Springer Theses

Recognizing Outstanding Ph.D. Research

For further volumes:
http://www.springer.com/series/8790

Aims and Scope

The series "Springer Theses" brings together a selection of the very best Ph.D. theses from around the world and across the physical sciences. Nominated and endorsed by two recognized specialists, each published volume has been selected for its scientific excellence and the high impact of its contents for the pertinent field of research. For greater accessibility to non-specialists, the published versions include an extended introduction, as well as a foreword by the student's supervisor explaining the special relevance of the work for the field. As a whole, the series will provide a valuable resource both for newcomers to the research fields described, and for other scientists seeking detailed background information on special questions. Finally, it provides an accredited documentation of the valuable contributions made by today's younger generation of scientists.

Theses are accepted into the series by invited nomination only and must fulfill all of the following criteria

- They must be written in good English.
- The topic should fall within the confines of Chemistry, Physics, Earth Sciences, Engineering and related interdisciplinary fields such as Materials, Nanoscience, Chemical Engineering, Complex Systems and Biophysics.
- The work reported in the thesis must represent a significant scientific advance.
- If the thesis includes previously published material, permission to reproduce this must be gained from the respective copyright holder.
- They must have been examined and passed during the 12 months prior to nomination.
- Each thesis should include a foreword by the supervisor outlining the significance of its content.
- The theses should have a clearly defined structure including an introduction accessible to scientists not expert in that particular field.

Alexander Zahnd

The Role of Renewable Energy Technology in Holistic Community Development

Doctoral Thesis accepted by
Murdoch University, Perth, Western Australia

Springer

Author
Dr. Alexander Zahnd
School of Engineering and Energy Studies
Murdoch University
Perth
Australia

Mechanical and Aerospace Engineering
University of California San Diego
La Jolla
USA

RIDS-Nepal
International Project Director/Advisor
Lalitpur
Nepal
www.rids-nepal.org

and

RIDS-Switzerland
International Project Director
Reitnau
Switzerland
www.rids-switzerland.org

e-mail: azahnd@rids-nepal.org

Supervisors
Prof. Dr. Philip Jennings
School of Engineering and Energy Studies
Murdoch University
Perth
Australia

Dr. Peter Freere
World Vision Australia
Melbourne
Australia

ISSN 2190-5053 ISSN 2190-5061 (electronic)
ISBN 978-3-319-03988-6 ISBN 978-3-319-03989-3 (eBook)
DOI 10.1007/978-3-319-03989-3
Springer Cham Heidelberg New York Dordrecht London

Library of Congress Control Number: 2013956336

Printed on acid-free paper

Springer is part of Springer Science+Business Media (www.springer.com)

Parts of this thesis have been published as journal articles and as peer reviewed conference papers as following:

Zahnd Alex, Jan Kleissl, Angel Clark, Wendy Cheung, Linda Zou, "Minimizing the Lead-Acid Battery Bank Capacity through a Solar PV—Wind Turbine Hybrid System for a high-altitude village in the Nepal Himalayas", ISES Solar World Congress, Cancun, Mexico, 3.–7.11.2013

Zahnd Alex, Philip Jennings, "The Role of Renewable Energy Technology in Holistic Community Development", Green Energy, chapter 5, pages 75–142, November 2012, Book editors: Dr. M. D. Tiwari, Dr. Anurika Vaish, River Publishers, PO box 1657, Algade 42, 9000 Aarborg Denmark

Zahnd Alex, The Role of Renewable Energy Technology in Holistic Community Development, ISES Solar World Congress 2011, Kassel, Germany, 28.8.–2.9.2011

Karaliunas M., P. Vitta, A. Žukauskas, Alex Zahnd, D. Bista, B. B. Chhetri and M. R. Updhyaya, "Characterization of Nepali Solid-state Lamps", Electronics and Electrical Engineering, 2009 No. 1(89) ISSN 1292–1215

Sanjel Nawaraj, Shah Malesh, Zahnd Alex, Upadhyaya Muniraj, "Power Generation Potential and Cost of a Roof Top Solar PV System in Kathmandu, Nepal", RETRUD 2009 (Renewable Energy Technologies for Rural Development), Kathmandu, Nepal, 15.10.–17.10.2009

Pandey Rojan, Alex Zahnd, "Identification and Evaluation of the Losses Occurring in a Solar PV System under Real Field Conditions for Rural Electrification in Humla, Nepal", ANZSES 2009 (Australia New Zealand Solar Energy Society), Townsville, Queensland, Australia, 29.09.–02.10.2009

Malla Avishek, Alex Zahnd, Haddix McKay Kimber, Ellul Alicia, Alex Zahnd, "The Importance of Monitoring and Performance Analysis of a Rural Solar PV Electrification Project", Engineers Australia, Queensland Regional Annual Conference 2009, Rockhampton, Queensland, Australia, 14.08.–15.08.2009

Zahnd Alex (key note paper), "Holistic Community Development and the Role of Contextualized & Renewable Energy Technologies in Improving Health Conditions in Rural Nepal" Engineers Australia, Queensland Regional Annual Conference 2009, Rockhampton, Queensland, Australia, 14.08.–15.08.2009

Fuller J. Robert, Lu Aye, Alex Zahnd and Sujit Thakuri, "Thermal evaluation of a greenhouse in a remote high altitude area of Nepal", International Energy Journal, Volume 10 Issue 2 June 2009

Fuller J. Robert, Alex Zahnd and Sujit Thakuri, "Improving comfort levels in a traditional high altitude Nepali house", Environment and Building (Elsevier), Volume 44 Issues 3 March 2009

Zahnd Alex, Kimber Haddix-McKay, "Benefits from a renewable energy village electrification system", Renewable Energy, Volume 34, Number 2, pages 362–368, February 2009

Haddix Mckay, Kimber, Zahnd Alex, "Holistic community development and the role of contextualized & renewable energy technologies in improving health conditions in rural Nepal", In: 3rd International Solar Energy Society Conference—Asia Pacific Region (ISES-AP-08) Incorporating the 46th ANZSES Conference, 25–28 November, Sydney, Australia 2008

Mulder Arjen, Zahnd Alex, "The porter problem in ecological economics Solar battery recycling in rural electrification":ISEE 2008 Conference, Nairobi, Kenya 4th July 2008

Kimber Haddix-McKay, Alex Zahnd, Catherine Sanders and Govinda Nepali, "Responses to Innovation in an Insecure Environment in Nepal", Mountain Research and Development (MRD), Vol 27, No 4, November 2007

Zahnd Alex, Kimber Haddix-McKay, "The Family of 4" (published in French with the title "Quatre Piliers pour un development holistique"), TRACÉS, Bulletin technique de la Suisse romande, pages 7–9, November 2007

Zahnd Alex, Kimber Haddix-McKay, "Pico-hydro Power Plant for elementary lighting as Part of a Holistic Community Development Project in a remote and impoverished Himalayan Village in Nepal", (published in French with the title "Eclairage sur mesure au Nepal"), TRACÉS, Bulletin technique de la Suisse romande, pages 11–13, November 2007

Bhusal Pramod, Alex Zahnd, M. Eloholma and L. Halonen, "Energy-efficient Innovative Lighting and Energy Supply Solutions in Developing Countries", International Review of Electrical Engineering (I.R.E.E.) Volume 2, No. 5, pages 665–670, October 2007

Zahnd Alex, Haddix Mckay, Kimber, (key note paper), "Problems Encountered with Solar PV Systems in Himalayan Villages and Possible Remedies", ANZSES 2007 (Australia New Zealand Solar Energy Society), Alice Springs, Australia, 02.10.– 06.10.2007

Fuller Robert, Schwede Dietmar, Malla Avishek, Zahnd Alex, "Prediction of Hot Water Usage in a Solar Heated Community Bathing Centre in Nepal", ISES Solar World Congress 2007, Solar Energy and Human Settlement. Beijing, China, 18–21st September, 2007

Bhusal Pramod, Alex Zahnd, M. Eloholma and L. Halonen, "Replacing Fuel based Lighting with Emitting Diodes in Developing Countries: Energy and Lighting in Rural Nepali Homes", LEUKOS, The Journal of the Illuminating Engineering Society of North America, Volume 3, Number 4, pages 277–291, April 2007

Zahnd Alex, Haddix McKay Kimber, "Benefits from a Renewable Energy Village Electrification System", WREN 2007 (World Renewable Energy Network), Perth, WA (Western Australia), Australia, 04.02.–07.02.2007

Zahnd Alex, "High Altitude Solar Water Heater Community Bathing Center", Solar Progress, Renewable Energy for Australia and New Zealand, Volume 27, Number 4, pages 21–24, December 2006

Thakuri Sujit, Zahnd Alex, "High Altitude Smokeless Metal Stove A Research, Development and Implementation Project through the Kathmandu University"; RETRUD 2006 (Renewable Energy Technology for Rural Development), Kathmandu, Nepal, 12.10.–14.10.2006

Zahnd Alex, Malla Avishek, "High Altitude Solar Water Heater Community Bathing Center: Renewable Energy Utilized for a Remote and Impoverished Himalayan Village in Humla, Nepal", ANZSES 2006 (Australia New Zealand Solar Energy Society), Australian National University Canberra, Australia, 13.09.–15.09.2006

Zahnd Alex, Haddix McKay Kimber, Komp Richard, "Renewable Energy Village Power Systems for Remote and Impoverished Himalayan Villages in Nepal", ICREDC06 (International Conference on Renewable Energy for Developing Countries 2006) in Washington DC, USA, 05.04.–07.0.4.2006

Rai Kanchan, Zahnd Alex, "High Altitude Smokeless Metal Stove Research and Development", Boiling Point 51, pages 37–8, January 2006

Zahnd Alex, Haddix McKay Kimber, "A Simple, Optimised PV System for a Remote Himalayan Village" ANZSES 2005 (Australia New Zealand Solar Energy Society), Otago University, Dunedin, New Zealand, 28.11.–30.11.2005

Zahnd Alex, Haddix McKay Kimber, "Renewable Energy Resources for Improved, Sustainable Livelihood A Case Study of a Holistic Community Development Project with a Remote and Poor Mountain Village in the Nepal Himalayas", 6th Asia Pacific Roundtable for Sustainable Consumption & Production (6APRSCP) Melbourne, Australia, 10.10.–12.10.2005

Zahnd Alex, "A Holistic Approach to Community Development", World Rivers Review, pages 8, 9, 15, February 2005

Zahnd Alex, "Handling Resources Sustainable and Responsible—Case Study of a Holistic Community Development Project with a Remote and Poor Mountain Village in the Nepal Himalayas, Utilizing the Available Renewable Energy Resources", Sustainable Resources Conference 2004 (SR04) Boulder Colorado, USA, 28.9.–5.10.2004

Zahnd Alex, "Handling (Managing, Administrating, Governing) Available Resources Responsibly—Case Study of a Renewable Energy Project with a Remote and Poor Mountain Village in the Nepal Himalayas", World Renewable Energy Congress VIII, Denver Colorado, USA, 28.8.–3.9.2004

Zahnd Alex, "Sustainable Renewable Energy Development of a Remote Community in Nepal" ANZSES 2003 (Australia New Zealand Solar Energy Society), Melbourne, Australia, 26.11.–29.11.2003

Zahnd Alex, "Appropriate Technology for the Poorest Mountain Communities in the Nepal Himalayas", Sustainable Resources Conference 2003 (SR03) Boulder Colorado, USA, 1.10.–4.10. 2003

Supervisor's Foreword

This thesis describes the main results of 20 years of theoretical and practical work in the remote, impoverished mountain communities of Nepal. The author, Alex Zahnd, did not set out to write an academic thesis when he began his community development work in Nepal in 1996, after he had been in Nepal since the mid-1980s. His goal was to bring some benefits from modern technology to the isolated and disadvantaged villagers who have a very low life expectancy and suffer from numerous illnesses that are rare now in industrialized countries. From the outset, Alex Zahnd sought to utilize local resources and appropriate technologies, developed for the local context, which the villagers would be able to understand and maintain. Rather than tell them what they needed, he sought to find out what their most pressing needs were and to design systems to assist them. Thus, he developed the concept of holistic community development (HCD), in which he attempted to address the core needs of the community in an integrated manner. This was a difficult task as he had to address social, cultural, economic, and technical issues with his system designs. Ultimately he succeeded in developing, deploying, and testing technologies to address the key needs of water supply, food for the long cold winter months, energy for cooking and lighting, sanitation, and community education.

This thesis has a strong theoretical basis, which Alex Zahnd has developed into a method or strategy for addressing community needs in developing countries. It includes details of the consultation processes, the system designs, the installation and education programs, the evaluation procedures, and the follow-up technologies to address further needs. This work has now been under way for 20 years and it has been continually evaluated and refined. The author now has a technique for sustainable community development in the third world and he wanted to share his knowledge with the international development community. Thus, he was persuaded in 2005 to start to write up the key aspects of his theory and its applications so that many others could benefit from his experience.

His work in Nepal is ongoing and he is continuing to apply his approach to holistic community development and new designs and insights continue to appear. However, it is now abundantly clear that Alex Zahnd has developed a versatile and robust approach to community development which should be applicable in many developing countries.

The thesis is a comprehensive account of the theory and application of HCD in the context of the impoverished districts of Humla and Jumla, located in the remote and hard to access northwest of Nepal. The Appendices contain details of many of the designs and educational programs that he has developed. They also highlight the importance of knowing the local culture and traditions, in order to have newly developed technologies and HCD concepts based on the local context, thus making them more likely to be acculturated, or sustainable, over the generations to come. My role as his main thesis supervisor was primarily to help him to tell his story in the form of an academic thesis, and it was a privilege to be associated with such important and beneficial work. I hope that this thesis will be widely read by community development practitioners and that it will lead to improved outcomes from international development programs.

Perth, Australia

Philip Jennings
Emeritus Professor of Physics
and Energy Studies Murdoch University
Perth, Western Australia

Preface

This Ph.D. thesis has not come together through the more traditional route; a predefined Ph.D. research project which emerged in the quest to find answers and solutions to problems/issues that have not been previously addressed and thus pose new challenges to be tackled. Rather, the present Ph.D. thesis is an amalgamation of a work in progress, initially started in 1996/1997 through realizing the urgent need for changes in the field of utilized and applied renewable energy resources and technologies, respectively, and community development, but without defined, satisfactory, or available solutions and tools to put into practice.

The author recognized this need by observing that the impoverished alpine communities of Nepal need solutions that are more holistic, applicable within their defined context, and with a much more practical angle compared to the current "standard" approach to community development projects.

Thus, from 1997 onwards, the author started to develop and apply new ideas and concepts of holistic community development. The concepts gradually included more and more projects that meet identified basic energy service demands through converting the locally available renewable energy resources.

The author was initially inexperienced in this field, and often had to make ad hoc decisions to do justice to the local circumstances and needs. The first few years were very much a phase of learning-by-doing. Many mistakes were made due to the limited understanding and knowledge at the time of how inclusive and holistic solutions need to be in order to bring forth long-term changes that are suitable for varying contexts and communities.

In particular, the decade from 2001 onwards was devoted to developing and implementing holistic community development projects in close partnership with the local communities through the NGO, RIDS-Nepal. The experience of working at a grass-roots level, and the initial results that the concept of applied renewable energy technologies, paired with long-term holistic community development brought forth, were most encouraging. Gradually, these projects brought substantial life improving changes to the small communities that RIDS-Nepal works with in the high-altitude Himalayan mountain valleys in Nepal.

Recognizing that there are many other impoverished communities in similar circumstances, in other areas in Nepal and indeed in other, similar needy countries, encouraged the author first to start writing papers and publications about the individual project parameters of the new holistic community development concepts of the “Family of 4” and the “Family of 4 PLUS”, and the reasons why the author has developed these new concepts, their outworking, and their impact among those with whom they have been applied.

Work experience, results, and impacts seen and experienced on the ground through the last 16 years of practical community development encouraged the author to put together all the papers and publications written since 2005, and to synthesize the results in the form of a Ph.D. thesis. It is his hope that other agencies, communities, and scholars may benefit from these experiences and apply them in contextualized ways within their own defined surroundings and communities.

Acknowledgments

It may not come as a surprise that there are many people I would like to thank and acknowledge for having been an active part in the long journey, since 1996, when I set out to learn and develop new concepts and understanding of how the needs and demands of impoverished communities can be addressed. With limited space available I can only mention some of the numerous people who supported and guided me with their experience, mind provoking ideas, cautions and advice through this passage of my life, which will continue after the submission of the Ph.D. thesis, with more understanding, wisdom, and even more vigor and determination.

In particular I want to acknowledge and thank Prof. Philip Jennings, my principal supervisor, for his invaluable guidance and advice throughout the process of the Ph.D. research program. I would like to extend my utmost gratitude to him for all his kind and continuous support, encouragement, and mentorship. He is one of the "real" Professors in the sense of the word, able to guide and lead younger professionals through his outstanding academic knowledge, expertise in industry, and international recognition, paired and enforced through his life-long experience as educationalist and expert of renewable energy and the needs of society. Special thanks I would like to extend as well to Dr. Peter Freere, my associate supervisor, for all his encouragement and valuable comments, particularly in the final stages of this thesis. His knowledge of Nepal and his practical experience of applied renewable energy technologies in the context of developing countries and the need for community development made him an ideal supervisor. He provided positive, critical advice to make the thesis applicable for a wider, professional interest community.

There would have been no opportunities for me to even start thinking about new ways of how to conduct projects without the village communities, motivated and interested to develop and implement with us in RIDS-Nepal, the new Holistic Community Development (HCD) concepts. Thus, I would like to acknowledge and thank all of the people of the 13 villages, one school and one health post with whom we have worked intensively since 2002 in the Humla district. They have allowed us to participate in their lives and become in many cases much more than mere project partners, but friends. They are: Chauganphaya Village and Chaughanphaya High School, Syada Health Post staff, Dhadhaphaya Village, Tulin Village, Kholsi Village, Pamlatum Village, Tangin Village, Hepka Village,

Darapori Village, Jharkholsi Village, Chala Village, Sata Village, Yangu Village, Lekha Village.

Likewise, there would have been no projects with these impoverished and needy communities without all the various donors from around the world, who supported our community-based implementation projects and research projects throughout the years. Thus, I would like to extend my gratitude and thanks to all of them, in particular to the ISIS Foundation (in particular Associate Prof. Kimber Haddix McKay, Leo Excel and Audette Excel, founder of the ISIS Foundation, who all became much more than project partners or donors, but close, life-long friends), Kadoorie Charitable Foundation (KCF), Light in Nepal (LiN), American Himalayan Foundation (AHF), Nepal Himalayan Mission (NHAM) and multiple individual people and enthusiasts who with their financial and moral support enabled me and RIDS-Nepal as a team, to carry out all the HCD projects in partnership with the local village communities. Their understanding and encouragement, particularly during times when things were very difficult, are greatly appreciated.

I would like to express my appreciation to the leadership of the Kathmandu University, Prof. Suresh Raj Sharma, our Vice Chancellor, Prof. Dr. Bhola Thapa, Dean of our School of Engineering and Prof. Dr. Bim Shrestha, my Head of Department of the Mechanical Engineering Campus, on whose staff team I have been a member since 2001. Kathmandu University (KU) provided the best environment within Nepal where I was able to start new student research projects each year that were relevant to the remote mountain communities. The results benefited not just the students, but also the end users in the remotest parts of Nepal. Over the years, seven of KU's graduated students have been my research assistants and worked with me for 3 years each at KU and through RIDS-Nepal, continuing to highlight the importance of applied practical student R&D projects and field-based research. They all have been as well to the remote villages, participating in the implementation of the projects with the mountain communities. Most of them have already finished their Master's Degrees in a University abroad and they are the ones that I hope will lead Nepal on the path of sustainable development in the years to come. Thus, I want to thank Kanchan Rai, Biraj Shrestha, Avishek Malla, Sujit Thakuri, Rojan Pandey, Nawaraj Sanjel, and Malesh Shah, for all their hard work and involvement in our projects during their time as my research assistants. I see them as future professional leaders of Nepal as they have embraced the vision for Nepal's needs and how they can be addressed.

Without the dedication and tireless effort of all our staff at RIDS-Nepal I would have not been able to bring ideas to fruition with our project partner villages. They have been faithfully working for years in this remote, harsh, and difficult working environment, with minimal equipment and facilities. Thus, I would like to extend my thanks and gratitude to the following staff who have been with me on this long journey of learning by doing and getting up after having fallen and continuing to focus on the vision and mission. Govinda Nepali, Haripal Nepali, Bam Bahadur Rokhaya, Sunita Budha, Sarita Shahi, Sher Bahadur Rokhaya, Gogan Rokhaya, Pradeep Shahi, Barka Shahi, Nayan Singh, Ram Raja Singh, Manish Pradhan,

Jhabendra Bhatterai, Bodh Raj Pandt, Palden Sunar, Sanjay Nidhi, Bishal Thapa, Prabin Pant, Bir Jung Shahi, Govinda Lama, Resham Katri, Uttam Khatri, Pradhuman Dhakal, and all the expatriate volunteers from around the globe who came for a defined time period to share their professional skills. With them I have been able to enjoy the smiles of the local end users which made up for all the hardship. I extend my thanks to Muni Raj Upadhaya, the owner of Pico Power Nepal (PPN), who is one of my earliest close Nepali friends and co-founder of RIDS-Nepal and with whom I have been working closely since 1997. He often worked with me relentlessly on new technical ideas, which initially have not found many supporters, to put them into practice, and to learn and improve them through experience.

Most of all I would like to thank my dear wife Sheila, who has always been my strong moral support with never ending patience and love during times when things just got "too" much. Also, I wish to thank my dear son Joel and my dear daughter Jessica, who showed kind understanding in times of extreme stress and time constraints. I cannot express enough thanks for their understanding and having enabled me to work in such remote and difficult areas, with extended absences from them. I am indebted to Murdoch University, from where I have previously graduated with my Renewable Energy Master's Degree as an external, online student in 2004. Through its relevant and motivating higher education programs I have benefited greatly for my professional career and services in Nepal. They provided a firm basis, paired with many years of practical experience, to undertake a Ph.D. research project. An Australian Government Endeavour Ph.D. scholarship, which was awarded to me, made it financially possible for me to undertake this Ph.D. research project.

Contents

Abbreviations and Acronyms

AD	Anno Domini; the number of years counted based on the traditionally reckoned year of the conception or birth of Jesus Christ of Nazareth
a-Si	Amorphous Silicon solar photovoltaic module material/technology
AC/DC	Alternating current/Direct Current
ACF	Action Contre La Faim, (a French INGO), or in English: Action against Hunger
ADB	Asian Development Bank
AEPC	Alternative Energy Promotion Centre (under the Nepali Government)
AGM	Absorbed Glass Mat–Battery (sealed and valve regulated lead-acid batteries which demand low maintenance and no topping up of distilled water)
Ah	Ampere-hour; value of a battery's energy storage capacity
ALRI	Acute Lower Respiratory Infection
AM	Air Mass
AMP	Amperes
Albedo	Fraction of shortwave solar radiation energy reflected from the earth's or a body's surface back into space/intercepted object
A.P.C.S.	Analog Process Control Services ltd in NSW Australia
ASHRAE	American Society of Heating, Refrigerating and Air Conditioning Engineers
AURI	Acute Upper Respiratory Infection
BC	Before Christ; the number of years counted before the traditionally reckoned year of the conception or birth of Jesus Christ of Nazareth
BOS	Balance of System
BS	Bikram Sambat
BTU	British Thermal Unit (1 BTU = 1.06 kJ, or 0.293 Wh)
°C	Degrees Celsius
CIBSE	Chartered Institute of Building Services Engineers
CC	Charge-/Discharge-Controller

CDM	Clean Development Mechanism
CDO	Chief District Officer (the government position with the highest political rank on a district level in Nepal)
CEM	Channel Extension Module
CFL	Compact Fluorescent Lamp
CIA	Central Intelligence Agency: The World Fact Book
CIE	Commission Internationale de l'Eclairage (French: International Commission on Illumination—standardization body)
CLO	Clothing Level. ASHRAE provides a CLO calculator to measure people's clothing level (by weight) to determine their living "comfort" level
COHb	Carboxyhemoglobin level (in %) of CO in the blood stream
C_{20}	Capacity (in Ah measured) of a battery discharged over 20 h
DANIDA	Danish International Development Agency
DDC	District Development Committee (a Nepal government appointed committee working in each district of Nepal)
DIY	Do It Yourself
DoD	Depth of Discharge, indicating the level of discharge of a battery/battery-bank. A DoD of 10 % means that the battery/battery-bank has been discharged by 10 % of its full capacity and thus remains 90 % full, or has a SOC of 90 %
DT605	Data logger from the company dataTaker, type 605
DT80	Data logger from the company dataTaker, type 80
EPA	Environmental Protection Agency (a USA Government Agency)
ESAP	Energy Sector Assistance Project (run by DANIDA and AEPC)
FFW	Food For Work
FY	Fiscal Year
GDP	Gross Domestic Product
GOBI	Growth monitoring; Oral rehydration therapy; Breastfeeding; Immunisation. Acronym for UNICEF's primary health care program
HASWH	High-Altitude Solar Water Heater
HASWHBC	High-Altitude Solar Water Heater Bathing Centre
HARS	High-Altitude Research Station (RIDS-Nepal's Simikot office and high-altitude research station, located, at 3,000 m above sea level)
HDI	Human Development Index
HMG	His Majesty Government of Nepal
IEA	International Energy Agency
IFA	International Fund for Agricultural Development (A UN financial institution)
INGO	International Non-Governmental Organisation
IPP	Independent Power Producer
ISIS	*The ISIS Foundation*
ISO	International Standard Organisation
ISPS	Institutional Solar PV Systems

K	Degree Kelvin
KLDP	Karnali Local Development Program, Jumla, Nepal
KU	Kathmandu University
km	Kilometre
kW	Kilowatt (power unit)
kWh	Kilowatt-hour (energy unit)
LED	Light Emitting Diode
LoL	Loss of Load, indicating the amount of time (in % of 24 hours a day) the load could not be provided because the system was designed too small, thus not able to provide the needed energy
l/s	Litres per second
LUTW	Light Up The World, Calgary, Canada
m	Metre
MOA	Ministry of Agriculture (a Nepali government Ministry)
m.a.s.l.	Metres above sea level
MDGs	Millennium Development Goals (to be reached in 2015)
MHP	Micro-Hydro Power Plant/Project
MJ	Megajoule (1 MJ = 0.278 kWh)
MOLD	Ministry of Local Development (a Nepali government Ministry)
Mongoloids	The Mongol people from East Asia, who conquered much of Eurasia in the thirteenth century, establishing the Mongol Empire
MOST	Ministry of Science and Technology (a Nepali government Ministry)
MPP	Maximum Power Point, indicating the point of highest power output
MW	Megawatt (1 MW = 1,000 kW, or 10^6 W)
MWh	Megawatt hour (1 MWh = 3.6 GJ)
NEA	Nepal Electricity Authority (Government owned), Kathmandu
NFE	Non-Formal Education classes mainly for women and out of school children, mostly girls
NGO	Non-Governmental Organisation (such as, e.g. RIDS-Nepal)
NHE	Nepal Hydro Electricity (a Nepali company)
NOTC	Nominal Operation Collector Temperature
NM	Nanometre (1 nm = 10^{-9} m)
NPV	Net Present Value
NREL	National Renewable Energy Laboratories, in Golden CO, USA
NRs	Nepali Rupees. The unit of currency in Nepal. One Rupee has 100 Paise
OECD	Organisations for Economic Cooperation and Development
O&M	Operation and Maintenance
PMV	Predicted Mean Vote
POA	Plain of the Array (means at the same angle as the equipment)
PPD	Predicted Percentage of Dissatisfaction
PPA	Power Purchase Agreement
PPN	Pico Power Nepal (a Nepali company)

PPP	Purchase Power Parity
PSH	Peak Sun Hours (number of sun hours per day with 1,000 W/m^2 global solar radiation intensity)
PV	Photovoltaic
PVSOL	Solar Photovoltaic System design software (Germany)
PVSyst 5.571	Solar Photovoltaic System design software version 5.57 (29th May 2012) develoepd by the University of Geneve in Switzerland
RAPS	Remote Area Power Supply
RET	Renewable Energy Technology
RETScreen	Renewable Energy Technology Screening software (Canada)
RIDS-Nepal	Rural Integrated Development Services-Nepal
RTD	Resistance Temperature Detector
Sanskritic	Much of the roots of Hinduism are in the Vedic civilisation which have been written in the Sanskrit language, relating Hinduism and Sanskrit inseparably
SHS	Solar Home System (small solar PV system for one home/family)
Si mono	Si mono-crystalline solar photovoltaic module material/technology
Si poly	Si poly-crystalline solar photovoltaic module material/technology
SINTEF	Selskapet for INdustriell og TEknisk Forskning ved norges tekniske hoegskole (The Foundation for Scientific and Industrial Research at the Norwegian Institute of Technology), http://www.sintef.no/Home/ (accessed July 30, 2012)
SOC	State of Charge, is the level of a battery or battery-bank's charging level. 100 % SOC means the battery/battery-bank is fully charged
STC	Standard Testing Conditions (1,000 W/m^2, 25 °C Ambient Temperature, 1.5 AM (1.5 Air Mass is the amount of air mass the sun's radiation passes through when reaching and penetrating earth's atmosphere at an approximate angle from the horizon of 48.2°))
SWOT	Strength—Weakness—Opportunities—Threats Analysis
TRNSYS	Transient Systems Simulation Program
TWh	Terawatt hour (1 TWh = 1,000 GWh, or 10^{12} Wh)
UA	UA (W/m^2 K) is the overall conductance value and is defined as the product of the overall heat transfer coefficient and the heat transfer area. It defines the heat transmission through a material or assembly of materials per unit surface area. The U-value of a material indicates how well heat passes through the material. The lower the U-value, the greater the resistance to heat and therefore the better this material insulates against heat losses. Thus, if a wall material has a UA value of 1 W/m^2 K, it means that for every degree of temperature difference between the inside and outside surface there would be 1 W of heat energy flowing through each m^2 of its surface
UN	United Nations

UNFCCC	United Nations Framework Convention on Climate Change
US	United States of America
UV	Ultra Violet (electromagnetic waves in the range of 100–400 nm)
V	Volts
VDC	Village Development Committee
W	Watt (power unit)
WB	World Bank
WECS	Water and Energy Commission Secretariat (in Nepal)
WHO	World Health Organization
Wh	Watt-hour (energy unit)
WLED	White Light Emitting Diode (a solid-state lighting technology using multi-layered semiconductor material, emitting light when connected to/powered by DC electricity)
Wp	Watt peak power output
W_R	Watt rated power output

Chapter 1
Introduction

1.1 Background and Context

Nepal is a unique country, in regard to its culture, people, geography, ecosystems and climate. It is situated in the lap of the Himalayas, landlocked between China to the north and India to the south, east and west. It was previously cut-off by thick forests and jungles infested with malaria and dangerous animals, and its culture developed in isolation from any foreign influence. The rich resources of Nepal, in particular the abundant water flowing from the Himalayas down to the Indian sub-continent, and its abundant solar energy, are heavily underutilised. Thus, ~75 % of Nepal's rural people are deprived of even the most basic energy services. They rely on traditional biomass such as firewood, agricultural residues and dung, for their daily cooking, heating and lighting purposes. With a human development index (HDI) of 0.458, Nepal ranks among the poorest developing countries.

Theory and practice reveal that in recent decades, community development addressed mostly the individual needs of projects' end-users, with often minimal interaction and participation of the receiver. This resulted in minimal long-term impact and/or new opportunities for the beneficiaries. The thesis aims to make a contribution to an improved approach to the community development of Nepal's people within their context and identified needs. It argues that appropriate and sustainable solutions for long-term development of local communities demand that people's self identified needs are recognised and addressed in holistic ways. This is because communities have multi-faceted needs, which are all dependent on, and interlinked with, each other, and thus can't be addressed through single-strand projects. Further, utilising the communities' locally available resources is crucial for long-term sustainable development.

Accordingly, the thesis' main research questions are:

- *What is the appropriate role of renewable energy technology in holistic community development?* and
- *Does the newly developed holistic community development concept of the "Family of 4" and the "Family of 4 PLUS" provide a sustainable way of improving the quality of life for remote and impoverished mountain communities in Nepal?*

A. Zahnd, *The Role of Renewable Energy Technology in Holistic Community Development*, Springer Theses, DOI: 10.1007/978-3-319-03989-3_1,

The research shows that tapping into the locally available renewable energy resources, and converting them through contextualised renewable energy technologies, has to be a central part of community development. This is done, through the study of practical examples of appropriate technologies for some of Nepal's remotest, high-altitude village communities, who live under the harshest imaginable conditions. The research shows that the most commonly expressed needs of the communities are for improved and easier access to basic energy services, and better hygiene and health conditions. Therefore the author developed the new Holistic Community Development (HCD) concept of the "Family of 4" (a Pit Latrine, a Stove, Basic Indoor Lighting and Clean Drinking Water). Once the "Family of 4" HCD is effectively implemented, the "Family of 4 PLUS", with various additional measures (such as greenhouse, solar drier, solar water heater, slow sand water filter) is introduced.

This thesis looks at the role of renewable energy technologies (RETs), designed for a specific context, to meet identified community needs embedded in the new HCD concepts. RETs such as solar PV, solar thermal (water heating, food cooking and drying), pico-hydro power, small wind turbines and biomass (firewood) cooking and heating stoves, are discussed within the context of implemented, village based, HCD projects. Impact data presented show encouraging results, and thus it can be claimed that the main research questions can be positively affirmed. The HCD approach produces synergies between the various components which provide additional benefits to the recipients at no extra cost. The newly developed RETs and HCD concepts include various new locally manufactured technologies, providing a contribution to the steady growth of the local economy. Numerous suggestions and recommendations for further R&D are presented.

1.1.1 A Glimpse at Nepal's History

Nepal's earliest recorded history dates back to the beginning of the first millennium BC. Ever since power struggles for political rulership through uprisings and wars took place. It was under the charismatic leadership of Prithvi Narayan Shah in 1768, who triumphed after 27 years of war, that the capital was moved to Kathmandu. After some disagreement with the British in 1814, Nepal had to hand over Sikkim, East of Nepal, and parts of the Terai plains as a result of the signing of the 1816 Sugauli Treaty. This established Nepal's present geographical border lines to the East and West.

Jung Bahadur seized control of Nepal in 1846 through another bloody massacre by butchering several hundred of the most important men while they assembled in the Kot courtyard. He assigned to himself the more prestigious title Rana, and thus the Rana dynasty was formed. He proclaimed himself the first Prime Minister of Nepal. The Rana regime, a harsh, strongly centralised autocracy, followed a policy of isolating Nepal from the surrounding world, so that no external or foreign influences were able to penetrate the society, purposely keeping people living in medieval conditions. This lasted till 1950 when a democratic government, operating under the scrutiny of the monarch, King Tribhuvan, took over and appointed him ruler in 1951. He stayed in full control till his son, King Birendra, succeeded him in 1972. Corruption and mismanagement did not improve, which led to the protest movement called "Jana Andolan" or "People's Movement" during 1989–1990. These violent protests, led to the dissolving of the King's cabinet, legalising political parties and thus making the panchaayat system obsolete. In the first free election, in May 1991, the Nepali Congress won the majority with the Communist Party as their main opposition.

Through the early nineties constant small uprisings, demonstrations and clashes between students and the government took place, which led one of the Maoist parties, in February 1996, to start the people's war, with the aim of replacing the monarchy with a Maoist government. This led Nepal into a 10 year long civil war, led by Dr. Baburam Bhattarai and Pushpa Kamal Dahal, better known in Nepal as Prachanda.

After the assassination of the King's family by the Crown Prince on 1 June 2001, the late King Birendra's brother, Prince Gyanendra inherited the throne. The political situation further deteriorated, with the monarch dismissing the government and assuming executive powers on 1 February 2005. Ongoing and intensifying strikes and ever more violent street protests in Kathmandu, forced the King to restore the parliament in April 2006.

In November 2006, after 10 years of brutal and violent war and civil unrest, the peace agreement between the government and the Maoists brought the long hoped for framework for possible peace. Elections followed on 10 April 2008, in which the Maoists emerged as the winners, securing a simple majority in Parliament and forming the new "Republic of Nepal" government. On 28 May 2008 Nepal was declared a Federal Democratic Republic, abolishing the 240 year old monarchy, forcing the last King of Nepal, King Gyanendra to leave the Royal Palace and become an ordinary citizen. However, this did not bring the hoped for changes and stability and Nepal continues to suffer in 2012 under its long-term, unstable and fast changing political power struggle. For more a more detailed historical background, see Appendix 3.

1.1.2 The Geography and Climate of Nepal

Nepal, with a land surface area of 147,181 km^2 (CIA 2009), lies between 26°22′–30°27′ northern latitudes, and 80°04′–88°12′ eastern longitudes. It is situated in the Gangetic plain and landlocked, with China/Tibet to the north and India to the south, east and west. It embraces unique climatic environments, from tropical to high alpine, in an unparalleled physical ecosystem, ranging between altitudes of 70–8,848 m.a.s.l. (Fig. 1.1).

Three distinct geographical zones cover Nepal. To the south, the mostly flat, heavy populated (50.2 %) Terai, ranging from 70–800 m.a.s.l., merging with the

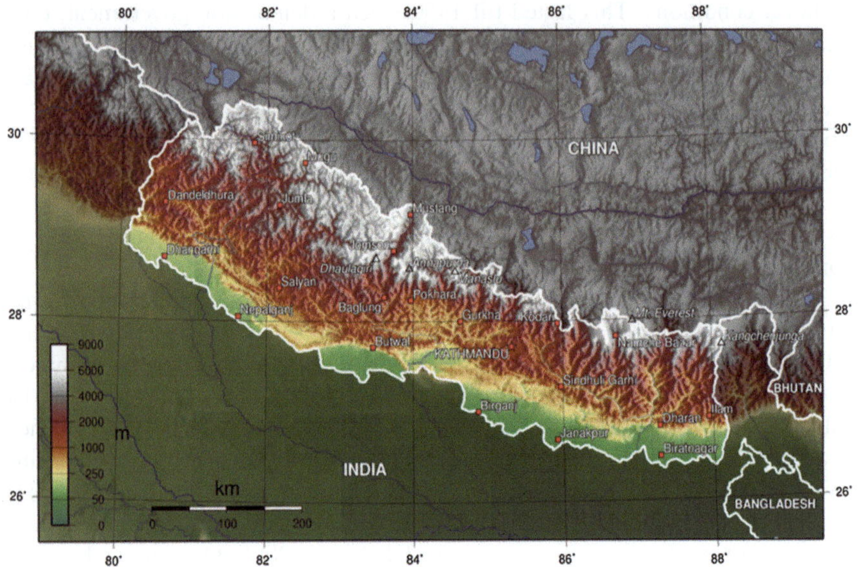

Fig. 1.1 Nepal Map (*from* http://en.wikipedia.org/wiki/File:Nepal_topo_en.jpg)

Indian border. In the middle lies the densely populated (43.1 %) hilly Siwalik and Mahabharata Range (~800–4,000 m.a.s.l.), with its fertile valleys such as Kathmandu and Pokhara. The sparsely populated (6.7 %), high-altitude Himalayas (~4,000–8,848 m.a.s.l.), include Mt. Everest and seven other mountains above 8,000 m[1], form a strong natural border with China/Tibet towards the north. The hard to cultivate high-altitude villages (~1,800 to ~3,500 m.a.s.l.) experience an alpine climate with cool summers and freezing winters. This distinctive geographical environment creates distinct seasons and different climatic zones, from the hot, humid tropics in the south, to the inhospitable, cold, arctic zone above 4,400 m.a.s.l.

Attributable to its exceptional topography and climates, Nepal is prone to various natural disasters. While windstorms and lightening strikes are experienced throughout the country, floods, drought and fires threaten the Terai, and heavy snowfalls and freezing conditions endanger the mountain communities. A strong population growth since 1960 has had a major impact on the natural resources, such as the forests. Landslides and deforestation form deep scars where once lush green forests thrived. With the Eurasian (north) and the Indian (south) Tectonic Plates in constant motion towards each other, with three major north dipping thrust

[1] Nepal has eight of the world's fourteen highest (>8,000 m high) mountains within its, or on its border line. They are: Mt. Everest (8,848 m), Kangchenjunga (8,586 m), Lhotse (8,516 m), Makalu (8,485 m), Cho Oyu (8,188 m), Dhaulagiri 1 (8,167 m), Manaslu (8,163 m) and Annapurna 1 (8,091 m). For more detailed information see: http://en.wikipedia.org/wiki/List_of_highest_mountains (accessed July 26, 2012).

faults "sandwiching" Nepal, earthquakes are a constant danger. For more details on this and Nepal's unique climatic and geographical conditions, please refer to Appendix 4.

1.1.3 A Glimpse at the Ethnicity and Religion in Nepal

Nepal's two main groups of people are also unique in their ways. Of the total population of ~26.6 million (June 2011 census (NPCS 2011a, b)), most Aryans live near the Indian borders and in urban centres, while the different tribes of Mongolians live predominantly in the hills and mountain regions. Approximately 75 % live in remote, difficult to access areas with the majority having no access to roads, clean drinking water, health facilities or electricity (Karki 2002). Fang (2007) also indicates that 42 of Nepal's 75 districts are considered food deficient, acutely food deficient and permanently food deficient areas. Nepal is also one of only a few countries with a lower female life-expectancy of 44–48 years in remote mountain areas, and 60.75 years (CIA 2009) in urban areas. Men live longer, 46–52 years on an average in remote areas[2], and 61.12 years (CIA 2009) in urban settlements. The high fertility rate of >4 children born per woman till 2007[3], feeds the high population growth rate of >2 %, with an annual urbanisation rate of 4.8 % (NPCS 2011a, b).

Officially (Nepal 2001 census) 92 different languages are spoken by 103 distinct castes and ethnic groups. Often, across a valley different languages are spoken, as the wild mountain streams have divided people groups for centuries. Most people in the mid- and high-altitude hill/mountain areas are subsistence farmers with little farm land, poor soil and harsh conditions, making a meagre living. Crops are vulnerable with an immediate impact on survival if natural calamities strike or weather patterns change. Due to the lack of roads and rough terrain in remote, high-altitude areas, markets are almost non existent and food has to be flown in. Consequently, staple food such as rice, costs up to 300 % more[4] compared to the Terai.

Agriculture is different in the flat, low-altitude Terai. Good soil, reliable water sources and milder weather permit the year round cultivation of crops such as rice, wheat, maize, sugar cane and vegetables (Fang 2007). Easier access to finances provides much greater food security (Fang 2007).

The Nepalese are family oriented and most tasks and events are believed to be connected with the spiritual world. Fortune and misfortune are widely accepted as

---

[2] The statistics for the remote areas are from the author's own survey data over a 16 year time span (1996–2012) working in the remotest parts of Nepal.

[3] http://www.indexmundi.com/g/g.aspx?c=np&v=31, and http://www.indexmundi.com/nepal/total_fertility_rate.html (both accessed July 26, 2012).

[4] These figures are from the author's own survey data over a 16 year time span (1996–2012) working in the remotest parts of Nepal such as Humla and Jumla.

the will of fate or *karma*. Most people believe in *bhoot* (ghosts), *pret* (evil spirits), *bokhsi* (witches), and *graha dasha* (a bad position of the planets), causing diseases in people/livestock, crop failures, or accidents. Incense, flowers, food, a rooster (for poor people) or a male goat (for richer people) are offered to pacify the spirits.

Nepal was a Hindu kingdom before 19 May 2006, when it became a secular state. The main religious beliefs are: Hindu 80.6 %, Buddhist 10.7 %, Muslim 4.2 %, Kirant[5] 3.6 %, Christians 0.5 %, others 0.4 %. A strong caste consciousness, historically divided into four main castes, deeply affects the way people behave towards each other. In Humla, mainly two religious beliefs are followed. Shamanism (a blend of Hinduism, Animism and witch craft), and Tibetan Buddhism. In each village *dhamis* and *jhankris* (witch doctors) are consulted as local healers and sorcerers. For more details on Nepal's unique ethnicity and religion see Appendix 5

1.1.4 A Glimpse at Nepali Culture

While numbers are useful to get a quick grasp of a country's location, size and economic condition, the more anthropological factors which make a culture work the way it does, are much more difficult to identify and understand. In order to engage in development work, it is important to first comprehend people's culture in depth, which is a time demanding and difficult task. This is even more essential in the case of Nepal, which was virtually isolated from the outside world for over two centuries until 1950. Therefore, a more in depth study is provided in Appendix 6, leading "behind" the curtain which masks the culture and habits and some of the main reasons why things are the way they are in Nepal. It aims to explain patterns of the society which are still at work today, dictating much of Nepal's culture, daily life and work. It is recommended for anyone involved, or going to be involved, in community development projects in Nepal. In the following, only a glimpse at some of Nepal's unspoken, peculiar and captivating behaviour patterns, invisible through the eyes of the short-term visitor to Nepal, is provided.

Nepal's heterogeneous society offers valuable social resources for its development. The life of subsistence farming brought a strong commitment to productive labour, with a high capacity for endurance and a mutually supportive and cooperative organisational style. Nepalese are known to be good traders and skilled craftsman. Bista (1991, 2–4) points out that the culture of fatalism, inherently unsupportive and in conflict with development and productivity, has become increasingly pervasive. This is reflected through the long-term, external infusion of donor funds for development projects. He goes on to say: "this deep belief in fatalism has had a devastating effect on the work ethic and achievement

[5] The Kirant religion is a blend of animism, such as ancestor worship and Shamanism, Shiva Hinduism (which is the oldest of the four main strands of Hinduism), and Tibetan Buddhism.

motivation, and through these on the Nepalese response to development. It has consequences for the sense of time, and in particular such things as the concept of planning, orientation to the future, sense of causality, human dignity and punctuality" (Bista 1991, 4). Bista (1991, 29–59) explains, that the class and caste system are "at the base of many of Nepal's development difficulties". Though introduced by the Brahmins arriving in Nepal during the medieval period, the rigid application of the caste system became apparent only through the ruling elite, the Ranas, in the twentieth century. He states that "any rigidities associated with an ascriptive system will have an adverse effect on development and Nepal suffers from these problems". Since then, absolute belief in fatalism has developed, with its emphatic sense that "one has no personal control over one's life circumstances, which are determined through a divine or powerful external agency".

Besides its relationship and connection with fatalism, dependency can be seen, though often in a more mature form, in the cooperative social behaviour of Nepalese indigenous groups. Nepalese are not strong individualists, but collectivism as such is not a hindrance to development. A deep rooted concept of collectivism is the concept of *afno manche*, meaning one's own people, belonging to the "inner circle" of a group of people. It encourages the problem of inclusion–exclusion as group members acquire favourable status through the trading of privileges/services rather than based on professional reasons. This can clearly impede project goals which need to be achieved. Bista (1991, 4) summarises it in one sentence, saying: "with *afno* manche one finds exclusionary tendencies, factionalism, failures in cooperation, and corruption in various forms leading to malfunctioning of development administration and dissatisfaction at every level". Thus, dependency and fatalism, deeply rooted and combined in the society's belief and cultural systems, pose significant impediments for sustainable development.

Another typical practice which is often observed in Nepali society is *chakari*, practiced by those with ambitions and interest in self-promotion. Bista (1991, 5) explains *chakari* is "a social activity which, in its most common form is simply being close to or in the presence of the person whose favour is desired. Instead of efficient fulfilment of duties and obligations, persistence in *chakari* is seen as merit, and with enough merit, favours may be granted. It is a passive form of instrumental behaviour whose object is to demonstrate dependency, with the aim of eventually eliciting the favour of the person depended upon". As one can imagine, to practice *chakari* requires persistence, requiring a life and work attitude which inherently creates fundamental barriers to successful development. Bista (1991, 93) summarises it succinctly: "Decisions are made, appointments determined because of pressing obligations formed through *chakari* and not as a result of an objective determination of what is best", and "the entire social apparatus then suffers as critical positions are filled and governmental decisions made as a result of *chakari*. It is a built-in guarantor of incompetence, inefficiency, and misplaced effort".

Education too is still associated with class/caste privileges and not with labour. It is a powerful symbol of status and class identity, rather than a stepping stone to professionalism. In this way education is not perceived as a means of acquiring skills for future productive use in the society, but rather as an end in itself which, once achieved, signifies higher status. This in turn automatically brings more privileges. Thus, once education has enabled the desired status, it is believed that one is "effectively removed from the workforce". In Bista's words: "within such an atmosphere a genuinely productive professional workforce can never be developed and professional activities must continue to be performed by the untrained" (Bista 1991, 6). Work is equated with *dhuka* or pain while the opposite of dhuka is *sukha* or bliss, which means to live without having to work, and this is considered to be fortunate.

Another impact of causality and fatalism on people's daily life in Nepal is the concept of *karma*, believing that one's circumstances have been determined in advance and that one's life has been fated. The course of future events is determined and thus people bear poverty and deprivation only with a promise of a possible better rebirth with little sense to attempt to influence the outcome. Such fatalistic belief affects the way projects are planned, problem solving activities are addressed and managerial responsibilities exercised. Fatalism also influences the concept and sense of time. Bista (1991, 84) says: "For Nepalese, time is like a flowing river,..., without a sense of beginning or end or particular divisions", and "Nepali culture does not have a sense of time as past, present or future; it is always an endless present". Thus planning often does not take place and a busy person is running around for someone higher up, thus he can not be of high caste or high status, because the more free time a person has the higher/more important he is considered to be.

Over the last two decades significant changes have started in Nepal's society, with many signs of opening up to the wider world and with increasingly more joint international projects. Many people know that there is plenty of work to be done for Nepal's development, but are still surrounded by a wider supportive society which considers work to be of low status. Thus Bista's wise words (1991, 9) "any culture needs to be scrutinised and its negative elements purged", and Nepal needs not only "to learn new things in order to progress, but must also unlearn old things" (1991, 5). These words are still very relevant today. See Appendix 6 for additional insights into Nepali culture.

1.1.5 A Short Overview of Nepal's Politics and Government

Since 28 May 2008 Nepal has been a Federal Democratic Republic, or "Sanghiya Loktantrik Ganatantra Nepal" in Nepali language, with Kathmandu as its capital. The democratic multi party system puts forward the politicians for the 601 seats in the Constituent Assembly, with the majority held by the two main Communist parties CPN-M (Maoists) and the CPN-UML (United Marxist-Leninist) and the main opposition the Nepal Congress (NC). The Supreme, Appeal and District courts provide the three-tier judiciary system.

Fig. 1.2 Nepal's 14 geographical regions called Anchal (http://www.un.org.np/reports/maps)

Nepal is divided into 5 development zones (Fig. 1.3), and 14 political zones (Fig. 1.2) for purposes of planning and development projects. Further, the country is sub-divided into 75 administrative districts (Fig. 1.3), each headed by a Chief District Officer (CDO), responsible for the district's general development. Each of the 75 districts has a number of VDCs (Village Development Communities), including 2–5 villages, represented by chosen members. Each village is sub divided into 3–6 Wards, with their representatives.

This complex political and administrative structure is so massive in size and slow in functioning that it developed a certain kind of organism with a life of its own mostly insensitive to the external environment. The National Planning Commission (NPC), chaired by the Prime Minister, oversees the general development of the country through its five-year plans. As it is written mainly by city based "economists, statisticians and geographers,..., the five-year plan document is worth very little. The NPC also tends to ignore the people, local institutions, and even the Ministries once they have made their initial recommendations." (Bista 1991, 107). These instruments of government are still in place in 2012.

The bloody and brutal civil war between the Maoists and the Government (1996–2006), killed an estimated 20,000 people, with many village families losing one member. From the Comprehensive Peace Accord (CPA) in 2006 onwards the Maoist party participated in the government, reducing the King to a figurehead, and later brought the monarchy to an end, diminishing Gyanendra, Nepal's last King, to the role of an ordinary citizen, forcing him to leave the Royal Palace in

Fig. 1.3 Geographical regions of Nepal. Nepal is divided into 5 zones from East to West, the Eastern, Central, Western, Mid-Western and Far-Western zones, 14 Anchal (a cluster of districts) and 75 districts. (http://www.un.org.np/reports/maps/npcgis/NatBio00002.jpg, available till May 2012)

June 2008. This ended the powers enjoyed by the royal Shah lineage over the Himalayan nation for 240 years.

The Maoist Party emerged as the leading party in the Democratic elections in April 2008, with their first Prime Minister in office in August 2008. Since then the political struggles for survival and democratisation have not diminished, and only time will tell how many years it will take to heal the deep wounds, inflicted through the decade long civil war. For additional information see Appendix 7.

1.1.6 A Brief Discourse on Poverty and Economy in Nepal

> "Poverty in Nepal is a deeply entrenched and complex phenomenon." (IFAD, n.d.)

Outdated as these government data are, they are still considered to be on the positive side of Nepal's actual poverty (<1.25 US$/day), as the decade long war (1996–2006) has hampered most efforts for development. According to the UN's Human Development Report 2011 (UN 2011), Nepal's HDI is 0.458 ranking at 157th out of 187 countries.

Table 1.1 Poverty by region in Nepal

Region	Population below poverty line (%)		
	Poor	Ultra-poor	Total
Mountain	29.3	26.7	56.0
Hills	21.3	19.7	41.0
Terai	28.7	13.3	42.0
Urban	13.2	9.8	23.0
Rural	26.4	17.6	44.0
National average	24.9	17.1	42.0

Source National Planning Commission. (1998), p. 203

Table 1.1 presents the figures from Nepal's 1996, nationwide poverty survey.

Nepal's population aged 15 years or less is 40 % and ~75 % of the population lives in rural, difficult to access areas, often with no road access within days of trekking and no grid electricity connection. IFAD, a UN financial institution dedicated to eradicating rural poverty in developing countries, states that approximately 40 % of Nepalese live below the poverty line of US$12 per person/per month (IFAD 2009). IFAD goes on to say: "while the overall poverty rate for Nepal is 31 %, this figure increases to 45 % in the Mid-Western region and 41 % in the Far-Western region"[6]. That shows that additional values, such as caste adherence, family size, remoteness and accessibility of villages need to be considered to define the number of people living under the poverty line. Already Lekhal (2005) and Hagen (1996) concluded that, "…poverty in rural areas in Nepal has a direct impact on environmental degradation and biodiversity loss. Alleviation of poverty in rural areas is therefore vital not only for economic reasons, but also for preserving the environment and biodiversity" (Lekhal 2005, 203). Thus, 30–45 % of the nation's population fall under the poverty line of US$1.25/day[7].

It is the author's experience that besides the use of figures, a good way to identify and define poverty, is to experience it first hand, because poverty, across cultures, manifests itself in the inability to meet basic human needs such as food, shelter, work, clothing, clean water, fuel, health care and education. It denies people chances, opportunities and choices for change and development. Further, it is widely accepted that "poverty alleviation and development depend on universal access to energy services that are affordable, reliable and of good quality" (Reddy 2002; Saghir 2005). According to the WEA report (WEA 2000), there are clear linkages between access to energy and reduced infant mortality and fertility rates and increased literacy and life-expectancy, as well as a clear relationship between poverty and remoteness. Schumacher (1973, 138) in his famous book, "small is beautiful" indicates that some of the primary causes of extreme poverty "lie in certain deficiencies in education, organisation and discipline". He goes on to say

[6] http://operations.ifad.org/web/guest/country/home/tags/nepal (accessed July 26, 2012).

[7] For a more detailed discussion on the IFAD 0.40US$ a day, the previous UN defined 1US$ a day and the now revised 1.25US$/day poverty line income, see Chap. Appendix 8

that in the context of poverty alleviation, "development does not start with goods; it starts with people and their education, organisation and discipline. Without these three, all resources remain latent, untapped potential". Thus besides the national poverty figure the author suggests to define poverty as well through a set of practical indicators such as:

- The use of traditional fuel,
- The burden of the reliance on traditional fuels,
- The effect of limited access to improved, modern energy services (electric, thermal, mechanical),
- Annual per capita electricity consumption,
- Infant mortality,
- Potential for sustainable energy service-producing projects,
- Level of education,
- Agricultural productivity and growth,
- Access to health services, and
- The Gini coefficient[8].

Despite the political instabilities and civil war, Nepal has made some progress towards poverty alleviation over the last decade from the mid 1990s onwards. Poverty rates declined across all of Nepal's development regions and ecological belts, and the World Bank, in its "Poverty Reduction—Poverty Nepal[9]" web site provides a summary of the poverty reduction achievements. Still Nepal's traditional fuel (biomass) dependence represents 85–90 % for all energy services nationwide and 100 % in the remote mountain areas. The annual per capita electricity consumption of ~70 kWh, accessible by ~25 %, ranks Nepal very low globally. Infant mortality is high, ranging from 53 to 86 per 1,000 live births for the richest 20 % and poorest 20 %, respectively. The Gini coefficient for Nepal has changed from FY95/96-FY06/07 from 0.342-0.473, (UNICEF 2010) indicating an increasing dichotomy between the poor and "better off" or "new rich" people in Nepal, feeding political tensions, that are common in many developing countries.

While the CIA, in 2007, ranked Nepal 197th out of 225 nations with an estimated US$1,500 GDP per capita, it down scaled that value to US$1,000 GDP[10] for the year 2008 (CIA 2009). In contrast, UNDP ranks Nepal as 68th out of 102

---

[8] The World Bank's Gini coefficient is a measure of statistical dispersion most prominently used as a measure of inequality of income distribution or inequality of wealth distribution. It is defined as a ratio with values between 0 and 1: A low Gini coefficient indicates more equal income or wealth distribution, while a high Gini coefficient indicates more unequal distribution. from: http://en.wikipedia.org/wiki/Gini_coefficient (accessed July 26, 2012).

[9] http://web.worldbank.org/WBSITE/EXTERNAL/COUNTRIES/SOUTHASIAEXT/EXTSAREGTOPPOVRED/0,,contentMDK:20574069~menuPK:493447~pagePK:34004173~piPK:34003707~theSitePK:493441,00.html (accessed July 26, 2012).

[10] https://www.cia.gov/library/publications/the-world-factbook/geos/np.html (accessed July 26, 2012).

developing countries with an annual GDP per capita of US$252 (UNDP 2006). The World Bank places Nepal, with an average per capita Gross GDP of US$260 for 2004[11], among the poorest countries in South Asia, ranking it the twelfth poorest country in the world. Further, urban Nepalese earn multiple times more than people in rural areas. Considering the Humla people's per capita GDP of only US$72 (KIRDRC 2002) it becomes even clearer, that while averaged national figures may be indicative for a country, they can distort the reality and needs of individual communities enormously.

Another, fast changing economic situation in Nepal is the significant increase over the last few years in remittances from the estimated 2–3 million Nepalese workers abroad. The World Bank indicates that the proportion of households in Nepal receiving remittances increased from 24 % in the FY95/96–32 % in the FY2003/04, and that the actual amount of remittance increased over the same time span by more than 80 %. This major, new financial income source has become a substantial component in Nepal's domestic economy. This carries potential dangers for the country's long-term economic condition, as it creates an artificial economic boom, fuelled by an unstable and artificially inflated source, subjected to the volatile international economy, as experienced since 2008.

Industrial activities are still small and include mainly the processing of agricultural products. Other important sub-sectors include trade with China and India, transport, communication and tourism, which collapsed during the civil war and are picking up again since the peace agreement in late 2006. Foreign investments in Nepal remain limited, due to the small size of its economy, its technological backwardness, its remoteness, its landlocked geographic location, its political instability, civil strife, and its susceptibility to natural disaster.

All of Nepal's fossil fuel needs are imported from India, which creates unstable supplies and political dependence. Nepal has significant, largely untapped renewable energy resources, in particular water and sunshine which can be converted into the needed energy services. The annual discharge of Nepal's ~6,000 rivers is ~236 billion m^3 (WECS 1999a), creating an 83,290 MW theoretical and 42,130 MW economically feasible, hydro-power potential (USAID 2003; Warnock 1989; WECS 1999b), able to generate annual ~300TWh. Till July 2005 only 546 MW (Regmi 2005), or 1.3 % had been exploited with a low average capacity factor of 47 % (Gautam 2005). Till August 2009 the power capacity increased to 635.845 MW (NEA 2009a) with an average capacity factor of 44 % for the year 2008–2009. Since 2000, due to long dry periods, rapid population growth and not enough increase in power generation capacity, forced the authorities to introduce ever increasing load shedding. This culminated in February 2009 with up to 20 h/day having no access to electricity in Kathmandu, Nepal's capital. Varying with the seasonal availability of rain water, load shedding will continue for the next

[11] http://web.worldbank.org/WBSITE/EXTERNAL/COUNTRIES/SOUTHASIAEXT/EXTSAREGTOPPOVRED/0,,contentMDK:20574069~menuPK:493447~pagePK:34004173~piPK:34003707~theSitePK:493441,00.html (accessed July 26, 2012).

decade. These conditions are borne by the local people and industries, as consequences of wrong political decisions taken 15 years ago, a long-term unstable government, civil war and no clear, realistic long-term vision for the country.

Lying at the +30° Northern latitude "solar belt"[12], with about 300 sunny days a year (Renewable Energy South Asia 2002; CRTN 2010), Nepal receives an average global solar radiation of 5.3–6.0 kWh/m^2/day. As a landlocked country, Nepal does not have ideal wind resources. But there are many north–south valleys, cut by rivers into the massive Himalayan mountain range, with strong winds blowing. So there are niche opportunities for wind power.

For additional information on Nepal's poverty, economy and renewable energy resources biomass, hydro, solar and wind, see Appendices 8 and 11.

1.1.7 A Brief Discourse on Development in Nepal

Nepal was a road less country in 1950, nevertheless ~17,282 km (World Bank 2010a) of paved roads have been built by 2007 all over the country (Fig. Appendix 9.1), despite the extreme topographical and meteorological conditions, which makes the development and maintenance of infrastructures, such as drivable roads, exceedingly difficult. This limits the delivery of social services to remote regions, slowing down/crippling any kind of development. Additionally, the high transport costs into the remote areas add to the hindrance of Nepal's development. Forty-two domestic and one international airport, in Kathmandu, serve the country, with many domestic airports still on green fields with minimal navigation systems, causing yearly detrimental outcomes, particularly during the monsoon season.

From an average 2.3 % GDP growth in the 1970s, it grew to 7 % in the late 1980s (through immense foreign aid and development projects), before it stabilised at ~5 % before the war in the early 1990 s. The war slowed economic growth to an average rate of 2 %, with a slow but steady increase since peace returned in late 2006.

Historically, Nepal has been self-sufficient, as all its citizens were basically subsistence farmers for centuries. Ill prepared to meet the challenge of development, especially after the long period of the Rana regime, who intentionally kept the country isolated, increasingly more foreign development projects started in the early 1980s. Suddenly the awareness of being a poor country, compared to the rest of the world, increased. Bista (1991, 133) remarks, that these changes began "to destroy the gracefulness, charm, generosity and hospitality even among the rural people", as the concept of community was there, but it was limited to the local village society and context rather than beyond. The government had no effective

[12] The convergence of heated air from the equatorial region in the Tropopause increases the air mass aloft around the 30° northern latitude (the Hadley cell). This in turn causes the air pressure at the surface to increase. Hence, at northern latitudes of around 30° a high pressure belt called "subtropical highs" (C. Donald Ahrens, Meteorology Today, 7th edition 2003, Brooks/Cole, Chap. 11, p. 286) is created, providing favorable weather to utilize the incoming solar radiation.

indigenous administration to determine the country's needs, or to manage and distribute the aid and no sense of direction for the future. Further, aid is not compatible with the fatalistic beliefs that all is determined, thus fundamentalists often oppose foreign aid which they believe interferes with what is determined for society (Bista 1991, 135). Though this radical belief is in the minority, it is a deeply rooted part of the culture, not verbally heard or uttered, but many behaviours, subconscious decisions and working patterns are tainted with this train of thought.

Within the Sanskritic tradition, education is divorced from any practical use and seen more as a way to distinctive social positions. This is not a conducive environment for the pragmatic approach education needs to bring to community and economic development (Bista 1991, 117–128). It is highlighted by Bista's (1991, 137–138) and the author's practical experience, that too often, Nepalese (almost all from high castes) who studied abroad and returned with new professional skills, higher degrees and with the needed exposure to western society/development/industry/economy, came back with lots of idealism, new ideas and values. But once confronted with the persistent old ways and traditions, built upon the principle of fatalism, they were defeated as few can withstand that strong, invisible force of culture and tradition.

Nepal is characterised by low levels of development and a high dependence on subsistence agriculture, with low productivity, land constraints, natural hazards and imported goods. Additionally, the high population growth causes short- and long-term food insecurity with 42 of Nepal's 75 districts experiencing food deficits (Fang et al. 2007). With $\sim$80 % of the population directly dependent on agricultural production, the potential and scope for agricultural development is enormous. But the professionals recruited to lead this development are high caste academics put in highly respected positions due to their education, and for whom agricultural labour is unthinkable. They have never worked the soil and thus what they learned never gets applied. The use of fertiliser and pesticides is limited, due to limited availability and high cost. There is an urgent need for investment in seed research, supply and irrigation facilities for the different climatic conditions. Food conservation techniques, such as drying and storing, are not well developed (Joshi et al. 2002). Between 20–30 % (MOA 1997) of agricultural production is lost at the post harvest stage.

It is very common that neither during the project's inception, design and planning, nor during the project implementation enough practical steps are undertaken towards local participatory partnership and increased project ownership. The knowledge and experience that public projects such as roads, hospitals and schools, involving several communities, villages or even districts, need as well to be "owned", cared for and maintained, is non existent even today. People and the local government bodies lack the sense of pride or ownership of such projects, compared to things they built through their own efforts. If no ownership feeling is present among the end-users, maintenance will not be done and thus the durability of the project is jeopardised and thus there is no justification to talk about sustainability.

As of 2005, Nepal's health sector facilities consisted of 84 government hospitals, 188 primary health centres, 698 health posts and 3,129 sub-health posts (WHO 2005), with most hospitals and health centres concentrated around urban and low-altitude areas. Remote mountain areas are mainly serviced by health posts and sub-health posts, where the availability of basic health equipment and drugs does not cover a whole year's needs (Emeriau 2006). Women in remote areas give birth in the unhygienic environment of their mud home, in the cow shed or in a specially dedicated "bleeding hut" where the pregnant woman stays for ~14 days, giving birth in total seclusion from anyone from the village, due to her "polluted" status.

Karki et al. (2002) says that only 18 % of the population has access to the national electricity grid (Fig. 1.4), most of which is concentrated in urban areas. In rural areas, rugged terrain makes grid expansion difficult and the combination of low population densities and low demand make it much less cost effective and economically unattractive for independent power producers (IPPs) and NEA. As a result, only 4 % of the population in rural areas have access to the national grid (Karki et al. 2002). The NEA forecasts the electricity needed up to the year 2020 (NEA 2009b). For the year 2009/10 the expected peak electricity demand was around 821 MW. The nationwide installed hydro-power capacity in August 2009 was 635.845MW_R (NEA 2009a). But with an estimated 500–520 MW only functional[13] and a capacity factor of 44 %[14], the average power generation capacity was only ~230 MW, which covered just ~28 % of the maximum peak load.

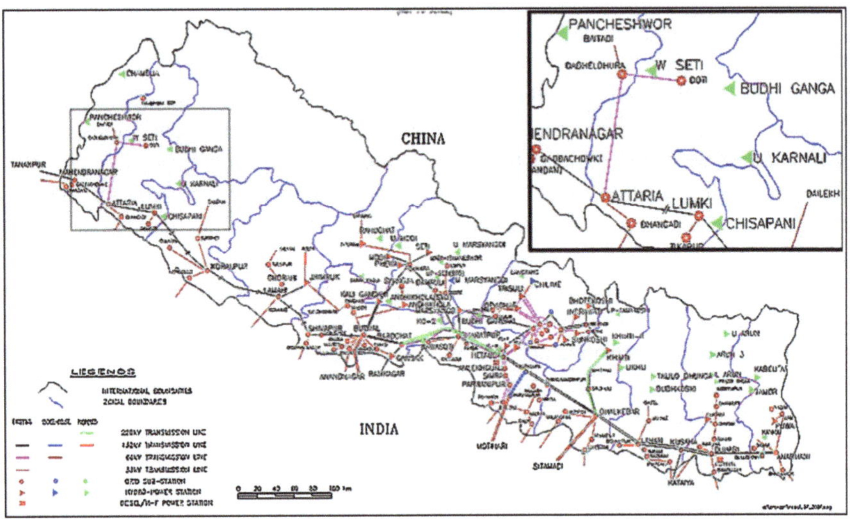

Fig. 1.4 Nepal's national electricity grid in 2009. In the legend the *green line* represents the 220 kV, the *red line* the 132 kV, the *brown line* the 66 kV and the *pink line* the 33 kV *transmission lines*. (http://www.geni.org/globalenergy/library/national_energy_grid/nepal/index.shtml, accessed August 10, 2012)

---

[13] Personal investigation and experience of the author of the hydro power situation and conditions in Nepal.

[14] Calculated from the NEA electricity generation data for 2008–2009, in *A Year in Review-Fiscal Year 2008/2009*, Annual Report, Chap. 3.3.2, p. 4

Studies conducted over the last decade about the relationship between poverty alleviation and access to clean, efficient energy sources and improved energy services basically all agree that there is a direct link between improved access to energy services and increased development and better living conditions (Goldemberg and Johansson 1995; Reddy 2000; Brew-Hammond et al. 2004). Consequently, poverty in Nepal is partially linked to the lack of access to improved energy services and energy infrastructures. The enormous, untapped potential of Nepal's renewable energy resources (1.1.6/Appendices 8/11) indicates that harnessing the locally available renewable energy resources provides great opportunities for Nepal's long-term, sustainable development.

Another point of importance to highlight is the issue of ongoing relevant monitoring of projects beyond their implementation time period as there is no government system of long-term monitoring of the progress of development. Thus, how the people have been actually affected by such projects, is simply not known.

The Maoist insurgency (1996–2006) further crippled economic activities, stalling development through destruction of existing infrastructure, forced migration and loss of life (Fang et al. 2007). Government funds were diverted from urgently needed development to pay for the war bills. Due to security concerns, foreign aid agencies suspended their activities. The war affected access to healthcare and education (UNICEF 2006), reduced agricultural production, and one of Nepal's main foreign income sources, tourism. Nepal will need foreign aid and support for years to come, as this huge annual income is an essential part of the national economy.

But the doctrine of fatalism inherently opposes development, incapacitating the country's social and economic advancement. One promising step forward, escaping the vicious circle of fatalism, could be the focus on increased training and empowerment of local people. People have thus far earned their means of living by putting their hands to work, be it on the land they live on or in the manufacturing or service industry. Bista puts it in the following way: "The Nepali population that has remained untouched by the Hindu caste principles is Nepal's greatest treasure" (Bista 1991, 151). Most of these people though live in remote rural areas. This means, that increasingly foreign aid programs have to make the extra effort to go out into the most remote places to live and work together with these people.

One such area, where people live and have always struggled to survive, is the Humla district, where the non-profit NGO, RIDS-Nepal (co-founded by the author) established in 2002 a long-term holistic community development program. One of the aims of this NGO is to start out afresh, with all above things in mind, on new paths, so that funds and support for village based projects get to those for whom they are intended.

For a more in depth study of the development of Nepal, including important key features of the state, conditions as well some anthropological and cultural mindsets of the Nepali society with regard to development, refer to Appendix 9.

1.1.8 A Glimpse into Humla

The title of the thesis is: "The Role of Renewable Energy Technology in Holistic Community Development". The thesis is based upon the author's practical experience of planning, designing, building/manufacturing, implementing, monitoring and following-up of installed HCD (Holistic Community Development) projects, in which applied, contextualised RETs (Renewable Energy Technologies) are included. All projects take place in partnership with 13 remote, high-altitude village communities in Humla, in a hostile and harsh environmental context. Thus, the thesis is an amalgamation of 16 years of theoretical learning and practical experience of living and working with these communities, designing, implementing and following-up RET projects embedded in long-term HCD programs.

In order to understand the central chapters of the thesis in context, it is important to provide initially some background information about the Humla district and its people. It aims to enable the reader to picture for himself the living conditions of the Humla people, identity, their main living conditions and their unique environmental context. They live so remotely that they can be reached only either by airplane and then by foot. If no airplane is available or the weather conditions do not allow it to fly, then it requires a 15–17 day trek from the next road head across and over some of the highest Himalayan passes and mountain trails.

Today, an estimated 75 % of Nepal's population lives in rural areas, while about half of this population lives in areas within a 2–17 days walking distance from the nearest road.

Humla (Fig. 1.3) is one of the 75 districts of Nepal. Humla, with Simikot as its main district town (Latitude: 29°58'20" North / Longitude: 81°49'06" East / Altitude: 3,000 m.a.s.l..), is remotely located in the far north-west of Nepal, bordering Tibet. It can easily be spotted with Google Earth with the above provided coordinates. Humla is elevated between 1,524–7,031 m.a.s.l. Lack of roads and extremely difficult air access are amongst the most important constraints for the local people's development and improvement of life and to tourism development in the area.

Humla is the home of two different ethnic groups. The Hindu Nepalis, inhabit mainly the villages in the southern and middle part of the Humla district, between 2,000–3,000 m.a.s.l. Their caste system, as described previously, is still today very strictly adhered to and a mostly unexpressed but clearly visible dichotomy exists between the different castes. In the more northern, higher altitude villages (between 2,500–4,000 m.a.s.l.), one finds a separate ethnic group. This group is collectively referred to as the Lama population, referring to its observance of Lamaism, or Mahayana Tibetan Buddhism. The ancestors of the Lama population migrated into these valleys from Tibet (today part of China) in the ancient past, with distinct populations in Humla tracing their descent from founding fathers of both Western and Eastern regions of Tibet. Both Kagyu and Nyingma lineages of

Fig. 1.5 Humla district has a rugged and harsh environment, making it very difficult to make a living from it. Deforestation increases land slides, further limiting the agricultural land. Already <1 % of the land is arable and can be used for growing people's staple food. In 6 of the 8 villages pictured in their mountainous and harsh surroundings, RIDS-Nepal runs HCD projects

Tibetan Buddhism may be found in practice among ethnic Tibetans of Humla (Zahnd anecdotal data) (Figs. 1.5, 1.6).

This remote area has remained under the powerful influence of superstition, and culturally conservative belief systems, giving Humla a reputation for social backwardness which contributes to its social and economic marginalisation. These conditions adversely affect the governance, human rights and the equitable functioning of Humli society. Women in particular suffer from the influence of superstitious beliefs, for example, Humli women are still routinely compelled to stay in filthy cowsheds whilst menstruating, and also during and for several days after childbirth.

Humla is not only rich in species biodiversity but also has a broad range of habitat types. The wild fauna is rich and includes several protected and endangered species in the Humla region, such as the snow leopard (Panthera uncia), the clouded leopard (Neofelis nebulosa), the musk deer (Moschus moschiferous), the gray wolf (Canis lupus) and the leopard cat (Felis bengalensis). It is the well known Yak's home and, in the upper Humla region, the Tibetan people use them for transporting their goods as well as to produce milk, from which they manufacture butter for their traditional butter tea, or Tibetan tea. In addition to the fauna, Humla alone has about 1,500 species of high-altitude plants, with a high concentration of different medicinal plants, which are one of the main income sources of the local people. Several plant species fall into the endangered plant category.

Fig. 1.6 Hepka village at 2,749 m.a.s.l., with 95 homes and Tangin village at 3,233 m.a.s.l., with 41 homes in upper Humla district, in its harsh, high-altitude environment. In both villages RIDS-Nepal has long-term HCD projects

According to the 2011 Human Development Report, the Nepalese HDI is 0.458. Within Nepal, the Humla region is amongst the poorest, with an HDI of 0.244 (DDC 2004). Its literacy rate of 27.1 % is also the lowest of the country (DDC 2004), with a women's literacy rate of ~5 %.

Humla is the most northern part of the Karnali Zone (Figs. 1.2, 1.3), which is in turn part of Nepal's mid-western development region. It is one of Nepal's larger, though sparsely populated districts with 51,008 people (2011 Census). It lies in the middle of the vast Himalayan mountain range between 29°35′–30°47′ northern latitude and 81°18′–82°10′ eastern longitude. The upper Humla villages lie at altitudes between 2,500–4,000 m.a.s.l., while their picturesque surroundings and mountain ranges reach altitudes up to 7,031 m.a.s.l. A harsh, high-altitude climate, with an average of 199 frost days a year (NASA 2003; Zahnd 2004a, 233), prevails. Most of the 5,355 km^2 Humla region is either covered with rocky mountain ranges, snow covered mountains, forests or high-altitude pastures. Only 50.2 km^2, or 0.89 % of the whole district area is suitable for farming (DDC 2004, 10).

The district lies in the rain shadow of the Himalayas and consequently receives much less precipitation during the monsoon season (June–September) compared to the rest of the country in the middle- and lower-altitudes. This effect increases with altitude, with settlements above 2,500 m.a.s.l entirely dependent on snow and glacial melt water. Yearly rainfall ranges from 800 mm in the north of Humla to 1,200 mm in the south (Emeriau 2006). This means that Humla does not share the same seasons as the rest of the country (Appendix 4), but has only three distinct

Fig. 1.7 Hepka village, a Tibetan village at 2,749 m.a.s.l. with 95 homes and 617 inhabitants in the upper part of Humla

Fig. 1.8 The harvesting time of the finger millet fields of Hepka village is *near*

seasons, the summer, with rains (June–September), the cold, sunny and dry winter (October–February) and a dry, sunny spring (March–May) (Figs. 1.7, 1.8).

While wheat, barley, 'pepper' (buck wheat), corn, potatoes, and kidney beans are grown as the staple food in single crops, apples, apricots and hard walnuts are the major fruits produced in the district. Though rich in natural resources such as water, sunshine, medical and aromatic herbs, these resources are not tapped into in appropriate or sustainable ways, because of lack of education, conservation awareness, basic infrastructure or economic opportunities.

Humla is known as a permanent food shortage area, and for most of the year food has to be flown in at inflated prices, due to an average food to transportation cost ratio of 1:6 (DDC 2003, 6)[15]. That is also reflected in the high rate of malnourished children <5 years of age, which is estimated to be 65 % (DDC 2003, 27, RIDS-Nepal 2007). The educational level of the Humla people is also low. Each person averages only 0.88 years of school education and the literacy rate for Humla women is just 4.8 % (KIRDRC 2002, 5). Thus it does not come as a surprise that Humla, out of Nepal's 75 districts, ranks second to last for poverty, deprivation, socioeconomic and infrastructural development, and female empowerment, and 72nd in terms of socioeconomic and infrastructural development (ICIMOD 2003).

During the decade long civil war (1996–2006) the political tension was very high, as the Maoists had many supporters and followers in such an impoverished region. The non-profit NGO RIDS-Nepal was the only local organisation working throughout these years in the villages and with the local communities in upper Humla. The political tension rose to such a high level at times that it was necessary to limit the days of work in some of the villages as there was explicit danger to life for the RIDS-Nepal staff (Fig. 1.9).

Almost all families and communities living in such remote places as upper Humla do not have access to improved energy services such as basic indoor lighting or improved cooking and heating stoves which carry the smoke outdoor. Further, their remoteness, the harsh, high-altitude Himalayan climate, with very limited agricultural opportunities, add to their poverty and widespread malnutrition. Beside Humla being one of Nepal's poorest regions, the long, harsh winters and a short growing season, make it subject to severe food insecurity, which is a hallmark of poverty in Humla.

Agriculture is difficult, due to the steep slopes (Figs. 1.5, 1.6), rocky and poor soil (Fig. 1.10), and the terrain limits accessibility. Additionally, the dry climate and unavailability of irrigation limits the potential for cultivation. Insufficient food can be grown to feed the population for the year and, due to the lack of road infrastructure, it is prohibitively expensive or impossible to import adequate food to the area. As a result, the region has high levels of chronic malnutrition and micronutrient deficiencies.

Pressure on forests for fuel, fodder and building materials has led to severe deforestation and drying up of local water sources. Reduced rain and snowfall over

[15] RIDS-Nepal paid NRs180/kg (or US$2.5/kg) during the project fiscal year 2009–2010 for each kg of equipment or food for the air transport from Nepalgunj in the south of Nepal to Simikot, Humla's only way to bring in goods into the district. The Simikot landing strip is Humla's only, till end of 2011 unsealed mud and stone airstrip (see: http://www.rids-nepal.org/images/google_earth/HARS_Simikot.kmz) (accessed July 26, 2012).

Fig. 1.9 Upper Dhadhaphaya village in Humla. All the Humli houses are built with stones and mud and wood beams. The roofs are covered with a thick layer (15–30 cm) of mud and around the edges with flat stones. The typical architecture of the area is to build the houses in clusters, close to each other and on top of each other. That is partly due to the shortage of flat and arable land

Fig. 1.10 Less than 1 % of Humla's land is suitable for agricultural work. In this harsh climate crops can be grown for only 3–4 months per year. The soil is meagre, nutrition deficient and stony. Also in Humla, the farmers still use only hand made wooden (hard, local "kharsu" wood) ploughs with no steel parts included

the last few years has further reduced crop and animal production. Landslides and heavy snowfall are the greatest natural hazards (Emeriau 2006). Thus Humla is amongst the very remote and fragile environments which are in a state of crisis due to increasing environmental exploitation and population pressure.

Consequently, Humla's extremely under developed condition, with severe food security issues, can partly be accounted for due to the difficult agricultural conditions and isolation. To address the widespread chronic nutritional deficiencies, development of local agriculture is needed to improve output and diversity of food production. That is further substantiated through RIDS-Nepal's surveys revealing that an estimated 65 % (RIDS-Nepal 2007) of Humla's children <5 years of age are chronically malnourished[16] according to the UN standards.

Night blindness prevalence in pre-school children, a sign of Vitamin A deficiency, was found to be 1.5 %, compared to the national average of 0.27 %. 20.8 % of women declared they had suffered night blindness during their last pregnancy, higher than both the national rate of 11 % and the mid- and far-western rate of 18.7 %. Further, micronutrient deficiencies were indicated by occurrences of anaemia and night blindness. Anaemia is related to a lack of iron and its prevalence among pre-school children was found to be 38.1 % with 2.2 % having severe anaemia (Emeriau 2006). All this data present a dire picture of the lives and living conditions of the Humla people.

The local diet in Humla is based around roti (unleavened bread made from grain) or rice[17] (imported) as the staple food, with dhal (lentils) and vegetables and fruits when they are available. Animal products such as meat and eggs are rarely consumed (Fang et al. 2007). Due to the lack of fresh vegetables and fruit, and reliance on unleavened bread, both Vitamin C and zinc deficiencies are common (Treese 2007).

Based on these findings, RIDS-Nepal started a malnutrition program in 2005 for children <5 years of age (as part of the "Family of 4 PLUS" program). It operates in those villages where people have identified their children to be malnourished

[16] Though there is no accepted international definition for "global chronic malnutrition", a possible description for it is: "the situation of a child who is stunted, underweight and experiencing nutrient deficiencies which have weakened them to a state from which they may never recover". Global chronic malnutrition is an indication of long-term food shortages and results in "stunting". The term "chronic" suggests that the malnutrition is a prolonged ordeal that is weakening the child's immune system, stunting their growth and impairing their intellectual capabilities which will affect them for the rest of their lives. It is most often the result of a combination of long-term inadequate dietary intake, infections and diseases. When children don't get enough to eat, their immune system can't fight off diseases and infections, resulting in greater severity and duration of sicknesses. Additional, long-term diseases speed up nutrient loss and suppress appetite, resulting in growth failure. Thus malnourished children's growth is hampered and they are shorter and lighter than they should be for their age according to the UN standards for weight and upper arm circumference measurements. Measuring weight and height of young, malnourished children and comparing them with the "standard" for the population is the most common way of assessing them (summarised from UNICEF 1998).

[17] Rice can not be grown in the villages in Upper Humla, thus people are fully dependant on the government rice subsidy program, through which rice is air transported to Humla. People are assigned a monthly rice quota they can purchase at a subsidized rate. This program has been running by the government since ~1980, making people of Humla fully dependant on rice instead of enabling them, through the teaching of new skills, to improve their ancient agricultural methods and techniques and plant and grow new crops suited to the local context.

Fig. 1.11 "Super porridge" or "sarbotham pito" (in Nepali) production from locally grown ingredients, is part of the learning for mothers with children <5 years of age, enrolled in the children's malnutrition program

and where RIDS-Nepal, in close partnership with the local village community, has already implemented the "Family of 4" program. As part in that program, RIDS-Nepal has conducted an extensive village survey, to determine what kind of food the various groups of people, distinguished by sex and age, eat on an average day each month of the year. That allowed us to identify the approximate daily energy intake from food for each person and group on an average day for each month. Further, the survey aimed to find out what kinds of average daily work and jobs each of the defined people groups is involved in each month. That allowed us to identify the approximate energy needed to perform these daily tasks. The aim was to find out the average food deficit intake for each person and group defined for their daily, normal work and jobs. That allowed us, for the first time, to quantitatively evaluate the often heard statement that "Humla is a permanent food shortage area, with people being malnourished". While that sounds harsh and needs to be addressed, only through time demanding field based research and data surveys could more accurate information be collected. Having done that, RIDS--Nepal could design a much more accurate, and applicable, malnutrition program for the malnourished children. This addresses food shortage and food deficiency as an integral part of a long-term HCD village program (Figs. 1.11, 1.12).

For six people groups/ages these surveys and data have been collected:

(1) Women ~15–50 years
(2) Men ~15–50 years
(3) Children 0–6 months
(4) Children 6–12 months
(5) Children 1–3 years

Fig. 1.12 To feed the malnourished child with "sarbotham pito", and to measure and weigh the child enrolled in the nutrition program every month, is part of RIDS-Nepal's malnutrition program

(6) Children 3–5 years

The following Fig. 1.13 presents a bar diagram of the collected RIDS-Nepal survey data for the average daily energy intake of the various food items eaten each day and the estimated average daily energy demand for all daily jobs, works and responsibilities. That resulted in the average daily energy deficiency for an average Humla person in the people group identified over the course of a year.

For the bars in Fig. 1.13, the required average daily energy intake needed for a female adult's daily works and jobs, was assumed to be 2,500 kcal or 10,467 kJ, over the year, based on the assumption of a heavy, daily workload.

Further, the lack of sufficient and diverse food, results in major nutritional deficiencies. A recent nutritional survey performed by ACF[18] in Humla and the neighbouring district of Mugu (Emeriau 2006) reported a prevalence of global acute malnutrition and severe acute nutrition equal to 12.3 and 3.3 % respectively. Global chronic malnutrition and severe chronic malnutrition prevalence were found to be equal to 63.9 and 37.1 % respectively. Acute malnutrition is related to immediate food shortages (crop failures, droughts) and results in "wasting", while global chronic malnutrition is an indication of long-term food shortages and results in "stunting". The global chronic malnutrition data for children <5 years of age from the RIDS-Nepal survey amounted to 65 %, and thus is comparable to the data

[18] ACF (Action Contre La Faim, or Action against Hunger) was founded in 1979 by a group of French people in response to the emergency in Afghanistan. The organisation is exclusively dedicated to ending hunger around the globe. http://www.actioncontrelafaim.org/en/content/acf-international, and http://www.actionagainsthunger.org/ (both accessed July 26, 2012).

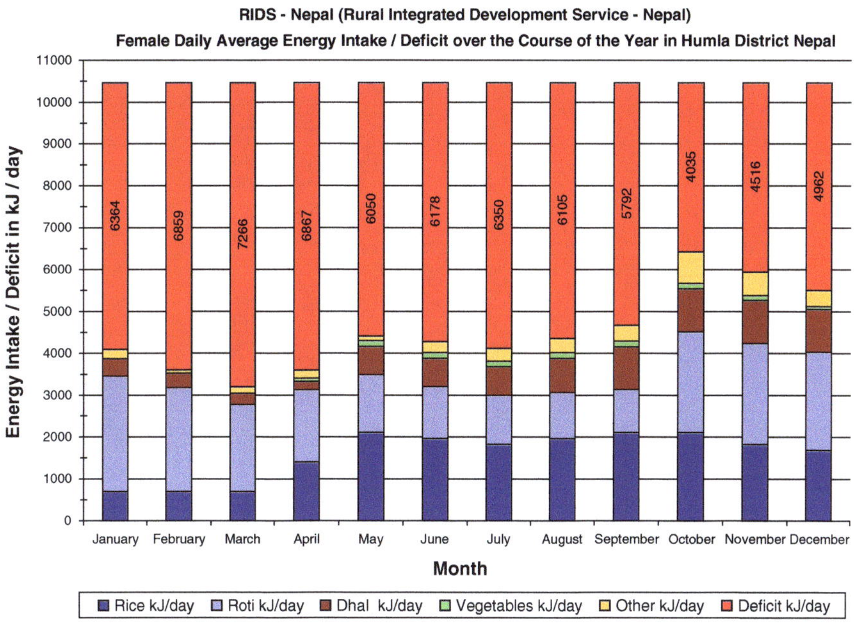

Fig. 1.13 Female daily average energy (food) intake/deficit over a year in Humla District, Nepal (RIDS-Nepal's survey 2007)

from ACF, though RIDS-Nepal's data were taken only in 13 villages during 2009 in the upper Humla district where RIDS-Nepal has been working since 2002.

A number of national and international organisations are active in providing food aid to affected areas in Nepal, the most prominent of which is the Nepal Food Corporation (NFC) and the World Food Programme (WFP). Both of these organisations focus on importing food supplies into disadvantaged areas, such as Humla. However, these heavily centralised programs are not well suited to a country like Nepal with extreme topography and lack of road infrastructure. High transportation costs reduce the amount of food that can be distributed in remote rural areas or prevent it from being distributed at all (Fang et al. 2007). Further, these programs are not long-term programs but usually run only for short-time periods, weeks rather than months, to cover some period of crop failure, which is additional to the permanent food shortages and would bring masses of people beyond the line of physical recovery. Necessary as such programs are, they do not address the long-term need to improve step by step the food situation in Humla, teaching people to learn new agricultural skills, improve agricultural tools, to start planting new crops, or build greenhouses and solar driers to produce and store more food. The prevalence of acute and chronic malnutrition was found to be higher than the national rate, and, in the case of severe chronic malnutrition, higher than in all western mountain regions of Nepal (Fig. 1.14).

A follow-up study by ACF in 2007 showed reductions in global and severe acute malnutrition from 12.3 and 3.3 % to 10.1 and 1.3 % respectively in the

Fig. 1.14 Child from Dhadhaphaya village, with bloated stomach, suffering from an intestinal parasite (worms) infection. This is a very common problem

initial study in 2005, indicating that the short-term food situation had improved. This study was conducted a year after the initial survey and just after NFC and WFP had conducted a short-term food relief action program because the 2006 crops failed terribly. However, global and severe chronic malnutrition prevalence were still 68.5 and 35.5 %, showing a small decline in the severe chronic malnutrition prevalence of 1.6 %. There was an increase of 4.6 % in the global chronic malnutrition prevalence, indicating that the long-term food situation, instead of improving through the food relief program, had worsened (Fournier 2007). This shows that only long-term solutions are able to improve the dire food situation of the Humla people in a sustainable and dignified way.

This short introduction to Humla and its people aimed to highlight just a few of the issues and circumstances the local people have been fighting all their lives. Though not comprehensive in any way, even the few facts and data provided are enough to show that there is a huge challenge ahead to relieve their poverty and malnutrition.

1.1.9 Contribution of This Research to New Knowledge

Discussions, studies, seminars and even postgraduate programs on community development are widely available. Most of them focus solely on intellectual research, followed by extensive written dissertations, "how to do" manuals and lectures. They generally produce sensible results and suggestions such as, highlighting the importance of participatory approaches to development projects; raising the problem of gender discrimination along with the need to address this inequality; and including the community in the project through consultation and practical hands on activities during implementation; plus other common conclusions. While these issues are all of crucial importance to any development project, and even more for any kind of community related development project, where not merely physical changes are aimed for, these issues are only part of the story. Unfortunately, there are few practical examples of the successful application of these principles to real life situations.

One of the main rational explanations for this is that most of the writings, discussions, seminars and lectures in Colleges and Universities, are around the basic philosophy of development. They emerge mainly through academic research rather than observation and participation in development activities and projects. The theory of community development is mostly based on intellectual anthropological studies, extracted from the published literature in the field, rather than extensive practical experience. This thesis is novel in that it aims to provide what has been missing in this field, a new approach to community development based on practical experience and academic research. It presents this new approach, based on literature research and practical experience, understanding and learning, gained from 16 years of field work in remote areas of Nepal, working amongst those to whom the development endeavour is directed.

Further, community development projects are usually strongly linked with technical projects. However, technical solutions have often been presented in terms of "standard" solutions and applications. For example, there is one kind of cooking stove available in the market. Thus, it is widely assumed that this stove is the solution for all communities and people, independent of where they live and what they cook. The question, of whether this stove is appropriate for the Humla people, who endure the terrible and life threatening indoor air pollution through their open-fire cooking/heating places, is not asked. In this thesis it is argued that a stove, in order to be relevant for a particular context, has to be designed and built according to the locally available fuel resource, the food people cook, the way they cook and the way they like their food to be cooked, plus any other energy service demands/needs, such as room heating and water heating that the people themselves have initially identified in the survey. This is called contextualised technology and only with this approach is it possible to solve some of the deep rooted development problems of these communities. This approach has to be applied to any technology considered for community development projects. While maybe an "off the shelf" technical solution goes some way, it is neither responsible nor fair towards the end-users, to talk of sustainability if the technology is not contextualised.

As previously stated, access to improved energy services is a key factor for better living conditions. With the important issues of global warming, depletion of fossil fuel resources and energy independence, the utilisation of renewable energy resources and technologies have come more and more to the forefront in the discussions about solutions. This is also true for communities who do not have access to the grid or a road. Renewable energy technologies are able to provide improved energy services, tapping into the locally available renewable resources, thus making the community energy independent. They do not contribute to oil depletion and they facilitate the solution to global warming. Thus, to join together the two main themes, a new approach to community development and a strong emphasis on the application of contextualised, renewable energy technologies, constitutes a new field of knowledge and research. This is what this thesis aspires to present. It aims to add novel thoughts, writing and practical examples to the theory and practice of community development as it starts out by first learning and understanding the issues people and communities face in their daily activities in a set context and how they identify and prioritise their own situations and needs.

This is only possible if time is invested to learn about the life and culture of the community in focus and to live and work with them for extended periods of their lives. This is one of the main reasons why studies in this field have rarely flowed on to practical outcomes. If time had been devoted to implementing even a few of the brilliant ideas scholars put forward, they could have changed the world already. Further, having recognised that improved access to energy services is of fundamental importance, renewable energy technologies must play an important role in that new approach to community development. Practical examples are needed of projects, which are monitored and refined as they evolve and over years and generations are more and more absorbed and finally merge with the community and become part of their culture. Evidence of practical, context related change is needed in order to talk responsibly about sustainable development. While such transformations have long time frames, no one can attempt to provide in a single piece of work the ultimate solution to community development. One life is just not enough, as such deep rooted changes at the cultural, anthropological level take generations, before they are entirely absorbed.

Thus this thesis aims to describe the experience and knowledge that contribute towards this objective. It claims to present a compact summary of 16 years of development work during which the present thinking emerged from grass-roots experience, wrestling with the status quo, thinking about development, working out new theories, applying them in practice, having monitored the failures and successes and thus refined them, only to start the cycle of learning all over again, multiple times. Thus the present work does not claim to have reached the end of the road but rather an important crossing in a vast road network which thus far has not been identified explicitly. The central part of the thesis, Chaps. 2–6 and the publications and papers presented in Appendix 15 demonstrate that the practical walk has not only started, but has already gone some distance, long enough to be sure that the road is new, appropriate, though not without difficulties but promising enough for the "main traffic" to switch on the indicator and move over.

Additionally, while the thesis does not explicitly deal with or discuss present policies for development in Nepal, nor in any other developing country, it has the potential to add new knowledge and understanding, based on real life experience, for policy makers and governments who are in the process of developing or revising their policies in regard to community development.

Finally it also rejuvenates a discussion which, according to the author's perception, has become liturgical in its practical approach among the experts. Too many have been preoccupied with discussing and presenting the different theories, resulting more in disagreements about theory, so that the actual intended end-user and beneficiary, the local person, family or community, has been pushed to the periphery and thus lost its central place of raison d'être.

1.2 Sustainability and Development

Over the course of the last four decades many summits (Appendix 10), conferences and seminars on "Sustainability and Development" have been held[19]. There have been many comprehensive resolutions written but much fewer successful examples of sustainable development brought forth as showpieces from which one can learn. The differences between developed and developing nations, the diversity of cultures and multi-national mixtures, and differences in living standards and economic power, make it impossible to come up with a unified, all embracing definition of "Sustainable Development".

But what has been learned over the course of time, through theory and practice, is that these complexities can be better understood if they are separated into more manageable, distinguishable categories. Thus some of the main lessons learned since the Stockholm Conference in 1972 and the Brundtland Report in 1987 are the following.

- Mankind lives within a limited ecosystem which, while it can bear to some extent the misuse of resources, it can only provide a limited supply of raw materials. The mantra of "steady economic growth for development", which has been venerated over the last few decades of the twentieth century, has now become questionable. A fast growing middle class around the globe has become aware that growth will lead sooner or later to a collapse, which may be due to the pollution of our habitat or the depletion of limited vital resources.
- No nation has more right to better living conditions than any other nation. Nor for that sake should any one nation be allowed to pollute the earth more than any other nation.
- The interconnectedness of these issues has been realised, also the time frame for addressing them has been understood as a generation rather than a few years.
- It has been recognised that each nation needs to wrestle within its own context and framework to define sustainable development and implement it accordingly. It is recognised that Governments are responsible for developing policies which are applicable within their defined social, political and environmental context.

[19] For a more detailed discussion on the last four decades of international summits and conferences on, or including the topic of "Sustainability and Development", the author has added an additional chapter, Appendix 10, in the Appendices.

Further, it has been acknowledged that any successful implementation of sustainable development policies and strategies requires the propagation and direct and active involvement of the local Government agency (LGA), responsible for their surrounding civil society and environment. The bottom-up involvement of NGOs and LGAs in the promotion and implementation of sustainable development projects must go hand in hand with the development of nationwide policies (top-down). The communities at the grass-roots level have to be involved in a participatory way through initial consultation and participatory design of the policies, and the implementation of the agreed actions.

- It has become apparent that involvement and participation of the local community in shaping their future living conditions is a central political value of democracy. Further, it also builds mutual respect and a healthy, strong ownership feeling for all the tasks and activities put in place. In addition, it draws upon the local community's experience and understanding of their own environment, resources, social and political structures. This is the most appropriate way to understand the local context and tackle the indigenous circumstances. This also impacts positively on local resource management as it draws upon long-term established local ecological and social knowledge, enabling sustainable development to grow from within the local community.
- Education for every person, independent of gender, age or ethnicity, has become a strong driver in the propagation and widespread dissemination of the need for changes from the present approach to development towards sustainable development.

Since the late 1990s the discussions about sustainable development have also increasingly intensified in Nepal. This is partly due to the unique, pristine and fragile environment of the Himalayas which influences the ecology way beyond Nepal's borders, and partly due to the increased collaborative development activities between the Government and foreign agencies. Thus it is not uncommon that related conferences, workshops or seminars take place in the capital Kathmandu. For example, in the opening remarks of the two-day seminar on "Education for Sustainable Development in Nepal", Dr. Trailokya Nath Upraity (Trailokya 2003, 5–6) from Japan paraphrased sustainable development, initially defined in the Brundtland report, as:

> Sustainable development, in a nutshell, is "living well within the means of nature" or, living in a way to create lasting economic prosperity, environmental health, and social justice for current and future generations. Hence, its goal is to develop a society in balance and improve the quality of human life. ...The concept of sustainability, in short, envisages the use of resources that replenish themselves over a certain period of time. ....In search of a balanced and sustainable view of development in any country, education has to play a major role to impress the minds of the growing generation that all efforts and measures undertaken for development are to be sustainable.

The Nepali Government's National Planning Commission presented in 2003 a document called, the "Sustainable Development Agenda for Nepal" in which it defines sustainable development in the following way (HMG 2003, 10):

> The over-arching goal of sustainable development in Nepal is to expedite a process that reduces poverty and provides to its citizens and successive generations not just the basic means of livelihood, but also the broadest of opportunities in the social, economic, political, cultural, and ecological aspects of their lives.

Many of the lessons learnt over recent decades are included in this concise definition of sustainable development for Nepal. Again it is not the theoretical definition which is faulty or not comprehensive enough, as it contains all the major words and factors it needs to embrace in order to be acceptable. Rather, it is the development of the necessary policies and even more the outworking of the policies, which in turn builds the basis for implementation. This is the actual challenge which is hard to address, and in the case of Nepal, this process is still to come. Thus there has up to now been plenty of discussion but not much visible outworking of the well verbalised objectives.

Today's knowledge and experience of "sustainable development" requires responsible policy makers to use the term "sustainability and development" with an increasingly wider perspective. They must flesh out the needed policies accordingly so that they include more types of development activities than in previous decades. The deliberate process of concurrently achieving social, political and ecological changes forces anyone active in that field to consider much broader approaches than hitherto.

1.2.1 The Need for Holistic Community Development

In keeping with the continuous trend of thought developed since the Brundtland report's first publication (Brundtland 1987), issues such as free/affordable quality education, social welfare, gender equality, skill development, people empowerment and available and affordable health care have to come more and more into the spotlight and must be integrated into the definition of "sustainable development". They have become increasingly a part of the emerging, more integrated, understanding of "sustainable development", reflecting the learning and thinking of the past 20 years. In addition, rather than aiming to produce the ultimate, universal, all inclusive definition for "sustainability and development", the trend of thought has progressed beyond that towards the more practical applications of "sustainability and development". It is the learning from real life experience through creative thinking, applied and relevant research, practical applications and project life-cycle evaluations with subsequent integration of the new knowledge into the ongoing development of the theories. It is through the progress of context related "learning by doing", together with retrospective reflections and revisions that the interface between "sustainability and development" takes on meaning and is defined and understood in its contemporary context. This is not a static, but an ongoing progress which is constantly evaluated and amended as learning takes place and new results are obtained. Or in short, "sustainable development" has to be understood primarily through the process of implementation and not through the development of new definitions only.

With a strong emphasis on grass-roots projects and "learning by doing", the author developed from 1996 to 2000, the first approach to a more comprehensive, local context related, community development approach in one of the remote, least developed Nepali Himalaya districts in the north-west of Nepal, called Jumla. As is normal in a project such as this, growing pains are part of it, thus the experience was very mixed. There were many positive and motivating results, various highlights, frequent eye openers as well as many nose dives, from which it was often not easy to recover. The problems faced were of many types. They ranged from simple technical failures due to an inexperienced and non professional approach to contextualised development of appropriate technologies, managerial shortcomings, to the inability of organisational structures to adjust fast enough to such a new approach, as well as to the rapidly changing environment and challenges in the world of "development".

The approach was very practically orientated, including the local community actively in each phase of every project. This strongly hands on exposure and experience provided the needed framework for the next step, the holistic community development (HCD) approach of the "Family of 4" and the "Family of 4 PLUS". For this emerging new HCD concept a new, non-profit NGO (non-governmental organisation), called RIDS-Nepal (www.rids-nepal.org) was co-founded by the author in 2002 and later, in 2005 registered by the Government. This was in order to have "new wine skins" for "new ways" as the implementation of the new HCD concept of the "Family of 4" and the "Family of 4 PLUS" demanded some different organisational structures and policies.

For example, it is paramount that the RIDS-Nepal field staff consists of local people trained and empowered through long-term on the job training and job responsibility related academic and practical training, if possible, provided within the field project area. Further, it is policy that each staff member, along with their daily work responsibilities, is either engaged in a non-formal education/training program, or enrolled in a formal educational program, aiming to improve his/her ability to fulfil the job responsibilities. Further, all the field staff have to live long-term in the area where they work, so that they become an integrated and active part of the local communities, with whom the projects are implemented in partnership.

This new concept of HCD is now being applied and implemented since 2002 in an even more remote and poverty stricken region of Nepal's north-western high-altitude area, the district of Humla. The "Family of 4" has been so named because the majority of villages and local communities partnered with since 1996 have repeatedly identified their own community development priorities to be:

(1) Pit latrines.
(2) Smokeless metal stove (metal is preferred in the high-altitude areas because indoor heating is required year around as well as improved cooking conditions).

(3) Basic, elementary indoor lighting (powered through RAPS (Remote Area Power Supply) systems which tap into the locally available renewable energy resources such as solar energy, the local stream or wind resource).
(4) Clean drinking water (from village based tap stands).

The project implementation phase of the "Family of 4" is usually ~2 years after the initial base-line survey has taken place in order to understand each family's (and the community's) initial context, situation and self-identified needs. Following this a mutual agreement is signed for a long-term project partnership between the local community and RIDS-Nepal. Once each family in the community/village has access to the improved living conditions the "Family of 4" enables, a follow-up survey is conducted among the same families to learn what impact the program has had in its initial implementation stage of ~2 years. During that follow-up survey the local people often identify new needs which they have now become aware of as they ascend the "ladder of development". Thus a second HCD concept, based on the "Family of 4" has been developed, called the "Family of 4 PLUS". It includes thus far (July 2012) the eight most often mentioned additional desired development activities/projects identified during the follow-up survey. These are:

(5) Greenhouse for increased and more secure, food production.
(6) Solar drier to dry the surplus vegetables, fruits or other food products (such as meat) in hygienic ways and with a moisture content allowing long-term storage.
(7) Slow Sand Water Filter (SSWF) installed inside the home, in order to store the collected and filtered drinking water from the tap stands in the village in a clean and safe manner.
(8) Nutrition program for malnourished children <5 years of age.
(9) NFE (non-formal education) programs for women and out-of-school children (mostly girls).
(10) Hot water (generated by a solar water heater system) to wash and take showers for improved personal hygiene.
(11) Solar Cooker to utilise the local solar energy resource and minimise the local biomass consumption.
(12) Scholarships for apprenticeship training in construction, midwifery, community nursing or agriculture in a technical school in Nepal.

These additional projects do not need to be implemented in every village, but are part of the main core of the "Family of 4 PLUS" and are implemented on a village by village basis depending on the local people's identified request for further development of their community.

These programs are envisioned to be part of the long-term ($\leq$2 generations or ~20 years) partnership between the NGO and the local community. It is important to think in such long time frames as any changes within a community, either social, technical, environmental or political, need years to have their full impact and become part of the community culture and way of life. It is a central part of the new HCD concept that such a long time frame is considered.

With the new NGO's organisational structure and policies this is more easily possible as all the field staff are local people, chosen and trained to live and work among their own communities. They have a strong interest to succeed as they benefit themselves at several different levels. They learn new practical and managerial skills, participate actively in the shaping and improving of their own local people, gaining the respect of the local community, which often results as well in gaining a better social status. In addition they are good motivating examples for the other local people.

Further, RIDS-Nepal's policies include the objective that "low caste" people, from the so called "Dalit" group, have a high priority to be employed. This, because these people are usually the poorest and most underprivileged in regard to education. Once a person from the Dalit caste is part of the RIDS-Nepal team, he/she has the opportunity to learn and grow on the job. The main Humla field project manager of RIDS-Nepal, from 2002 to 2010, Govinda Nepali, was from this group. He took his opportunities and worked hard and successfully passed the class 12 school leaving certificate, majoring in Science. He then enrolled in a 3 year Bachelor degree in social science and English, graduating in 2010, all alongside his daily work responsibilities. In 2011 he enrolled for a full-time Masters Degree in Rural Development. Needless to say, this demands much more from someone than just fulfilling the expected job responsibilities.

Since the initiation of the Humla programs in 2002, he and the NGO have three times been awarded by the Government the "Best Project and NGO of the Year" award for the Humla district. Thus it is obvious that an important gap, the unjust division of society through the caste system, which has direct implications on their opportunities in life, has been bridged with dignity and honour.

Further, 80 % of all 16 RIDS-Nepal field staff have passed successfully one or more formal educational courses since they started to work with RIDS-Nepal. These people, who have learned to work hard and earn their rewards, will be the future leaders of their local communities. They are respected among their own people because they have their roots in the local community.

In this way the new concepts of the "Family of 4" and the "Family of 4 PLUS" are embedded in an appropriate organisational framework, and carefully applied on a long-term basis in the local region. They are holistic and integrate people's identified development needs and thus have the potential to be important stepping stones towards sustainable development at the village level. The local people have gained the knowledge of what is needed and the skills to put their hands to work. They know from experience that hard work will bring forth positive changes. The "Family of 4" and "Family of 4 PLUS" provide healthy ownership and pride in the program from within the community and that all contributes to "sustainable development". This present piece of work aims to highlight a possible approach to sustainable development for a defined context. It provides a model that can be adapted to other cultures, communities and situations.

1.3 Research Questions and Objectives

1.3.1 Research Questions

The previous sections of this chapter, as well as the associated chapters in the Appendices have provided the background and detailed information required to realise the complexity of Nepal's culture with its ethnic and social diversities. These are deeply interlinked with the political development of the State, as well as the fragile and unique environment of the Himalayas, within which Nepal is embedded. Further, Appendix 10 lists the main summits and conferences dealing with development issues and their implications on a global scale. Summaries of their outcomes and impacts and the decades' long debates on sustainable development, its meaning, how it has been understood and implemented in general are discussed.

Already in 1965 Miss Lucy Morgan (Morgan 1965) wrote:

> Each [community development] program is bound by its own culture and each is faced with different social, economic, political, educational, and religious barriers. Population explosions, personnel shortages, lack of adequate training facilities, new government structures, outmoded administrative patterns, sagging economies, pressure for results, and imbalance in budgets present but a few of the daily situations that must be faced if community development programs are to move forward.

With this statement she identified already in the 1960s, even before the mainstream sustainable development debates started, the many facets and some of the key issues to be considered for sustainable community development. The next four decades have added plenty of details in regard to definitions and the understanding of the interconnections between the various levels of social, economic, political and environmental issues. In particular, the issue of energy resource depletion and energy consumption patterns has become one of the hot issues of the debates. Hardly anyone today would deny that we do have a significant problem on hand with the present, unequal and unsustainable energy conversion pattern in the limited environment of the earth's ecosystem.

Thus, since the early 1990s, more and more voices have been saying that the world's nations have to switch over from the present pattern of using non-renewable, mainly fossil and nuclear energy resources, to the more widely and equally distributed utilisation of renewable energy resources through applied renewable energy technologies. However, there is a remarkable lack of practical, real life examples of sustainable development, be it in the field of community development, applied renewable energy technologies or the combination of both. The amalgamation of both fields is an area not yet widely touched on, but believed to be of crucial importance in the present and even more in the future ongoing development of the understanding of sustainable development.

Thus the above analysis gave rise to the two main research questions for this thesis:

- *What is the appropriate role of renewable energy technology in holistic community development?*
- *Does the newly developed holistic community development concept of the "Family of 4" and the "Family of 4 PLUS" provide a sustainable way of improving the quality of life for remote and impoverished mountain communities in Nepal?*

1.3.2 Primary Objective

The Human Development Report (2006) stated in its foreword to the summary chapter that:

> Human development is first and foremost about allowing people to lead a life that they value and enabling them to realise their potential as human beings. (HDR 2006, V)

To live life in dignity, and allow each individual to develop their full potential, is still for many people a dream rather than reality. Development of people's living conditions, widening their chances and opportunities to become the people they are meant to be, requires the integration of social, physical, spiritual, economic and environmental development. That means a holistic approach to all development is essential in order to pay justice to mankind's multi-faceted nature and needs. Further, it enables synergistic benefits and impacts among multi-pronged, complementary projects. Such an approach can not be undertaken by an individual entity or body, but demands a strong partnership with all stakeholders. Thus, awareness and education of the participants is central and at the forefront of any project activities. Educating the local communities about sustainable community development through non-formal education and participatory hands on jobs and projects is a good practical strategy. In this way the end-users are not only the implementers but are also the agents of change, rooted within their cultural and environmental context. This can take place through indigenous NGOs with locally employed and trained staff and partnering with like minded national and international groups of people who have access to, and can mobilise the needed, financial resources. Thus the resulting primary objective for this thesis is:

- *To design, implement and evaluate an integrated approach to holistic community development, called the "Family of 4" and the "Family of 4 PLUS", in high-altitude areas of Nepal.*

1.3.3 Additional Specific Objectives

Additionally, there are some specific objectives which deserve attention as they are an intrinsic part of the new HCD concept and become apparent due to the interlinked nature of the "Family of 4" and the "Family of 4 PLUS". These are the

Base-Line & Follow-Up Surveys and synergistic benefits gained due to the holistic nature of the HCD concepts. In the following a short introduction to each of the three specific objectives is provided. Examples of both the survey questionnaires can be found in Appendix 14, and the first indicators, results and impressions regarding the potential synergistic benefits are discussed in more detail in the discussion Chap. 7.

As previously mentioned in 1.2, significantly more emphasis has to be given to the process of hands on implementation of "sustainable development" and not merely to the development of new definitions. Ongoing learning from real life project experience demands a cyclical project life-cycle. It's the first and the last step of the whole project cycle which we need to focus on in order to keep on learning. Therefore:

- The **first specific objective** is *to understand the initial circumstances and context of the remote mountain villages of Nepal, in order to identify the needs and define possible projects together with all stakeholders.*
- The **second specific objective** is *the evaluation and identification of the quantitative and qualitative impacts of the implemented project activities.*

Embedded within the identification and evaluation, the first and the last steps of a cyclical project cycle, are all the other, more managerial, hands on implementation and practical project activities.

It is paramount for the new HCD concept of the "Family of 4" and the "Family of 4 PLUS" that before any project activities are planned a Base-Line Survey takes place in the community or village interested to work in partnership with RIDS-Nepal. Paying due justice to the nature of holism and people's multiple needs and ingrained desire for improved living conditions, the base-line survey has been developed by a professional task force. Thus an anthropologist, a medical doctor, a community development specialist (the author) and an experienced project implementer (the RIDS-Nepal field manager, trained by the author) put their heads and experience together and over the course of a year developed a 56 question Base-Line Survey questionnaire and 52 question Follow-Up Survey questionnaire (Appendices 14.1, 14.2 contain the full set of questions of each questionnaire).

While each of the "Family of 4" and the "Family of 4 PLUS" project parameters enable life changing improvements, it is harder to measure the extent of the potential synergistic benefits that the fully integrated approach of this new HCD concept sets in motion. Thus,

- The **third specific objective** in this thesis is *to demonstrate that the holistic long-term approach to community development brings significant synergistic benefits.*

They are difficult to identify and associate with an individual project, as all the "Family of 4" and "Family of 4 PLUS" project components do support and enhance each others' impact and thus benefit the local end-user community. Further, in order to identify synergistic benefits and impacts on changed living conditions requires not just a few weeks or months, but rather years. Thus, time, in the sense of long-term participation and monitoring, needs to be taken once more into consideration as a crucial parameter. Not 1–2 years but at least 5–10 years and more of intensive, local presence and active partnership with the local community. However, while this does not lie within the time span of this thesis, the first findings of possible synergistic benefits have been extracted from the interpretation of interviews and collected data from follow-up surveys after the first 2 years. These are presented and discussed in Chap. 7.

1.3.4 Significance of the Research

Over the course of the last four decades the number of impoverished people and communities has not declined as expected, despite the enormous effort of international and governmental organisations. Many high level international summits, round table talks and conferences have taken place to address the need for new approaches to development. Huge amounts of financial investment and support have been provided by all nations for cross cultural as well as for national development projects.

Despite this, an estimated 2.4 billion people, or more than a third of the world's present population, live with no elementary electrical indoor lighting, a stove which reduces the harmful indoor air pollution, a pit latrine, no educational facilities or access to clean drinking water near their home or village. Thus they all can be considered to live in "poverty", without access to the basic necessities and unable to meet their daily minimum needs.

Thus, the beginning of the twenty first century sees more poverty stricken people with minimal or no hope for a better future and opportunities than the world's total population in Edison's time, when the great agent of change, industrialisation, began to take hold. The shear difference in living standards between nations is staggering and can not be justified in any responsible way. No doubt, the last four decades of effort, despite the massive interventions, has not brought forth the expected changes and results. Thus the question arises, "has the world community of development professionals, aid organisations and governments been on the right track in addressing this unjust situation?" Again, this question has to be asked and answered with practical solutions.

Thus the present thesis argues that the past four decades of development work, carried out by the world's nations and aid organisations, have not brought forth the expected and indeed desired results. It argues, that the current, single-project, short term approach to community development can not be justified and that an urgent paradigm shift is needed. It goes on to present one possible new way forward, the Holistic Community Development approach of the "Family of 4" and the "Family of 4 PLUS". This new approach considers the learned experience and necessary changes, contextualising them for a defined community.

Over the last two decades in particular it has emerged that access to improved energy services for individuals, families and indeed whole communities, is one of the most important aspects of enabling development to become sustainable (Reddy 2000; WEA 2000; Saghir 2005). Such is the nature of many impoverished communities. They have no direct access to their country's grid network, nor often even any road access. Thus, the question arises whether the conversion of the locally available renewable energy resources, through contextualized RETs, meet the needed energy services of impoverished communities? If so, do RETs play an important role in HCD?

The thesis presents an approach that was developed on the basis of the learning which took place over the course of 16 years of practical experience of living and working among local communities struggling to meet their daily needs rather than a merely mentally derived new paradigm. It does not claim to be the "silver bullet" to address the urgent need for changes in the approach to sustainable development. However, it claims to include some insights that are vital for a redirection, to enable development and improved living conditions for our neighbours in need, in close partnership, with mutual understanding and respect.

1.4 Chapter Formulation

The thesis contains nine chapters. The first chapter is introductory and presents Nepal as a country by a walk through its history, culture, ethnicity and current conditions in regard to development, politics and economy. Also it describes the main project and field research area, the district of Humla and its people. It discusses the uniqueness of Nepal's geographical and climatic conditions. A closer look at poverty within Nepal reveals that simple figures do not tell the full story of such a diverse society with deep rooted cultural traditions still strongly influencing people's daily life.

Chapter 2, called, Theory and Context, discusses the theoretical basis of HCD and its need for a defined context such as Nepal's remote mountain communities. It demonstrates through case studies the benefits of basic village electrification systems, powered by utilising the locally available renewable energy resources.

Chapter 3, called Power/Energy Generation and Lighting. It places some emphasis on the power/energy generation for basic indoor lighting for remote villages in the high-altitude Himalayan Mountains through solar PV systems and pico-hydro power plants. It highlights the importance of going beyond the mere technical aspects of a rural village electrification project, by considering as well the social parameters in the design of a village electrification project. Technical and non-technical problems encountered in implemented projects, as well as possible remedies, are discussed. Through case studies the importance of long--term data monitoring is emphasised, so that shortcomings and losses can be identified and considered in the design of new village electrification projects. A HDI case study points out the possible life improvements over a decade with basic indoor lighting.

Chapter 4, called Thermal and Processing Technologies, presents some of the newly developed technologies as part of the novel HCD concept. The contextualised technologies discussed include the smokeless metal stove, the high-altitude greenhouse, the solar drier and the high-altitude solar water heater bathing centre. Results of their monitored performance in the village context are presented and discussed.

Chapter 5, called Thermal Comfort in the Home, looks into the potential for improving the comfort level in a traditional high-altitude Nepali home through measurements and suggested improvements.

Chapter 6, called Social and Environmental Issues, proposes possible ways of recycling old batteries from solar PV systems in order to avoid any potential ecological danger. It also discusses the social and political hardship and insecurities the people in Nepal faced during the 10 year civil war from 1996 to 2006 and how this related to, and affected the community development projects.

The Discussion chapter, Chap. 7, follows in this logical sequence. It addresses the achievements, successes and failures of the HCD projects in which applied RETs play a leading role in the overall new approach to development. The research question and objectives are revisited and discussed. The role of RETs within HCD projects is discussed in the light of the research and practical experience presented in the publications/conference papers. This chapter highlights the main findings.

Chapter 8, Conclusions, summarises the findings, the contribution of the thesis to the wider debate about HCD and RETs and topics for further research projects.

Chapter 9, References, includes all the literature which has been used and consulted for the PhD research project and writing of the thesis.

The Back Matter (comprising all the Appendix chapters) contains all additional, relevant themes and issues that enhance the understanding of the context within which the thesis took place. It is meant to be a kaleidoscope of tools for readers interested in more depth information, increase in understanding and application of the new HCD concepts, in order to contextualise the approach for their own, unique situation. They include among others: a comparison of the Nepali versus the western calendar; a history of the fluctuations and changes of the Nepali currency against the US$ currency over the last two decades; the titles of all the published articles and presented papers which provide the bulk of the thesis' central Chaps. 2–6. Nepal's identified main renewable energy resources and the extent they have been utilised until today is discussed, as well as results of traditional, fuel based lighting and light emitting diode lamps powered by solar PV systems. The Back Matter chapters also provide examples of some of the major HCD project related awareness raising and educational tools, developed by the author and RIDS-Nepal. These include the brochures, posters and various awareness raising and teaching videos. Further it includes the title page of the Solar PV operation and maintenance training handbook, exclusively written for the context. An example of the RIDS-Nepal Non-Formal Education work books and games, written specifically for the Humla village students who never had the privilege to join a school. Also the titles of the songs, which were initiated by the author and written and composed by RIDS-Nepal staff and sung by local people in their own language, are included. They present another creative awareness raising tool for HCD project topics. The titles of the various dissertations, in which the author acted as local supervisor, with topics related to one or the other HCD and RET projects in Humla, as well as the titles of the gallery of pictures of the projects of the new HCD concepts of the "Family of 4" and "Family of 4 PLUS", developed for and applied in Humla, are included. The full content of all these "holistic community development tools", all the published articles and conference papers, all the awareness raising and teaching vidoes, the full scope of the various gallery pictures of the "Family of 4" and "Family of 4 PLUS", all songs, posters, brochures, the full content of the Solar PV operation and maintenance training handbook and the various Non-Formal Education work books and games, and copies of the supervised dissertations can due to obvious reasons not be included in the printed and eBook version of the thesis. However, all these "tools" are available upon request on a DVD from the author. The RIDS-Nepal Internet web site address, which is constantly updated and expanded, is provided. On the web site all the past and current implementation and research projects can be visited. One of the highlights on the web site are the Google Earth links through which all the villages, where long-term HCD with applied RETs projects are implemented, can be visited. Another highlight is the link to the monitoring data site, which allows registered users to create their own data table and graphs of interest for various village electrification systems, the two hot water village bathing centres and the high-altitude greenhouses. All these tools, instruments and teaching materials have been developed by the author and RIDS-Nepal.

References

Bista, D. B. (1991). *Fatalism and development—Nepal's struggle for modernization* (2nd ed.). 5th Impression, Calcutta: Orient Longman (ISBN 0 86331 245 5, 1994).

Brew-Hammond, A. et al. (2004). *Reducing rural poverty through increased access to energy services: a review of the multi-functional platform in Mali*. New York: United Nation Development Program, Environment and Energy. Retrieved July 21, 2012, from http://www.undp.org/content/undp/en/home/librarypage/environment-energy/sustainable_energy/reducing_rural_povertythroughincreasedaccesstoenergyservicesmppi.html

Brundtland, G. H. (1987). *Our common future,* United Nations World Commission on Environment and Development (WCED), 4th August 1987, ISBN 0-19-282080-X. Oxford: Oxford University Press. Retrieved July 30, 2012, from http://www.un-documents.net/wced-ocf.htm

CIA. (2009). *The world factbook, Nepal*. Retrieved July 20, 2012, from https://www.cia.gov/library/publications/the-world-factbook/geos/np.html

CRTN. (2010). *Solar parabolic cooker*. Retrieved July 20, 2012, from http://www.crtnepal.org/?option=technology&tid=3735333030

DDC (District Development Committee) Humla. (2003). *Periodic district development plan, Humla, first five year plan 2002/2003–2006/7*, Government of Nepal Kathmandu, Nepal.

DDC. (2004). *District profile of Humla (DPH)* (in Nepali language). Government of Nepal, Simikot, Nepal.

Emeriau, S. (2006). *Nutritional exploratory mission-Humla and Mugu districts, Nepal*. Kathmandu: Action Contre le Faim (ACF).

Fang, C., et al. (2007). *FAO/WFP food security assessment mission to Nepal*. Rome: Food and Agriculture Organisation/World Food Programme.

Fournier, A. (2007). *Action Contre la Faim: Anthropometric and mortality surveys*, Mugu and Humla districts, April 2007, Kathmandu, Nepal.

Gautam, U. et al. (2005). Nepal: Thermal energy for export. *South Asia Journal* July–September 2005. Retrieved April 13, 2010, from http://www.southasianmedia.net/Magazine/Journal/9-nepal_thermal_energy.htm (November 1, 2011 not available any more).

Goldemberg, J., & Johansson, T. B. (1995). *Energy as an instrument for socio-economic development*. New York: UNDP/BDP Energy and Environment Group. Retrieved November 1, 2010, from http://www.energyandenvironment.undp.org/undp/indexAction.cfm?module=Library&action=GetFile&DocumentAttachmentID=1025 (July 25, 2012 not available any more).

Hagen, T. (1996). Environment and biodiversity seen with the eyes of the farmers. *Environment and biodiversity in the context of South Asia* (pp. 1–11). Nepal: Ecological Society.

HMG, (2003). *Sustainable development agenda for Nepal*, His Majesty's Government of Nepal, National Planning Commission, 2003. Retrieved May 10, 2009, from http://www.scdp.org.np/pub/SDAN_eng.pdf (November 1, 2011 not available any more).

ICIMOD (The International Centre for Integrated Mountain Development). (2003). *Table of composite index of poverty, deprivation, socioeconomic, infrastructural development and women empowerment index*. Kathmandu, Nepal. Retrieved July 20, 2012, from http://www.icimod.org/ and http://www.hdihumla.org.np/remote-districts-of-nepal.htm

IFAD (International Fund for Agricultural Development). (n.d.). *Rural poverty in Nepal*, UN agency. Retrieved July 19, 2012, from http://operations.ifad.org/web/guest/country/home/tags/nepal

Joshi, C. B., et al. (2002). Development and dissemination of solar drying systems in Nepal: Problems and prospects. *International Energy Journal,3*(2), 53–73.

Karki, A., et al. (2002). Micro-hydro power in Nepal: Access to electricity for isolated rural population in the hills and mountains. *International Energy Journal,3*(2), 89–97.

KIRDRC (Karnali Integrated Rural Development and Research Centre). (2002). *Governance in the Karnali, an exploratory study*. August 2002, Jumla, Nepal.

Lekhak, H. D., & Lekhak, B. (2005). *Natural resource conservation and sustainable development* (2nd ed.). Kathmandu: Kshitiz Publication.

MOA (Ministry of Agriculture). (1997). *Progress report 1980/81-1995/96,* Lalitpur, Post Harvest Loss Reduction Division, Ministry of Agriculture (MOA).

Morgan, L. (1965). Community development: Observation around the world. *Paper presented at the American School Health Association and the Public Health Education Section of the American Public, April 1965,* University of North, Carolina School of Public Health, Chapel Hill.

NASA. (2003). *Processing, archiving, and distributing earth science data,* NASA Langley Research Centre. Retrieved July 21, 2012, from http://eosweb.larc.nasa.gov/

NEA (Nepal Electricity Authority). (2009a). *A year in review-fiscal year 2008/2009*, Annual Report, pages 9, 83–84, Kathmandu, Nepal. Retrieved March 23, 2010, from http://www.nea.org.np//reports/annualReports/YjqRDwQ9E3anrep09.pdf (November 1, 2011 not available any more).

NEA. (2009b). *Electricity demand forecast,(load forecast) 2003–2020*, Chap. 4, Table 1. Retrieved March 23, 2010, from http://www.nea.org.np/corporateplan.php (November 1, 2011 not available any more).

NPC (National Planning Commission). (1998). *The incidence of poverty by region in Nepal*. HMG (His Majesty Government) of Nepal, Kathmandu, Nepal, pp. 203

NPCS. (2011a). *National population and housing census 2011*. Central Bureau of Statistics, Government of Nepal, June 2011. Retrieved July 30, 2012, from http://www.census.gov.np/

NPCS. (2011b). *Preliminary leaflet 2011.* Central Bureau of Statistics, Government of Nepal, June 2011. Retrieved July 30, 2012, from http://www.census.gov.np/images/pdf/Preliminary%20Leaflet%202011.pdf

Reddy, A. K. N. (2000). *Energy and social issues-world energy assessment: Energy and the challenge of sustainability*, Chap. 2, (pp. 39–60). New York: United Nations Development Program (UNDP).

Reddy, A. K. N. (2002). Energy technologies and policies for rural development. In: T. B. Johansson, & J. Goldemberg (Eds.), *Energy for sustainable development: A policy agenda*, Chap. 4, (pp. 115–136). New York: UNDP.

Regmi, B. R. (2005). Sustainable development of hydro power potential in Nepal. *Kathmandu Hydro Power Conference June 2005,* Nepal.

Renewable Energy in South Asia. (2002). *Renewable energy South Asia*, Chap. 1.2.2. http://www.worldenergy.org/wec-geis/publications/reports/renewable/overview/chap_1_2.asp

RIDS-Nepal. (2007). Nutrition data survey in Humla. Nepal: Unpublished RIDS-Nepal Nutrition Survey in Upper Humla.

Saghir, J. (2005). *Energy and poverty: Myths, links, and policy issues.* Energy Working Notes No. 4. Energy and Mining Sector Board of the World Bank Group, Washington, No. 4 May, 2005.

Schumacher, E. F. (1973). *Small is beautiful-a study of economics as if people mattered*, Blond & Briggs Ltd. ISBN 0 009 922561 1.

Trailokya, N. U. (2003). *Education for sustainable development in Nepal: Views and visions.* Kathmandu, Nepal. Retrieved July 23, 2012, from http://enviroscope.iges.or.jp/modules/envirolib/upload/1515/attach/report9_intro.pdf

Treese, J. (2007). *New approach to improving nutrition in Humla, Nepal.* Kathmandu: RIDS–Nepal.

UN. (2011). *Human development report 2011-sustainability and equity: A better future for all* (p. 129). Retrieved July 24, 2012, from http://hdr.undp.org/en/media/HDR_2011_EN_Complete.pdf

UNDP. (2006). *Human development report 2006-beyond scarcity; Power, poverty and the global water crisis* (p. 294), United Nations Development Program, New York, USA. Retrieved July 24, 2012, from http://hdr.undp.org/en/media/HDR06-complete.pdf

UNICEF. (2006). *Situation of women and children in Nepal 2006.* Kathmandu: United Nations Children's Fund (UNICEF).

UNICEF. (1998). *The state of the world's children 1998*, UNICEF, Focus on Nutrition, Oxford press UK. http://www.unicef.org/sowc98/sowc98.pdf. Accessed July 24, 2012

UNICEF. (2010). *Child poverty and disparities in Nepal.* Nepal Report 2010 Overview, UNICEF, Kathmandu, Nepal. Retrieved July 24, 2012, from http://www.unicef.org/rosa/Child_Poverty_Study-_Overview%281%29.pdf

USAID. (2003). *Regional hydro power resources: Summary and analysis of selected SARI data,* USAID-SARI/Energy Program, Kathmandu, Nepal, November 2003. Retrieved July 24, 2012, from http://www.sari-energy.org/ProjectReports/RegionalHydroPowerReport.pdf

Warnock, J. B. (1989). *The hydro resources of Nepal.* U.K.: Water Power and Dam Construction.

WEA (World Energy Assessment). (2000). *Energy and the challenge of sustainability*, UNDP, New York, USA. Retrieved July 24, 2012, from http://www.undp.org/content/undp/en/home/librarypage/environment-energy/sustainable_energy/world_energy_assessmentenergyandthechallengeofsustainability.html

WECS (Water, Energy Commission Secretariat). (1999a). *WECS bulletin no. 10.* Kathmandu: HMG of Nepal.

WECS. (1999). *Annual REPORT 1999.* Kathmandu: HMG of Nepal.

WHO. (2005). *Distribution of health facilities in Nepal.* Nepal: World Health Organisation Kathmandu.

World Bank. (2010). *Transport in South Asia, Nepal transport sector.* Retrieved July 24, 2012, from http://web.worldbank.org/WBSITE/EXTERNAL/COUNTRIES/SOUTHASIAEXT/EXTSARREGTOPTRANSPORT/0,,contentMDK:20560914~menuPK:869038~pagePK:34004173~piPK:34003707~theSitePK:579598,00.html

Zahnd, Alexander. (2004). *Case study of a solar photovoltaic elementary lighting system for a poor and remote mountain village in Nepal*, MSc RETs Dissertation, Murdoch University, Perth, Western Australia. Retrieved July 24, 2012, from http://www.rids-nepal.org/index.php?option=com_docman&task=doc_download&gid=2&Itemid=104 and http://www.rids-nepal.org/index.php?option=com_docman&task=doc_download&gid=3&Itemid=104

Chapter 2
Theory and Context

2.1 Introduction

Appendix 10 provides a summary of the major conferences and Summits held with regard to the wider issues of development of the living and environmental conditions of the world's poor, developing and developed countries. Section 1.2 discusses the main points recognised, and the lessons learned since Stockholm 1972, highlighting that all the various issues connected with development are complex and thus it is not possible to work with a single overarching theoretical model of sustainable development. Further, it stressed that in order to achieve concurrently, social, political and ecological changes in the real life settings of the end-users of development projects, anyone active in the field must consider much broader approaches than previously applied.

Though it is not often spoken or explicitly written about, there is a close connection between technical and environmental development projects and health improvement of the beneficiary community. However, most of the development activities and projects address improvement of people's health and living conditions, which also improve health, well being and the physical condition of individual people as well as whole communities. Therefore it is justified to say that technical projects can be considered as well to be active preventative health care measures.

Initially, when the concept of "primary health care for all" emerged in the 1970s, as a concept, it was integrative and multi-level. WHO gave health the holistic meaning of "the physical, mental and social well being of the individual" (Alma Ata 1978). With its "Health for All by the Year 2000" initiative, the WHO sought to ensure that "as a minimum, all people in all countries should have at least such a level of health that they are capable of working productively and of participating actively in the social life of the community in which they live" (Rifkin and Walt 1986, 561). Projects inspired by this thinking were multi-pronged, aimed at, for example, decreasing diarrheal disease by improving hygiene and sanitation with infrastructure improvements to drinking water supplies. At that time it was axiomatic among community development professionals and theorists of development that projects should be holistic, multi-pronged, and involve the efforts of a multi

A. Zahnd, *The Role of Renewable Energy Technology in Holistic Community Development*, Springer Theses, DOI: 10.1007/978-3-319-03989-3_2,

disciplinary team, acting in concert with local people. Later, due in large part to shifting economic conditions and a greater emphasis by donor agencies on quantifiable results and outcomes, a more narrow definition of primary health care emerged. This definition is often associated with the somewhat contentious "selective approach", discussed later. Thus, the emphasis on multi-pronged projects fell by the way side in the 1980s and 1990s, when international funding for development aid came under increasing scrutiny, and the scope of projects narrowed in order to meet ever tightening control over reporting and transparency, and an overwhelming need to show results—preferably within one fiscal year.

Though selective approaches to development (projects with only single components such as indoor lighting or a village drinking water system) are effective in achieving carefully de-limited goals, selective approaches cannot produce the critical synergistic benefits of a multi-pronged, holistic project framework.

In the 1970s the concept of primary health care began to occupy the world's attention. The WHO published a crucial document in 1973 called, "Organisational Study on Methods and Promoting the Development of Basic Health Services" (Perpich 1975; UN 1997; WHO 2008), in which it laid out some of the main issues. It identified the fact that the roots of the health services crisis facing the developing world did not lie exclusively in the domain of the health sector. Instead, the authors recognised that health had to be situated in a social and economic context and that political structures and priorities had to be considered as well.

This was the climate out of which the Alma Ata conference in 1978 developed. Holistic community development as a strategy for improving health outcomes emerged in a period during which there was significant debate about the nature of poverty and the need to understand the structural changes that would be necessary to improve the lives of the impoverished. The foundation of this approach relied upon the idea that improvements in health would always be contingent in part upon changes in the social, political and economic factors that ultimately determine the quality of life of the World's poorest.

2.2 Comprehensive Approaches to Community Development

Advocates of a comprehensive or holistic approach to improving livelihood and health emphasised the fact that many issues affect health (Magnussen et al. 2004). The focus was on the fact that in most project sites in the rural developing world, the physical infrastructure (lack of roads and bridges), the absence of running water and electricity, the difficulties associated with getting people to accept and use such innovations as latrines, improved cooking/heating stoves, and lack of transport (among many other social and economic factors) all affect health outcomes. According to this assumption, the assistance of a variety of professionals in broad, multi-sectoral programs to improve health status in a target population was considered necessary. Health was viewed not simply as a problem of disease but one of several factors contributing to lower the quality of life that the poor experience. Some examples of comprehensive primary health care programs would include projects that improve health awareness and training, community public health education, improved access to affordable energy services, drinking water taps, capacity to grow more nutritious food, improvements in the safety and cleanliness of living areas, building health posts and furnishing them with staff, supplies, and on-going training.

It is important to situate this approach in its appropriate historical context. Until the 1980s there was a strong sense of optimism in discussions and debates about international development. The world economy was growing and the social climate, particularly in the late 1960s and 1970s was favourable for projects that emphasised holism, community, and social equality. Investments in international development were high and expected to continue increasing. Multi-sectoral, broad based comprehensive primary health care was "in".

Fast forward a few years to the 1980s. The world economy looked very different—economic recession, debt, the oil crisis and an increasingly unfavourable trade climate had taken their toll, and the world economy was not growing as hoped. Development projects across sectors had failed and many people had lost confidence in the process of international development. Projects focused on health were particularly concerning to analysts of development, because the possibility that health was not a development sector that was going to be able to be "turned over" to some governments in the developing world any time in the next 50 years was beginning to be understood (LaFond 1995). As LaFond (1995) notes, weakening public support for overseas assistance in the 1980s strongly affected international donor agencies' budgets to programs that could not show "results". Health focused UN agencies (e.g., UNICEF) and the WHO were affected by heightening public scrutiny and criticism as well, and were increasingly governed by the need to show results that were simple to display, easy to understand, and obtainable within a defined funding cycle (often only 1 year). This dramatically changed the nature of the projects that were promoted and funded.

2.3 Hard to Measure and Lacking "Results" Triggers Donors' Response

As donors became aware of the complexity of the problems affecting health outcomes, they were also realising that it was difficult to measure immediate health impacts for many of the components of the holistic community development programs that had been initiated in the 1970s. For example, it is difficult to measure the short-term health impact of a water tap system, because although it may improve hygiene and sanitation initially, other factors can intervene and depress the benefits of these improvements on health. In response to the perception of these measurement and assessment problems, many agencies started to scale back in the nature and scope of the programs in which they invested.

Thus emerged selective primary health care, an approach that "espoused the mobilisation of health services to attack the most prevalent disease problems" (Rifkin and Walt 1986, 562). It is centred on short-term, single-goal oriented interventions, and cost effectiveness. Vaccination programs are an example of this approach. These are simple, effective and affordable interventions that prevent devastating, often fatal diseases experienced by many of the rural poor (e.g., measles, pertussis, tetanus, poliomyelitis, diphtheria and tuberculosis). The hallmark of the selective primary health care approach is UNICEF's "GOBI" program. GOBI[1] is an acronym for Growth monitoring, Oral rehydration therapy, Breastfeeding and Immunisation. These are four medical interventions that well-exemplify the selective approach: they are simple, low technology, single-sector, quantifiable interventions thought to have a low cost, high impact effect on child survival. GOBI was part of the much touted "Child Survival and Development Revolution". A quote from a UNICEF official illustrates the appeal of Child Survival and the GOBI approach:

> The disastrous consequences of the recession prompted UNICEF to re-examine its work with a view to exploring and reordering its priorities at times when resources are limited amidst increasing demands and needs. This led to the adoption of low cost, high impact measures which would contribute to child survival and development…and [are] viewed as the core of a cost effective and high impact programme (quoted in LaFond 1995, 26).

Unger and Killingsworth (1986) assert that in the mid 1980s, selective primary health care was the darling of bilateral co-operation agencies, foundations, academics and research institutions, and international agencies. They suggest that "rather than health factors, the major determinants of this adoption have been political and economical constraints acting upon decision makers" (Unger and Killingsworth 1986, 1001).

2.4 Lessons from the Field

Comprehensive primary health care is fairly well studied, and available studies allow us to identify the approach's successes and weaknesses. A review of the impact of a variety of interventions under the comprehensive approach, suggested that the evidence for the impact of community based health care programs in poor countries on the mortality rates of young children is weak (Ewbank and Gribble 1993).

More recently, a study was conducted in the Gambia that analysed the influence of village level primary health care on mortality rates in children (Hill et al. 2000). The Gambian study compared mortality rates over 15 years among 40 villages. All of the villages had access to maternal and child health programs and vaccinations from a local health centre, but only some of the villages had primary health care. Primary health care was defined as the presence of one paid community health

---

[1] The website for more information about the UNICEF's GOBI program can be found at: http://rehydrate.org/facts/gobi-fff.htm. Accessed 26 July 2012.

nurse for approximately every five villages, in addition to a village health worker and a trained traditional birth attendant.

During the 15 years studied, funding and support for the primary health care program grew, was maintained, and then dropped. The authors found that while the primary health care program was well funded and supported, the mortality rates of 1–4 year olds were significantly lower in the villages with primary health care. For children ($\geq$5 years old), the results were less clear cut, though for both age groups, mortality rates rose after support for the primary health care staff weakened (Ewbank and Gribble 1993; Hill et al. 2000).

Other analyses of specific community based interventions aimed at improving hygiene, sanitation and indoor air quality show that the introduction of certain technologies brings positive results and can have a strong effect on improvements in health, as described below.

Indoor air pollution (IAP) kills one person every 20 seconds in developing countries (Warwick and Doig 2004), and there is no doubt that IAP contributes to the incidence of respiratory disease in the villages where the author works. Reductions in IAP are known to be very effective in improving the health status of people, particularly among women and children under the age of five. The many negative health effects of excessive IAP are well understood (Smith et al. 2000), and the health benefits of transitioning from heating and cooking methods that produce heavy IAP loads to cleaner methods have been documented. Ezzati and Kammen (2002) estimate that in their study population in rural Kenya, interventions that reduced IAP, caused a reduction in acute upper respiratory infections by 24–64 %, and acute lower respiratory infections by 21–44 %.

Through RIDS-Nepal we conduct 24 hours IAP tests in Humli households with traditional, smoky lighting and open-fire cooking/heating systems (Figs. 2.1 and 2.3.). In homes with the RIDS-Nepal solar PV systems and smokeless metal stoves (Fig. 2.2), the TSP, PM_{10}, $PM_{2.5}$ and CO values[2] were much lower, thus showing that these technologies improve indoor air conditions very significantly. Following solar lighting and smokeless metal stove (SMS) installation, households had IAP levels measured[3] that were 10–50 times lower than the levels we measured[4] in traditional, open-fire conditions. This was true for all measured values, including TSP, PM_{10}, $PM_{2.5}$ and CO values. Preliminary analyses of this health data suggest that this reduction in IAP has substantially reduced the incidence of symptoms of respiratory disease.

[2] TSP stands for "Total Suspended Particles" and includes all particles suspended in the air generated by a polluting source. PM_{10}, $PM_{2.5}$ are particulate matter of <10 or <2.5 μm (micrometres) in diameter and thus able to enter the respiratory system. CO is carbon monoxide.

[3] Since 2006, RIDS-Nepal has used an EPAM-5000 Environmental Particulate Air Monitor to measure the $PM_{2.5}$, PM_{10} and TSP indoor air pollution. This portable EPAM-5000 includes a gravimetric and photoelectric detector. http://www.hazdust.com/epam5000.php. Accessed 26 July 2012.

[4] These average improved IAP levels are based on the RIDS-Nepal's recorded (unpublished) measurements in over 50 households. Each IAP test is over a 24 h time period over the years 2006–2010, in homes without and with the installed smokeless metal stove over the years 2006–2010.

Fig. 2.1 "Odhan", an open-fire place with three stones or a simple three footed iron holder is the traditional cooking technology used by all family households in Humla, creating enormous health hazards through very high indoor air pollution

Fig. 2.2 With the smokeless metal stove, up to half the firewood can be saved, food is cooked all at one time and the indoor air is clean

Other interventions including safe stool disposal have been shown to be effective in reducing diarrheal disease rates (Tables 7.3, 7.4 and 7.5). A study of domestic hygiene and diarrhoea in Bangladesh showed that post neonatal mortality rates were 68 % lower in families with latrines than in those without (Rahaman et al. 1985). Gorter et al. (1998) showed that Nicaraguan families with children whose faecal matter was properly disposed of by using diapers were at reduced risk of diarrhoea compared with those families in which children went without diapers.

Fig. 2.3 Previous to the elementary indoor lighting the use of "*jharro*" (a resin soaked pine wood tree stick) was the traditional way of generating a dim, smoky indoor light

Hand washing after defecation has been found to be uncommon among people in the developing world (Han et al. 1986; Huttly et al. 1994). In order to wash effectively enough to prevent disease transmission, one study in Guatemala found that mothers had to wash their hands an average of 32 times per day, using an additional 20 liters of water (Graeff et al. 1993). A review of studies of hand washing suggests that "the promotion of hand washing with soap is an intervention that appears to be both highly effective, reducing diarrheal incidence by between 27–89 %, and feasible" (Curtis et al. 2000). The ability of people to wash their hands frequently and effectively depends upon readily available water, which is a significant problem for many poor communities.

Potable water is often thought to be absolutely critical in preventing diarrheal disease. However reviews of 67 studies in 28 countries in 1986 and another 17 studies in 1991 "concluded that improvements in water availability were probably more important than in water quality" (Curtis et al. 2000, 27). In fact, it appears that the health effects of having clean (drinking and cooking) water are swamped by the benefits associated with having enough water to increase hygiene and sanitation practices, in particular dishwashing and hand washing.

Flies have been linked to diarrheal incidence in a number of studies. One study of Israeli soldiers demonstrated a significant reduction in diarrhoea when yeast baited fly traps were introduced (Cohen et al. 1991). Another study showed that spraying insecticide was effective in killing houseflies and reduced the incidence of diarrhoea by nearly 25 % compared to villages where the insecticide was not sprayed (Emerson et al. 1999). Unfortunately, control of flies in a village setting is extremely expensive and difficult. However, safe stool disposal is an achievable goal. Preliminary analysis of our data shows a reduction in the incidence of diarrhoea in villages where pit latrines have been built, which is believed at least partly attributable to reductions in the probability that flies are contacting stool, which would limit disease spread.

Many studies have shown some effect of a variety of interventions falling under the umbrella of comprehensive primary health care. However, measuring the effects on health of some of these interventions can be very difficult, particularly on short time scales. This is because the negative effects of other problems that the target community faces can cancel out the health effects of any one of these interventions. Thus, measurement of the benefits of comprehensive programs has to be extremely careful and for validity must also be compared with controls.

2.5 Comprehensive Versus Selective Approaches Needs in Humla, Nepal

The author's practical experience over 16 years in the field has convinced him that a selective approach alone cannot effectively and responsibly improve the livelihood and health outcomes of Humli people. Currently, it is widely recognised that acute respiratory infections (ARIs), diarrheal diseases and malnutrition are the main problems that need immediate and sustained attention in rural Nepal (Winrock 2004; Benguigui 1999; Pandey et al. 1989b). These problems cannot be addressed without making deep and significant changes to infrastructure and behaviour patterns. While selective projects, targeting diseases like polio and measles are critical, and do show results that are easily quantified, they cannot be the only approach to tackling the serious health problems facing Humli people in the struggle for improved overall living conditions. A study in central Nepal, of interventions aimed specifically at reducing ARIs, showed a 60 % drop in ARIs with education, immunisation and antibiotic treatment of children showing signs of pneumonia, but the authors concluded that this reduction was overshadowed by the still unacceptably high rates of mortality from the other killers, primarily diarrhoea, and malnutrition (Pandey et al. 1989a). Despite the fact that this realisation may be shared by other people in the development field, project models still tend not to be as holistic as the author recommends.

In our time in the villages, we have seen positive synergistic effects of holistic community development programs, when project components are chosen by villagers based on their needs assessment and when these components dove tail together to improve the overall hygiene, sanitation and access to elementary energy and health services in the village. Factual data with regard to synergistic benefits are very difficult to obtain and sort out, particularly in regard to their source, magnitude and quality. This is because most benefits are related to long-term improved living conditions and increased preventative health care from the various, concurrent projects and these changes are hard to identify and express in merely numerical terms. Further, changes need to be tracked and observed over a long time period. However, the author and RIDS-Nepal have gathered many anecdotal testimonies, shared by the end-users. They include statements such as: "since the SMS has been installed, the women and children of the homes have been freed from their chronic coughs and chest pains as well as their previously resilient eye infections and heart pains"; "The pit latrine each family has built as a precondition for an SMS has enabled the whole village to become much cleaner and more hygienic"; "The people, and in particular the children, experience far fewer cases of diarrhoea with the consequent dehydration, loss of valuable nutrients, subsequent weakness and inability to work or learn".

These improved health conditions, in combination with the basic indoor lighting, enable the women and out of school children to participate in the daily NFE classes. To be able to read and write, along with the basic numeric skills, brings invaluable long-term personal as well as community benefits, which would have not been possible without the HCD approach.

This is what inspires us to continue to improve upon a model we call the "Family of 4" and to expand it into the "Family of 4 PLUS" over the years. The outline of this approach was originally designed by the author in 2000 after having worked with, and lived among the local communities for over 4.5 years. Over the years the author and his team have developed, improved upon and added new project components, based on participant observation and direct consultation with villagers. We find that the empowerment associated with the teaching/learning, equipment ownership, and project design and implementation uplifts individuals and groups of individuals within the village in a fashion that could never be achieved by a selective approach to improving overall living conditions and health outcomes (Zahnd and Malla 2006). This experience has convinced us that a return to holistic community development, with projects designed after detailed needs assessment is conducted in concert with villagers themselves, is not only good and successful; it is imperative. Further, history has brought forth example after example of development projects worldwide which have demonstrated that local people have to grow into technological and subsequent behavioural changes slowly, for the simple, and perhaps obvious—but rarely admitted—fact that traditions and cultures change much more slowly than new technologies can be introduced (Zahnd and Haddix 2007b, c). Therefore, the newly developed HCD concept is set in a time frame of two generations, during which the various projects are defined, implemented and followed-up. This reflects about a 20 year time span in the present context of Humla.

It is crucial to understand that the local community is at the centre of any holistic development project and that the technologies applied are to serve and support their struggle for a better life. Therefore, any project has to be based on a thorough understanding of the local context and culture, and must include an understanding of the "invisible" causes of poverty, and the impact on the community of decades of deprivation. This approach demands time, compassion and dedication. These more "human" aspects of a development project are crucial factors that need to go alongside the technical aspects. In this way the people are recognised from the beginning as equal partners and not just as receivers of imposed ideas. This time intensive, often frustrating process is central to a holistic development project.

2.6 The Role of Renewable Energy Technologies in HCD

Almost all of the identified 1.6–2 billion people without access to electricity (Mills 2002), 1.1 billion without safe drinking water, 2.3 billion suffering from water related diseases (with over 2 million children dying each year), 2.4 billion without adequate sanitation (TEAR 2002) and 2.4 billion relying on traditional biomass for their daily energy services (IEA 2002), live in developing countries, and four out of five live in rural areas (IEA 2002). Lack of electricity and heavy reliance on traditional biomass are hallmarks of poverty in developing countries (IEA 2002). Deforestation in Nepal is widespread and the once picturesque, biodiversity rich forests and valleys are stripped of their valuable resources in unstainable ways.

Today, over 100 years after Edison's visionary statement that: "we will make electricity so cheap that only the rich will burn candles" (Mills and Johnson 2002), this is only true for the industrialised countries. More people today do not enjoy the luxury of having light in their homes in the developing world than the world's population in Edison's time.

There is a clear relationship between poverty and access to electricity (IEA 2002; Saghir 2005). Poverty levels increase the more remote and inaccessible the communities are, while costs for electrification increase due to transport and maintenance costs. It is also absolutely essential that projects are sustainable, locally appropriate, and are developed in close partnership with the local community. The availability of elementary energy services is a crucial agent of long-term community development. The author argues that tapping locally available renewable energy sources, through applied renewable energy technologies, developed for a defined geographical, cultural and climatic context, is central to project success. These technologies, in concert with the other contextualised project components described in this thesis, form the backbone of our holistic approach to community development.

Tapping into locally available renewable energy resources to provide the necessary energy services in appropriate, affordable and culturally sensitive ways, in conjunction with projects addressing health, food, hygiene and educational needs, result in synergistic benefits. The approach embraced in this collaborative effort between RIDS-Nepal, its donor partners and the local people, is that the combined outcome of a comprehensive community development project bears more sustainable benefits than the sum of each individual project. This is particularly true with respect to the four critical pieces of community development in remote Nepal: pit latrines, stoves, lights, and clean water, which the author calls the "Family of 4". We believe in the synergistic effect of a project such as the "Family of 4" and thus have created projects that energise villagers' faith in, and enthusiasm for, making the model work as a whole.

Approximately 75 % of Nepal's 26.6 million people live in rural areas, with around half of these so remote that neither a road nor the national grid is likely to ever reach them. While Nepal has no fossil fuel resources, it is a country that is rich in renewable energy resources such as hydro power and solar energy. These abundant and locally available renewable energy resources can be tapped into with locally developed technologies that the community has helped adapt to the local conditions.

Poverty has many faces, and cannot be defined simply with economic values and figures, as previously discussed. One clear way to improve upon this situation is to work toward providing people living in poverty with access to, and control over sustainable energy producing projects. It is widely accepted that "poverty alleviation and development depend on universal access to energy services that are affordable, reliable and of good quality" (Reddy 2002; Saghir 2005). Access to energy services is a key in satisfying the daily needs of human beings, and there are clear linkages between access to energy and reduced infant mortality and fertility rates and increased literacy and life-expectancy, as the WEA report explicitly illustrates (WEA 2000).

While links between health and access to energy are illustrated in the studies cited above, there is an important gap in household level studies of the role of renewable energy technologies in improving health status. Many authors have described the effect of reducing IAP for improving respiratory health (Ezzati and Kammen 2002), but most if not all of those gains relate to the introduction of stoves. As described above, the very substantial reductions in IAP we see in the homes in our project villages is attributable to the combined effect of changing both heating/cooking methods (with smokeless metal stoves) and lighting methods (with solar PV/pico-hydro systems).

The author would argue then, that for those people living in remote areas, far away from mainstream development and the national grid, renewable energy resources present some very promising opportunities. They can make clear and definitive headway towards improved livelihood and health outcomes for the whole family through improved access to energy services, if they are responsibly and sustainable utilised. It is also clear that local universities, such as Kathmandu University in Nepal, can play a vital part in the development of contextualised RETs, locally developed and adjusted for the geographic, meteorological and cultural conditions in which they will be expected to function. The author's reasoning is, that a RET is/can be called "contextualised" when its design has emerged after considering the end-users' energy service demands, their living conditions, economic power and ability to learn and apply new technical skills, as well as the long-term sustainability of the system.

2.7 Basic Village Electrification System for People in Humla

Humla is the most northern district of the Karnali Zone, which is part of Nepal's western regional development area. This Zone is isolated from the mainstream of Nepal's economic development initiatives due to its remoteness, and because it is the only Zone with no connection to a road. In fact, a 16 day walk from the next road head is required to reach Humla's district centre, Simikot. The alternative is to book a 1 hour flight in an old Twin-Otter from Nepalgunj, in the very south of Nepal at the border with India. This flight lands in Simikot, before 2012 on a gravel landing strip, but comes irregularly, depending upon the weather and the routine diversion of the air fleet to other parts of the country depending upon politics and business.

The entire Zone is known for its permanent food shortage, environmental degradation, low productivity, regular disease epidemics among people and livestock, a harsh and unforgiving climate, weak education system, negligible employment opportunities, severe gender inequities in workload, health and education, and a growing trend toward emigration.

Humla lies also 16 days walk from the next grid connection, in the most difficult terrain, with high, snow covered peaks, forests, wild rivers and a harsh climate, with freezing winters, stormy springs and the annual monsoon rain. In these circumstances just to think of the shear cost of a grid connection is prohibitive. These features help explain the dire living conditions in the villages of the Humla people where our projects are located. Here people's lives are marked by poverty and suffering that is almost untouched by the modest but significant development gains experienced elsewhere in Nepal.

Lighting is often the first use of electricity in a developing country, and people are willing to invest in home or village electrification once they understand the potential health improvements, the increased educational opportunities for their children, and the possible financial savings (Mills and Johnson 2002) for their families. Our anecdotal testimonies show, that women, the primary users and organisers of household energy, place a high value on improved energy services such as lighting and an efficient cooking and heating stove. Thus the most appropriate solution is embedded power generation, through the utilisation of the locally available renewable energy resources. The remoteness makes it obvious that non-renewable fuels (such as petrol or diesel for a generator) are not affordable, having almost quadrupled in price by the time they reach Humla, due to the transport costs. Thus it is the available renewable energy resources, such as the streams, the sun and the wind, which are feasible and sustainable resources for power generation.

When we think of the availability of power, we mostly think of the availability of electricity by the flick of a switch at any time, and for almost any kind of application, independent of the power demand. This is in part because we either live in a developed country, or in an urban place in a developing country where the power grid is a reality. In the case of Humla this will probably never be the case, as embedded power generation demands considerable financial investment and careful long-term planning. That is why, for a remote and impoverished end-user community, which never had the experience of power access in the form of electricity, the first step into that new realm of experience is often a basic village electrification system. This is a simple power generation system for a few, low power consuming, long lasting lights, such as WLED (white light emitting diode) lamps (Figs. 2.4 and 2.5), utilising the locally available renewable energy resource.

In recent years Remote Area Power Supply (RAPS) systems have become a very suitable and appropriate power generation approach for remote communities. A RAPS system is a power generation system that is clearly defined in its scope,

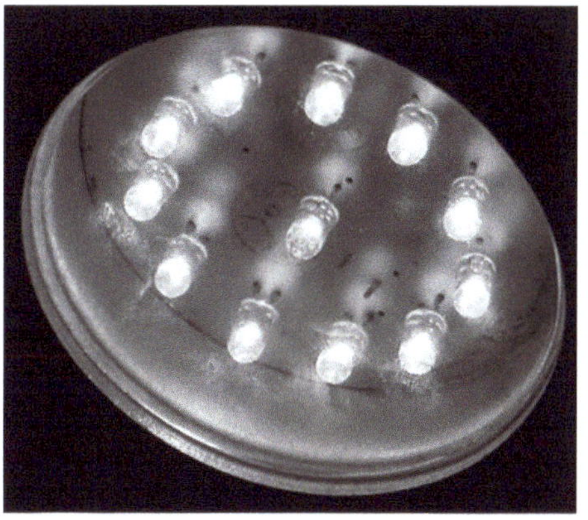

Fig. 2.4 WLED lamp (AC or DC) with 12 NSWP510DS diodes, consuming a total of 1.1 W. These WLEDs have a life-expectancy of >50,000 h, which is about 20 years for an average household's use

Fig. 2.5 Humli family in their main room with two WLED lamps. The lamps provide enough lighting for people to see one another and to do daily indoor tasks such as cooking, cleaning, socialising and school homework

capability and life expectation. It utilises either a single or several renewable energy resources—the latter case is called a hybrid system (Sect. 3.2.2).

The author's reading, practical and anecdotal experiences show that improved access to energy services is one of the main steps to the fulfilment of the Millennium Development Goals (MDGs). People do not want energy resources as such but rather the energy services they can provide through the exploitation and conversion of available renewable energy resources. Almost all of our local communities, with no access to modern energy services, identify lighting, cooking, heating and clean drinking water as their main needs for improved living conditions. We have noted that lighting, for brighter illuminance and cleaner indoor air conditions, for reading/studying and socialising, is the energy service which ranks most often at the top of the wish list of people in Humla (Figs. 2.6, 2.7, 2.8 and 2.9). It often marks the first milestone on the path of a community's development towards an improved living standard. But, as the EnPoGe report (Cecelski 2002) also recognises:

> Rural electrification is generally the costliest and structurally the most loss making activity for power utilities. It mostly brings in micro consumers, which worsens the load factor, and [puts] politically sensitive pressure on tariffs (Madan 2003).

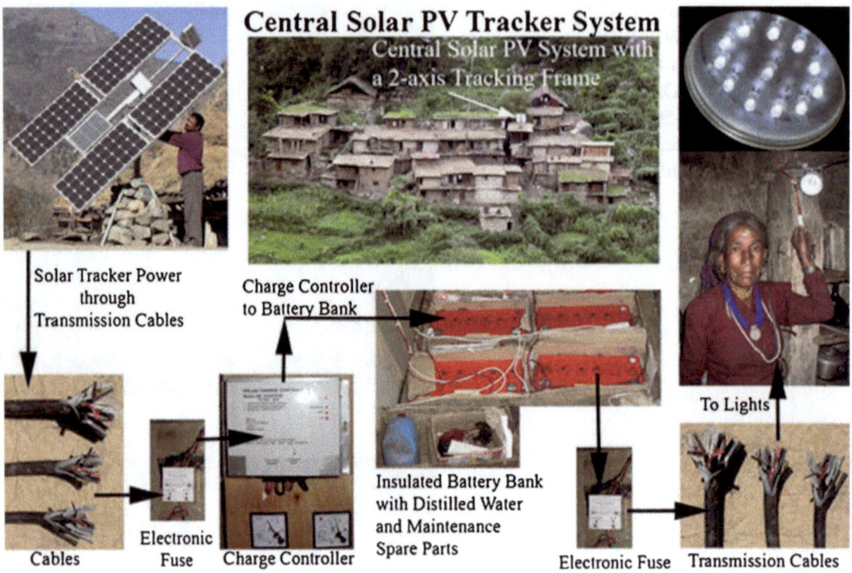

Fig. 2.6 RAPS system: The central solar pv village tracking system as installed in the village of Tulin in the RIDS-Nepal Humla project area (Sect. 3.3.2.3, Figs. 3.49 and 3.50)

Fig. 2.7 WLED lamps for minimal but locally appropriate, long lasting light services. Alongside is also an efficient smokeless metal stove, designed for the particular local cooking and heating needs, installed in each home

Fig. 2.8 Education is known to be a powerful tool in development, and benefits powerfully from the introduction of indoor lighting. Evening non-formal education classes can now be held

Fig. 2.9 Proud and participatory owner of a pico-hydro power plant. The community has chosen one of their own people to be specially trained to operate and maintain the pico-hydro power plant

This does not provide a good starting point to consider the electrification of the remotest and economically weak communities, living in some of the most hostile and inaccessible areas of the world. The worldwide trend to privatise utilities, to have electricity generation, transmission and distribution under the direct influence of the free market, aiming for maximum profits and ever increasing market shares, favours the fulfilment of the "wants" of the rich rather than the "needs" of the poor.

2.8 Restoring the Importance of Holistic Community Development: The "Family of 4" and the "Family of 4 PLUS"

The author's extensive experience working and living in the villages in this region of Nepal, in combination with the teaching and reading of the development literature, have convinced him that a return to a HCD model is the most likely approach to bring real and sustainable life changes with positive health improvements for local people. The author began implementing the current model of HCD in Humla with RIDS-Nepal and its donor partners in 2002, after baseline studies of the target communities, carried out in consultation with members of each household and in group discussions between the local people and the project team, identified a latrine, a stove, electric indoor lighting and clean drinking water as their most urgent needs to be addressed. This confirmed the author's previous experience of working and living for years in remote communities in Nepal (Zahnd 2003, 2004b, c). Hence, these most pressing needs, understood to be deeply interlinked with each other, have been defined as the "Family of 4", intrinsically intertwined and thus compelled to be applied and implemented concurrently (Zahnd and Haddix 2005). At the baseline, none of the households in these villages has indoor lighting, a smokeless stove for heating/cooking (householders still heat and cook with open-fires), a place for human waste disposal other than the fields surrounding the village, or clean drinking water. Thus, after the RIDS-Nepal team begins the first 2 years phase of a "Family of 4" HCD program partnership with the village, each household has access to the following "Family of 4" program's components (Fig. 2.10):

- A pit latrine.
- A smokeless metal stove with a hot water stainless steel tank.
- A basic indoor lighting system (solar PV/Pico-hydro/Wind Turbine, with three 1 W white LED lights per household).
- Access to clean drinking water from tap stands, from a community owned spring.

It is an essential part of the "Family of 4" program that each of the four "pillars" are continually developed, based on the ongoing learning of implemented projects and the resulting feedback of the end-user communities. Thus the various components of the "Family of 4" undergo ongoing improvements, based on monitored field research data and anecdotal end-user experience. While the

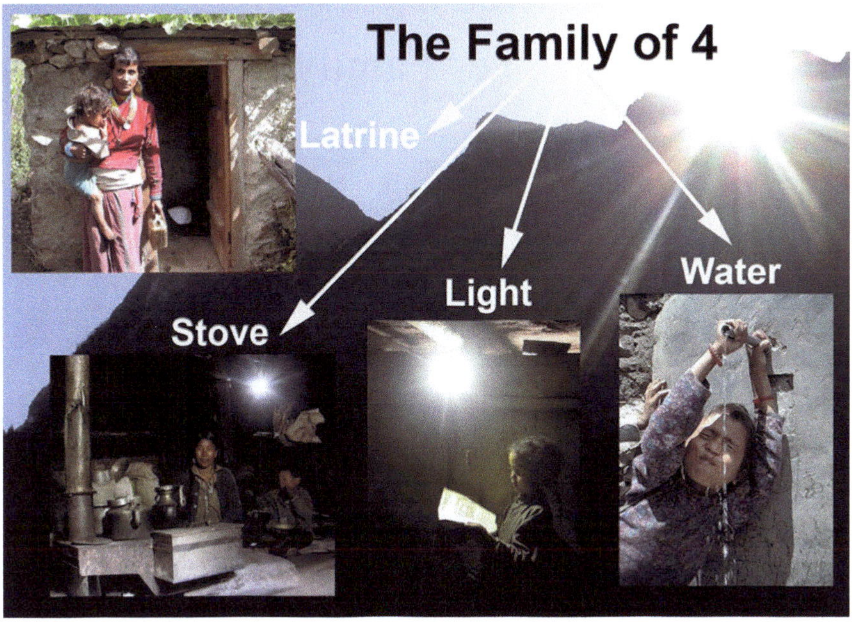

Fig. 2.10 The "Family of 4" consists of a pit latrine, a smokeless metal stove, basic indoor lighting for each household and access for families to clean drinking water from village tap stands

design of the pit latrine, the smokeless metal stove[5] and the gravity fed drinking water systems have already reached a high level of appropriate design and local manufacturing, the WLED lamps experience almost a yearly redesign due to the fast development of the relatively new WLED technology and thus consequent availability of improved diodes in the market.[6]

---

5 In 2006, the Nepal government announced a nation wide forthcoming, long-term subsidised stove program. Thus a 2 year evaluation of the major available stoves in Nepal was launched, which resulted in the approval of the author's designed and developed "Jumla design" smokeless metal stove (SMS) as the National Standard stove model. Thus, since 2009, a nation wide stove subsidy program (subsidising 50 % of the manufacturing cost) with annual production of 50,000 SMS, is running.

6 Since the initial development of white light emitting diodes in the mid 1990s by Dr. S. Nakamura from Nichia in Japan, the light output of the diodes has constantly improved, from initially 14 lm/watt in the author's first WLED electrification project in Jumla in 1999, to 83 lm/watt in RIDS-Nepal's 2009/2010 village electrification system. With this value, the WLED technology has reached good quality CFL's luminescence. Leading researchers in this field, such as Prof. A. Zukauskus from the University of Vilnius in Lithuania, as well as literature (see references 5 and 6 in Appendix 15.13) show that till 2020 WLED lights will be on the market with a luminosity of 200 lm/watt. Thus this compels RIDS-Nepal to continue to invest effort into the constant redesigning and development of the WLED indoor lighting technology applied for the Humla communities.

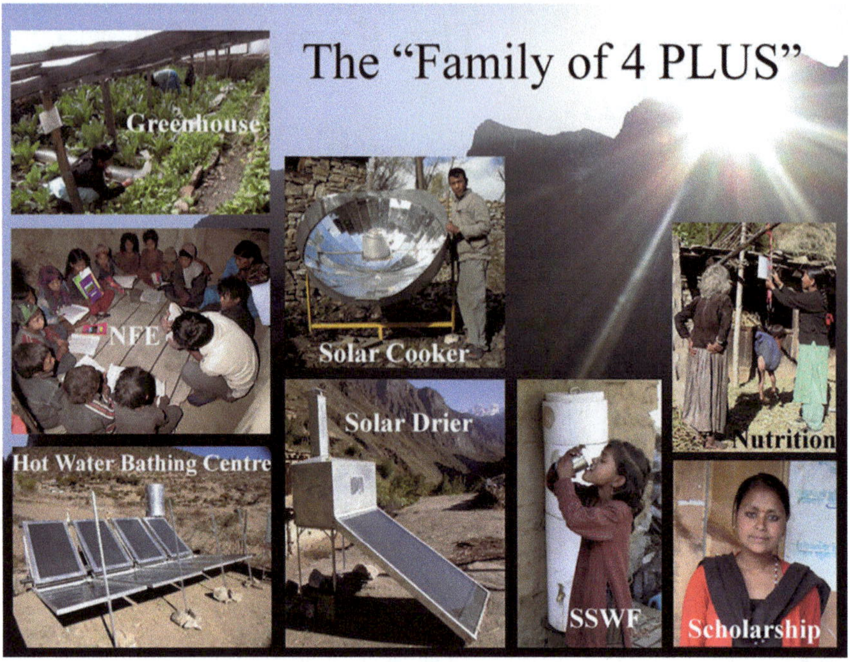

Fig. 2.11 The "Family of 4 PLUS" consists of greenhouse(s), solar drier(s), NFE (Non-Formal Education) for mothers and out-of-school children, a solar water heater bathing centre, an indoor slow sand water filter for each home, a mal-nutrition program for children <5 years of age, solar cooker(s) and a scholarships for apprenticeships for young Humla women/men

With the "Family of 4" being implemented for each family in a village, the people experience positive changes over time. They become increasingly aware of additional needs they would like to be addressed for their family's/community's long-term development. Thus, once the "Family of 4" program has been running for about 2 years, the second phase, the "Family of 4 PLUS" program (Fig. 2.11), again derived from anecdotal experience collected from the "Family of 4" program end-users and RIDS-Nepal's experience of working and living with the communities, can be launched. As the main basic needs have been addressed in the "Family of 4" program, the "Family of 4 PLUS" program includes a wider range of identified needs, which can change from village to village, according to their circumstances. Again, each project is part of ongoing development process through field research, data monitoring and the periodically conducted follow-up surveys in the villages. Thus, till 2012 eight components of the "Family of 4 PLUS" program have been defined, developed and, if identified by the local community, are implemented in partnership with the villages.

- Greenhouse for high-altitude villages.[7]
- Solar drier in the village.[8]
- Slow Sand Water Filter (SSWF).[9]
- High-Altitude Solar Water Heating (HASWH) bathing centre.[10]
- Access to a nutrition program for malnourished children <5 years of age.
- Non-Formal Education (NFE) classes for mothers and out-of-school children.[11]
- Solar Cooker for cooking during the day.[12]
- Karnali Technical School scholarship for a 2.5-year hands on apprenticeship.[13]

---

[7] A PhD student from the University of Melbourne, AUS, with the author's co-supervision, is supporting the further development of the initial RIDS-Nepal high-altitude greenhouse concept and design.

[8] Since 2007 the Nepali government is providing a 50 % subsidy on the solar drier, developed by the author through Kathmandu University student research projects from 2002 to 2004.

[9] The SSWF, developed by the author, is since 2008 available on the local market.

[10] The HASWH is since 2006 available on the local market. It has been developed by the author through Kathmandu University student research projects from 2003 to 2005.

[11] RIDS-Nepal developed under the leadership and guidance of the author, till end of 2011 four NFE work books for the women and out of school students of the RIDS-Nepal NFE classes. The four work books (chapter Appendix 17) are also available for other local NGOs and interest groups in indigenous literacy programs from the RIDS-Nepal web site at: http://www.rids-nepal.org/index.php/Non-Formal_Education/View_category.html. Accessed 26 July 2012.

[12] Another alternative is the use of solar cookers (e.g. the parabolic SK14 solar cooker), utilising the Sun's power to cook, heat water for hygienic purpose, and to drink. However, good and proven this technology is, the promotion and use of a solar cooker poses particularly enormous obstacles to indigenise and make it widely used and appreciated in the local community. Issues such as the cooking hours, which are limited to sunny days and day hours, when the women are mostly engaged in field and crop work, or to have women cook outside the house, where they can be seen and where evil spirits have easy access to the food, are just some of the reasons why it is very hard to convince the local people that the solar cooker is a realistic alternative, alongside the smokeless metal stove. This, because the solar cooker demands significant changes and alterations to their prevailing culture and belief systems, which are deeply rooted in the local society and hard to change. However, RIDS-Nepal has in all its offices at least one SK14 solar cooker installed and in daily use, to demonstrate that it is possible to use a solar cooker in this environment and that we can save precious firewood resources. Local people are invited to come and share the experience and see for themselves that the food can be cooked with a solar cooker and that it tastes the same if not even better and is in no way "cursed" by any evil spirit. In these practical ways, living and working alongside local communities we hope to slowly penetrate the thick layer of culture through the practical examples of making the solar cooker an integrated part of our life and work.

[13] RIDS-Nepal's scholarship program provides a unique opportunity to young women and men from remote Humla villages for a Technical and Vocational Education and Training (TEVT) at the Karnali Technical School (KTS), in Jumla, a 7 day trek south of Humla. Till end of 2010, three young people from Humla villages have successfully completed their TEVT courses and are employed in different governmental and non-governmental development organizations in Humla. Since 2009, a further four Humla students (two women and two men) have enrolled in this scholarship program at KTS. The students, in consultation with RIDS-Nepal, can choose the

We believe in the synergistic effect of a project such as our "Family of 4" and "Family of 4 PLUS" because we have seen individuals and groups (e.g., women) rally together in ways they never did before, and take a qualitatively different and new approach to problems facing them. For instance, the NFE classes targeting women bring literacy to members of this community who have never previously been able to read posters or brochures with health messages. Two short examples illustrate the point. Now, not only can women read them, but this ability seems to reinforce their interest in building and maintaining their family latrine, and in the protection of the cleanliness of their water source. Emboldened by new knowledge and skills, we see women seeking out more information and innovation to improve other elements of their lives in a fashion that is hard to imagine if we had limited our scope to providing, say, latrines alone.

Thus, the "Family of 4" and "Family of 4 PLUS" HCD concepts enable villagers to pursue their efforts for the changes and development they eagerly want to see taking place in their families and villages. It is easy for people to appreciate the benefits to themselves, their families, and the larger group from participating in such a project.

2.9 Sociocultural Features of the Villages and Our Field Staff Team

In RIDS-Nepal we believe that people are the centre of each project, and applied technologies serve and support them toward improved living conditions. This approach demands that the local context, language and culture have to be learned and understood in order to comprehend the unspoken and invisible "software" issues of the community. This demands time, compassion and dedication—crucial parts of a project, difficult to identify and judge and even more difficult to budget and "sell" to some donor agencies.

To a very large degree, the success of the RIDS-Nepal projects in Humla is based upon the strengths and character of the team members on the ground. In addition to the author's background in applied renewable energy engineering, we have local team members who work full-time, year round on the projects and who come from Humla or the neighbouring district, Jumla. The Jumli team members have completed extensive hands on training as well as certificate programs in

(Footnote 13 continued)
profession they want to study. RIDS-Nepal is supporting this program through funds raised from philanthropic organisations and individuals. The TEVT Scholarship Program is important as it provides local young women and men with a unique opportunity to learn a trade (construction sub-overseer, agriculturist) or profession (nurse, midwife), which are urgently needed in such remote and impoverished areas as Humla. With this scholarship program, the required skilled workforce for Humla can be developed over the coming years, empowering local individuals with the right tools and education and thus they can increasingly take the responsibility of development of their own communities into their own hands.

sustainable technology at the Karnali Technical School, and each of them is matched by a Humli counterpart whom they personally trained. RIDS-Nepal also has close ties with the Kathmandu University (where the author is a lecturer and researcher since 2001) and other international Universities, with whom joint research projects are conducted every year. Our field staff contains both male and female employees who are well trained in the applied technologies, and are articulate, motivated, and of middle and lower castes.

Although Nepal has formally abandoned the caste system, features of the system still prevail in most parts of the country and dictate social discourse and patterns of interaction. In Humla, villages are usually mixed in terms of caste composition, ranging from the lowest (Dalit) to the second highest (Chettri) caste. Humlis tend to be humble people who can be ill at ease with high caste urban Nepalis. Our team is composed of approachable individuals, each of whom has a special mix of lower/middle caste position, good education, excellent linguistic skills and all are native to the rural areas of Western Nepal. In addition to the transparency of the project budgets and accounting and the quality of the work, we know that the social skills and capacities of our team make them highly effective on the ground. RIDS-Nepal was the only development organisation permitted to work in Upper-Humla undisturbed by the Maoists during the recent insurrection. This is remarkable, given that during the same period (2002–2007) many offices involved in community development throughout Nepal were shut down, project activities ceased and some offices burned or ransacked.

In the sections below some of the main project components, central to our long-term HCD model, as well as some of the "soft" features that distinguish our experience with them, are briefly described. Each of the pieces of equipment that we install is developed locally, bearing in mind the cultural, meteorological, social and economic contexts and the technical limitations of working in this valley. For example, we needed to balance such considerations as the materials available in the villages to build the latrines, accommodating the heating needs and culinary preferences of local people in the design of the smokeless metal stove, the architecture of the houses in the design of the stove pipe, the amount of solar irradiation available in every household/cluster of households in each village considered to be electrified with a solar PV system, and so forth.

This level of detailed research, development and customising of technologies is part of what we call "contextualised" technology. It demands significant forethought and long-term commitment to working in the villages, "tweaking" the technologies as new demand or behaviour patterns evolve, and resources to put into baseline needs assessment and follow-up research lasting for at least one generation ($\sim$10–15 years).

An example illustrating how important it is to contextualise technologies comes from our experience with stoves in Humla. While most of the local families cook and heat on open-fires, there are a variety of types of stoves to be seen around the valley. Some function only as benches or counter space, seeming to have been designed with profit (alone) in mind, as they do not come anywhere close to meeting the energy service demands of the people. What local people told us that they want is a cooking and heating system that allows them to prepare multiple dishes simultaneously, to produce the national dish of *dal bhat tarkari* (lentils, rice and vegetables) twice a day, hot all at the same time, whilst also warming the room in the winter. Humli people also like to eat a type of unleavened bread (*roti*) that has to be toasted against hot coals to produce the desired taste. These demands are tough to manage on a single burner stove or open-fire place without an air flow regulator or exhaust damper, in order to control the combustion process according to the meal being prepared or the heating demands of the season.

The RIDS-Nepal smokeless metal stove, designed by the author, in contrast, is designed to meet these preferences, with an easily adjustable air intake and exhaust pipe valve, three burners on which food can be simultaneously cooked, and a toaster slot to toast *roti* against the coals (Figs. 2.2, 2.5, 2.7, 2.14, 4.7, 4.10 and 4.11). All of these tasks can be accomplished without opening the front of the stove, which is critical, since operating a stove with the door open allows smoke into the room, and causes the combustion rate to flare out of control, increasing the rate at which firewood is consumed. Each of our stoves has a stainless steel water tank abutting one side, where water can be easily boiled and stored for drinking, washing, or other needs. Further, each stove has a unique number so that we can follow it up for years to come. This has proven important for quality monitoring and trouble shooting.

In Sect. 2.10 we describe, in detail, each of the HCD components in the "Family of 4" model and in Sect. 2.11 the "Family of 4 PLUS" components.

2.10 The "Family of 4"

The "Family of 4" is a set of innovations that are installed, as a group, into each home in a target village. It includes a pit latrine, a smokeless metal stove, basic indoor lighting (through a locally available and readily utilised renewable energy resource such as solar, hydro or wind), and access to a safe drinking water system.

Common and easily treated conditions prevail in the target communities, and make life in this already challenging biophysical environment utterly miserable for many villagers. Some of the primary conditions affecting people in the remote mountain communities where we work are: scabies and other skin conditions, due to unhygienic living conditions; chronic and often severe upper and lower respiratory chest infections, due primarily to indoor air pollution from cooking over open-fires; gastro-intestinal worms and other parasites due to the lack of human waste disposal systems; and dysentery and Giardia infections from polluted drinking water.

To address only one of these problems with a technical solution is often done by donors with a limited mandate, time frame, or budget. While such limitations are a reality for many donors, experience shows that a single-pronged approach is neither sustainable nor beneficial in the long-term. The lure of the single-pronged approach—its simplicity, the possibility of completing the project within a single fiscal year for results to be reported back to the donor, and so on—must be resisted.

The "Family of 4" HCD approach addresses the key features of village life which are responsible for primary health problems. The synergistic benefits of the components are consequently many times more powerful than individual projects, such as "just" lighting, or "just" clean water, or "just" better sanitary conditions when implemented alone.

2.10.1 Pit Latrine for a More Hygienic and Private Environment (1st in the "Family of 4")

Usually, people in these villages defecate wherever a free and private place can be found. Due to lack of awareness of the importance of hygiene and sanitation, a shortage of land, and with no local examples to imitate, the pit latrine is not a traditional part of the infrastructure for a household. In the "Family of 4" model, the pit latrine is the first component to be built (Figs. 2.12 and 2.13). Work on the smokeless stove and the indoor lighting (solar PV/pico-hydro systems) does not proceed until the less exciting, lower prestige work of building the latrine has been finished. Because human waste is considered to be "polluting" in the local ideology, people hesitate to be associated with the building of a latrine or its maintenance.

We also know from other studies that women like to go in small groups to guard one another and to use the opportunity to socialise with their peers away from the scrutiny of their mothers in law (Dhakal, personal communication, 2008). Encouraging people to overcome this set of beliefs and associated habits has been one of the largest challenges we face.

An approach that we have found useful is to increase awareness and education about the issues surrounding hygiene and sanitation with posters/brochures/flip charts we designed as well as with songs written in the local dialect, using images familiar to people from their own valley (Appendices 20, 21, 23 and 24). We also emphasise these messages using the same materials (and more) in the NFE classes (Appendix 17). Gradually, people see the need to use and properly maintain their pit latrine. As a result, the walking paths, the surrounding village fields, and the streams are cleaner than ever before, greatly diminishing the risk of the spread of diarrhoea and other diseases.

2.10.2 Smokeless Metal Stove for High-Altitude (2nd in the "Family of 4")

In terms of energy, Nepal's traditional biomass fuel consumption provides for 85–90 % of all energy services nationwide and 100 % in remote mountain areas such as Humla. An open-fireplace with no chimney and a home full of smoke is "normal" in Humla. Often, the daily firewood consumption is 20–40 kg (Zahnd

Fig. 2.12 The first "pillar" of the "Family of 4" is always the Pit Latrine (PL), one per family, as preventative health care through improved hygienic conditions and environment is a crucial pre-requisite for sustainable development

Fig. 2.13 Children are the main ones to be taught how to use the PL daily, as they are the ones who get used in the most natural way, and thus it becomes part of their new culture

1998, unpublished data), and the indoor air pollution has a direct damaging impact on the health of women and children in particular (ITDG n.d.). This causes respiratory diseases, asthma, blindness and heart disease (IEA 2002), resulting in the extremely low life-expectancy for women and the high death rate of children <5 years of age in Nepal (Warwick 2004; IEA 2002). Increasing deforestation results in a scarcity of local firewood, forcing women and children to spend 7–8 hours gathering fuel wood further afield every second day (IEA 2002; Haddix and Zahnd 2005).

Now, each household has an efficient smokeless (i.e. with a chimney flue) metal stove, with a firewood consumption rate of nearly half previous levels (Fig. 2.14). As described above, the stove is designed for local culinary preferences and heating needs. The stove is also time efficient, as it allows women to cook *dal bhat*

Fig. 2.14 The Smokeless Metal Stove (SMS), designed by the author in 1998. Up to 2011 an estimated 25,000 homes (through RIDS-Nepal, the government subsidy program and other NGOs) in the remote districts of Nepal have had this SMS now installed. It cooks the traditional food all at one time, uses ~40 % less firewood, provides appropriate room heating and 9 liters of hot water for drinking the local “butter tea” as well as facilitating an improvement in personal hygiene. This is called “contextualised” technology, with people, their culture and way of living in the centre and technology developed to suit them in their context

tarkari all at one time, and it provides hot water for drinking and washing in an attached 9 liters stainless steel water tank. The stove ensures a smoke free, cleaner, and safer home environment, where children are no longer at risk of falling into the open-fire. Thus, it comes as no surprise that women, the main users and organisers of the family’s energy demands, esteem the SMS.

2.10.3 Solar PV System (3rd in the “Family of 4”)

The conceptual basis of HCD projects is that various needs, as identified by the target population, cannot be addressed simply using a single approach. Rather, primary health and resource dependent needs have to be addressed in a comprehensive, multi-pronged approach. Villagers articulate their primary needs and help to design projects that efficiently, sustainable and holistically address them. Thus, electricity is “just” one of the four pillars of the “Family of 4” concept.

However, by only attending to one piece of the overall picture, other critical needs can be neglected, such as a lack of a human waste removal system, and contaminated drinking water. In such an eventuality, the installation of lights will be of marginal benefit and the potential health benefits resulting from the new lighting system could not be clearly demonstrated.

In order to understand the local population's need for indoor lighting, it is important to understand how homes were lit previously and what activities occurred in homes after dark. In Humla, all families traditionally use *jharro*—a resin rich wooden stick from high-altitude pine trees (Fig. 2.15)—to light indoor living spaces. Burning *jharro* provides very smoky and minimal indoor lighting (Fig. 2.16).

Traditionally, every person in the Humla villages uses the open-fireplace for lighting their home interiors, in addition to cooking and heating. In this culture, women and children are most likely to suffer from the heavy load of indoor smoke pollution (Warwick and Doig 2004), causing respiratory diseases, asthma, blindness and heart disease (IEA 2002). While the US-EPA (Environment Protection Agency) PM_{10} 24 h standard is set at 150 $\mu g/m^3$, not to be exceeded more than once per year on average over 3 years, indoor open-fire places create PM_{10} levels $\geq$20,000 $\mu g/m^3$ (Warwick 2004, RIDS-Nepal unpublished data) on a daily basis, for hours at a time.

Thus, once the potential health improvements, the improved social gatherings and the better educational opportunities for their children through basic indoor lighting are understood, it is clear that it is frequently on the top of people's lists of the needs to be addressed.

In order to design a solar PV village system that will reliably light a house over its life span, the following issues were important for our team to identify and monitor:

- The solar irradiation ($kWh/m^2/day$) for the location of the solar PV system.
- The village's population, annual growth, its load distribution and growth pattern.
- The sustainability, ease of installation, and maintenance of components.
- Feasibility and reliability of locally developed and manufactured products.
- Trade off between sustainability/cost versus high efficiency.
- Participation of all stakeholders in every project step, including in economic terms.
- Culturally appropriate training, hand over, operation and maintenance.
- Minimal or no ecological impact during installation and operation.

In Sect. 3.3.2, the three types of solar PV systems (single home, cluster, and centralised village systems), which the author developed and installed in Humla through RIDS-Nepal, are described in detail, (Zahnd and Haddix 2007a). Here, a cluster system is briefly described. Using the checklist described above, 18 clusters of 4–12 households were defined in RIDS-Nepal's project partnering village Dhadhaphaya.

Each cluster has a central house, chosen by the community, on whose rooftop a 75 W_p PV module is mounted on a seasonally adjustable aluminium frame (Fig. 2.18). In that house, usually in the kitchen, is a 12 V DC battery-bank, consisting of 2 deep-cycle, 12 V DC solar batteries, each with 100 Ah capacity. They are in a locally made wooden box, insulated with the initial packing material

Fig. 2.15 *"jharro"* production from high-altitude pine trees kills the trees slowly

Fig. 2.16 *"jharro"* the traditional means to generate light in the home, or while walking in the village in the night. Strong black smoke, soot and a very dim light is the output

(Figs. 2.17, 3.35, 3.43, 3.91, 3.92 and 3.104) and with locally available silver birch trcc bark and pinc nccdlcs.

Each houschold gcts thrcc 1 W WLED lights, conncctcd to thc clustcr battcry-bank through armoured underground cables. These 1 W lights, with 83 lm/watt[14] have a life-expectancy of up to 100,000 hours (Craine 2004). The lamps are developed and manufactured in Nepal by Pico Power Nepal.[15] They provide

[14] Each WLED lamp consists of 12 high quality Nichia NSWP510DS WLED diodes with a 50° light angle.

[15] Pico Power Nepal (PPN) can be contacted through Mr. Muni Raja Upadhaya, at: muniraj@rids-nepal.org or through the author.

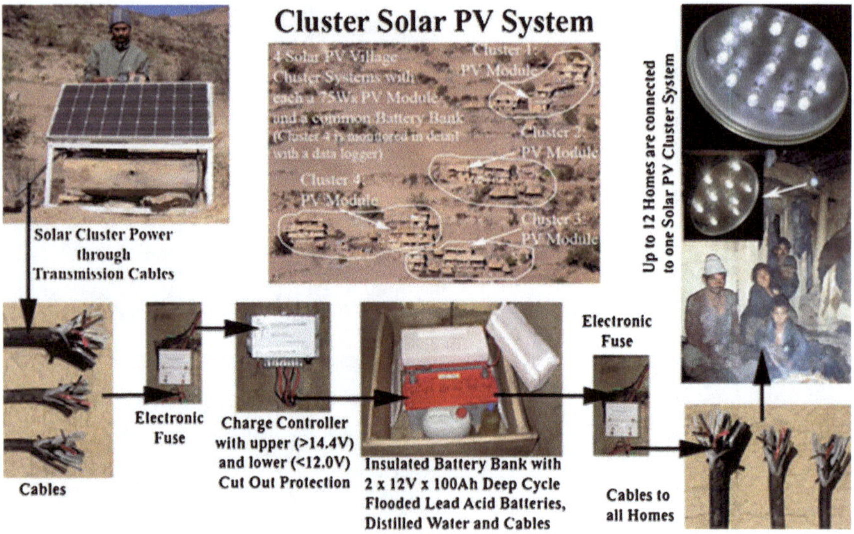

Fig. 2.17 Cluster Solar PV System Schemata as installed for four clusters in Pamlatum village in Humla through a RIDS-Nepal's long-term HCD project

Fig. 2.18 A solar PV cluster system for up to 12 homes. One 75 W_p PV module (*right side*) generates power which is stored in the battery-bank installed inside the home. All other cluster homes are connected through underground cables to the battery-bank, which provides up to 3 days of power without sunshine

enough light to socialise, read, and carry out other daily tasks, thus eliminating the need for an open-fire or *jharro* for lighting.

Some of the "soft" factors associated with a project like this are monitored by our team and compared with data collected in our baseline and follow-up studies, in which we ask members of every household about social and attitudinal expectations and changes associated with each project component. For instance, we carefully choose the household in which the battery-bank for each cluster system is located. This needs to be the household of an individual who is well respected, honest, and sociable.

We do not advocate cross caste user systems—each cluster of households drawing power from the same battery system should be comprised of families from

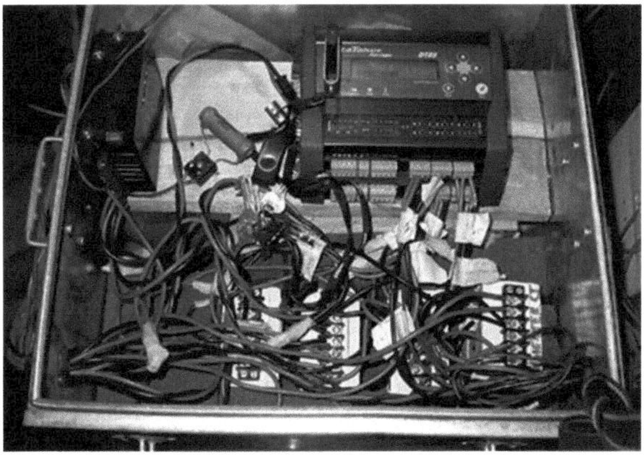

Fig. 2.19 A sophisticated dataTaker DT80 data logger monitors and records 22 different parameters for one of each of the three types of solar PV village systems the author developed. This provides us with detailed data for long-term performance

the same caste. This cuts down on conflict and disputes over usage load and maintenance duties.

Sophisticated data monitoring and recording systems installed in several typical PV system types in the villages (Fig. 2.19), monitor the daily solar radiation, the energy generated and stored and the consumers' usage and load over the course of the day, week, month and year. Using the data generated by these systems, we can see what the usage pattern looks like, how the people learn for the first time to live with electrical light inside their home and whether villagers are connecting other electricity using devices (other than the lights) into the wires. Further, they provide important data to understand the long-term performance of a PV system under real conditions. These data are crucial for the ongoing development of our projects, and help us make improvements to the solar PV systems. Our team is in and out of the households year-round, troubleshooting and training/re-training and helping address issues that arise until all the glitches have been removed.

2.10.4 Clean Drinking Water from a Community Owned Spring (4th in the "Family of 4")

The main rivers in Humla, contaminated by upstream villages, human waste, washing and disposal of dead animals, used to be the daily drinking water supply for most of the villages where we work. Now, in our project villages, water flows from tap stands that deliver clean water from the community owned springs high on the slopes above the villages. From the springs, water is brought through gravity flow to the villages through 50–90 cm deep buried underground polyethylene pipes to cemented tap stands, providing clean and fresh drinking water to clusters of households via local water tap stands in each village (Figs. 2.20 and 2.21).

This is often a very technically challenging project component, since the terrain is steep, rocky, covered with snow during the winters and dangerous in the areas

Fig. 2.20 Clean and pure water from the water tap for the whole family

Fig. 2.21 The water is brought via underground buried polyethylene pipes from the springs high above via gravity flow down to the village. In the village several water tap stands, defined by the local community, provide clean and fresh water 24/7, preventing unnecessary diseases

where the springs originate. Additionally, there can be some social tension associated with the spring and the location of the tap stands, and the users of each one. Members of higher castes prefer not to draw water from the same tap as members of the lowest castes. They make these preferences known to our team members in a way that, given some of our team members' own lower caste positions, can be hurtful and discouraging.

This is a feature of a caste bound society like Nepal's that can factor into projects in ways that can be damaging to team spirit and the ability of groups of people to work together. Despite the experience of frequent frustrations such as these, we have been able to successfully navigate these difficult issues to date with commitment and diplomacy—mainly because we are aware of the issues and invest the time and effort to confront and address them.

Fig. 2.22 The High-Altitude Research Station (HARS) which is also the RIDS-Nepal Simikot field office. Here the newly developed RETs, often emerging out of collaborative projects between RIDS-Nepal and universities around the world, are field tested. Problems can be identified and technologies improved, before they are considered ready to be implemented in the project villages. This is important for developing contextualised technologies and long-term sustainable HCD

2.11 The "Family of 4 PLUS"

2.11.1 Greenhousing and Solar Driers

With 199 days of frost in a year (NASA 2003; Zahnd 2004a), only 3–4 months (DDC Humla 2003) of agricultural work is possible. Thus many Humlis suffer permanent food shortages with high levels of malnutrition, especially in children. A low cost greenhouse prototype, constructed using local stones, wooden beams and UV stabilised plastic from Kathmandu, was built at the RIDS-Nepal High-Altitude Research Station (HARS) in Simikot, Humla in 2005 (Fig. 2.22). The greenhouse now produces vegetables for 10 months per year. This is absolutely critical as our data show that an astonishing 67 % of Upper-Humli children under the age of 5 have stunted growth, a sign of serious long-term malnutrition (Haddix and Zahnd 2005).

Further, in order to store the greenhouse's product yields in clean and hygienic ways, a solar drier was developed in tandem with Kathmandu University, as part of another practical student research project, led and guided by the author. These driers are now becoming popular in Humla, and allow families to efficiently and properly dry and preserve their precious foods for consumption in times of food shortage.

2.11.2 Non-Formal Education Classes for Mothers and Out-of-School Children

With a literacy rate for women as low as 4.8 % in Humla (KIRDRC 2002), booklets and brochures purchased in the distant cities are not appropriate for awareness raising and education, at least not initially. Thus we developed a novel, Humla specific NFE program for mothers and out-of-school children, with topics

Fig. 2.23 Women enrolled in the NFE evening classes meet 6 times a week to learn how to read, write and communicate. Teaching material is developed locally and the main themes are the "Family of 4"/ "Family of 4 PLUS" village project components

that support the "Family of 4" and "Family of 4 PLUS" projects, including pit latrine usage, cooking stoves, solar lighting, safe drinking water, greenhouses, solar driers, nutrition and improved hygiene (Appendix 17). These themes are taught in evening classes, initially through our pictorially based posters, games and songs (Appendices 19, 20, 22 and 23). As students' literacy increases, other locally relevant reading materials are introduced and used under the guidance of the RIDS-Nepal trained NFE facilitator (Fig. 2.23). In this way, the participants are immediately involved with subjects relating to the other project components which have been or are being implemented in their village, at a level appropriate to their educational experience, leading them to discover the benefits of functional literacy, and ultimately to participate in the development of their own teaching materials.

2.11.3 Nutrition for Malnourished Children <5 years of age

As indicated above, a very high number (67 %) of children <5 years old are malnourished in our project villages. This reality is the result of unproductive land, the harsh environment, unpredictable, recurring natural calamities like landslides and droughts, remoteness and lack of knowledge of how to utilise the minimal available foods for maximum nutritional benefit. To begin to help this situation, we developed a nutrition program for the most malnourished children <5 years of age and their mothers, in the project villages. Up to 10 mothers per village are instructed in basic nutrition, and trained how to mix and prepare a super porridge called "*sarbotham pitho*". This is made from locally grown lentils, wheat, soya beans and corn. This is added to the diet of the most severely malnourished children in the village. Weekly, the mothers of these children are visited by one of our female team members, for counselling and to answer questions and help with problems (Fig. 2.25). On a monthly basis, each child is weighed and measured (Fig. 2.24), and data are recorded so that each child's growth and health condition can be tracked.

Fig. 2.24 Malnourished children identified by the village and <5 years of age are enrolled in the RIDS-Nepal mal-nourished children's program. Each child is weighed and its data recorded monthly

Fig. 2.25 Mothers, enrolled with their children in the malnourished RIDS-Nepal program, are counselled on a weekly basis. They are taught how to care and cook for the children. Everyday problems are raised and discussed

2.11.4 Solar Heated Bathing Centre

The rivers in this high elevation area have been found to be warmest from June to August, measured as 12–16 °C. The rest of the year they are between 4 and 12 °C. Thus, water for bathing needs to be heated by wood fires, and wood collection is already a huge burden for women. Thus, a commonly owned high-altitude solar heated bathing centre for women and men has been designed and built (Zahnd and Malla 2006).

The solar water heaters are based on the thermosiphon principle, and are designed and manufactured in Nepal, with hot water storage tanks and insulation to protect the system from freezing in the winter. These heaters were developed through a collaborative research project with the author's students at Kathmandu University. One bathing centre unit, consisting of four flat plate solar absorbers and

Fig. 2.26 The first (world wide) high-altitude bathing centre for a village community in Darapori village in Humla, Nepal at 2400 m altitude above sea level, developed as a joint Kathmandu University student research project with RIDS-Nepal, was commissioned in May 2009

one hot water storage tank, allows up to 300 people to enjoy hot showers (calculated at 10 l, 50 °C water per person) once every 2 weeks, addressing the pressing need to improve local hygiene. The population of a village defines the number of units needed. A "village bathing centre committee" is responsible for keeping track of and maintaining the infrastructure. A data monitoring system, recording the incoming solar irradiation, the intake water, the absorber, the hot water storage tank temperatures and the daily hot water consumption, was installed in the Darapori village bathing centre in May 2009 (Sect. 4.4.1.7, Figs. 4.66, 4.67, 4.68, 4.69, 4.70, 4.71, 4.72, 4.73, 4.74, 4.75 and 4.76), as well as in the first prototype high-altitude bathing centre in Simikot in October 2005 (Appendix 26.6, Figs. HASW_11–77). It provides valuable feedback for evaluation and future improvements (Fig. 2.26).

2.11.5 Slow Sand Water Filter (SSWF)

Although the villages with running "Family of 4" projects have a village based clean drinking water system, it is often the case that the drinking water gets contaminated by the way people store and handle the water in their homes.

Unhygienic conditions inside the home, attract flies, rats and other animals such as chickens and dogs, and pose a potential danger of pollution to the drinking water. Thus, the author developed an indoor Slow Sand Water Filter (SSWF) for the average household family size in the Humla district of six people (Figs. 2.27, 2.28 and 2.29).

The SSWF is filled with two kinds of sand, in order to first filter out the rough, more visual parts, before the actual biological process through the

Fig. 2.27 Teaching how to install and how to use the indoor SSWF by a RIDS-Nepal staff member

Fig. 2.28 Filling up the SSWF by a home owner for his restaurant

Fig. 2.29 Each family member benefits from the pure drinking water

"Schmutzdecke"[16] takes place in the main part of the SSWF. A 9 litre water tank, containing the purified water, is accessed by a brass tap. RIDS-Nepal's ongoing faecal e-coliform tests have confirmed that up to 98 % pure water can be achieved and maintained for consumption with the SSWF, which meets the Nepali standard.

2.11.6 Solar Cooker

With firewood being the only fuel resource to provide the family's basic cooking and heating services, it comes as no surprise, that deforestation in Nepal is wide spread (Figs. 4.4 and 4.5), causing annually landslides and the loss of land and lives. The RIDS-Nepal smokeless metal stove will slow down this devastating deforestation in the years ahead. However, it will take decades of measurements and limitations to firewood collection, with rigorous monitoring of the proper use of the stoves, to mitigate the severe destruction of Nepal's unique forests.

[16] "Schmutzdecke" is the German word for the fine layer of bacteria, which is responsible to "purify" the water running through the SSWF. This layer of bacteria is developed on top of the fine sand layer over the course of the first 10–14 days of use (filling the SSWF with water twice a day and emptying the filtered water). Because the "Schmutzdecke" consists of bacteria, the SSWF has to be filled daily so that the bacteria remain in a moist environment, as otherwise they would die and the filtration of the water would not take place.

An alternative, alongside the SMS, is the use of solar cookers (Appendix 26.5) to cook, heat water for hygienic purposes and for drinks. However, good and proven this technology is, the promotion and use of solar cookers poses particularly enormous obstacles to indigenisation in the local community. Issues such as the cooking hours, which are limited to sunny days and day hours, when the women are mostly engaged in field and crop work, or to have women cook outside the house, where they can be seen and where evil spirits have easy access to the food, are just some of the reasons why it is very hard to convince the local people that the solar cooker is a realistic alternative, to the smokeless metal stove. This is because the solar cooker demands significant changes and alterations to their prevailing culture and belief systems, which are deeply rooted in the local society and hard to change.

2.11.7 Karnali Technical School

RIDS-Nepal's scholarship program provides a unique opportunity to young women and men from remote Humla villages for a Technical and Vocational Education and Training (TEVT) at the Karnali Technical School (KTS), in Jumla, a 7-day trek south of Humla. So far, three young people from Humla villages have successfully completed their TEVT courses and are employed in different governmental and non-governmental development organisations in Humla. Since 2009, a further four Humla students (two women and two men) have enrolled in this scholarship program at KTS. In consultation with RIDS-Nepal, the students choose the profession they want to study.

RIDS-Nepal is supporting this program through funds raised from philanthropic organisations and individuals. The TEVT Scholarship Program is important as it provides local young women and men with a unique opportunity to learn a trade (construction sub-overseer, agriculturist) or profession (nurse, midwife), which are urgently needed in such remote and impoverished areas as Humla. With this scholarship program, the required skilled workforce for Humla can be developed over the coming years, empowering local individuals with the right tools and education to increasingly take the responsibility of development of their own communities into their own hands.

2.12 Expected Results

If the hypothesis is correct, that a holistic project, in close partnership with the community, will have a more sustainable long-term impact than individual projects, planned periodical impact surveys should identify this. Beyond the quantitative changes currently (Haddix and Zahnd 2005) being monitored in villagers' time usage, morbidity, and nutritional status, a range of qualitative impacts are being monitored, including:

- Overall improvement in hygienic conditions inside and around the homes.
- Decrease in firewood consumption for cooking, heating and lighting purposes.
- Increase in women's literacy rate, with more girls joining the local school.
- Increase in social interaction, now possible in community gatherings after dark.
- Increased community based, community initiated development projects.
- Less breakdown times, improved use and maintenance of the new technologies.

The experience of working in the front line of holistic community development shows, that the MDGs will be hard or impossible to achieve in the set time frame. This is partly because they were defined primarily by professionals who had limited or no exposure to, or lived among the people experiencing the realities the MDGs were designed to change.

The anthropological insight and practical experience gained over the last decades makes clear that the lack of dignified human living conditions for individuals, communities, even whole districts or regions cannot be addressed with one off, single-pronged, approaches within a fiscal year. The author argues that development efforts must be designed from the bottom-up, from the grass-roots level, according to the end-users' perceptions of their most urgent needs, rather than according to a donor's development agenda.

This is not an entirely new concept, and, for the field of international development in the health sector, was well developed and quite adequately articulated at Alma Ata in 1978. Developments in foreign aid around the globe since then have dramatically shifted project design away from comprehensive and grass-roots approaches towards selective, single-pronged projects designed from the top-down, which it is argued have undermined the efforts themselves as well as people's faith in the overall endeavour. But people's concerns, needs and problems are never simple, nor simply solved. All human needs are couched in complexly nuanced physical, social, mental and spiritual environments. Issues to address in development projects must always be understood, discussed collaboratively and analysed in the specific context of the ultimate beneficiaries. By definition this means that project planning should always unfold collaboratively with the target community and that the planning process should fit as comfortably as possible within local traditions, beliefs, and behaviour patterns, as well as the geographical and climatic context. Thus, unless a paradigm shift toward retooling development efforts in the direction of small scale, bottom-up, grass-roots based, holistic community development takes place, the author believes that there is little hope of reaching the noble but heretofore unobtainable MDGs.

In addition to the value placed on the holistic approach to community development projects targeting health outcomes, the author has learned to pay due respect and importance to relevant, cultural milieu. Project planners and implementers must spend time living in the target community. The amount of time shared living with local end-users or beneficiaries of the development effort does not necessarily have to be protracted. But it is critical for project planners and implementers to have a solid sense of local behaviour patterns, social hierarchies, expectations of the development effort and barriers to success.

Hard experience has taught us that there is no short cut to sustainable and appropriate development. Living with the people targeted by the project, and learning to understand and respect the logic behind local culture and ancient tradition is crucial to the relevance of any HCD project.

In an earlier latrine installation effort, we were initially flummoxed by the fact that despite the appearance of abundant open space in which to place the latrines, no villager seemed to believe their household could afford the space for a latrine. The logic to their protestations slowly emerged—that piece of land was inhabited by a spirit that could not be polluted; another was the subject of a three generation old dispute over ownership; a third was too close to rice, a sacred crop, not to be contaminated by proximity to human waste (Zahnd and Haddix 2007b). These nuances cannot be learned from books nor downloaded from the Internet—and without patience and a deep respect for local people and their perceptions of reality, issues such as these can develop into complete impasses, or can lead to projects that quickly unravel after the implementers depart for the capital city. Real life experience has to be the basis for context relevant, respectful and dignified sustainable development.

Our experience in the field compels us to argue that only a long-term HCD approach can sustainable and responsibly bring about serious change in terms of health outcomes and livelihood improvement for people living in remote impoverished conditions such as in Humla. We would like to see a wide scale paradigm shift, back to long-term, holistic, multi-pronged projects that are reliant upon villager led needs assessment and rigorous, annual, household level studies of the social, attitudinal, technical and health outcomes associated with the studies.

Ideally, projects should have long-term donor commitment for the follow-up and maintenance of projects. Projects need not be large scale, as in the case of the exemplary Millennium Village Project[17], which shows already encouraging results if the vision of holism becomes reality as reported by the Economist for several villages in Western Kenya (The Economist 2006). The commitment needs to be long-term, because the positive effects to be experienced as a result of this kind of effort will not be tangible in the short-term. Donors and project staff members also need to be aware of the fact that by and large this type of work is not generally profit

[17] The Millennium Project was commissioned by the United Nations Secretary General in 2002 to develop a concrete action plan for the world to achieve the Millennium Development Goals and to reverse the grinding poverty, hunger and disease affecting billions of people. The Millennium Villages are based on a single powerful idea—impoverished villages can transform themselves and meet the Millennium Development Goals if they are empowered with proven, powerful, practical technologies. By investing in health, food production, education, access to clean water, and essential infrastructure, these community led interventions will enable impoverished villages to escape extreme poverty, something that currently affects over 1 billion people worldwide. http://www.unmillenniumproject.org/mv/index.htm. Accessed 26 July 2012.

making in the short-term. At some point, small shops and businesses may, and should, indigenously emerge to service the long-term equipment needs to maintain and improve different project components. This can represent an important development for local people to help them to engage, if they wish, with the cash economy.

Finally, it is important to remind the reader that change in behaviour may be slow, and may take up to two generations to occur. This is especially true for project components such as latrines, which are so heavily laden with symbolic meaning that may, at first, be invisible to newcomers to the cultural and social system. Working with local people with patience, compassion, and a shared understanding of the commitment each side has made to tackle the challenges that will inevitably arise, is necessary and will bear fruit that will nourish all participants, in the longer term.

References

Alma Ata (1978). Declaration of Alma-Ata. *International Conference on Primary Health Care*, Alma-Ata, USSR, 612 September, 1978. http://www.searo.who.int/LinkFiles/Health_Systems_declaration_almaata.pdf. Accessed 31 July 2012.

Benguigui, Y. (1999). *Technical guidelines for the prevention, diagnosis, treatment, and control of ARI at the primary care level, Section IV: Prevention and control* (Chap. 16, pp. 313, 323). Washington, D.C.: Pan American Health Organization (1999). http://www.paho.org/English/AD/DPC/CD/AIEPI-1-4.16.pdf. Accessed 27 July 2012.

Cecelski, E. (2002). Enabling equitable access to rural electrification: Current thinking on energy, poverty, and gender. *ENERGIA News,* Vol. 5, No. 3 (2002). http://www.energia.org/fileadmin/files/media/EN112002_cecelski.pdf. Accessed 21 July 2012.

Cohen, D., Green, M., Block, C., et al. (1991). Reduction of transmission of shigellosis by control of houseflies. *Lancet, 337*, 993–997. doi: 10.1016/0140-6736(91)92657-N. Accessed 28 July 2012.

Craine, S. (2004). *Technical and economic optimization of village electrification systems* (pp. 9, 11, 14, 36). MSc English Dissertation, Curtin University, WA.

Curtis, V., Cairncross, S., & Yonli, R. (2000). Domestic hygiene and diarrhoea—Pinpointing the problem. *Tropical Medicine and International Health, 5*, 22–32.

DDC (District Development Committee) Humla (2003). *Periodic District Development Plan, Humla, First Five Year Plan 2002/2003–2006/2007*. Nepal: Government of Nepal Kathmandu.

Dhakal, (2008) personal communication with author

Emerson, P., Lindsay, S., Walraven, G., et al. (1999). Effect of fly control on trachoma and diarrhoea. *Lancet, 353*(9162), 1401–1403.

Ewbank, D., & Gribble,J. (Eds.). (1993). *Effects of health programs on child mortality in Sub-Saharan Africa*. Washington DC: National Academy Press.

Ezzati, M., & Kammen, D. (2002). Evaluating the health benefits of transitions in household energy technologies in Kenya. *Energy Policy, 30*, 815–826.

Gorter, A. C., Sandiford, P., Paun, J., et al. (1998). Hygiene behaviour in rural Nicaragua in relation to diarrhoea. *International Journal of Epidemiology, 27*(6), 1090–1100.

Graeff, J., Elder, J., & Booth, E. (1993). *Communication for health and behaviour change: A developing country perspective*. San Francisco: Jossey Bass.

Haddix, M. K., & Zahnd, A. (2005). *A social and health impact assessment in Chauganphaya and Kholsi, Phase One*. Kathmandu: The ISIS Foundation & RIDS-Nepal.

Han, A., Oo, K., Aye, T., & Hlaing, T. (1986). Personal toilet after defecation and the degree of hand contamination according to different methods used. *Journal of Tropical Medicine and Hygiene, 89*, 237–241.

Hill, A., MacLeod, W., Joof, D., et al. (2000). Decline of mortality in children in rural Gambia: The influence of village-level primary health care. *Tropical Medicine and International Health, 5*, 107–118.

Huttly, S., Lanata, C., Gonzales, H., et al. (1994). Observations on handwashing and defecation practices in a shanty town of Lima Peru. *Journal of Diarrhoeal Diseases Research, 12*, 14–18.

IEA (2002). *Energy and poverty* (Chap. 13). World Energy Outlook.

ITDG (n.d.). *Reducing exposure to indoor air pollution* (p. 14). ITDG Publication. http://www.itdg.org/docs/smoke/itdg%20smoke%20-%20part%203.pdf. Accessed 20 July 2012.

KIRDRC (Karnali Integrated Rural Development and Research Centre) (2002). *Governance in the Karnali, an exploratory study*. Jumla, Nepal, August 2002.

LaFond, A. (1995). *Sustaining primary health care*. New York: Save the Children Fund.

Madan, G. (2003). *Impacts of rural electrification on poverty and gender in Indonesia.* Vol. 1: Facts, *analysis, and recommendations* (p. XIX). Washington, D.C.: Energy, Poverty and Gender (GEPOGE).

Magnussen, L., Ehiri, J., & Jolly, P. (2004). Comprehensive versus selective primary health care: Lessons for global health policy. *Health Affairs, 23*(3), 167–176. http://content.healthaffairs.org/cgi/content/full/23/3/167. Accessed 1 November 2011.

Mills, E., & Johnson, S. (2002). *A dramatic opportunity for technology leapfrogging in the developing world.* Lawrence Berkeley National Laboratory.

Mills, E. (2002). Why we're here: The $230-billion global lighting energy bill. *Proceedings of the 5th International Conference on Energy-Efficient Lighting* (pp. 369–385). France: Nice.

NASA (2003). *Processing, archiving, and distributing Earth science data.* NASA Langley Research Centre. http://eosweb.larc.nasa.gov/. Accessed 21 July 2012.

Pandey, M. R., et al. (1989a). Impact of a pilot acute respiratory infection (ARI) control programme in a rural community of the hill region of Nepal. *Annals of Tropical Paediatrics, 9*, 212–220.

Pandey, M. R., et al. (1989b). Domestic smoke pollution and acute respiratory infections in a rural community of the hill region of Nepal. *Environment International, 15*(1), 337–340.

Perpich, J. (Ed.). (1975). Health in community development. *International Health Conference*, 20–22 October, 1976, National Academy of Sciences, Washington USA (pp. 6–7). http://books.google.com.au/books?id=c20rAAAAYAAJ&pg=PA6&lpg=PA6&dq#v=onepage&q=&f=false. Accessed 19 July 2012.

Rahaman, M., Rahaman, M., Wojtniak, B., & Aziz, K. (1985). Impact of environmental sanitation and crowding on infant mortality in rural Bangladesh. *Lancet, 2*, 28–31.

Reddy, A.K.N. (2002). Energy technologies and policies for rural development. In T.B. Johansson, & J. Goldemberg (Eds.), *Energy for sustainable development: A policy agenda* (Chap. 4, pp. 115 136). New York: UNDP.

Rifkin, S., & Walt, G. (1986). Why health improves: Defining the issues concerning 'comprehensive primary health care' and 'selective primary health care'. *Social Science and Medicine, 23*, 559–566.

Saghir, J. (2005). *Energy and poverty: Myths, links, and policy issues.* Energy Working Notes, World Bank: Energy and Mining Sector Board, No. 4, May 2005.

Smith, K., Samet, J., Romieu, I., & Bruce, N. (2000). Indoor air pollution in developing countries and acute lower respiratory infections in children. *Thorax, 55*, 518–532.

TEAR Australia (2002). *TARGET No. 2,* February 2002.

The Economist (27 April 2006). The magnificent seven—How a few simple reforms can lift African villages out of poverty, SAURI, Kenya. *Economist* 2006. http://www.unmillenniumproject.org/documents/Economist042706.pdf. Accessed 23 July 2012.

UN (1997). *Community involvement in primary health care: A study of the process of community motivation and continued participation* (p. 9). Economic and Social Council. http://www.cf-hst.net/UNICEF-TEMP/Doc-Repository/doc%5Cdoc362050.PDF. Accessed 23 July 2012.

Unger, J., & Killingsworth, J. (1986). Selective primary health care: A critical review of methods and results. *Social Science and Medicine, 22*, 1001–1013.

Warwick, H., & Doig, A. (2004). *Smoke—The killer in the kitchen*. ITDG Publication. ISBN 1 85339 588 9. http://practicalaction.org/docs/smoke/itdg%20smoke%20report.pdf. Accessed 24 July 2012.

WEA (World Energy Assessment) (2000). *Energy and the challenge of sustainability*. New York: UNDP. http://www.undp.org/content/undp/en/home/librarypage/environment-energy/sustainable_energy/world_energy_assessmentenergyandthechallengeofsustainability.html. Accessed 24 July 2012.

WHO (2008). *Address to the 61st World Health Assembly*, 20 May 2008. http://www.who.int/mediacentre/events/2008/wha61/hafdan_mahler_speech/en/index.html. Accessed 24 July 2012.

Winrock (2004). *Household energy, indoor air pollution and health impacts—Status report for Nepal*. Kathmandu: Winrock International Nepal. http://www.pciaonline.org/files/Nepal_Household_Energy_and_Health_Overview.pdf. Accessed 24 July 2012.

Zahnd, A. (1998). *Firewood consumption survey in Jumla villages*, Jumla, Nepal. Unpublished data survey.

Zahnd, A. (2003). Appropriate technology for the poorest mountain communities in the Nepal Himalayas. *Sustainable Resources Conference*, 1–4 Oct 2003, Boulder, Colorado, USA.

Zahnd, A. (2004a). *Case study of a solar photovoltaic elementary lighting system for a poor and remote mountain village in Nepal*. MSc RETs Dissertation, Murdoch University, Perth, Western Australia. http://www.rids-nepal.org/index.php?option=com_docman&task=doc_download&gid=2&Itemid=104 and http://www.rids-nepal.org/index.php?option=com_docman&task=doc_download&gid=3&Itemid=104. Accessed (both) 24 July 2012.

Zahnd, A. (2004b). Community development through sustainable energy use and conversion in harmony with the environment in remote districts of the Central Himalayan Mountains Nepal. *Great Himalayas Conference*, 12–15 Dec 2004, Kathmandu University, Nepal.

Zahnd, A. (2004c). Handling (managing, administrating, governing) available resources responsibly—Case study of a renewable energy project with a remote and poor mountain village in the Nepal Himalayas. *World Renewable Energy Congress VIII*, Denver Colorado, USA, 28 Aug–3 Sept 2004. http://www.rids-nepal.org/index.php?option=com_docman&task=doc_download&gid=9&Itemid=104. Accessed 24 July 2012.

Zahnd, A., & Haddix, M.K. (2005). Renewable energy resources for improved, sustainable livelihood: A case study of a holistic community development project with a remote and poor mountain village in the Nepal Himalayas. *6th Asia Pacific Roundtable for Sustainable Consumption & Production (6APRSCP)*, Melbourne, Australia, 10–12 Oct 2005. http://www.rids-nepal.org/index.php?option=com_docman&task=doc_download&gid=7&Itemid=104. Accessed 24 July 2012.

Zahnd, A., & Malla, A. (2006). High-altitude solar water heater community bathing centre—Renewable energy utilised for a remote and impoverished Himalayan village in Humla Nepal. *ANZSES 2006*, 13–15 Sept 2006, Canberra, Australia. http://www.rids-nepal.org/index.php?option=com_docman&task=doc_download&gid=11&Itemid=104. Accessed 24 July 2012.

Zahnd, A., & Haddix, M.K. (2007a). Solar PV systems in Himalayan villages: Problems and possible solutions. In *Proceedings of the 45th Annual ANZSES Conference*, 2–6 Oct 2007, Alice Springs, Australia. http://www.rids-nepal.org/index.php?option=com_docman&task=doc_download&gid=61&Itemid=104. Accessed 24 July 2012.

Zahnd, A., & Haddix, M.K. (2007b). Eclairage sur mesure au Nepal (translation of the English manuscript: Pico-hydro power plant for elementary lighting as part of a holistic community development project in a remote and impoverished Himalayan village in Nepal) (pp. 11–13). *TRACÉS 24, Bulletin technique de la Suisse romande*, Nov 2007. http://retro.seals.ch/digbib/view?rid=bts-004:2007:133::1549&id=browse&id2=browse5&id3=2. Accessed 24 July 2012.

Zahnd, A., & Haddix, M.K. (2007c). A mountain to climb? How pico-hydro helps rural development in the Himalayas. *Renewable Energy World*, *11*(2), 118–123. http://www.renewableenergyworld.com/rea/magazine/archives?tech=&page=1&keywords=A+mountain+to+climb&x=21&y=14. Accessed 24 July 2012.

Chapter 3
Power/Energy Generation and Lighting

3.1 Nepal's Energy Consumption Pattern

An interesting trend in the energy consumption pattern and growth for Nepal's non-renewable and renewable biogas and micro-hydro energy resources over the years 1990–2002 (presented in TJ (10^{12} J)), can be seen in Table 3.1.

While in 1990 Nepal's traditional fuels made up 96 % of the total annual energy consumption, in 2002 they are still providing 86 % of all (households/ industry/transport) energy services. In actual mass the consumption increased by 27.5 %, while Nepal's forests shrunk over the monitored time period by ~22 % (World Bank 2010). The per capita traditional fuel consumption decreased from 1,072 kg per year in 1990 (with a population of 18.1 million and an assumed mixed traditional fuel energy value of 12 MJ/kg) to 987 kg (i.e. by 8 %). This is a result of the sharply increased consumption of imported petroleum fuels in urban areas (the typical climbing up on the "energy ladder" (ITDG n.d.) in developing countries). Compared to the average world wide per capita biomass consumption between 1994 and 2005 of 1.9 kg/day (average population of 6 billion people and a total annual biomass consumption of ~50.7 EJ,[1] with an assumed mixed traditional fuel energy value of 12 MJ/kg), Nepal still has a 52 % higher biomass per capita consumption of 2.9 kg/day (23.5 million people and 12 MJ/kg).

During the same time period there has been an almost five fold growth in the consumption of non-renewable petroleum energy resources, mostly for the transport sector. With 57 %, or 183,402, of all of Nepal's registered vehicles in 2001 being in use in the Kathmandu valley (CEN 2003; Shakya 2003), it is clear why Kathmandu suffers from serious air pollution problems. Also the massive increase in coal consumption, in particular from 1999 to 2000, with an almost four fold increase, indicates that the industrial use of heat energy (especially for the brick kiln

---

[1] Averaged from Janet Ramage *Energy A Guidebook*, Oxford University Press, 1997; WEC 2005, *2004 Survey of Energy Resources*, pages 247–252, 259, http://www.worldenergy.org/documents/ser2004.pdf (accessed July 26, 2012),. WEC 2007, *2007 Survey of Energy Resources* pages 333–334, and http://www.worldenergy.org/documents/ser2007_final_online_version_1.pdf (accessed July 26, 2012).

A. Zahnd, *The Role of Renewable Energy Technology in Holistic Community Development*, Springer Theses, DOI: 10.1007/978-3-319-03989-3_3,

Table 3.1 National energy consumption data for Nepal 1990–2002 in TJ (10^{12} J) (WECS 2003)

Fueltype	1990	1991	1992	1993	1994	1995	1996	1997	1998	1999	2000	2001	2002
Traditional	232727	237614	242604	247699	252901	258212	263634	267138	272893	278748	284735	290659	298208
Fuelwood	207887	212253	216710	221261	225907	230651	235494	237555	242687	247884	253199	258636	264985
AgriResidue	9332	9528	9728	9932	10141	10354	10571	11645	11893	12166	12446	12732	13148
Animal dung	15508	15833	16166	16506	16853	17207	17569	17938	18313	18698	19090	19491	20076
Commercial	**10095**	**14830**	**17848**	**18322**	**20580**	**24784**	**27759**	**29441**	**32740**	**34576**	**44515**	**46701**	**49826**
Petroleum	7899	10933	13040	14842	16679	19119	21615	23623	26619	28180	30224	31286	32317
LPG	0	197	281	379	481	643	916	1075	1131	1232	1508	1975	2401
Motor spirit	601	822	897	968	1040	1172	1380	1497	1572	1674	1862	1984	2119
Air turbine fuel	332	687	878	1039	1094	1357	1469	1731	1860	2009	2056	2283	1716
Kerosene	3129	3544	4437	5206	5877	6559	7568	8841	10226	10696	12006	11472	14018
High Speed diesel	3752	5142	6196	6724	7436	8597	9501	9783	11402	11978	11780	12367	10857
Light diesel oil	87	119	100	60	117	149	174	78	38	21	156	134	94
Fuel oil	0	253	86	280	424	406	341	320	54	189	428	482	569
Other	0	168	165	186	209	236	266	299	337	380	428	482	543
Coal	307	1777	2461	1090	1359	2839	3085	2540	2579	2618	10064	10803	12355
Electricity	1889	2120	2347	2390	2542	2826	3059	3278	3542	3778	4227	4612	5154
Renewables	**160**	**178**	**221**	**249**	**251**	**339**	**455**	**631**	**897**	**1061**	**1236**	**1454**	**1678**
Biogas	128	144	185	212	212	298	412	536	678	826	981	1179	1327
Micro hydro	32	34	36	37	39	41	43	46	48	51	56	60	62
Solar	0	0	0	0	0	0	0	0	0	0	0	0	0
others	0	0	0	0	0	0	0	49	171	184	199	215	289
Total	242982	252622	260673	266270	273732	283335	291848	297210	306530	314385	330486	339014	349712

industry) is sharply increasing, due to the increased urbanisation and building boom.

Electricity consumption, though almost exclusively generated with hydro power in Nepal, increased almost threefold over the monitoring period, with an average annual increase of ~10 %. While that indicates an infrastructural challenge to meet such a demand growth, the amount of electricity consumed in the whole energy mix was still only 1.47 % of the total in 2002.

The energy services generated through new renewables (not including hydro power plants >1 MW), show a widely fluctuating picture year by year with growth rates between 10 and 30 %, and more than a ten fold overall increase during the recording period. But the actual energy generated by renewables (including hydro power plants >1 MW) is still very low, and contributes ~1.9 %[2] of the whole energy generation mix.

The high rate of growth in energy consumption indicates the need for clearly defined long-term energy policies, which take the unique and fragile Himalayan ecology into consideration. These policies must be sensitive to the delicate environment, allowing people to utilise the locally available renewable energy resources on a competitive basis with the non-renewable energy resources. That would allow Nepal to protect what can still be protected, and may help to restore to some extent what has already been harmed through the uncontrolled growth in consumption of fossil fuels and inefficient combustion processes.

3.2 Approaches for Improved Energy Services for the Poorest of the Poor in Nepal's Remote Himalayan Villages

While significant progress during the last two decades has been made in rural electrification through extending electricity services to rural villages, it also became clear that with traditional electrification methods such as grid connection alone, most of the remaining, isolated rural villages in the developing world will not be reached within the foreseeable future. Geographical and climatic conditions, in combination with low population density, minimal energy demand levels and low growth potential, are some of the reasons that make rural electrification costs often prohibitive. The continuously increasing rates of migration and urbanisation, in particular in developing countries, often support the decision by governments to direct their effort to improving and extending urban electrification systems.

[2] Considering that ~95–98 % of all electricity is generated by hydro power plants > 1 MW in Nepal, with ~2–5 % of the nation's electricity generated with diesel generators, an additional ~5,000 TJ Electricity generation can be added to the NEW Renewables' Energy mix of 1,678 TJ, thus making up a total of around ~1.9 % of Nepal's total energy consumption.

3.2.1 Grid Connection

Nepal's national grid (total length 4,346 km) is designed, built and maintained by the government owned NEA, providing ~7 million customers with grid electricity. These consumers are mostly located in the few urban areas and highly populated flat, subtropical southern part of Nepal (Fig. 3.1). The few existing grid extensions into the more rural areas are characterised by frequent overloading, poor reliability, high line losses and power theft. The lack of grid extensions is understandable in the context of Nepal, as they are usually very costly, long-term investments. Further, the remote and mountainous areas are not only a challenge from the technical and geological point of view, but are sparsely populated, with few consumers per square km, with low energy demands and low expected load growth. Often, in rural areas, consumers are charged a flat rate, if at all, which provides no incentive to conserve electricity (Fig. 3.2).

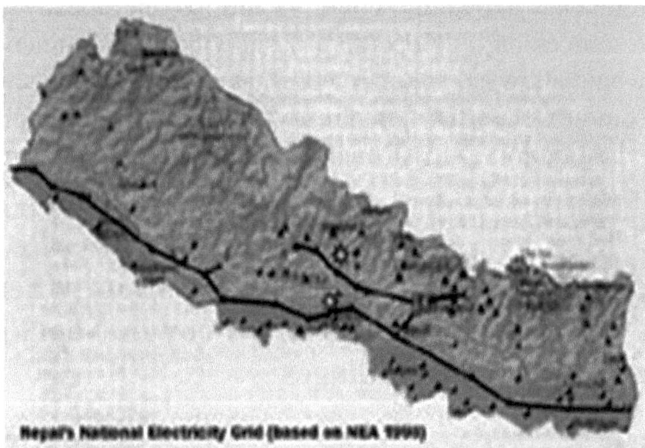

Fig. 3.1 NEA's national electricity grid, which is mostly connecting urban areas in the flat low-altitude areas of Nepal

Fig. 3.2 Though this resident near Nepalgunj had to provide land for transmission poles, it has not enabled the electrification of the home

All of these conditions have placed great pressure on NEA, and the company has been restructured in 2005/6 into five business units, the generation, transmission, electrification, engineering services and distribution and customers services (retail) (NEA 2009, 9). These units aim to be profit making businesses. Thus there is now even less hope for future rural electrification projects through NEA, because it is not profitable to serve the poor and remote mountain communities. Further, existing rural power generation plants are being sub-contracted either to interested private companies or to the communities themselves, to run them as local businesses. But most of the micro-hydro power plants for rural village electrification projects have been originally designed and built with the traditional approach of "100 W per household" for incandescent bulbs, making them too big and too costly for the local communities to maintain and repair.

3.2.2 RAPS Systems

Nepal's unique geographical and topographical conditions demand new approaches and technologies in order to reach the ~20 million people living in difficult and remote mountain areas without electricity. A viable option to grid extension is the generation of electricity within, or near to, a community, using locally available renewable energy resources. One solution that is growing in popularity is to generate the required power locally through a Remote Area Power Supply (RAPS) system. A RAPS system can be defined as a power generation system, generating electricity for rural homes and communities. Such systems are small scale (usually $<50\ kW_p$) self-contained units, providing electricity independent of the main electricity grid or mini grid network. RAPS systems range from small petrol generators, able to power appliances directly, to more complex installations using either renewable energy only, or in combination with diesel or petrol generators.

A RAPS system that has a combination of energy sources, such as a wind generator, solar panels, petrol or diesel generator, battery charge control system, battery storage and inverter, is called a hybrid RAPS system. For the remote mountain areas of Nepal a suitable RAPS system could contain solar PV panels, a small micro-/pico-hydro power plant, or a small wind generator. A petrol powered back-up generator is not a feasible solution for the remote/poor mountain communities, as the transport cost for fossil fuels (carried by porters for days or weeks) is unrealistic. When the renewable energy resource is not available for prolonged periods of time, the way to ensure an appropriate back-up system is to design a battery-bank, able to provide the required energy for a realistic span of time.

RAPS systems are often designed with 20 years life-expectancy. This demands specific conditions to provide the right foundation for the design. To define these in the initial stage of a project is by no means an easy task, as they will form the "outer frame" and limits of the RAPS system, and are not easily changed afterwards. It is good engineering practice to include in the life-cycle cost[3] all the necessary operational, maintenance, repair and spare parts costs. This provides a realistic energy unit (kWh) price that each consumer has to pay, in order to keep the RAPS system operational. The locally available energy resources, the community's average daily energy service demands (including a realistic future load demand growth) and their affordable budget will determine the kind and size of the RAPS system. Thus in the following, a basic village electrification system in the context of a Humla village, is a power generation system, generating small amounts of energy through a RAPS system.

The following outline contains the procedure that is followed once a village community has requested a joint village electrification project with RIDS-Nepal:

- The local community has to initiate the request for the implementation of a HCD project in their village, which includes a basic village electrification project for lighting.
- First a house by house base-line survey has to be carried out, in order to have the local people contemplate upon and identify their personal living conditions.
- The local community has to be introduced to the WLED technology and lighting services it can provide through practical examples of operational systems, before the project is designed.
- The local community's financial capability to participate in the initial project costs as well as in the operating and maintaining of the project over its life-cycle, has to be agreed upon and signed, before the project is designed.
- A written and signed (by thumb print) agreement by all stakeholders, on the kind of voluntary participation, initial cost participation, monthly fees, initial training for installation, operation and maintenance, has to be in place.
- The definition of the project scope and design occurs in close partnership and consultation with the local community, increasing the ownership responsibility.
- Along with the elementary lighting project, a minimum of three other HCD project components, a pit latrine, a smokeless metal stove, and clean drinking water from village taps, have to be implemented alongside each other. Future inclusions of NFE (non-formal education for adults and out-of-school children), greenhouses and solar driers should be considered.
- A carefully defined and faithfully executed follow-up program (>10 years) has to be followed, once the initially installed HCD programs are implemented.

[3] Life-cycle costs (LCC) include all costs of a project (cradle to grave), from its initial start, including planning, procurement of equipment, interest and inflation rates, transport, to infrastructure building, installation, operation, repair/maintenance, decommissioning and salvage value at the end of the project's useful life time. The LLC calculation is an economic model, spanning over the entire project's life time. LLC is expressed in economic terms/values, often as cost per unit of service/energy. The LLC analysis helps to choose the most, long-term cost competitive solution of ownership of a project.

- The follow-up program should also include a data monitoring and recording program, as well as periodic village surveys, to record factual and anecdotal experience of the changes occurring over time. It is often difficult to convince sponsors to fund this important aspect of the project.

Nepal's rich resources of hydro power and sunshine can be utilised locally through appropriate renewable energy technologies such as micro-hydro (5–10 kW) and pico-hydro (0.1–5 kW) power plants (Zahnd 2004a, 22), or solar photovoltaic modules and arrays, to generate the desired electrical power to meet basic energy demands. RAPS systems can provide a realistic and appropriate solution for the vast majority of Nepal's population according to a community's identified needs and financial capability within their foreseeable lifetime. Further, it is often the community that owns a RAPS system. Thus in order to have strong ownership for the RAPS system, the community has to participate actively in each project step. That starts with the initial planning (defining and setting all the parameters of a RAPS system) and design stage, the building and installation of the RAPS system, as well as operation and maintenance. Local people are trained in the basic theory and on the job in all practical installation and operating work, assuring a high degree of ownership and thus willingness to maintain and repair the RAPS system. With such an approach and implementation, a RAPS system can provide basic electricity needs, which are often limited to elementary lighting, for the most remote communities.

3.2.3 Basic Rural Village Electrification

One of the first questions to ask for any village electrification system design is: "What is the most important energy service demand of the family?" And most of the time in remote villages the answer is: Light. In order to design an appropriate electrification solution for lighting purposes, it is important to understand the methods and technology people have thus far used to fulfil their lighting needs.

For example in the high altitude areas in Humla, in order to provide light in the otherwise dark homes (as most have only very small unglazed windows which are mostly closed due to the cold), every home creates light inside their home by burning the traditional "*jharro*", a stick of resin soaked pine wood (Figs. 3.3, 3.81, 3.82, 3.101). They cook with firewood on an open-fire, with either three stones, or a three legged metal piece, called an "odhan" (Figs. 3.3, 3.4, 3.82, 3.101). Needless to say, the open-fire place and the "*jharro*" burning all evening, create very unhealthy living conditions. In this context, what options are available other than the smoky "*jharro*" for providing light for each home, light that is necessary for cooking, evening gatherings and children's study times?

The bulky equipment for a 10 kW or larger micro-hydro power plant weighs several 1000 kg. Once installed, experience shows that it is almost impossible to dismantle it, when it is in need of repair due to failure. The technical expertise is

Fig. 3.3 Open-fire place cooking and "*jharro*" burning. The traditional way of having a small dim light inside the home is the "*jharro*" (resin soaked pine wood stick burning to the left, with a one evening ration ready to be burned at the *bottom left*). Burning "*jharro*" emits thick black smoke and soot which adds to the already unhealthy conditions created by open-fire place cooking

Fig. 3.4 Open-fire place cooking is the traditional and common way to cook the daily food. The mother and often the children sit around the smoky fire as the meals are prepared. Women and children are most likely to suffer from exposure to the indoor smoke pollution (open-fires create PM_{10} levels $\geq$20,000 $\mu g/m^3$. US EPA 24 hours average limit is not to exceed 150 $\mu g/m^3$ more than 3 times a year, *Smoke, Health and Household Energy,* ITDG, September 2002), causing health hazards such as respiratory diseases, asthma, blindness and heart disease (IEA 2002, 367–368)

not available in the village and the high transport cost for porters and air freight are prohibitive for most of the poor mountain communities.[4] These issues raise serious questions of appropriateness and sustainability of this otherwise widely accepted technology.

---

[4] During 1996–2000 the author lived in the remote mountain area of Jumla, where initially 23 micro-hydro power plants have been installed over the course of a decade. Out of these 23 power plants, only 3 were functional, providing the local communities with indoor lighting. All other micro-hydro power plants were not operational due to mostly simple technical and socio-economical problems.

But what kind of electrification technology and system is appropriate where the local people have identified home lighting as their most urgent need? Considering the local community's needs, demands and financial capability (most of them are below the poverty line), a basic lighting system, that is affordable and reliable, is what they need. In order to provide a possible solution for lighting in this context, it is important to start improvements at both ends, the power generation and the lighting power consumption. If less power is needed for the lights, less power needs to be generated. Smaller systems are a lower financial burden for the local community, and thus more affordable without an external sponsor. Further, they are normally easier to build, operate, maintain and especially to repair, if they fail. They can also be designed in closer partnership with the local community, making them the proud owner of the system. Such a system is called a basic village electrification system.

3.2.4 Appropriate Lighting Technologies

In most homes, which are powered by a micro-hydro power plant, and sometimes even by a solar PV system, in the rural mountain areas of Nepal, 2–3 incandescent bulbs, each 25–60 W, are installed. Thus a village with 50 homes needs a 10 kW micro-hydro power plant. The local people do not know any other lighting technology, thus they have never questioned its appropriateness. The incandescent bulb's standard life span is 1,000 hours.[5] They are easily breakable, and to get a new bulb in the remote mountain areas is difficult, time consuming and expensive. Every single light bulb has to be flown in by airplane, and then carried by porters to the main bazaar, before one can purchase one at an exorbitant price. If a high quality CFL (compact fluorescent light) bulb, with a high Power Factor (PF) of >0.9 is used (Fig. 3.5), which consumes 7–11 W (and is comparable with a

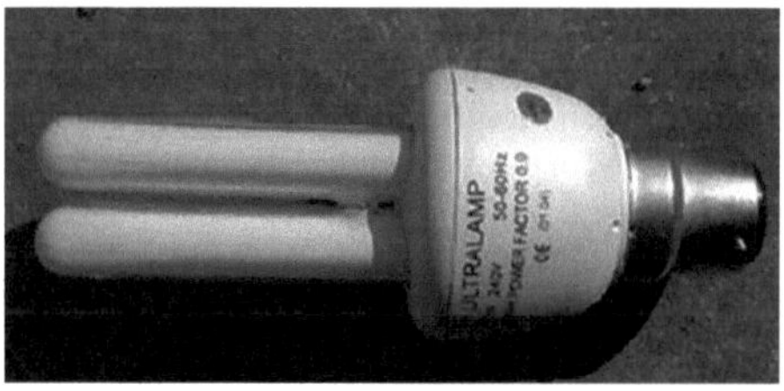

Fig. 3.5 A high quality, 11 W CFL Lamp (manufactured by Ultralamp, an Australian-Chinese collaboration), with a high Power Factor (PF) of >0.9, and with a life-expectancy of 8,000–12,000 h. Its light output of ~80 lm/W can be compared with the luminance of a traditional 55 W incandescent bulb

[5] Wikipedia provides a very good, detailed historical and technical discourse of the traditional incandescent light bulb at: http://en.wikipedia.org/wiki/Incandescent_light_bulb (accessed July 26, 2012).

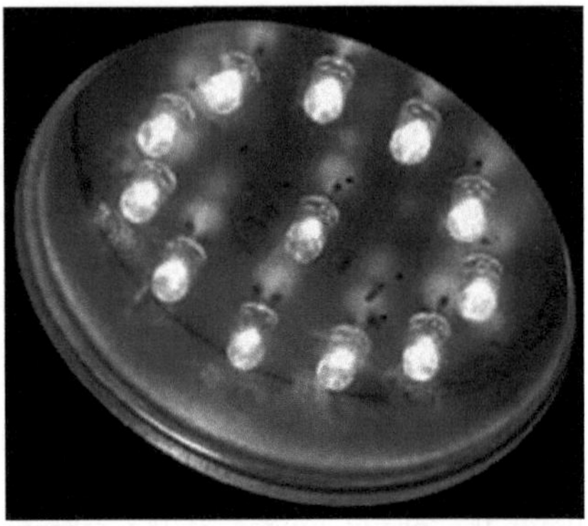

Fig. 3.6 Nichia NSPW510DS (12 diodes with a 50° light angle) WLED lamp, consuming 1.1 W, with ~83 lm/W and a life-expectancy of >50,000 h

35–55 W incandescent bulb with respect to its light output), the power generation can be 5 times smaller for the same amount of homes and lights. Further, CFL lamps have life expectancies of 8,000–12,000 hours. Thus, a 2 kW pico-hydro power plant would be enough to provide the same lighting services.

Even less power needs to be generated if 1 W WLED lights are used. Further, the expected life-time of WLEDs is between 50,000 and 100,000 h according to the manufacturers. Since early 2009 Nichia has had WLEDs on the market with ~83 lm/W (Fig. 3.6), making the technology competitive[6] with good quality CFL lamps, with around 80 lm/W.

---

[6] The word "competitive" is used, because by providing merely the value of lm/W (lumens per watt), called Luminous Efficacy, indicating the ratio of luminous flux to power for a lighting technology, is not sufficient to provide enough information for the users to get a clear understanding or a "feeling" for the brightness of different lamps. It is important to include also the angle of radiation that light is emitted by a source. To know this important specific parameter for a light source means that one knows how much of the radiated light is usefully intercepted or applied. For example, an incandescent bulb, with a viewing angle of 360°, emits light all around, with a high percentage of it ending up to be in non-useful directions (e.g. upwards), while LEDs have often a much narrower viewing angle between 15 and 120°. Thus, an LED with a 30° viewing angle means that the light emitted from the LED forms the shape of a cone with a 15° angle around the centre axis of the cone. By changing the emitting angle of the same LED, e.g. from 30° to >30°, changes (decreases) the luminous intensity of the LED, called candela (cd) but not the luminous flux, called lumen (lm), and vice versa. Lumen and Candela measure different things and thus are not easily and direct comparable. Lumens, measuring the power of light perceived by the human eye, measure light output at the source. Lumens are defined as the luminous flux emitted into one steradian (sr) by an isotropic point source having a luminous intensity of 1 cd. Candela on the other hand measure the light emitted per unit of solid angle, a quantity that does not vary with the distance from the source of the light. Thus, a 1 cd (1 lm/sr) light source will produce 1 lm/m^2 at the distance of 1 m from the light source. The candela (1 cd = 1,000 mcd (millicandelas)) is the luminous intensity, in a given direction of a source that emits light that has a wavelength of approx. 555 nm (yellowish-green and the most sensitive wavelength for the human eye) and that has a radiant intensity in that direction of 1/683 W per steradian. As the candela value is independent of distance, one can think of it as the emission from the lamp without being interested in what happens to the photons it has ejected. The candela is mostly used when dealing with focused light such as LEDs and flashlights. By taking into

WLEDs long life-expectancy (>20 years) and their ruggedness (light emitting diodes are almost unbreakable compared to CFL and incandescent bulbs), are important factors for sustainability, and this makes them very suitable for remotely located, rural village electrification projects. Thus, WLED lamps are the most appropriate and sustainable solution to fulfil the local people's identified demands for minimal lighting in the design of a basic rural village electrification system for remote mountain communities in Nepal.

3.3 Technologies Appropriate for Improved Energy Services in Nepal's Remote Himalayan Villages

Among the various available renewable energy technologies, very small hydro power plants, solar PV systems and small scale wind turbines work out to be the most applicable technologies. It is important in order to increase the possible success of a project, that the most appropriate and sustainable technology for a defined context is chosen. This leads to initial questions such as:

- Is the technology chosen based on: Least Cost, Preferred by the Community?
- Has sustainability been considered before efficiency?
- Will the power system be technically, economically, socially and environmentally sustainable?
- Can the consumer afford the energy services over the life-cycle time of the system?
- Can the end-users participate in all steps of the projects and can they be trained to operate and maintain them?
- Is the technology also culturally appropriate and acceptable to the end-users?

(Footnote 6 continued)
account both, the viewing angles and the lumen output, the comparison between a CREE white LED (CREE X-RE R2) with about 100 cd and a viewing angle of 90°, and an incandescent bulb, with about 105 cd and a viewing angle of 360°, as well as some other, more known and traditional light sources, looks like the following table, copied from http://www.ledrise.com/shop_content.php?coID=17 (accessed July 26, 2012).

Light source	lm	lm/W	Viewing angle	Useful viewing angle[a]	Useful lm
CREE X-RE R2 White LED	242	92	90	100 %	242
100 W incandescent	1200	12	360	33 %	396
25 W Halogen	260	9	360	33 %	85.8
15 W T8 neon	1350	90	360	33 %	445.5

[a] Without case or reflector

- Can needed spare parts be made available from the local/national market in due time and at affordable prices?
- Can all stakeholders' (end-users, project implementer, donor) expectations be met?
- Will new activities and possible income generating projects result?
- Will the overall living conditions of the villagers be improved?

These questions have to be asked and answered satisfactorily by all stakeholders of a project. This is time consuming, and often involves additional costs, which are hard to justify to donors.

In the following we will look into the three major applicable and locally available RETs for Nepal's remote mountain villages, hydro power, solar PV systems and small scale wind turbines.

3.3.1 Hydro Power Plants

With Nepal's vast, economically feasible hydro power potential of 42,130 MW, it is obvious that this technology is the one to consider first. The hydro power industry in Nepal distinguishes between large-scale ($\geq$10 MW), medium-scale ($\geq$1 MW, $\leq$10 MW), small-scale ($\geq$100 kW, $\leq$1 MW), micro-hydro ($\geq$5 kW, $\leq$100 kW), and pico-hydro ($\leq$5 kW). As the context of our study is very small scale power generation, for basic lighting services only, the focus of this chapter is limited to pico hydro power plants.

3.3.1.1 Pico-Hydro Power Plants and Case Study

As the name indicates, pico hydro power plants are intended to generate small amounts of power (<5 kW) from a small stream/river. They are mainly intended as power generation plants for remote villages which have identified their energy service demands as basic lighting, as part of their first exposure to electricity. In the mid 1990s, NHE (Nepal Hydro Electricity) developed, with input and field test results from the author, the first pico-hydro power plant, generating 200 W. It was mandatory that only material and equipment that is available, or could be manufactured, in Nepal was used in making this plant. The principle of using a motor as a generator, according to the book "Motors as Generators for Micro-Hydro Power", by Nigel Smith (1997), was the basis of this approach, as small induction motors are widely and cheaply available in Nepal. Using a motor as a generator does not achieve the highest efficiencies, but it is an appropriate technical solution for the context in which they will be used. The technology is easy to understand, operate and maintain, and tough enough to survive the harsh and difficult conditions under which they have to work for years, with a low failure rate.

Fig. 3.7 The Thalpi village people built the whole power house and stone/wood water canal, as part of their project input, thus being rightly proud of their pico-hydro power plant

Fig. 3.8 The power house and the water canal of the Godhigau village 165 W pico-hydro power plant are built with local wood, with no cement, because cement in this remote area it is prohibitively expensive

A first 200 W (though in the field it produced only 165 W) pico-hydro power propeller turbine with a resistive air-heater as dump-load, was installed in the village of Thalpi (Fig. 3.7), in the remote Jumla district of north-western Nepal in 1998, followed by the second in the neighbouring village of Godhigaun in 1999 (Fig. 3.8). The pico-hydro power plant is designed with a low negative head, and uses 25 liters of water per second. The negative head means, that the pico-hydro plant has a conical shaft of 8° after the propeller turbine. That shaft is 2.1 m long, and can be installed in between the farmers' terraced fields, which are frequently watered. In this way, no additional water needs to be diverted from the fields to run the pico-hydro power plant. As each pico-power plant is a self-contained unit, it is easy to add further plants in series if the energy demand of the end-users increases. For Thalpi village, with 30 households (and 245 people), three WLED lamps (each consuming 1 W) were installed for each household. There was no detailed life-cycle cost analysis carried out (as most of the local materials used have no trade value and thus would need to be costed first). Every household paid monthly 20NRs (0.4AUD, 2000 value), towards the maintenance, security and repair costs, to make the system sustainable.

Fig. 3.9 165 W pico-power plant

RIDS-Nepal buries all the transmission cables (armoured cables) underground, despite the substantially increased cost. This is justified because of the enormous amount of forest degradation in Nepal's high-altitude Himalayan areas, and thus no further trees have to be cut for transmission poles throughout the lifetime of the power plant. Further, underground cables are protected from the often harsh weather conditions, such as snow, torrential rains, storms and lightning. In both the following case studies from the first two pico-hydro power plant installations in the Jumla district (Figs. 3.7, 3.8), no cement has been used either for the power houses or for the water canals, which get their water from diverted small streams (Figs. 3.9, 3.10, 3.11, 3.12, 3.13, 3.14, 3.15).

In each village 2–4 people, chosen by the village pico-hydro power plant committee, have been trained to operate and maintain the pico-power plants, and to inform the project implementer of problems they can not handle. One of them is also a watch man, responsible for the security of the power plant. He is paid a small monthly salary (300NRs) from the fees collected each month from the users. Thus the pico-hydro power plants are fully managed by the local village and the sense of ownership is high, as they all have put a lot of hard work into carrying the equipment, organizing all the local building materials, and building the power plant.

The first 1.1 kW pico-hydro power plant (Figs. 3.16, 3.17, 3.18, 3.19, 3.20, with a water flow of 83 l/s, a 3.1 m negative head and a maximum efficiency of 43.6 %) was installed and inaugurated on the 6th December 2006 in a remote village called Kholsi in Humla.

The 75 families and the primary school have together a total of 230 AC WLED lights installed, consuming a maximum of ~250 W. As the lights are usually on

Fig. 3.10 Godhigau pico-hydro wooden water canal

Fig. 3.11 Installation of the Thalpi pico-hydro in the wooden canal

only for several hours in the morning and evening, the power plant has a low load factor. As this village has a high-altitude climate, they are always in need of warm/hot water. Thus a "water heating dump load" was designed for the power over-production, heating 500 l of water in a polyethylene storage tank (Fig. 3.18), insulated with locally available materials.

The warm/hot water generated by this system is used for improved personal hygiene and showering. It can also be used to prepare meals, reducing firewood consumption even more than already reduced by usage of the efficient smokeless metal stoves (~40 % if properly used) installed in each family home, as part of the long-term "Family of 4" HCD approach. What are some of the advantages and disadvantages of a pico-hydro power plant in this situation?

Fig. 3.12 First test if light can be generated. What a life experience!

Fig. 3.13 Even under harsh climatic conditions (down to −20 °C) the pico-hydro power plant was performing flawlessly

Advantages:

- Local available, maintainable technology, cheap to run and easy to operate.
- It runs under harsh and low maintenance conditions.
- The generator is so small that it can be carried by only one person.
- Several pico-hydro power plants can be installed in series once the power demand grows.
- No cement is used for the construction of the power house and water canals, as carrying it by air plane and then by porter to the high-altitude area, increases the price thirteen times.

Fig. 3.14 Now there is no need anymore to burn "*jharro*" to have a small light inside the home, and the air is much cleaner

Fig. 3.15 Besides basic literacy and socializing, education is a major part of increasing the people's awareness

Fig. 3.16 Beside the generator, the turbine and roof, the pico-hydro power plant infrastructure is built, using only locally available materials such as stone and wood. No cement at all is used. That creates a strong ownership among the users

Fig. 3.17 Inside the power house of the pico-hydro power plant. The 1.1 kW generator is seen, with the wooden water canal, which cleans and streamlines the water for the propeller turbine's intake

Fig. 3.18 With 230 WLED lamps running, another minimum 700 W of the 1.1 kW generated at all time are diverted to heat water for hot showers and hot water to cook food

Fig. 3.19 The generator's 225 VAC are transformed to 615 VAC and via underground cable sent to the village

Fig. 3.20 For the WLED lamps the voltage is again transformed to 225 VAC in the village

- The local end-users can be involved in each project step, especially in the building phase, as they have to organise all the local materials such as wood, stones and mud.
- Through the intensive participation of the local community a strong sense of ownership is created, which is important for long-term sustainability.

Disadvantages:

- The induction motor and simple propeller blades (not airfoil shaped) are not very efficient.
- Designed for a "first time" rural village electrification, mainly for lighting purposes.
- The pico generators are induction motors, thus starting a motor could lead to difficulties in regard to severe voltage drop or even loss of excitation.

3.3.2 Solar PV Systems

The average solar irradiation for Nepal is 4.8–6.0 kWh/m^2/day (Appendix 11.5). In the high-altitude areas the solar radiation resource is in most places at the upper end of the scale (Blumenthaler et al. 1997), if the location is not shaded from surrounding mountain ranges or in a deep valley. That highlights the importance of having exact and reliable data for the local prevailing conditions. Again, one of the crucial points for each solar PV village project is to use and install as much as possible locally available or locally manufactured equipment. That increases the appropriateness of the project, and helps the local industry to grow in a more independent way, creating new income for the society, as well as teaching new skills to craftsmen—all important parts of a holistic community development project.

Further, over the last decade, the author, together with RIDS-Nepal and the Kathmandu University, developed and installed three different approaches to solar PV village electrification systems, the Solar Home System (SHS), the Solar PV Village Cluster System and the Solar PV Central Village System. Each approach depends on a village's context in regard to geography, village cohesion and locally identified needs, which are in most cases basic lighting services. In collaboration with PPN (Pico-Power Nepal) 1 W WLED lamps were developed.

It is important to underline, that any rural village electrification project is never implemented as an individual project, but as part of a much wider long-term holistic community development project in that particular village. Every rural village HCD project includes at least a pit latrine and a smokeless metal stove for each family, access to clean drinking water from various taps in the village, and often a greenhouse and non-formal education (NFE) classes for mothers and out-of-school children. In this way synergistic benefits occur and increase the overall positive development outcome for each village, family and person substantially beyond those of a single, individually implemented project.

3.3.2.1 Solar PV Home System (SHS) and Case Study

The most straightforward approach is to install a small scale solar PV system for each individual home, called a solar home system, or SHS. This was also in the mind of the policy makers for the first Nepal government solar PV home system subsidy program. On each house one solar PV module is installed, mostly between 20 and 40 W rated power output, with up to three 10–20 W fluorescent tubes. Each system has a 12 VDC, solar deep-cycle, battery of 40–75 Ah capacity and a simple charge- and discharge-controller, to protect the battery from too high discharge and charging rates. The market price (2009) for such a SHS is high, around NRs 20,000–25,000 (~US$278–347), and thus is not affordable by anyone in the remote areas without a subsidy.

In 1998 the Nepali government, supported by foreign investment, started the first solar PV subsidy program, making it possible for rural communities to purchase SHSs initially at 50 % of the market price. Then the first 5 years subsidy program started in 2001 through the government owned and foreign sponsored AEPC (Alternative Energy Promotion Centre).[7] The subsidy was gradually reduced by 10 % per year, so that after the first 5 years phase ended, the SHS set price reflected the going market price.

Unsurprisingly, fewer and fewer rural Nepalis were able to purchase a SHS for their families over the course of the subsidy program, since hoped for improvements in their economic status over the years did not keep pace with the reduction of the subsidy. In Nepal, there is a direct relationship between remoteness and

[7] The AEPC (Alternative Energy Promotion Centre) web site is: http://www.aepc.gov.np/ (accessed July 26, 2012).

poverty. The more remote a community is, the poorer its people (WEA 2000). This is further highlighted by contemplating the higher transport costs and increased effort required to build, operate and maintain power plants in these areas.

Thus a second 5 years AEPC led SHS subsidy program was introduced, with new subsidy rates, which were revised after the reception of an additional EU (European Union) grant. The subsidy levels were defined according to the remoteness of the end-users. For very remote mountain communities such as Humla, US$107 (in NRs equivalent, July 2007 rate) per SHS with a 10–18 W_p solar PV module system and US$154 per >18 W_p solar PV module system are available. Further, 75 % subsidy for the PV system cost is granted for community, health post or school buildings.

These government solar PV subsidy programs caused a mushrooming of solar PV companies in the country's capital, Kathmandu. Instead of the three solar PV companies existing in 1997, now 56 registered companies help meet the needs of the population and benefit from the subsidy programs. While the development of this renewable energy industry is encouraging and can only be welcomed, the author argues that the standard of the majority of the solar PV systems installed is questionable from a sustainability and end-user satisfaction point of view.

The increased availability of solar PV equipment has also increased price competition. Because national standards for solar PV systems are not comprehensively developed in Nepal, it is not surprising that the quality of available PV systems has dropped as market competitiveness intensified. Retailers' small price margins forced system installations to be done under time pressure, without appropriate training for end-users to learn how to run and maintain their newly purchased solar PV system. Further, follow-up visits are rare due to their expense, especially for remote places such as Humla District.

There are now many cheaper SHS available on the growing SHS market, but their quality and performance is often questionable and thus they are not considered to be appropriate for remote areas (nor in fact for anywhere else), and therefore they are not considered further in this study. The present SHS subsidy program (2007–2012) runs under the leadership of the government's AEPC is defined as follows:

1. Subsidy is provided to SHS of 10, 20 and 30 W_p or more.
2. The maximum subsidy for SHS $\geq 30W_p$ capacity is NRs 8,000 per system.
3. Additional 50 and 2.5 %[8] subsidy per SHS system is provided to the users in remote village development committees (VDC).
4. The level of subsidy is reduced each year at the rate of 10 %.
5. The subsidy for SHS used by public institutions such as the VDC buildings, schools, clubs, health posts/centres etc. is as high as 75 % of the cost.

[8] The AEPC (Alternative Energy Promotion Centre) solar PV system subsidy policy does not state for what the additional 50 and 2.5 % subsidies are, but the author interprets it as the 50 %, or maximum 4,000 NRs, due to the remoteness and the 2.5 % due to the increased transport cost to the remote village of the end-user.

Fig. 3.21 For the Darapori village individual SHS are the most appropriate lighting solution. Each SHS has one 16 W_p solar PV module (Kyocera KC16T), and a protected and insulated sealed lead-acid battery-bank (12 VDC with a capacity of 14.4 Ah), powering 3 WLED lamps (Figs. 3.6, 3.22)

Poor families will not have the economic capability to purchase a SHS once the subsidy has run for a few years, as the purchase cost is still too high, and there is no substantial cost decrease in the foreseeable future for solar PV equipment. Further there is no clear follow-up program in place despite the requirement that the installers' are paid the last 10 % of the total subsidy only after the SHS has been running for 1 year. That is not enough of an incentive to provide the end-users with appropriate quality equipment which will last far longer. Thus most of the installers do not really count on the last 10 % and there is very little fore-thought as to how the installed SHS can be followed up and maintained beyond the first year.

Additionally, the author recommends that SHS projects should be run as part of a long-term HCD project, with the end-users as active partners, participating through their voluntary work as well as some financial contribution, together with the implementer (who works on a long-term basis in that area) and the donor agency (which provides the main project funding according to an agreed project proposal and budget). Individual SHS (16 W) are appropriate when a village has its houses randomly scattered as in the RIDS-Nepal project partner village of Darapori (Fig. 3.21).

Each house has one solar PV module mounted on an aluminium frame (adjustable or fixed angle), its own battery-bank, providing up to 5 days of lighting services for the 3 WLED lamps per household without sunshine (Figs. 3.22, 3.23, 3.24).

If a village consists of several different castes and considerable distances between the houses, the individual SHS approach provides a sustainable solution. This approach requires that each household owner is responsible for his own system. Frequently, we have observed that a healthy competition arises due to each

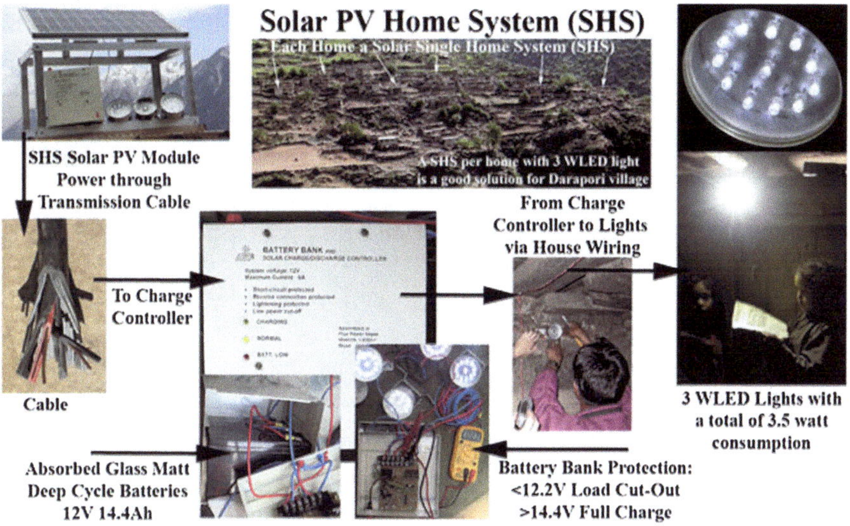

Fig. 3.22 Typical RIDS-Nepal Single Home Solar System (SHS) Schemata as installed in individual households if the houses are built separately from each other as e.g. in Darapori village

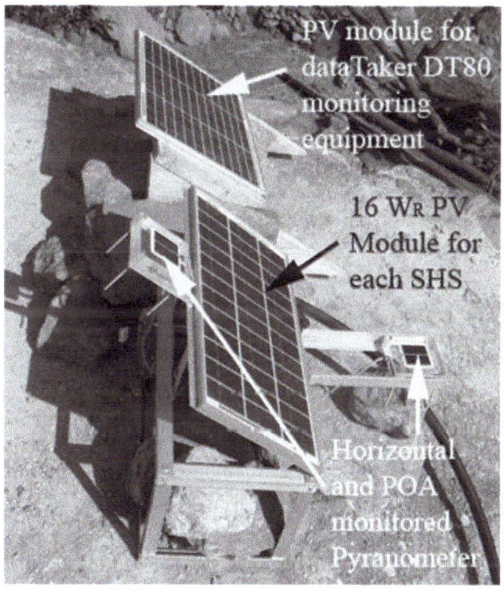

Fig. 3.23 SHS with a 1-axis, seasonally adjustable or a fixed angle aluminium frame. 22 parameters of this particular single SHS system are monitored 24/7 and recorded with a dataTaker DT80. This provides detailed system performance

owner's pride in their own SHS. That increases the likelihood that more care and maintenance takes place, increasing the SHS's reliability and life-expectancy.

Figures 3.25, 3.26, 3.27, 3.28, 3.29, and 3.30 show SHSs, as part of a HCD project, in Khaladig village in Jumla,

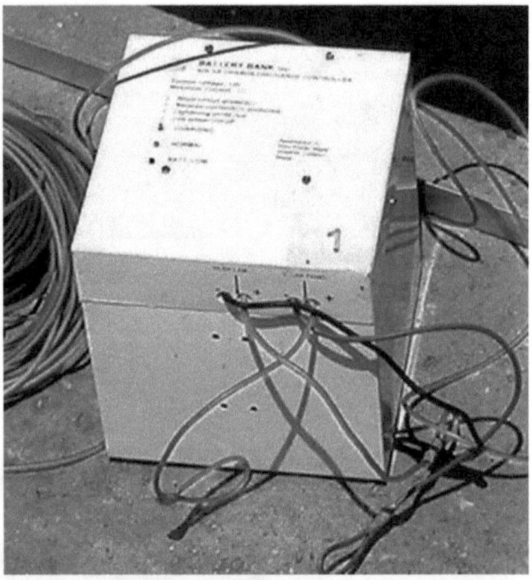

Fig. 3.24 Each SHS has its own, insulated battery box. Included is the battery-bank's charge-/discharge-controller, protecting the batteries from being overcharged (>14.4 VDC) or too heavily discharged (<12.2 VDC)

Fig. 3.25 The first step: raise awareness in the community and enquire and define what their needs for energy services are

Fig. 3.26 They grow now up with indoor light

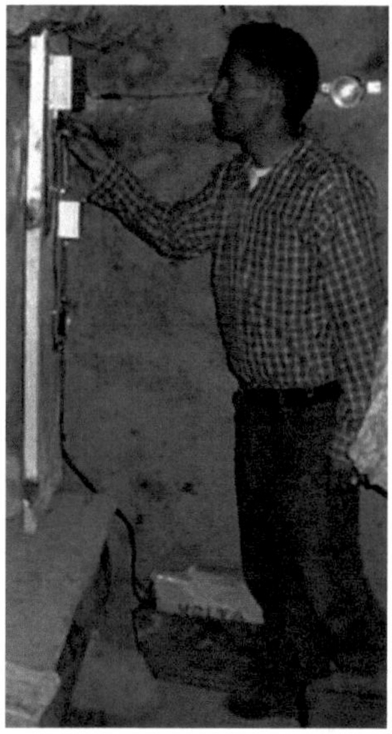

Fig. 3.27 Checking the battery and charger

Fig. 3.28 Training the end-users to maintain and conduct basic repairs is crucial for sustainability of every solar energy project

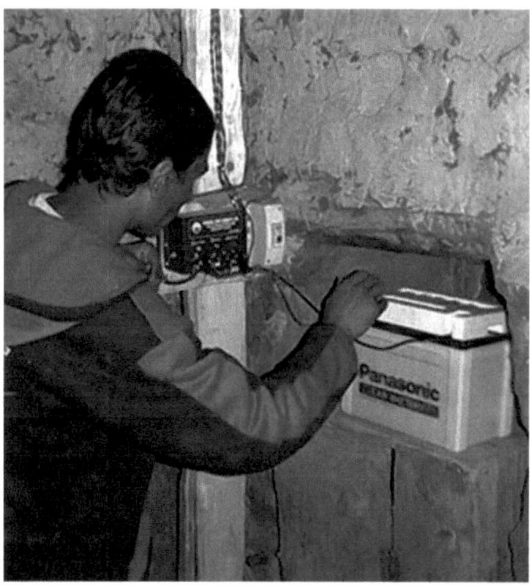

Fig. 3.29 On the job training through hands on installation

Fig. 3.30 One 40 W solar PV module for 3 lights

3.3.2.2 Solar PV Village Cluster System and Case Study

While SHSs are a very appropriate technology to bring light into the homes in remote villages, we learned that another appropriate approach is to get the whole village motivated to apply as a community for a basic rural village electrification project. That brings the dynamic, skills and work force of the whole village community into the project. Further, it creates a challenge and responsibility for each person of the village to participate in the project, thus including the poorest, marginalised and handicapped. The common planning, designing and building of the village power system creates a strong sense of community project ownership.

Over the years, the author has developed two approaches to electrify a whole village. They are the village cluster solar PV system and the village central solar PV system. Which of the two approaches is chosen, depends mostly on the village's situation in regard to the positioning of the houses and the village's geographical position compared to the sun's path over the year. The "cluster village solar PV system" (Figs. 3.31, 3.32) is most suitable if the homes in a village are built separately, in small groups or clusters at some distance from each other. In such cases, the cluster solar PV system approach provides a technically and economically viable solution.

A cluster is made up of 4–12 homes, sharing one solar PV module (often 75 W_p), mounted on a 1-axis adjustable aluminium frame and a common battery-bank. The interconnection of all the cluster homes to the solar PV cluster system is through armoured underground cables, powering three 1 W WLED lamps per home for 7–8 hours a day. In Nepal, a clustered spatial arrangement of homes often also maps caste, with people of different castes residing in different clusters.

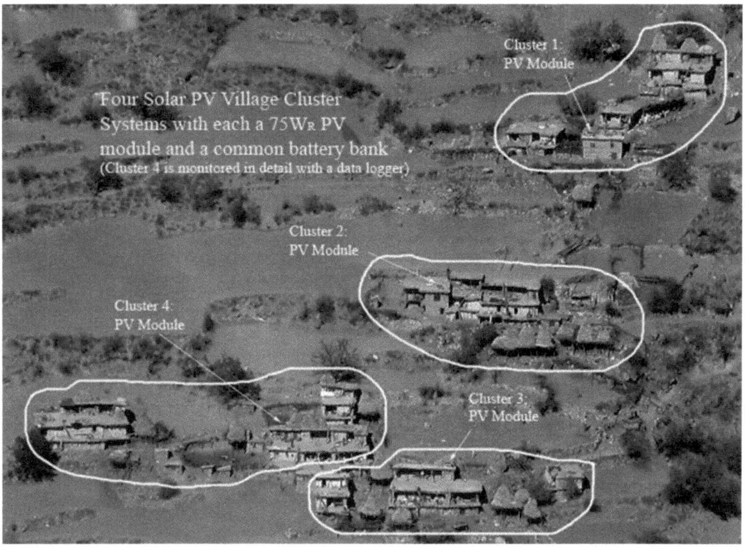

Fig. 3.31 Pamlatum village Cluster Solar PV System with 4 clusters. Each cluster (4–8 homes), has one 75 W_p BP275F solar PV module and a battery-bank (12 VDC, 200 Ah capacity)

Fig. 3.32 Each cluster has a 1-axis, seasonally adjustable aluminium frame, generating ~15 % more energy, compared to a fixed angle installation. All cluster homes are connected with the power house through armoured underground cabling

The cluster PV system takes advantage of cultural coherence within caste groupings in management and maintenance of each cluster in the system.

In a cluster solar PV system, the house in the middle is the cluster's power house with one 75 W_p (such as a BP275F[9] Si mono-crystalline) solar PV module

[9] The BP275F specification sheet is available at: http://www.oksolar.com/pdfiles/Solar%20 Panels%20bp_75.pdf (accessed July 26, 2012).

Fig. 3.33 Clusters, each with 4–12 homes connected

Fig. 3.34 One 75 W_p module, 1-axis seasonally adjustable aluminium frame, fixed with stones

on the flat mud rooftop. Further, one charge-/discharge-controller and an appropriately sized and well insulated battery-bank are installed inside the house. Each house has 3 WLED lamps. Thus up to 12 homes, with 36 WLED lamps are connected to one cluster, proving 7–8 h of light per day.

Each cluster has a newly developed electronic fuse to protect the system from misuse and overloads. Figures 3.33, 3.34, 3.35, 3.36, and 3.37 show and explain the village cluster solar PV system in Dhadhaphaya village in Humla, installed as part of a HCD project from January 2005 on. Since September 2005, 18 clusters

Fig. 3.35 Cluster power house with charge-controller and insulated battery-bank

Fig. 3.36 Project staff and user install the solar PV systems together, creating knowledge, skills and ownership

Fig. 3.37 Three WLED lamps per household for clean indoor air

(total of 1,182 W_p of installed solar PV modules),[10] for 170 homes (housing a total of 1,067 people), with 510 WLED lights have been in operation. Each day the trained local operators provide power via the cluster power houses' off/on switch to each cluster house for a total of 7–8 hours, 3–4 hours in the morning and 3–4 hours in the evening.

3.3.2.3 Solar PV Central Village System and Case Study

If the geographical conditions are favourable and the houses are built closely together, as in many villages in Humla (Figs. 3.38, 3.50), and the community is made up of mostly the same caste, the basic village solar PV system can be built as a central system, with a central village power house containing the battery-bank and the charge-/discharge-controller for the whole village.

On top of the central power house a 2-axis self tracking frame, developed by the author and built in Nepal, holds four 75 W_p solar PV modules, enough to power a village with up to 70 homes. The community as a whole takes the responsibility to participate in the building of the village system, maintenance of the system by trained local people, and the collection of monthly user fees.

In Chauganphaya village (Figs. 3.38, 3.39, 3.40, 3.41, 3.42, 3.43, 3.44, 3.45, 3.46, 3.47, 3.48), with a mere ~3 W per household energy demand, the 2-axis tracking frame with 4 × 75 W_p modules provides 63 homes with 7–8 hours lighting per day, with a minimal battery-bank DoD of ~20 %. This central village solar PV system was installed in January 2004.

All 63 homes are interconnected to the power house in the shortest possible way via 20–40 cm deep buried underground armoured cables. Three sizes of armoured copper cables are used, 4 mm^2 for distances of 20–50 m, 2.5 mm^2 for distances between 10–20 m and 1.5 mm^2 for distances $\leq$10 m. The whole system is protected with a central electronic fuse, so that any misuse or overloading of the system causes the immediate shut down of the system. An electronic fuse is used, because experience shows that many SHS have failed because the charge-controller glass fuse broke and the users either did not know about the fuse, as they were not trained to maintain the system, or it was not locally available.

Some months later, when the battery has not been recharged it fails too and normally that's the end of the SHS and the dream of light inside the home. Thus, the electronic fuse was developed because it is not possible, in such remote places, to get any glass fuses, which are commonly used in government subsidised solar PV systems.

Also in the village of Tulin, a 300 W_p central 2-axis tracker solar PV system (Figs. 3.49, 3.50) has been installed in 2006, providing power to each of the 29

[10] Of these, 15 clusters have one 75 W PV module, and three clusters one 19 W PV modules, as these three small clusters have only 4–6 homes per cluster. Thus for 170 homes (in Sept. 2005), including an average annual 3 % population growth over 10 years, a total 1,182W_R of PV modules have been installed, 7 W per household with each 3 WLED lights consuming each 1 W.

Fig. 3.38 In the middle of the village of Chauganphaya the local people have provided the land to build the central solar PV system power house. From the power house all homes are connected via armoured underground cable, buried at a depth of 20–40 cm

Fig. 3.39 The local people built the power house with their own materials. It contains the battery-bank, which provides up to 5 days power for all the lights during no-sun days, and a charge-/load-controller (Figs. 3.42, 3.43)

Fig. 3.40 The north–south axis can be changed from 5° to 60°, according to the sun's seasonal position

Fig. 3.41 The tracker follows the daily sun path from east–west, and back in the morning to east, by itself

Fig. 3.42 Charge-/ discharge-controller, with separate load controller, inside the power house

Fig. 3.43 The battery-bank well insulated in a locally made wooden box

Fig. 3.44 The lighting project is never on its own, the metal stove is always part of it

Fig. 3.45 A total of 189 WLED lamps are installed in 63 homes in Chauganphaya

Fig. 3.46 Lighting

Fig. 3.47 Smokeless metal stove

Fig. 3.48 Pit latrine

Fig. 3.49 A central 2-axis tracking solar PV system provides 30–35 % more daily energy generation compared to a fixed angle installation over the course of a year

Fig. 3.50 The 29 homes (195 people) in Tulin village are closely situated. This constellation provides a feasible solution for a central solar PV system with a 2-axis tracking frame (300 W_p), and a central battery-bank (24 VDC with a capacity of 300 Ah). All homes are connected through armoured underground cabling, saving precious trees

Fig. 3.51 Pit latrine

homes' 3 WLED lights via buried underground cables connected to the central power house.

In each, long-term HCD project, light inside the home goes along with a pit latrine, a smokeless metal stove, clean drinking water, and often greenhouses as above shown in Figs. 3.51, 3.52, and 3.53.

Fig. 3.52 Greenhouse

Fig. 3.53 Clean water

3.3.3 Small Scale Wind Generator System and Case Study

The utilisation of wind energy through ever increasing sizes of wind generators and wind farms has become the major new renewable energy technology on the international market. The wind technology and the manufacturing industry are mature and new development and creative designs are occurring at an amazing rate.

Through wind farms the generated electricity has become highly competitive with the major fossil and nuclear power plants in many countries. However, Nepal will not generate its major energy needs from large wind generators or wind farms in the foreseeable future. Rather, the valuable local wind resources of particular geographical places can be utilised for small scale applications, for single homes or rural villages.

Figures 3.54, 3.55, 3.56, 3.57, and 3.58 show the self made 2-bladed wind turbine, which was installed in Jumla in 1998, as part of the author's first small

Fig. 3.54 Self made two-bladed wooden wind turbine, wooden tail with metal frame, and a Bosch car alternator as power generator. The wooden blades are hand carved airfoil shaped blades with a blade diameter of 205 cm. From ~300 rpm onwards the wind generator provides 12 VDC to charge the battery-bank

Fig. 3.55 Bosch K1 14 V, 35 A car alternator, 1:4 wheels with belt set system, brakes, capacitors, electronic circuit plate

scale wind test project. It ran, and provided power for one home for 4.5 years (Fig. 3.56). The wind turbine was installed in a valley in close proximity to the house. Therefore the predicted power output of 180 W, was seldom reached. The aim of this project was to put to the test a self made wind turbine and to see if it could survive the harsh high-altitude (≥2,500 m.a.s.l) Himalayan wind conditions and climate. Thus, there is a future potential to manufacture small wind turbine blades from locally available wood.

The 18 m steel tower (Fig. 3.58) was screwed together with three six metre long pieces of Ø125 mm, Ø100 mm, and Ø75 mm from the bottom-up. At heights of 9 and 16 m, three steel cables (Ø8 mm) were fixed to the pole and anchored to the ground, in order to stabilise and hold the tower firmly in place in any wind condition. The bottom of the steel tower was anchored in a big stone pile, without any cement. During the 4.5 years of operation the wind turbine performed satisfactorily. It did not need to be taken down once for repairs.

While there is still a long way to go in regard to technical issues, such as an improved and appropriate design for the blades (in order to manufacture them by hand by local craftsmen), and the right choice of the generator with a direct drive,

Fig. 3.56 The self made wind turbine charges a 12 VDC battery-bank of 150 Ah capacity, enough for one home. For 4.5 years the author tested this wind turbine

Fig. 3.57 Power generation of maximum 180 W can be achieved, enough for several homes

Fig. 3.58 Wind turbine at 18 m height with steel cables to anchor the pole

encouraging results have been achieved. The test phase has shown that there are ample reasons to pursue the research and development for such small scale wind turbines, which can be manufactured and maintained by the local industry. They can provide one more appropriate and sustainable renewable energy technology to meet the needs of rural villages and communities, providing them with improved energy services for the overall development of their living conditions.

3.4 The Need for Ongoing Field-Based Research, Equipment Development and Field Based Testing

It has been the author's experience, that mostly field based research and new equipment development has to take place on an ongoing basis, in order to develop equipment which is locally manufactured. This equipment is generally easier to maintain, and can in general cope better with the local conditions and climate. To implement newly developed equipment and technologies is always a risk, thus in the case of the RIDS-Nepal Humla HCD projects, the High-Altitude Research Station (HARS) has been built (Fig. 3.59).

HARS serves as a research station, where newly developed, Humla specific technologies are put to test under real field conditions and stringent monitoring. The data we collect and the "real life" applications, provide valuable feedback for the new technologies. Necessary technical modifications and improvements can take place, enabling these new, locally developed technologies to be more appropriate and sustainable for the defined context. An example is the high-altitude SWH for the Darapori village, which has been in use since November 2005 under rigorous tests at HARS (Figs. 3.59, 4.53, 4.54, 4.55, 4.56, 4.57, 4.58, 4.59, 4.60, 4.61, 4.62, 4.63, 4.64, 4.65). This quality assurance process will minimize the failure rates and system downtimes in the villages. Experience shows, that sustainability and appropriateness are key aspects of a project that must be addressed in order to deliver the intended energy services to the beneficiaries. That means we must keep the end-users at the centre of our focus and

Fig. 3.59 HARS in Simikot Humla at 30° northern latitude and 3,000 m altitude. Here the various new developed solar PV home systems, solar PV central systems, high-altitude solar water heating systems, solar drier, parabolic solar cooker, greenhouse, and smokeless metal stoves are tested and monitored under real and long-term field conditions

develop and install renewable energy technologies that can deliver what is expected by all stakeholders.

3.5 Design Approach for a Contextualised Solar PV Village Electrification System for a Remote Himalayan Village

While easy access to electricity by the flick of a switch is an important goal and a means to increase living standards, it is important to recognise it is only the means, and not the end in itself. Thus rural electrification, in order to be relevant and sustainable, has to be embedded in projects addressing also health, food security, drinking water, indoor air pollution and education issues. While this is a basic necessity to the approach of sustainable development, this chapter concentrates solely on rural village electrification through PV systems. Issues such as identifying and specifying the social and technical parameters to define an elementary lighting service, tools to gain access to the available energy resource, and suggested software tools to simulate PV systems appropriately will each be highlighted and discussed.

Under the heading of rural electrification, or RAPS systems, a wide range of power generation systems can be defined. In Nepal the traditional thinking among government and rural electrification agencies is that each household needs access to a minimum power rating of 100 W (Craine 2004). Thus homes in the rural mountain areas, powered by a micro-hydro power plant (8–20 kW), and sometimes even by PV systems, have 2–3 incandescent bulbs per household, each consuming 25–60 W. That approach denies remote mountain villages any possibility for a basic electrification system. The relationship between poverty, remoteness and access to energy services can be further understood by contemplating the higher transport costs and increased effort required to build, operate and maintain a power plant in these areas. These circumstances demand new approaches, with new appropriate technologies to reach these villages. The initial electrification of a community is often best addressed with a step by step approach, starting with minimal energy services, initially providing only light inside the homes (Reddy 2000, 41). In order to cut down on the equipment size and weight, saving on infrastructure and transport costs, only the minimum lighting service is considered. Thus a basic village electrification system can be defined as a small embedded power generation unit, providing DC (direct current) electricity, just for minimal lighting purposes inside homes, for a few hours a day. It utilises and converts the locally available renewable energy resources into electric power through solar PV systems, or a pico-hydro power plant[11] (<1 kW) (Figs. 3.60, 3.61).

In the context of remote mountain villages in Nepal, the author designed a basic village electrification system, with a power generation system utilising a locally

[11] A pico-hydro power plant is considered as a power generation system with <5 kW power output.

Fig. 3.60 Three 1 W WLED lamps in the home, just sufficient for basic needs

Fig. 3.61 Old and young remain in the remote mountain villages, while the working age people migrate to the cities

available renewable energy resource, to provide energy for low power (1 W) WLED lamps. The light service provided is enough for people to see one another and to do daily indoor tasks such as cooking, cleaning, socialising and school homework. This is considered the first step on the path charted from a traditional energy use system (using only biomass) to introduce, learn to use and live with a commercial, cleaner, and more convenient energy source.

3.5.1 Important Parameters to be Identified and Specified

A basic rural electrification system aims to provide improved lighting services for the community, thus improving overall living conditions. This is a long-term process, as cultural and living conditions change at a far slower pace than such a project's implementation. That means that the new energy services need to be reliable, able to cope with the assumed growth (load and population), maintainable, and affordable for the local people.

It is crucial to understand that the local community is at the centre of any HCD project and that the technologies applied are to serve and support their struggle for a better life. Therefore, any project has to be based on a thorough understanding of the local context and culture, and must include an understanding of the "invisible" causes of poverty, and the impact on the community of decades of deprivation. This approach demands time, compassion and dedication. These more "human" aspects of a development project are crucial factors that need to go alongside the technical aspects. In this way the people are recognised from the beginning as equal partners and not as receivers of imposed ideas. This time intensive, often frustrating process is central to a holistic development project.

Thus, besides the technical specification for a design of a rural power generation system, the local context and conditions of the end consumers also play important roles. Thus "social parameters", such as the level of education, economic status and growth of the community, as well as more technical parameters, such as the available solar energy resource, initial and future load demand and days of independence, need to be identified.

3.5.2 Social Parameters

3.5.2.1 Population Growth

Many developing countries are continuing to experience high urbanisation rates. High unemployment rates, poor education facilities, a shortage of arable and irrigated land, and political instability. These are some of the main reasons that people (usually the young) leave their rural heritage for the "brighter future" promised by an urban environment.

That leaves many rural areas with a population distribution deficient in the important middle age, working and income generating group. For example, in Humla, the annual average population growth between 1981 and 1991 was 5.41 %, while the annual average population growth between 1991 and 2001 was 1.71 % (Zahnd 2005b). Increasing food shortage through natural calamities, customary land distribution among family heirs, political instability and civil war are local reasons supporting migration to urban areas in Nepal and India.

A rural village electrification scheme is often designed for a 20 years life-expectancy. Thus the end-user community's population growth and age distribution over this time needs to be known. It influences the system's size, the kind and quality of lighting services, and its ability and need for expansion.

3.5.2.2 Economic Growth

Rural village electrification is expensive, and therefore the economic side of each project has to be defined before a project can go ahead. The initial and ongoing project costs (for generation, maintenance and repairs) are usually higher per kWh of generated energy than for urban dwellings, due to remoteness, low capacity level and the need for high sustainability. With increased migration of the working age group, the already economically weak rural communities are increasingly likely to lack access to cash. This demands a clearly defined financial plan for the life-cycle cost of a rural village electrification project, in order to provide the community with realistic services according to their economic capacity. It is necessary to also include any long-term government subsidies, grants or loans (based on the present and emerging policies) in a project's economic plan, as well as initial project, installation, training, maintenance and repair costs throughout the power plant's lifetime. Issues such as future access to a road or other transport or communication infrastructure can have an important influence on the local income economy, and thus have to be realistically evaluated. Another factor to account for is the "external" earned revenue sent back to the families in the village. In the case of Nepal, the revenue earned through migratory workers in foreign countries and sent back periodically to their family members is estimated to be higher than Nepal's government revenues from its two main income sources, agriculture and tourism, combined (Tiwari 2006; Basnyat 2006). Thus, also for rural villages, the factor of economic capacity cannot be neglected.

3.5.2.3 Educational Experience

While the literacy rate for Nepal nationwide is 40–60 %, the literacy rate for women in Humla is only 4.8 %, and the average years of schooling is just 0.88 years per person (KIRDRC 2002). Studies show consistently that improved access to energy services (in particular lighting) increases education, which has a direct influence on improved health of the whole family, the productivity of people, their income generating capacity, and gender equality (Saghir 2005; Choe 2001; Goldemberg 2000), increasing the people's overall living standard on a long-term basis. Thus the initial education level of a community will determine also the need for appropriate awareness raising and non-formal education programs, alongside rural electrification (Zahnd 2005a). Further, the existing education level indicates the approach to be used in the initial instruction and training programs for the village's electrification scheme (Figs. 3.62, 3.63).[12]

[12] See in Appendix 18 the Solar PV System Training Manual we developed for local village users.

Fig. 3.62 A WLED light, consuming 1 W, enough for reading and homework

Fig. 3.63 Two WLED lights in the living room with a smokeless metal stove

3.5.2.4 Lighting Services Needed

What is the appropriate light output needed in the homes of the remote mountain villages for different tasks, and what is the primary function of lighting? In summary, the Australian Government defined the primary functions of a lighting system in the home and the work place as follows (Energy Programs 1994):

- To provide a safe visual environment.
- To make it possible to easily see the task.

Table 3.2 The rungs of the proposed "Lighting Service Ladder" for a basic rural village electrification system

No	Lighting technology	Units	lm/W	Watt used	Life-expectancy (h)	Price/ lamp US$[a]	Remarks
1	WLED	9	60–80	0.85 W	50,000–100,000	22.14	Nichia NSPW510DS diodes[b]
2	WLED	12	60–80	1.1 W	50,000–100,000	29.52	Including always 15 % wire losses
3	WLED	18	60–80	1.7 W	50,000–100,000	44.29	Prices in Nepal for local manufacturing
4	CFL	1	70–80	7 W	~8,000–12,000	5.43	High quality Compact Fluorescent Lamps
5	CFL	1	70–80	11 W	~8,000–12,000	6.43	High Power Factor PF 0.95 CFLs
6	CFL	1	70–80	15 W	~8,000–12,000	6.86	All CFL from Ultralamp AC & DC available[c]

[a] These prices are for the year 2008/9

[b] As used in all RIDS-Nepal Humla basic rural village electrification projects, see Figs. 3.6, 3.14, 3.44, 3.46, 3.60, 3.62, 3.63. Detailed technical specification for the Nichia NPSW510DS WLED diodes are available at: http://www.nichia.co.jp/specification/en/product/led/NSPW510DS-E.pdf (accessed July 26, 2012)

[c] As used in Humla and other projects for office and higher light output needs. http://www.ultralamp.com/cfl.html (accessed March 30, 2010, on November 2, 2011 not available any more)

- To provide a comfortable, pleasant visual environment.
- To achieve the lighting function as efficiently and cost effectively as possible.
- To be easy to clean and maintain.
- People must be able to safely orientate themselves and move about within buildings.

While for developed countries, defined levels of lighting according to the task are available (AS1680.1-1990[13]), no such standard is available for rural villages. Applied local knowledge, experience and realistic expectations allow the specification of the lighting level to be recommended for a defined context (Zahnd and Kimber 2005).

The following rungs of a "Lighting Service Ladder", identified according to the lighting services needed, are proposed by the author for a basic rural village electrification system with future capacity expansion possibilities (Table 3.2).

The calculated life-cycle cost (AS3595-1990[14]) is decisive at which rung of the light service ladder the community can start to climb, as this cost has to be covered

[13] The Australian Standard for "Interior Lighting" AS 1680.1-1990 is available for purchase at: http://www.saiglobal.com/PDFTemp/Previews/OSH/As/as1000/1600/16801.pdf (accessed July 26, 2012).

[14] The Australian Standard for "Energy Management Programs—Guidelines for Financial Evaluation of a Project" AS 3595-1990 is available for purchase at: http://www.saiglobal.com/PDFTemp/Previews/OSH/As/as3000/3500/3595.pdf (accessed July 26, 2012).

and maintained on a long-term basis by the community, in order to achieve a sustainable project with lasting benefits.

3.5.2.5 Ownership

Another important factor is the ownership of the power generation plant. Is it a private home, using daily light mostly in the evening and early morning, or a public building, with classes during the day which do not really need light, but where evening community gatherings or presentations take place on occasion, demanding higher light output. Further, publicly owned projects need organised management in order to be sustainable, as cooperative responsibility for a project is often not taken as seriously as the responsibility for a privately owned project.

3.5.3 Technical Parameters

3.5.3.1 Load Demand and Load Growth

The size of a basic rural village solar PV system depends directly on the load demand and number of homes. In the projects we design, all homes have the same number of WLED lights. In central or cluster solar PV systems, the trained operator often provides a defined and affordable number of hours of light per day through a centrally located on/off switch. In discussions with the village eldership, the historical population growth over the last 20 years was identified and averaged. Experience shows that it is critical to include population and load demand growth over the coming 5–10 years for the system design. The 10 years end-limit is identified because the battery-bank,[15] consisting of locally available batteries, typically needs replacement after a maximum of 10 years. Therefore the initial battery-bank's designed capacity has to reflect the realistic daily load demand during the next 10 years period. It must also be able to provide the energy needed by the community to get through a predictable number of cloudy days.

3.5.3.2 Life-Expectancy

The solar PV modules are often the most expensive equipment in a PV system. Thus, it is important that they generate maximum power, with as little down time as possible. Today, manufacturers give a 25 years life-expectancy guarantee

---

[15] As a rule of thumb, once a battery-bank is a few ($\leq$3) months in use, no additional batteries should be added, even if they are new, from the same brand and with the same capacity (Ah).

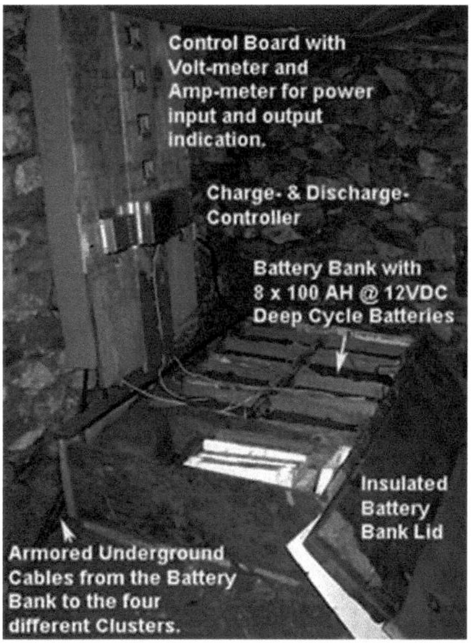

Fig. 3.64 Battery-bank well insulated in a locally made wooden box, designed to last 8–10 years, including the load growth

($\geq$80 % of rated power output). Accordingly all the peripheral equipment has to be designed to either match that, or some multiple of it, for the duration of the project. This is important for remote and difficult to access villages, and a mark of good professional design. Periodic maintenance, check-ups and power plant journaling, helps to identify potential shortfalls and needs for parts to be serviced or exchanged in time. The power plant journal allows different operators to know the exact status of the power plant. That keeps the down-time minimal, the services provided reliable, the capacity factor of the power plant high, and the end consumers satisfied. Approximate lifetime expectancies for some of the more important equipment are:

- Solar PV modules and PV array frame: 25 years.
- Battery charger and discharger: 8–10 years (Figs. 3.42, 3.89, 3.90, 3.94, 3.98).
- Battery-bank with conservative design (Sect. 3.5.3.4): 5–10 years (Figs. 3.35, 3.43, 3.64, 3.68).
- Lights (dependent on the technology and use (Sect. 3.5.2.4)): 4–25 years.
- Transmission cables: Underground 25 years (Fig. 3.135), untreated wooden poles: 2–3 years (Figs. 3.70, 3.71).
- Switches: 3–5 years (Figs. 3.92, 3.94, 3.99).

Fig. 3.65 Three Pyranometers with different angles to the sun measuring and recording the solar radiation

3.5.3.3 Solar Energy Resource Assessment

Without proper knowledge and assessment of the locally available renewable solar energy resource, a PV system is destined to disappoint all stakeholders, in particular the end-users. But, for remote and rural villages, carefully recorded solar irradiation data (Fig. 3.65) are rarely available. What follows is a short description of available tools, software and information required to assess the locally available solar irradiation resources:

- Average number of sunny days a year, over at least 5 years. Often available at the national meteorological institute. Needs adjustment to the local geographical and climatic conditions.
- National solar radiation data bank (usually from a weather station installed at the national airport), adjusted for the local village's geographical and climatic condition.
- RETScreen (http://www.retscreen.net). This is a Canadian government sponsored renewable energy technology project/system design tool which includes all the major renewable energy technologies and is freely available on the Internet. It is an Excel based spreadsheet that directly accesses NASA satellite solar radiation data, integrating the data in its calculation spreadsheet.
- SWERA (http://en.openei.org/apps/SWERA/), initiated by NREL (National Renewable Energy Laboratory). It assesses the local solar radiation and wind resources. Solar radiation data (40 km × 40 km surface area) for Nepal are freely available.
- NASA satellite solar radiation data are freely available for any place on the globe with the input of the latitude and longitude for the location in question,

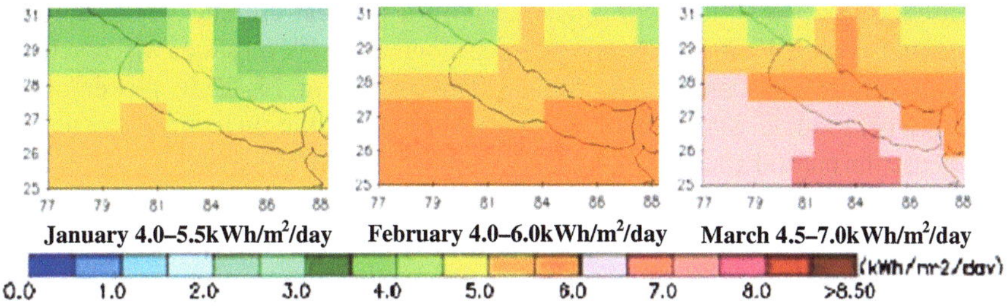

Fig. 3.66 Average Solar Irradiation on a flat surface tilted at 30° towards Equator for 1983–1993 for Nepal

from: http://eosweb.larc.nasa.gov/. The NASA satellite data bank generates monthly solar irradiation data, averaged over 10 years (Fig. 3.66).

- METEONORM 7 (Version 7.0.22 November 2013), a commercial software tool that provides hourly and monthly solar irradiation data for any defined place in the world (http://meteonorm.com/download/software/).
- Solar Radiation Calculator SolRC version 1.3. An older, but a simple, easy to use and very practical Excel spreadsheet providing solar irradiation calculation for a defined place. It is free available from the Internet.[16]
- Measurement of the locally available solar radiation on surfaces with different angles to the sun, as done in our project area. This is the most reliable, though also a time demanding and expensive way. Pyranometers need to be installed and the data recorded and interpreted over at least 3–5 years (Fig. 3.65).

Two additional parameters to consider for the design of a solar PV village system are the geographical and climatic conditions. A village in a deep valley, surrounded by high mountain ranges, has significantly shorter sunshine hours per day, particularly during the shorter winter months, with lower solar altitude angles. While METEONORM is able to include the surrounding mountain ranges (360° horizon) in its simulation, in most other tools a realistic factor (e.g. 0.9–0.7) has to be chosen in generated solar irradiation data, to account for the shorter sunshine hours. The climatic conditions need to be known, because a solar PV modules' power output is directly related to the modules' temperature, which can be estimated by knowing the ambient temperatures throughout the seasons. Si mono-crystalline and Si poly-crystalline PV modules have an average power output reduction of 0.4–0.5 % per

[16] SolRC version 1.3 (2001) is available in an Excel worksheet at: http://www.google.com.au/url?sa=t&source=web/ct=res/cd=1/ved=0CAgQFjAA&url=https%3A%2F%2Fclassshares.student.usp.ac.fj%2FPH301%2F2010-Dr.Raturi%2FLecture%2520Notes-First%2520part-Intro%2Csemi-comductors%2CSolar%2CPV%2FSolar%2520radiation%2520calculator.xls&rct=j&q=SolRC+version+1.3+&ei=MsSrS5OLHIGMtAP90833Cw&usg=AFQjCNG9bk8FgHnq3s2liHIUYvYqCsD0IA(accessed March 30, 2010, on November 2, 2011 not available any more)

increased °C >25 °C.[17] Thus if a village is in a tropical climate where ambient temperatures reach 40 °C during the summer months, the PV modules will reach 60–70 °C under direct solar beam radiation. That results in 15–22 % power output reduction per module. Smart design, positioning and installation techniques can help to increase the natural air and wind flow between the solar PV modules on the array to cool the PV modules down (Figs. 3.34, 3.41, 3.65, 3.87, 3.116, 3.118). In the high-altitude Himalayan villages, the ambient temperature during the sunny winter months is around the freezing point. If the PV modules' temperatures reach 9 °C under direct solar beam radiation, their power output increases by up to 8 % compared to the manufacturer's rated power output. These significant differences have to be recognised, identified and accounted for.[18] They have to be included in the available local solar energy resource on a monthly basis, so that the PV system is able to generate the expected energy for the defined lighting services.

3.5.3.4 Availability, Reliability, Days of Independence

A basic rural village electrification system is often not designed to deliver energy services at the discretion of each consumer. Rather, in the initial project steps the local villagers themselves decide how many hours a day they want light and are

[17] That means for every degree >25 °C (the standard temperature at which all PV modules are rated under laboratory conditions) these modules lose 0.4–0.5 % of their rated power output.

[18] For example, it is important that the defined (developed and locally built or externally purchased) solar PV charge controller is able to handle the higher input voltages possible for a solar PV module, or a serially connected string of solar PV modules, when they are very cold, <5 °C, in particular if a higher system voltage, such as 24 V, 48 V or higher is chosen. This, because the voltage of silicon based solar PV modules, in general, increases proportionally compared to the indicated product specification by ~0.35 % per decrease in °C < 25 °C. Thus, while the BP 275 Si mono-polycrystalline solar PV module, as used in most RIDS-Nepal solar PV village systems in Humla, is rated with a V_{OC} of 21.4 V and a V_{MPP} of 17.4 V at 25 °C, producing 77.5 W power output at 1,000 W/m^2 (STC), it generates a V_{OC} of 23.3 V and a V_{MPP} of 19.2 V at 0 °C, producing 83.9 W power output, or even a V_{OC} of 24.0 V and a V_{MPP} of 20.0 V at −10 °C, producing 87.1 W power output, which is 12.4 % higher than at STC. Thus, for a 48 V solar PV system that means that instead of a system voltage of ~69.4 V at 25 °C with a power output of 310 W, the charge controller has to cope and handle a system voltage of ~80 V at −10 °C with a power output of 348.4 W. Further, at RIDS-Nepal's HARS station (Fig. 3.59) we measure every year, particularly during the clear, blue sky months during the cold winter season, global solar radiation values of up to 1,200 W/m^2, which is 20 % above the STC conditions for which each solar PV module has been rated. As the current generated is proportional to the intercepted global solar radiation, a significantly higher current is flowing through the solar PV system and through the charge controller, generating a whopping 104.9 W from a 75 W PV module! Taking these two parameters (lower temperature and higher global solar radiation) together, the 48 V solar PV system generates instead of 310 W, 419.6 W, or 35.4 % more power. Extreme as these conditions may be, they do occur in the high-altitude Himalayas and thus generate such a significant range of system voltages and power output increase. Thus the local context has to be understood in detail and has to be considered in the development and manufacturing, or purchasing of, equipment in the designing stage of the solar PV system.

able to afford it, as this directly impacts the size and cost of the energy generation system. For the initial introduction of electricity into the village it is desirable to provide the power reliably and have the lights available as required.

Reliability and availability depend on the quality and size of the system, which is directly related to its costs, and the personnel operating and maintaining the system. For a remote village it is good practice to aim for high sustainability rather than highest possible efficiency with new, untested and unreliable equipment. Thus it is advisable to keep the energy services to a minimum, though consistently provided, if costs are the limiting factor.

Being most of the time a community owned system, a basic rural village electrification system needs trained people to operate, service and maintain it. They will be trained in the basics of solar PV and practical installation and maintenance skills before the project installation. They are chosen by the village elders or the village electrification committee.

Another important factor the local end-users have to decide at the initial project step is, how many days of lighting services do they want, and can afford, if the sun is not shining? These are the days of independence, when the battery-bank has to provide the whole energy without being charged and without being too highly discharged, leading to a shorter life-expectancy.

An advised maximum DoD for a battery-bank for remote areas is 0.30 (30 % of its full capacity).[19] Thus the more days of independence, the higher the battery-

Fig. 3.67 A project staff member trains the chosen local people in how to install and maintain the WLED lights, switches and house wires

[19] The DoD defines the capacity (Ah) drawn in percentage from the battery's full capacity. For remotely installed battery-banks it is advised to limit the DoD rates to 30 % of its full capacity, as the number of times and the level of DoD a battery is discharged, is related to its life-cycle expectancy. A battery-bank for a rural system should be designed to last between 5 and 10 years, as transport costs are very high, system down time should be kept low, and recycling opportunities are not yet available in Nepal. The DoD 30 % value is based on the author's 16 years of practical experience with the locally available batteries for solar PV systems, so that they can perform their best for an expected reasonable life-time of 5–10 years.

Fig. 3.68 A project staff member teaches the local trainees how to check, clean and maintain the battery-bank

bank's capacity needs to be, hence also higher costs. In areas such as Humla, where the annual monsoon covers the sky for days at a time, one must know beforehand what services are available, and what restrictions are designed into the system. The real state of a battery's charge can reliably only be defined through the battery's sulphuric acid specific gravity, the voltage and the internal battery temperature. As this is technically difficult and expensive to do, the battery-bank is protected through a defined minimum voltage, in-built into the discharge-controller. Once reached, it cuts the power to the consumers.

Another factor related to high availability and reliability is that the whole system is protected at various crucial points against short cuts and overloads due to unauthorised misuse. Thus for the installed systems in our projects in Humla, an electronic fuse was developed (Figs. 3.35, 3.89, 3.94, 3.105) which triggers when a pre-defined maximum current flows. Thus the battery-bank, the DC transmission lines, and the WLED lights are reliably protected.

3.5.3.5 Maintenance and Repair

Clearly defined system parameters, the use of appropriate and reliable material, whenever possible locally manufactured, are important parts of a long-term sustainable and successful basic rural village electrification system. Once the system is installed, checked and in operation, it needs ongoing maintenance and repair. Local people, chosen by the system committee, are trained by the project staff (Figs. 3.67, 3.68). They participate in the initial installation, which provides them with due respect in the community as the ones in control. Further, the operating and maintenance team are provided with the following tools, to which they were introduced as part of the initial skills training:

- Operation list to be filled in periodically with the most important system parameters: battery-bank voltage; specific gravity; when, why and which fuse

triggered; when solar PV modules were cleaned and lights cleaned/exchanged. All maintenance and repair work is recorded in the power plant journal.
- Basic tools, including also a voltage meter and battery specific gravity measurement tube. A set of spare WLED lights in case some fail, or for newly built homes. A roll of house wire and underground cable,[20] in case a new house (due to population growth) has to be electrified.
- A defined communication procedure (through letters or mail runners) between the system operator or maintenance staff and the project designer (RIDS-Nepal). That allows information about failures or untimely breakdowns, which cannot be fixed or handled by the locally trained staff, to be communicated.[21]
- Monthly visits by the project designer,[22] and a periodical[23] comprehensive family survey is carried out in the village. That provides crucial long-term feedback and information on the system performance and how to change/adapt future approaches and technologies for context related increased appropriateness and sustainability. It also provides crucial information on the long-term impact that lighting in the home has on the whole life and development of the people.

3.5.4 The Optimised Basic Rural Village Solar PV System

When all the different social and technical parameters are identified in the end-user community, they form the basis for the design of the actual rural village PV system. Knowledge of these parameters is essential for the design of village power systems that suit the local context with ample provision for the future. Consideration of the social context, or "soft" issues of a project, along with the important technical issues, pays respect to the owners' and consumers' living context. It also

---

[20] In the realised cluster and central village solar PV system projects in Humla, Nepal, only armored underground cabling is used. The main reasons for that are the harsh climatic conditions with snow, storms, torrential rains, and floods. Additionally, the area suffers from dramatic deforestation due to heavy reliance on forest resources, and in any case the soft Himalayan pine tree poles, would last only for a few years due to rotting. These, plus the aim for high reliability and sustainability of the system justifies the substantially higher costs for underground cabling. The armored underground cables are initially defined so that they are able to carry twice to three times the initial current, dependent on the systems' load demand, anticipated load growth over the next 10 years and the defined lighting services the users want and still can pay for.

[21] In the case of the systems installed in Humla there is a well established project office and high-altitude research station in Simikot, Humla's district centre, able to be reached from any place in Humla within maximum 2–4 days of walk.

[22] One of the project's main aims is to gain long-term experience with basic rural village electrification schemes, thus an initial baseline survey before the project starts, followed by the follow-up re-survey in the 2nd, 5th, 10th and 20th year after the project implementation is part of the design of each village system, which is itself part of the village's long-term HCD program.

[23] The periodical comprehensive family survey is carried out as part of the follow-up survey questionnaire (Appendix 14.2) which is carried out in the 2nd, 5th, 10th and 20th year of a "Family of 4" or "Family of 4 PLUS" HCD program.

recognises and in fact honours their choice to remain in the village and work on improving their living conditions. Lighting of their homes is one of the most important aspects of a holistic community project. Our work is embedded within the wider scope of a HCD project approach, which also addresses health, education, indoor air pollution, clean drinking water, and food security. These are all urgent and interrelated issues of remote mountain communities.

3.5.4.1 Tools for the Simulation and Design of Basic Rural Village Solar PV Systems

For the actual calculation and design of the village PV system, various approved software tools are available. They require various levels of detailed input of technical and financial data. Whichever "tool" is used, what makes the "optimised" design different from the standard design method is that defined numeric values of the social parameters, identified in Sect. 3.5.2, are also input. The following recommended software programs have been used for the Humla PV village systems design:

- PVSyst6.12 (01-11-2013) (http://www.pvsyst.com/) is a commercially available, very detailed sizing, simulation and data analysis tool for complete PV systems design. It can import solar irradiation data generated by METEONORM and TMY2 (Typical Meteorological Year, version 2) from NREL. It offers an extended and detailed data bank with the technical specification of the important PV system components. It can simulate stand-alone and grid-connected PV systems. It allows in-depth technical specification input for most of the equipment needed for the design simulation, and the detailed financial module included provides a thorough treatment of investment, payback period and price of the generated energy. Its sheer scope and ability to present data in graphical and report form is praiseworthy. A free, 15 days, full edition version, can be downloaded.
- RETScreen4 (11-2013) (http://www.retscreen.net). More than a mere solar PV system calculation and design program it is a "clean energy awareness, decision support and capacity building tool" (pvresources 2005). The energy production, the greenhouse gas emission reduction and the life-cycle cost of a solar PV system can be simulated according to the input data. It has a link to the NASA solar radiation databank and thus can calculate values for any system worldwide. It is based on an Excel spreadsheet with a good user manual, and is freely available.
- PVSol 6 (06-2012) (http://www.valentin.de/index_en) is a commercial solar PV systems simulation and design software tool for RAPS as well as for grid-connected solar PV systems. It includes a detailed economic efficiency simulation module, allowing it to simulate the life-cycle costs under various defined conditions and assumptions.

- PV-CHART 3.5 (11-2013) (http://fchart.com/pvfchart/). This very practical interface provides a comprehensive solar PV system analysis and design program for professionals and students. Based on algorithms developed by the University Wisconsin, USA, it can simulate monthly average performances for each hour of the day. The calculations use solar radiation utilisability[24] to account for statistical variation of the intercepted global radiation and the assumed/defined load.

3.5.5 Case Study: The Dhadhaphaya Village Solar PV System

Dhadhaphaya village[25] is a poor, undeveloped mountain village in the remote district of Humla, Nepal. Based on the village's request for a HCD project, meetings took place between the village elders and the project staff of the RIDS-Nepal Humla project. A detailed HCD project plan was developed, including the following components:

- A pit latrine for each household;
- A smokeless metal stove for each household;
- A basic rural village solar PV electrification scheme for 3 WLED lights per household;
- A village drinking water scheme;
- Five greenhouses to grow vegetables; and
- Two non-formal education classes, one for mothers and one for out-of-school children.

In the following only the basic rural village PV electrification project in Dhadhaphaya is discussed, as part of the village's larger HCD project. The following parameters for the solar PV system design were identified and deemed critical to the success of the electrification project:

---

[24] The solar radiation "utilizability" is defined as the fraction of global solar radiation that is incident on a surface exceeding a specified threshold or critical level. It is a simple and elegant way of estimating the long-term effect of the global solar radiation on any solar process. Initially applied in the calculations of the performance of flat plate solar thermal collector it is now also used for thermal concentrators, passive and solar PV systems. For more detailed information and explanation see: Beckman W.A and S.A. Klein. 1986. "Solar Radiation Utilizability", http://sel.me.wisc.edu/publications/journals/rosru84.pdf (accessed July 26, 2012).

[25] Dhadhaphaya village, in the Humla district, is located at 29°59′ Northern Latitude, 81°57′ Eastern Longitude, and at 2,550 m above sea level. See the village in Google Earth via the link on the RIDS-Nepal web site: http://www.rids-nepal.org/images/stories/explore_nepal/google_earth/Dhadhaphaya.kmz (accessed July 26, 2012).

Fig. 3.69 A Cluster Solar PV System, generating enough energy to power up to 12 homes each with 3 WLED lights, each consuming ~1 W

- During the detailed survey (August 2004), Dhadhaphaya village had 167 homes and 1,067 people. The average historical population growth is 2.2–3 %, or 3–5 additional homes built each year.
- With the METEONORM software and the NASA Internet based solar radiation data bank, the locally available solar irradiation has been simulated in hourly and monthly values. Further, at 3 h walk away, at the RIDS-Nepal HARS in Simikot, the local solar radiation has been measured on three different surfaces (Fig. 3.65) since May 2004. These data provide a comparable set to the METENORM and NASA simulated data. Thus the solar irradiation data used for the Dhadhaphaya village was an average of the three data sets, multiplied by a factor of 0.8, as a high mountain range towards the East, shortens the morning sunshine hours significantly.
- The village is surrounded by high mountain ranges, thus a 360° horizon silhouette (with the NASA 90 m horizon tool[26]) was created of all the mountain peaks around and included in the PVSyst simulation design. Thus due respect was paid to the shorter daily sunshine hours in the village.
- The average ambient temperature during the day ranges from 0 to 35 °C, thus the assumed solar PV module temperature under direct beam radiation is estimated to vary from 10 to 65 °C.
- Due to the layout of the 170 homes (September 2005) in the village, we used a "cluster" PV system approach (Figs. 3.33, 3.34, 3.35, 3.36, 3.37, 3.69), rather than a central solar PV system with 2-axis self-tracking frames (Figs. 3.41, 3.49,

[26] The "NASA90-m Horizon tool" is a digital terrain model (DTM) tool, tiled with a 90 × 90 m resolution, called NASA's Space Shuttle Radar Topography Mission (SRTM). It uses satellite data of the surrounding area of a defined geographical location (latitude, longitude and altitude) and thus can identify the 360° surrounding mountain ranges of the defined geographical location (village). Because the model considers the altitudes of the 360° surrounding area (mountain ranges) it provides the defined geographical location's (village's) horizon angle of the incoming/intercepted solar radiation. This tool is not available to the public, but to the developer of the METEONORM software tool, who provided the author with the 360° surrounding horizon for Dhadhaphaya village.

3.50, 3.59). Thus various clusters, each with 4–12 homes were defined, with most clusters having one BP275F, 75 W_p Si mono-crystalline solar PV module with its own battery charge-/discharge-controller and battery-bank.

- From each central cluster home with the power system, armored underground cables lead the power to individual cluster homes, powering the 3 WLED lights in each home.
- A total of 18 clusters (with 1,182 W_p),[27] for 170 homes, with 510 WLED lights were installed.
- The armored underground cables are able to cope with a tripling of the power in the years to come. They have a life-expectancy of 25 years, similar to the solar PV modules.
- The battery-banks are designed to cope with a doubling of the daily load, and an independency of up to 5 days. Its life-expectancy is designed to be 8–10 years. It is installed in a well insulated and aerated, locally made wooden box, so that the battery temperature does not rise above 25 °C, or drop below 10 °C, even during the harsh winter months (Figs. 3.91, 3.92).
- The charge-/discharge-controller is defined for a life-expectancy of ~10 years, and able to cope with a doubling of the power input (if the cluster community decides they want to add another solar PV module) and power output (load growth due to population and consumption growth).
- Due to the extreme poverty of villagers in this community, the initial project costs were raised by RIDS-Nepal through external donors. The local community participated in each project step, from the initial project definition to the project installation stage. They carried all the equipment to the village, dug the canals for the underground cables, and provided all required locally available materials (stones and wood). Thus their overall participation in the whole project, expressed in economic terms, amounts to ~25 % of total costs, thus creating a strong feeling of community ownership for the project.
- Being community owned, each family has to raise a monthly amount of 25NRs (~US$0.36), after the initial payment of 500NRs per cluster for the maintenance fund. These payments have been approved by the village elders, based on the needs for maintenance and basic repairs.
- For each cluster, a person was selected by the community and trained to operate and maintain the system. This person is paid a small salary from the monthly fees raised from the users.

[27] Of these, 15 clusters have one 75 W PV module, and three clusters one 19 W PV module, as these three small clusters have only 4–6 homes per cluster. Thus for 170 homes (in Sept. 2005), including an average annual 3 % population growth over 10 years, a total 1,182 W_R of PV modules have been installed, 7 W per household with each having 3 WLED lights consuming each 1 W.

3.5.6 Lessons Learned

Some valuable experience was derived from the implemented basic rural village PV electrification systems. Some of the crucial social and technical parameters for the design of a village PV system are:

- Alongside each project (since 2002) constant development of all locally manufactured equipment took place. This has proved to be of key importance. It increases the community's ability to maintain and repair equipment locally, while also allowing local people to learn new skills and improve the local economy. Specifically, the WLED lights, solar battery charge-/discharge-controller, the electronic overload protection, and the 2-axis self-tracking frame were all developed and manufactured in Nepal.
- Since 2002, every 1–2 years a new, more appropriate WLED light design and solar battery charge-/discharge-controller have been installed. While in most of the previous years this worked out well, some of the 2005 WLED diodes were purchased (due to time constraints) from a dealer in India, and not directly imported from the manufacturer in Japan. That turned out to be a mistake, as the dealer mixed obviously lower quality diodes with the high quality ones. Several of those WLED lamps have been found to be malfunctioning within the first month of their operation. These malfunctioning lights have now been exchanged with new, original diodes.
- The newly developed 2005 battery charge-/discharge-controller was installed for the first time with a microprocessor, allowing various battery charging/discharging and periodical gasification procedures. However, the designer was not fully aware of the harsh mountain conditions where the controller was to be used in our project. Thus, considerable failure rates in these units forced the project team to withdraw all 18 charge-/discharge-controllers and exchange them for the approved 2004 model. Since then, no complaints have come in.

These were valuable lessons. Research and new equipment development has to take place on an ongoing basis, while failure rates and system downtimes have to be avoided as much as possible. And there is often a difficult choice between these two paths. These experiences have shown, that sustainability and appropriateness are key aspects of a project that aims to deliver the intended services to the beneficiaries. To keep that in mind means to keep the end-users at the centre of our interest, and to develop and install technologies that can be applied to achieve that goal.

3.6 Solar PV Systems in Remote Himalayan Villages in Nepal: Problems Encountered and Practical Solutions

Over the last decade Nepal has experienced a steady growth in solar PV system installations in remote mountainous communities. The reason for this is the growing understanding that grid connected infrastructure will never be feasible for these inaccessible parts of the country. Government subsidy programs, funded by foreign governments and INGOs, have caused a mushrooming of solar PV companies in the country's capital, Kathmandu. But, as mentioned above, the standard of the majority of the solar PV systems installed is questionable from a sustainability point of view. The designers, manufacturers' and policy makers' view of rural village electrification through solar PV systems is driven and biased by an urban, academic and business oriented perspective. Their perception does not take account of the end-users' local context and needs. Likewise the end-users lack awareness and education about the use of renewable energy systems, which results in inappropriate and unrealistic expectations. These two mismatched worldviews threaten the long-term relevance of the local renewable energy industry and the improvement in living standards of these remote mountain communities.

The major problems fall into two categories, technical and non-technical. In the following sections the range of these problems is explored and suggestions for possible remedies are provided. Through the examples of installed solar PV systems these remedies are reflected in practice in Humla, one of the remotest areas of Nepal. The case studies are from three villages, in each of which a different solar PV system approach has been followed, according to differing geographical and cultural conditions. Each village has one system specific data monitoring system, recording data for detailed, long-term understanding of the solar PV system's performance and interaction with the users. This chapter reflects on 5 years of monitoring results and the practical experience gained since 1996 of installing/ living with solar PV systems. Recommendations and strategies for improved and context related solar PV village system projects are provided. These aim to prevent premature and unnecessary failure of future installations. These ideas are presented through case studies, based on practical field experience and the evaluation of published reports of other similar projects.

3.7 Technical Problems

The issues described above created a business environment for solar PV systems in Nepal that has not rewarded quality. Cheap materials and equipment easily entered this open, unregulated market. The pressure to take advantage of available subsidies resulted in rapid turn over of system components, and quick solar PV system dissemination into the rural areas. The lack of monitoring of solar PV systems entering the market, combined with little or no follow-up trouble shooting in the villages aggravates these problems and undermines end-user confidence in the new technology. Despite the fact that we strive to avoid these pitfalls by using high quality materials, extensive training and trouble shooting support, as well as on going research into system design, we still experience a variety of challenges. Major technical problems and/or shortcomings, which are outlined below, are often experienced in SHSs installed by Kathmandu based, commercial solar PV companies in Humla, some 17 days walk away from the next road head. Many of these SHS were installed for some of the poorest and most disadvantaged groups (e.g., among Dalits,[28] the 'untouchables'), and were partly funded by the government subsidy scheme. All of these systems use DC power only for lighting and there are no inverters used.

3.7.1 Solar PV Module

In most of the ~300, government subsidised SHS studied in this project, the solar PV module was the piece of equipment that performed the best. In Nepal, all available solar PV modules are imported from one of the major PV module manufacturing countries, including Japan, China, India and Germany. Significantly, SHS module failures were most commonly caused by no name Chinese PV modules, which performed poorly when compared to branded PV modules. The most common technical fault that we identified in damaged PV modules was the broken and corroded connections of the serially connected PV cells and the connection box.

3.7.2 Solar PV Module Frame

Most of the SHS that we inspected are installed at a fixed angle, usually inclined at 30–45° south (Figs. 3.70, 3.71). That angle is appropriate for Humla's northern latitude (30°). The PV module frames are usually a simple steel construction clamped to a wooden pole. The soft Himalayan pine wood tree is in high demand for firewood and substantial deforestation is occurring in Humla. It is also a very soft wood, and needs replacement every 2–3 years. For these reasons, the wooden poles, though chosen because they are locally available at no cost, are not ideal to mount PV modules.

---

[28] Dalits, the "untouchables", are the lowest caste in the hierarchy of the caste system. More relevant and detailed information is available at: http://en.wikipedia.org/wiki/Dalit (accessed July 26, 2012).

Fig. 3.70 Unstable frame installations on soft tree poles can not maintain appropriate long-term PV module positioning for year round highest energy yields

Fig. 3.71 Due to the softness of the wood, this frame position cannot withstand strong winds and snow

3.7.3 Solar PV Module to Charge-Controller

In many cases, the solar PV module extension cable to the charge-controller is non UV-stabilised, normal house wiring cable. The sun's high UV content in this high-altitude area (Blumenthaler 1997) makes such cable brittle in less than 2 years, exposing the copper wire to the elements. In a few instances, we have seen the wires being used for clothes drying, adding increased weight and moisture to the cable mantling, which was already weakened by the weather.

3.7.4 Charge-/Discharge-Controller

The charge-/discharge-controller[29] (CC) is a central piece of equipment of any solar PV system. It is required to charge and discharge the battery-bank in a regulated fashion. An overcharge protection prevents the battery from excessive charging and gassing, which would otherwise result in loss of distilled water and

[29] A charge/discharge controller is the piece of equipment in a solar PV system responsible for charging and discharging the battery-bank, as well as providing the demanded users' load according to the solar PV module(s)/arrays' power generation.

Fig. 3.72 The soot of open-fire places inside homes creates a harmful environment for solar PV equipment, especially for the CC. Unprofessional installation can lead to unexpected short cuts, damaging the CC

Fig. 3.73 The damaged CC forced the user to connect the battery directly to the PV module and load. After that, the battery lasted only 2 more months

increased temperature, increasing the corrosion of the batteries' positive grid. Likewise it must protect the battery-bank from being excessively discharged. That is most easily achieved by cutting the load off after reaching a certain battery voltage. Measurements of available charge-controllers on the market have shown that there is reason for serious concern in regard to the deep-cycle discharge protection of SHS battery-banks. Battery voltage cut-off values as low as 10.2 VDC have been measured. That coincides with the findings in the Sukatani report (Reinders et al. 1999) which identified the same concerns. Further, as the Sukatani report recognised, Humli users also by-passed the charge-controller and connected the battery directly to the load (Figs. 3.72, 3.73, 3.76). While that initially provided longer periods of energy for their lights, it shortened the battery's life-expectancy significantly because the battery was discharged daily far below the allowed DoD recommended by the battery manufacturer.

3.7.5 Battery-Bank

Due to the intermittently available solar energy, as well as the time mismatch between its availability during the day, and the main use of electricity in a SHS for

Fig. 3.74 Some common problems with batteries in the field: dirt, uneven placement, low distilled water level, no climate protection, corroded connections

lighting in the early morning and evenings, the solar energy generated during the day has to be stored. This is still done most economically with deep-cycle, flooded lead-acid batteries, even though this storage technology has been in use now for over 150 years. It stores and delivers energy directly as DC electricity. The widespread use of lead-acid batteries and the moderate cost per unit of energy stored, combined with reasonable energy efficiency and the ease with which they can be recycled (if the infrastructure is available), account for this battery's popularity. However, as there is no official lead-acid battery recycling[30] program in Nepal, it is of great importance that the solar PV system battery-bank has a long life-expectancy.

We have observed that inappropriate battery-bank design, installation and use are the main reasons for the premature failure of solar PV systems. In order to achieve a long battery life-expectancy, certain installations and operational criteria have to be followed. The main issues are:

- Charge/discharge protection through voltage and ampere limitations.
- Appropriate current charge/discharge (to avoid excessive temperature build-up and corrosion).
- Correct sulphuric acid density (which is normally 1,250 g/l, though it can be between 1,200 and 1,280 g/l for flooded lead-acid batteries, depending on their use and the expected life-cycle) and periodic topping up with distilled/rain water.
- Operation in the correct temperature range (best between 15 and 25 °C or maximum range between 10 and 30 °C).
- Shallow daily DoD cycling (best <10 % DoD, maximum 30 % DoD).
- Clean and grease battery connections periodically.
- Correct battery connection cable sizes (according to charge/discharge current).
- The battery-bank capacity needs to match the load demand and the load growth over at least half (~4–5 years in Humla) of its expected lifetime.

The installation settings shown in Figs. 3.73, 3.74, 3.75, 3.76 demonstrate conditions often observed in the field. The lifetimes of these batteries were

---

[30] The author started a prototype battery recycling and testing program in 2009 with RIDS-Nepal.

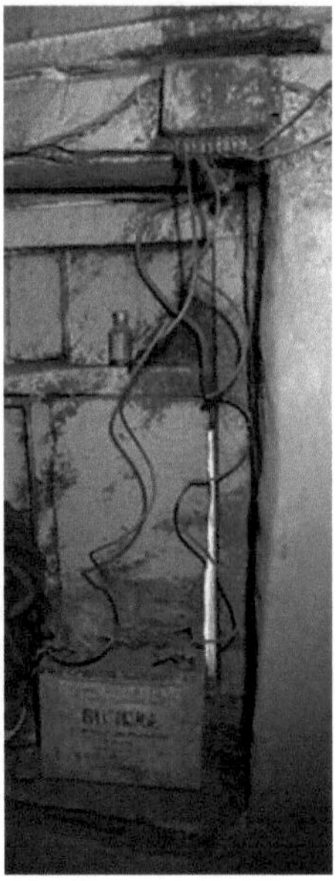

Fig. 3.75 No temperature protection

Fig. 3.76 Another commonly reported issue: Battery is directly connected to the PV module and to the load

shortened because they were not protected against the varying climatic conditions, or against the high indoor air pollution in the average Humla home where people cook and heat on an open-fire. Further, we observed that the electrolyte within the battery is often not maintained at a level high enough to cover all the battery's cells. Thus it does not come as a surprise when the SHS users interviewed mentioned that their batteries lasted only 6–9 months and in some cases even less.

3.7.6 House Wiring

Humla homes are made of stones, wooden beams and mud. Most homes have their food grain storage inside, which attracts mice. The mice eat everything they can get hold of, and they have a special fondness for PVC copper wire mantling. Further, homes have open cooking and heating places, creating enormous amounts of smoke, soot and particulate matter, endangering people's health as well as potentially harming indoor solar PV system equipment. These circumstances demand understanding of the local conditions and thus special care in the way house wiring is carried out (Figs. 3.77, 3.78).

Fig. 3.77 Loose house wiring is soon damaged and can pose a danger for children

Fig. 3.78 Mice find it easy to nibble on the house wiring and battery cables. Further, it is not permissible (Batteries store and release energy through a chemical reaction which takes place through the charging/discharging current, which dictates the battery internal temperature, the ambient temperature, the internal construction (numbers of lead plates and thickness of lead plates) and use of materials for the two different plates, the negative lead grid plate, alloyed with antimony or calcium to improve the mechanical characteristics, and the positive lead oxide, the "spongy" plate. That means that each battery is "unique" in itself and thus cannot be paired with other, different batteries, if the best performance is aimed for. The only time, several batteries can be joined to increase the battery-bank's capacity, is at the beginning of a project, with the same branded, new, capacity size and technology) to join batteries of different technologies, ages, brands or capacities

3.7.7 Indoor Lighting

SHS are PV systems generating very small amounts of energy. Relative to the amount of energy generated, they are expensive to purchase. In order to get the highest benefits from a given SHS, it is thus important that lights with low power consumption and long life-expectancy are used.

Many of the SHS that we inspected in the field have 10 W fluorescent tube lights, three for each home, powered by a 10 W_p solar PV module with a 20 Ah capacity battery. It can be expected that the battery is deeply discharged every day, and, if a charge-controller is used, the lights can be used only for a very short time per day (Figs. 3.79, 3.80).

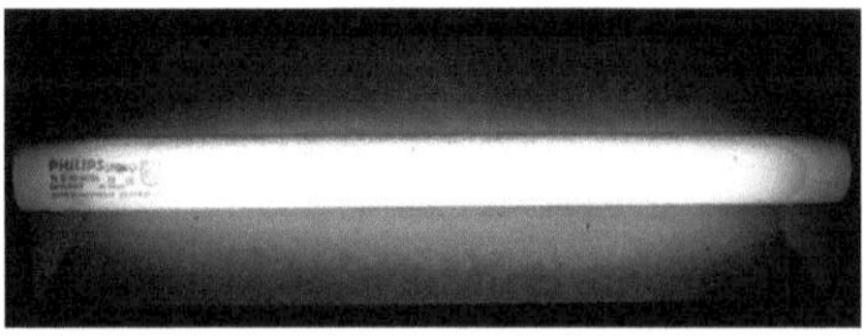

Fig. 3.79 10 W tube light, horizontally installed at the wall, allowing mud and dirt from the home's ceiling to fall on the tube and reflector

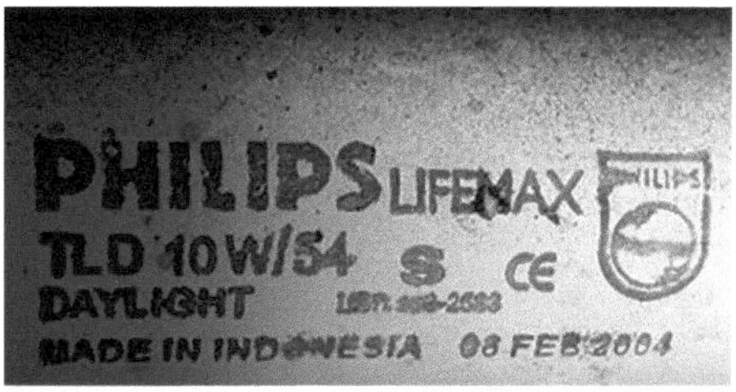

Fig. 3.80 Many SHS have a 10 W_p solar PV module, 3 × 10 W fluorescent tube lights and a 20 Ah, 12 VDC battery

3.8 Non-Technical Problems

Projects are not designed, implemented and operated in a vacuum, but in set cultural contexts for particular groups of people. For projects to succeed, planners and implementers must recognise how culture has defined and shaped local subsistence patterns and customs. In a context such as Nepal (Sects. 1.1.1–1.1.9, Appendices 3–9), the author believes it is critical that the cultural context, the "software" issues of a project, are taken into consideration from the beginning of a project. This demands that the local context, language and customs have to be learned and understood by the project partners before the project is initiated, in order to comprehend the unspoken and invisible "software" issues of the community. This demands time, compassion and dedication –crucial parts of a project.

Such invisible "software" issues are often the cause of unforeseen problems during the project implementation phase, causing delays, massively increased costs, and even project failures, because the uptake of innovations by local people usually occurs at a far slower pace than scheduled project implementations.

While in some villages uptake of innovations and willingness to learn new behaviour patterns is quick and uniform, in other villages change is slowly accepted and/or actively resisted by some members of the population. From experience of implementing projects and observing other projects, it appears that villages comprised of certain castes or mixes of castes experience innovation uptake at different rates from others. For instance, in our own experience in Humla, villages comprised of members of Hindu castes known for their cultural conservatism or risk aversion, have a slower and patchier uptake pattern of innovations than villages comprised of Buddhist villagers (Haddix and Zahnd 2007).

Thus, besides the technical specification for the design of a solar PV system, the local context and conditions of the end-users play an absolutely central role in the success of a project. These issues are often subtle and require excellent knowledge of the local populations. Thus, for every project site it is important to identify and plan for ways to adapt project implementation strategies to the local cultural context, and to budget and "sell" the importance of this piece of the project to partnering donor agencies.

In the following discussion, some important social issues (Sects. 3.8.1–3.8.5), and lessons learned about how to address them (Sects. 3.10.1–3.10.5), are identified.

Fig. 3.81 Dim and smoky light from *jharro*. Particulate matter produced by burning this resin rich wood poses a real danger to respiratory health

3.8.1 Energy Demand/Need Assessment

Solar PV systems for remote communities are designed primarily to provide lighting services that they have not previously had. In order to understand the local population's need for indoor lighting, one has first to recognise how homes were lit previously. In Upper-Humla, all families traditionally use *jharro* to light indoor living spaces. *Jharro* is a resin rich wooden stick from the high elevation Himalayan pine[31] tree whose flame provides smoky but minimally adequate indoor lighting (Figs. 3.81, 3.82). *Jharro* is gathered by inflicting a deep cut on the soft pine tree, forcing it to produce resin in order to "cure" the wound (Figs. 2.15, 6.3/ Appendix 25.3-Light_5-9). This resin rich layer of wood is cut away after a week and chips of this wood are burned to generate light.

To understand the traditional use of *jharro*, we measured illuminance levels in homes where it was in use. Illuminance levels of just 2 lx on the floor of the home surrounding (>1–2 m) the fire pit/cooking stove where *jharro* is burned were standard. This lighting level is only enough to move around the room and see the people and items in it. At an illuminance level of 2 lx, it was not possible to read. In order to get enough light to read, one had to sit right beside the smoky *jharro* (<0.5 m), and hold the reading material at just the right angle. The illuminance level on the cooking stove/or fire pit varied between 5 and 15 lx depending on its height (0–0.5 m) (Figs. 3.4, 3.82, 3.101).

---

[31] *Pinus wallichiana* is a pine native to the Himalayas and Hindu Kush region from north-eastern Afghanistan onwards across through north Pakistan, India, Nepal and Bhutan. It grows in high-altitude mountain valleys (1,500–4,300 m.a.s.l.) and can grow to a height of 50 m. The wood is not very hard, but durable, if it is not put under the strenuous cycle of getting frequently wet/ soaked and dried through rain and sunshine. It is highly resinous, the main reason why it is used to produce "*jharro*", for the traditional indoor lighting (Figs. 3.3, 3.81, 3.101). It burns well and quickly, with an estimated average calorific value of ~14 MJ/kg, and has a strong resinous smoke, one of the main reasons for widespread respiratory diseases. It grows well in a more temperate climate with dry winters and wet summers. For more information see: http://en.wikipedia.org/ wiki/Pinus_wallichiana (accessed July 26, 2012).

Fig. 3.82 Families in Humla traditionally burn *jharro* for indoor lighting, as candles or kerosene are typically unavailable or unaffordable

3.8.2 Awareness Raising of the Need for Improved Lighting Services

Having understood and measured the traditional indoor light source, it becomes clear that it is important to alert local people to the damaging health consequences associated with *jharro* use (Bhusal et al. 2007). It is also helpful to explain to people that *jharro* is an extremely inefficient type of lighting. This is true whether we are thinking in terms of the energy expended by humans in procuring *jharro* from the forest, or in terms of the amount of luminous flux produced by *jharro* (usually measured in lumens), as a ratio of the amount of energy consumed to produce it (usually measured in watts). Measurements indicated that the luminous efficacy of *jharro* (0.04 lm/W) is half of the efficacy of a kerosene fuel based lamp (0.08 lm/W; Mills 2005) and more than 300 times less than the 9 diode WLED lamps used in the villages with solar PV systems installed by RIDS-Nepal from 1999 to 2006, and even less with the new 12 diode WLED lamps (Fig. 3.83).

Since 2006, RIDS-Nepal has shifted from the 9 diode (Nichia NSWP510BS with ~40 lm/W) WLED lamps to 12 diode lamps, using Nichia's most recently released diodes (Nichia NSPW510DS with ~80 lm/W (Figs. 3.83, 3.84)). This allows us to produce lamps with about double the luminous intensity. Further, in collaboration with local users, we determined that the baseline illuminance level that a solar PV system has to provide is 5–15 lx for general tasks, and 25 lx for reading. This lighting level can easily be provided with RIDS-Nepal's 9 diode Nichia WLED lamps (Bhusal et al. 2007), and is even more readily achieved with our new 12 diode lamps.

Based on our experience, we would argue that especially since 2009, higher power consuming lamp technologies such as CFL, tube lights or even incandescent bulbs, are not appropriate for solar PV systems. Providing the right lighting level, using a technology that is affordable and sustainable has proven to be crucial for context appropriate design and cost effective solar PV systems in these rural areas.

Fig. 3.83 RIDS-Nepal's manufactured 12 VDC WLED lamp with 12 NSWP510DS diodes, consuming a total of 1.1 W

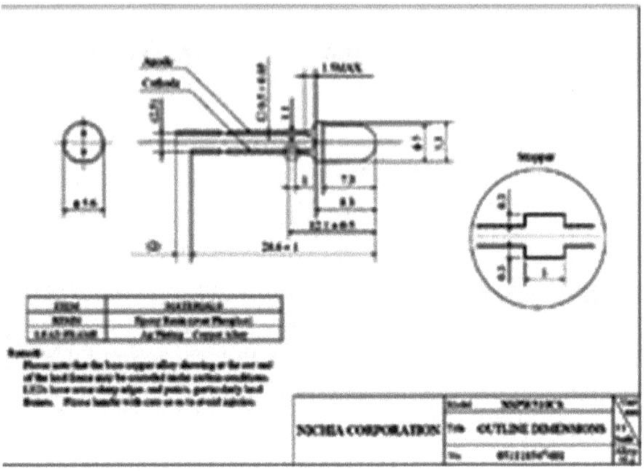

Fig. 3.84 Technical specification of the NSWP510DS white LED, with a life-expectancy of ≥50,000 h, or ≥20 years

3.8.3 Training for Solar PV System Operation and Maintenance

Under the government subsidy scheme, solar PV system operation or maintenance training is not offered, nor are end-users typically trained during the installation phase of their system.

While people are happy to have electric light in the mornings and evenings once the system is in place, without training they are unprepared to operate their system appropriately. Basic maintenance of batteries, such as topping them up with distilled/rain water, cleaning the battery-bank's terminals, checking the charge-controller connections, fuse and indicators, and cleaning the solar PV module during longer periods without rain are neglected. Villagers are not to be blamed for this, however, since in most Nepalese SHS projects training is not funded.

Often system owners try to add features to their system that will damage its integrity, for example other lights, radios, or battery chargers. Due to lack of

Fig. 3.85 Without training on maintenance and appropriate use, SHSs often break down rapidly, leaving the users with unfulfilled hopes and debts

Fig. 3.86 The three blackened WLED lamps after 6 months in a home without a smokeless stove where the family still cooks and heats using an open-fire. In this condition, the lamps fail to provide the intended illuminance. This shows the importance of applying the "Family of 4" HCD concept

training, they could not have known this would jeopardize the system's life-expectancy and performance, or even damage the system. Also due to a lack of training, short circuits are frequently caused by bad DC wire connections, which cause the charge-controller's glass fuse to burst. As there are no spare glass fuses sent along with the government subsidised solar PV systems and new glass fuses are not available in Humla, the end-users usually remedy the situation by taking the charge controller out and joining the solar PV directly to the battery (Figs. 3.73, 3.75, 3.76, 3.77, 3.85, 3.86). Without the charge-controller, the battery rapidly becomes so deeply discharged that it can not be re-charged again (Figs. 3.73, 3.77, 3.78).

3.8.4 Installation and Follow-Up

Villagers in Humla frequently reported that the company they bought the subsidised solar PV system from, installed it, charged them extra for the installation, and left as soon as the system was put in. End-users were not given contact information in case a technical problem arose, or if a piece of equipment needed to be repaired or exchanged.

Previously, one of the main Kathmandu based solar companies had a solar PV shop in Simikot, Humla's district centre. But due to low turnover, high staff and transport costs the company closed the shop. Currently, there is no support or technical staff in all of Humla to follow-up the many solar PV systems installed by a variety of companies. RIDS Nepal's full time, district project staff do support, maintain and follow-up their projects but are not able to provide support to all of the others.

3.8.5 Performance Monitoring

"Out of sight out of mind". That is the harsh reality of most of the village based solar PV systems, in particular those installed under the government subsidy program. There is a preventative measure working against this failure, in the form of a clause in the subsidy program, which says that the last 10 % of the subsidy shall be paid to the selling company only after the first annual inspection. But 10 % does not provide enough incentive to actually maintain and troubleshoot for the installed PV systems. Instead, the margins are calculated in such a way that the 90 % subsidy plus the end-user's installation cash payment makes a reasonably profitable business.

It is not meant to unfairly heap blame upon the retailer since, to be fair, one has to recognise the time and cost demanded by ongoing system follow-up, due to the days or even weeks of hard trekking away from the next road head. Thus, aside from the RIDS-Nepal projects, there is not one longitudinally monitored solar PV system in the Humla region, known to the author, for which field data are available. This makes comparisons and evaluations of the various systems' performances difficult.

3.9 Technical Remedies

The experiences and circumstances described above demand a fresh look at the procedures and implementation of rural village solar PV systems. Some of the issues outlined, such as the lack of training typically provided by SHS companies, suggest that the basic approach to this type of community development has to be re-examined. Toward this end, possible remedies for the technical and non-technical problems identified in the Sects. 3.9 and 3.10 are discussed next.

3.9.1 Solar PV Module

As indicated above, our experience with solar PV systems in rural Nepal shows that the fewest problems are caused by the solar PV modules. This industry is well established with international standards/quality controls, and the good record of the modules should persist if retailers continue to import quality solar PV modules. RIDS-Nepal uses two solar PV module manufacturers for our three different solar PV system approaches. The first was until 2007 the BP275F (75 W_p as in Figs. 3.41, 3.49, 3.59, 3.65, 3.69, 3.88) Si mono-crystalline PV module from BP, and since 2008 the Si poly-crystalline BP375F[32] 75 W_p module, manufactured in their highly automated production plant in Bangalore, South India. Both these PV Modules are used for the central (Figs. 3.38, 3.39, 3.41, 3.49) as well as for the cluster PV system approaches (Figs. 3.33, 3.34, 3.69, 3.87, 3.88). Since 2003, not one of these panels has been defective or otherwise the cause of any PV systems failure. The other module we use is the 16 W_p, Si poly-crystalline PV module KC16T[33] from the Japanese manufacturer Kyocera (Figs. 3.22, 3.23), with equally excellent performance.

As with all of the equipment we install, investment in high quality, standardised solar PV products is worth the additional cost, especially due to the remoteness of the target communities and the need for high durability and minimal maintenance. The investment will easily pay for itself over the life-cycle of a village solar PV system (Zahnd 2004b).

The efficiencies of solar PV modules in real field applications are most often lower than their STC[34] performance under laboratory conditions. Our field experience since 1998 and the data we have collected on efficiency since 2005 show that if all of the generated power is fed into the battery-bank, a ~2–3 % efficiency drop from STC performance for a high quality, single installed PV

[32] The BP375F solar PV specification sheet is available at: http://www.bp.com/liveassets/bp_internet/solar/bp_solar_usa/STAGING/local_assets/downloads_pdfs/pq/BP380_9-09.pdf (accessed July 26, 2012).

[33] The technical specification sheet for the Kyocera KC16T poly crystalline solar PV module can be consulted at: http://www.kyocerasolar.eu/index/products/download/English-cps-33141-files-62466-File.cpsdownload.tmp/KC16T_21T_32T_ENG_July_2009.pdf (accessed July 26, 2012).

[34] Standard Test Conditions (STC) are the testing conditions to measure solar photovoltaic (PV) cells' or modules' nominal output power under defined laboratory conditions. These are: Global solar radiation of 1,000 W/m^2, with a standard spectral distribution received on the solar PV module and the sun rays entering the earth's atmosphere at an angle of ~48.2° (1/1.5 Invcosθ) to the horizon/horizontal PV module, corresponding to an Air Mass (AM) of 1.5. The PV cell/module junction temperature is held at constant 25 °C during the measurement. More details about the STC (Standard Test Conditions) can be found at: http://en.wikipedia.org/wiki/Solar_panel#Module_performance_and_lifetime (accessed July 26, 2012).

module (SHS or Cluster PV system), and a ~3–4 %[35] drop can be expected for serial and parallel connected PV modules, as installed in a central PV solar system. Our data agree with Reinders et al.'s results (1999), which show that the single 40 W_p solar PV modules installed in that project have a STC efficiency of 11.1 %, while in the field they achieved only 8.1 % when all the generated power was fed into the battery. The main reasons for this drop are (1) the PV module is operating above 25 °C, (2) the PV module was not operating at the MPP (Maximum Power Point), (3) the intercepted global solar radiation is below the STC defined 1,000 W/m^2, and (4) there are ohmic losses in the cables. Our experience also showed that additional losses occurred through PV module mismatch[36] in solar PV arrays (with serial and parallel connected PV modules), such as those used in the central solar PV village approach.

3.9.2 Solar PV Module Frame

Alongside the three solar PV village electrification approaches described above, we developed also three solar PV module frame concepts, to maximize utilisation of the available daily solar radiation, and minimise the cost for the frame. The first is a 2-axis self-tracking system (rotating daily from East to West), with a manual bi-weekly North–South adjustment to accommodate the sun's seasonal path. Figure 3.59 shows three such 2-axis tracking frames, installed at the RIDS-Nepal Simikot HARS since 2004. Figure 3.49 shows the central solar PV village system of Tulin village, powering 29 homes for 195 people. This system produces a measured average of 30–35 % more energy than a similar system with a traditional fixed angle frame. The installation of the 2-axis tracking frame does not require cement, which is advantageous in a place like Humla, where cement is very expensive and often unavailable. The frame we designed uses locally available stones (Figs. 3.40, 3.41, 3.49) to secure it against wind and snow.

---

[35] While a ~3–4 % drop in power conversion efficiency of serial and parallel connected solar PV modules may not sound significant, this has to be seen in context. A high quality, Si monocrystalline solar PV module, such as the BP275, has a rated intercepted global solar radiation conversion to DC electricity at STC of 12.04 %, indicated by the manufacturer. Thus, a 3–4 % drop due to non-STC conditions and mismatch of other solar PV modules installed in the same PV array means, that the whole solar PV array converts the intercepted global solar radiation at a rate of 8–9 % only, instead of 12 %. This 3–4 % drop has to be considered in respect to the solar PV module's 12 % efficiency, thus a drop of 25–33 %, cannot be neglected in the initial design and development of a solar PV system.

[36] When several solar PV modules are serial or parallel connected, so called "mismatch" losses have to be included in the system's calculation. These losses are related to the fact that the PV modules in the array do not exactly possess all the same I/V characteristics (with slightly different short circuit current Isc, open circuit voltage Voc and maximum power point MPP values), because each PV module usually consists 36 individual solar PV cells (for a 12 V module), which do not have exactly the same characteristics. With various solar PV modules connected this fact is accentuated and thus increased losses occur which have to be accounted for in the design.

Fig. 3.91 The Cluster PV system battery-bank consists of 2 equal (brand, capacity and age) flooded lead-acid batteries, well insulated and protected. Spare distilled water (clean rain water) is included in the battery box. This 200 Ah solar PV cluster battery-bank, with 8 homes and total of 24, 1 W WLED lamps for up to 7–8 hours a day, has an average daily DoD of 6–8 %, providing up to 5 days the needed energy without sunshine with a maximum DoD of ~30 %

While such shallow cycling demands a bigger battery-bank capacity, and adds initially to the project's budget, it pays back in the long run, as the number of cycles is not proportionally declining with increasing DoD but, assumes a parabolic shape. Thus it is crucial that one knows all the manufacturer's specifications for a chosen battery technology and brand before the whole system design is finalised. This is another important factor for the sustainable operation and long lasting performance of a solar PV system in a remote area like Humla, where all the equipment has to be air lifted and then carried by porters into the project sites.

The examples in Figs. 3.91 and 3.92 show locally made wooden boxes used to house battery-banks, with Styrofoam covering the batteries and additional external insulation used for the cold winter months. These boxes keep the battery-banks within an acceptable temperature range at all times. With one or two spare bottles of distilled water (Fig. 3.91), appropriate maintenance for the batteries is made simple. In Humla, rain water serves this purpose. Periodic cleaning of the battery connections, greasing of the connectors and storing rain water for the batteries in a cool, dark place, are easy steps which positively enhance a battery-bank's service and life-expectancy. Last, but not least, a blank data sheet for each battery-bank, allowing users to chart when the battery was topped up and serviced, helps us and users to understand the frequency and time needed to maintain the energy storage system.

Fig. 3.92 The battery-bank should be fully covered and protected e.g. in a locally made wooden box. In our projects, the wooden box is sealed with cow dung and covered with additional heavy jute sacks to protect it from mice and dirt

Fig. 3.93 The house cable enters the home via an open ended polyethylene pipe. Cables are mounted underneath the wood beams

3.9.6 House Wiring

The house wiring comes into the room through an open ended polyethylene pipe (Fig. 3.93). In order that the mice cannot get to the PVC copper wire insulation, the cables are fixed underneath the overhanging wooden beams that carry the roof (Figs. 3.93, 3.95). Wires, switches and connection points have to be high enough so that children cannot reach them (Figs. 3.94, 3.95).

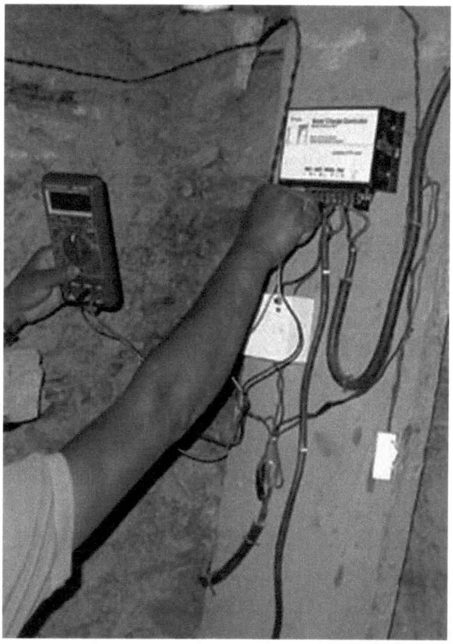

Fig. 3.94 Armoured cables from the PV module to the CC, PVC insulation house wiring inside, placed out of reach of children

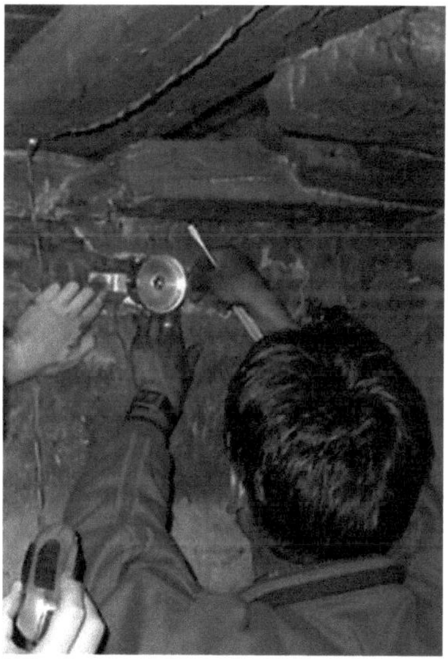

Fig. 3.95 The WLED lights are installed in a place where they can be turned on 2-axes so that they can be used for different tasks at different times

If available, wiring clips (Fig. 3.93) are useful and handy to fix the cables underneath the wooden beams. These are small issues which do not add substantial cost, but can be crucial to the long-term sustainability of a solar PV system. Finding and fixing mouse damage can be very time consuming and inconvenient to repair, thus it is well worth preventing through proper wiring installation.

3.9.7 Indoor Lighting

Each home electrified with a solar PV system has three WLED lamps installed, usually in two rooms. We like to locate two in the main cooking, living and sleeping room, and one in the store room. The three WLED lamps consist of two Nichia WLED lamps, each with 12 NSPW510DS (since 2009) white WLEDs, providing ~80 lm/W (Fig. 3.83), consuming 1.1 W each, and one 1 W Luxeon WLED (~30 lm/W) lamp from Lumileds, consuming 1.3 W (Fig. 3.95/Appendix 12.5.2 right lamp). That amounts to a total of 3.5 W power consumption per household.

These WLED lamps provide an illuminance level that is sufficient for the daily indoor tasks and reading as defined by Bhusal et al. (2007). WLEDs are known to have a long life-expectancy if they are operated under normal operating conditions (not over heated during soldering, and limited to 20 mA of current). Typically, a life-expectancy of >50,000 h[37] can be expected before brightness begins to substantially reduce. With an average of 7–8 hours per day this is 20 years of use, after which, considering the fast development of the WLED technologies, there will be new WLEDs available (Figs. 3.96, 3.97, 3.98).

Considering the fact that WLEDs, available on the markets, reached 100 lm/W in 2010, providing a life-expectancy of $\geq$100,000 h (Žukauskas 2007), there is a hopeful future for WLED lamps. Thus for the present and coming decade, one can say with some certainty that the WLED technology will continue to provide a very appropriate first time indoor lighting solution for the communities of Humla (Figs. 3.99, 3.100).

It is important to repeat that in our projects, a solar PV system for elementary lighting in a village or home is never implemented alone. Rather, solar PV systems are always introduced alongside the "Family of 4" HCD concept that the author developed and implements with RIDS-Nepal in Humli villages (Sect. 2.10). The negative health impacts of indoor air pollution caused by open-fire cooking and heating are so enormous that electric lights alone can not make a substantial positive change to quality of life. Additionally, to introduce the WLEDs into homes that continue to use open-fires for heating and cooking would massively shorten the life-expectancy of the solar PV system equipment, due to the damaging impact of the particulate matter of open-fires upon the optimal functioning of the system (Fig. 3.86).

[37] The author has two AC powered NSWP510BS diode WLED lamps, each with 12 diodes, under long term testing, one since 1st October 2006 and one since 13th July 2007. Thus, by the 30th June 2012 the first has ~50,000 and the second ~43,000 h. The visual illuminance drop measured by the author is ~50–55 % and ~40–45 % respectively. These, compares favourably with cheaper WLED diodes, with measured illuminance drops of ~75–85 % after 2,000 h, during tests conducted by Rudolfo Peon (University of Calgary, Canada) from 2004 onwards. The data and results are presented and documented in: *15x5 mm WLEDs Array Light Output Degradation vs Time* (2006), Advanced Solid State Lighting Laboratory, Calgary University, Canada.

Fig. 3.96 Reading under a WLED lamp

Fig. 3.97 First time indoor lighting

Fig. 3.98 All WLED lamps are tested and the CCs' cut-in/out voltage adjusted

Fig. 3.99 Two adjustable WLED lamps are installed in the main room

Fig. 3.100 A solar PV lighting project with WLED lamps is never implemented alone. It is always installed alongside a pit latrine, a smokeless metal stove project, and a safe drinking water system, as part of the HCD approach, the "Family of 4"

3.10 Non-Technical Remedies

As previously described (Sect. 3.8), projects unfold within a cultural context, where behavioural patterns, local expectations, and personal and community goals precede the arrival of the innovations we introduce. These factors have to be taken into consideration, and constitute what we call the critical "software" issues of a project.

People's education, expectations of the project's outcome, their economic conditions and ability or inability to maintain the system, are just some of the important "software" issues to know and understand in order to define the scope of a project. It is crucial that the community or household is able to take full ownership and pride in the implemented project, and for that they have to participate in all stages of a project. From the very beginning of a project to the point when control is completely handed over to the community, the end-user has to play an active role.

Thus in a solar PV project, local people have to participate in defining the lighting service they feel that they need and can afford. They have to be willing to send several people per village central system, one person per cluster system and if possible one person for 3–4 single SHS for the 10 day solar PV training course RIDS-Nepal developed and holds before every project's implementation. In the early stages of project planning, end-users have to understand, and sign an agreement to pay a per family amount per month in order to maintain the system.

3.10.1 Energy Demand/Need Assessment

As described in Sect. 3.8.2, an illuminance level of 25 lx for reading tasks is recommended for an elementary lighting service, while an illuminance of about 5–15 lx is recommended (Fig. 3.101) for general purpose tasks (Bhusal et al. 2007). With the locally developed WLED lamps these illuminance levels can be achieved (Fig. 3.102).

Fig. 3.101 From a smoke filled, open cooking fire place and use of *jharro* lighting to

Fig. 3.102 Clean indoor air with a smokeless metal stove and WLED lights

The long life-expectancy, low power consumption and 2 year guarantee provided by PPN for WLED lamps make them very affordable in regard to their running costs, once the initial purchase cost has been paid by the project. Thereafter, a monthly fee payment of 25NRs (~0.4AUD, 2007 value) per family is sufficient to maintain the solar PV system.

3.10.2 Awareness Raising, Lighting Technology and the "Family of 4" Concept

By demonstrating the brightness and illuminance of the WLED lamps, the local people are introduced to their new indoor lighting. They are also made aware that if they do not change their method of cooking and heating by purchasing and installing a highly subsidised smokeless metal stove, the new electric lights will not be able to provide the intended services (Fig. 3.86) and thus the HCD approach to improving health and quality of life will be undermined.

Further we are careful to emphasise the negative health impacts that the traditional open-fire and use of *jharro* has, especially on small children and women. We explain, many times, if necessary, the importance of the holistic approach to community development and the long-term benefits and living standard improvements that come along with it.

Therefore, along with the research, development, installation and trouble shooting that our on the ground team do, an enormous amount of time and effort is put into teaching about the "Family of 4" concept. This approach bundles the solar PV powered indoor lighting with WLED lamps together with a smokeless metal stove, a pit latrine for each household, and the building of a communal safe drinking water system (Sect. 2.10), as depicted in Fig. 3.103.

Fig. 3.103 Awareness raising about the reasons for and benefits of the holistic "Family of 4" project in a village community

3.10.3 Training for Solar PV System Operation and Maintenance

We have observed that understanding and maintaining one's own solar PV system creates a sense of pride and ownership in end-users, and also has the beneficial impact of teaching valuable skills and knowledge that can be useful in and outside the community. These are important steps toward a new technology being accepted and integrated into the local community over time.

However, ensuring that local users have the required knowledge demands theoretical and practical training, which is time consuming and costly. We recognise that these costs are a critical part of a sustainable solar PV village electrification system, and are as important as the hardware and transport costs. The cost effectiveness of an investment in these sorts of "software" features may not be obvious initially, but can be seen in the mature stage of a project, when systems are operated and maintained in the way they are designed and built for.

Understanding of a solar PV system's working principles, scope and limits of service provision helps the owner to use the system properly. The majority of Humlis are illiterate. However, our experience shows that although many local people never had the privilege of going to school, they are quick and eager learners. Thus some basic theory, hands on practical training and inclusion in the

Fig. 3.104 Periodic checks of the battery-bank's voltage with a multimeter and topping up of the battery with distilled water (rain water) should be done by the person who owns or is responsible for the solar PV system

Fig. 3.105 In our projects, practical sessions for local people on how to maintain their solar PV system are provided as part of the initial 10 days RIDS-Nepal solar PV training program. How to operate the system, what to check periodically, and how to identify and communicate problems to the RIDS-Nepal staff are part of the curriculum. This provides experience and confidence for locally trained people and increases the sustainability of the solar PV system

planning and installation of the PV system is fruitful, satisfying to everyone, and always needs to be a core part of a project (Figs. 3.104, 3.105).

3.10.4 Installation and Follow-Up

With the basic training course that we provide, it is possible for local people to participate in the solar PV system installations. This provides the local trainee with

Fig. 3.106 Learning by doing. Joint installation of the solar PV system after the 10 days of solar PV training. The training creates experience for the trainee and confidence in the users. The trainees also earn a small salary from the monthly users' fees for their services

Fig. 3.107 Participatory installation of the charge-controller and the battery-bank, putting theory into practice. This teaches new skills and creates a strong ownership feeling in the SHS users. It also engenders respect in the local "specialist"

the knowledge of where each cable goes and where the main connections are, which are critical to understanding a solar central or cluster PV village system.

Our project staff periodically follow-up every system that we install, interviewing the owner of the system as well as the trainee who is responsible for the system's operation and maintenance. Every system has a serial number and the results of the interview are recorded in the project's data bank. This gives us a simple and easy system for tracking the performance of the SHS, and helps us identify any shortcomings which have to be addressed. In this way ongoing local equipment development can occur, improving the system's performance and sustainability (Figs. 3.106, 3.107).

3.10.5 Performance Monitoring

Having recognised the lack of monitored, followed-up or even anecdotal field experience of installed solar PV systems, RIDS-Nepal made it a policy from the beginning of its long term Humla based HCD project in 2002 that all the staff have to live in the project area. We also require that each solar PV system installed in the village has a number and is listed in RIDS-Nepal's data bank with relevant owner and household data.

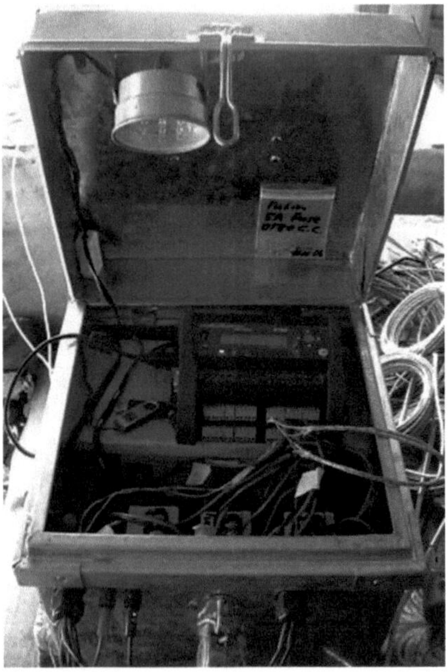

Fig. 3.108 DT80 Data Monitoring Box with independent Solar PV Power System (Figs. 3.23, 3.49, 3.116)

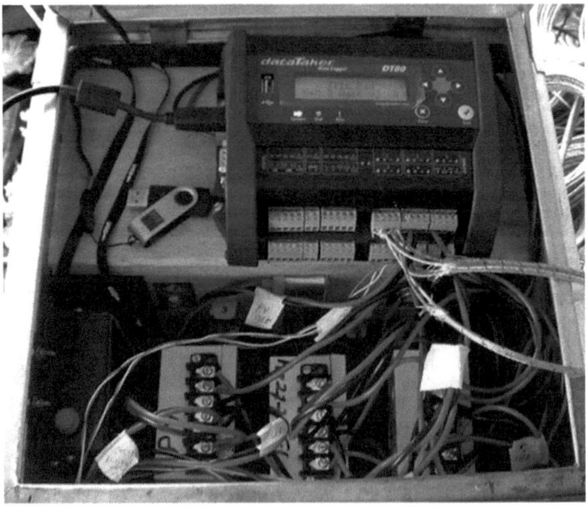

Fig. 3.109 24/7 the DT80 data logger records and calculates 22 parameters in three different time schedules for all three different solar PV system approaches

This ensures that each system is followed-up, inspected and reported upon as a regular part of the ongoing work schedule. Detailed data monitoring reveals a system's performance, which in turn is invaluable in the revolving learning process of improved solar PV system design and manufacturing. It is such an important and neglected aspect in most projects, that a whole section of this chapter, Sect. 3.11, is dedicated to that.

3.10.6 Summary

Our field experience and observation of other projects has shown that the majority of the solar PV systems installed by other implementers are not fulfilling the initial design demands in regard to equipment quality, performance and sustainability. End-users are often left with additional debts from the initial installation, beside their shattered and short-lived dream of improved indoor living conditions through electrification. As identified and discussed above, there are technical and non-technical reasons for the high failure and user dissatisfaction rate, which should be treated with the same importance and respect. This is supported by other published literature, with a classic case presented in the solar PV–Wind hybrid system in Xcalak village in Southeast Mexico. The report writer notes that "the technical results of the PV lighting system efforts had proven acceptable, but the resulting social issues in the case of Xcalak have presented much more of a challenge than PV lighting systems". Further, the writer recognised that "since the users were not paying for electricity [as there was no mechanism set up to allow for tariff collection], the system saw tremendous load growth [53 % within the first 18 months] as more electric appliances were added" (Foster et al. 1999).

The author's 16 years of experience of living and working with solar PV powered energy, in rural as well as urban settings has been priceless for generating an understanding of the need for sustainable and context appropriate electrification solutions. This experience helped shape an appreciation of the quality and value of solar generated electricity, where even miniscule amounts of energy, such as lighting through WLED lamps, can bring forth enormous changes for whole families and communities. The author's experience and anecdotal data show that these changes are even more beneficial when lighting projects are integrated in an HCD project using, for example, the "Family of 4" approach.

The following recommendations are derived from practical experience with installed and monitored solar PV systems by the author in Humla and Jumla. They summarise the main lessons learned over the last 16 years. They are supported by studies of other systems analysed in a similar fashion, such as in Xcalak, Mexico, and the findings published by the IEA (1999). The 15 most important recommendations are:

- Before a solar village PV project is even planned, a base-line survey of the community's current living conditions and self-identified needs has to take place. Social issues (the "software" issues in a project) need to be given the same importance as technical issues.

- A solar village PV system project should always be an integrated part of a wider HCD project approach such as the "Family of 4".
- The end-users have to take the initiative and explicitly express their interest in starting and partnering in a solar PV village system project.
- The available technologies need to be presented, demonstrated and explained to the end-users, with attention paid to services and life-cycle costs, prior to the project's final planning stage.
- The end-users have to take part in defining how the system will be used when it is installed (such as load demand, load growth, fee payment and prohibited loads), and in approving of the project's conditions and partnership with project implementers and third parties.
- The end-users need to form a solar PV system committee, responsible for appointing operators, overseeing the maintenance schedule, bringing in the regularly scheduled fee payments and taking the responsibility for implementing the rules that the community agrees upon.
- Professional design of the PV system, based on actual local geographical, climatic, meteorological and cultural data and conditions, is essential.
- Locally developed and manufactured equipment is preferred to imported equipment. Equipment has to be checked for appropriate quality, reliability and ability to withstand the local environmental conditions. This is crucial to achieve a high system capacity factor.
- Training the end-users on solar PV basics and hands on system installation has to take place before and during the project's implementation phase.
- After a system's installation, it has to undergo a careful check before a formal handover to the users takes place. This is an intrinsic and compulsory part of the project.
- All users need to be made aware and trained to use the available electricity in the most energy efficient way and to use only energy efficient and approved appliances.
- Each installed PV system should be numbered and registered for monitoring in a follow-up program (usually with spreadsheet based data input).
- Initially, for the first 2 years, intensive and then less intensive, but still systematic, periodic system follow-up needs to take place.
- Long-term data acquisition and system performance monitoring of a typical PV system approach is highly recommended, and indeed required for larger projects, for design evaluation, improvement and technical and economic calculations for best performance.
- The integration of a rural solar PV village project with a University based academic research program (as is the case with RIDS-Nepal and the Kathmandu University) can help to bridge the existing rural/undeveloped urban/developed divide, as well as creating new knowledge and insights through technical and social science analysis.

3.11 The Importance of Monitoring of Performance: Analysis of a Rural Solar PV Electrification Project

Unfortunately many solar PV village projects are not able to provide the energy services expected by the end-users, nor achieve their projected life-expectancy. Thus they fall short of their intended aim. Long-term monitoring of solar PV systems in their installed context, the analysis of their detailed performance over time and local seasonal shifts is almost unheard of, especially in developing countries. The vital information gained through monitoring is essential for understanding a system's actual field performance. This leads to more appropriate system design, considering the local context in regard to the prevailing meteorological conditions and energy service demands.

Since 2006 we have collected data to monitor the performance of the central solar PV system approach and the cluster solar PV system approach. Since April 2007 we have also been monitoring the single SHS approach. The functioning of each system is monitored in great detail by a data monitoring system (Figs. 3.108, 3.109, 3.112, 3.113). For each of the three solar PV system approaches, a DT80 dataTaker,[38] records and calculates data on 22 parameters. Data is collected every 10 seconds, averaged over 1 min and recorded in three separate, time averaged schedules, 10 min, hourly and daily (24 hours average).

Understanding these data is vital for the evaluation and re-design of more efficient, economical, and sustainable solar PV systems. The following 20/22 most important parameters[39] are recorded in each of the three monitored representative solar PV system approaches:

- Global Solar Radiation measured with a calibrated silicon cell pyranometer on a horizontal surface (international standard).
- Global Solar Radiation measured with a calibrated silicon cell pyranometer on the POA (plane of the array) of the solar PV module.
- Generated solar PV module voltage (directly measured with the DT80).
- Generated solar PV module current (measured and transformed into voltage by a calibrated current transducer).
- Power generated by the solar PV module (calculated).
- Solar PV module efficiency under the measured conditions (calculated).
- The solar PV module manufacturer's efficiency minus the measured "field" efficiency (calculated).
- Current flowing from the charge-controller into the battery (measured and transformed into voltage by a calibrated current transducer).

[38] For more details and the technical specifications of the DT80 data logger from DataTaker see: http://www.datataker.com/DT80.php (accessed July 26, 2012). For reliability purpose, all DT80/DT605 data loggers are powered by their own, dedicated solar PV system.

[39] The 2 parameters not listed are related to the DT80 data-logger: Internal DT80 temperature and the DT80 internal memory back-up battery voltage.

- Battery charging voltage, measured between the charge-controller and the battery (directly measured with the DT80).
- Power flowing into the battery-bank (calculated).
- Current flowing from the charge-controller to the load (measured and transformed into voltage by a calibrated current transducer).
- Voltage measured between the charge-controller and the load (directly measured with the DT80).
- Power demanded by the load (calculated).
- Current flowing from the charge-controller to the air heater dump load (measured and transformed into voltage by a calibrated current transducer).
- Voltage measured between the charge controller and the air heater dump load (directly measured with the DT80).
- Power flowing to the air heater dump load (calculated).
- Solar PV module temperature (T-type thermocouple[40] directly measured with the DT80).
- Ambient air temperature (T-type thermocouple directly measured with the DT80).
- Temperature difference between the solar PV module and the ambient temperature (calculated parameter used for the solar PV module efficiency calculation).
- Battery-bank temperature (T-type thermocouple directly measured with the DT80).

Data are downloaded from each DT80 and saved in the field on external USB memory sticks, hand carried to the RIDS-Nepal Simikot office and loaded onto the HARS PC. They are presented in tabular and graphical forms, as in Figs. 3.110, 3.114, 3.119, 3.120, 3.121, and 3.122.

The data monitoring system was designed with the guidelines proposed by IEC (1998), which suggests, for example, that measurement of generated PV power instead of just current (due to the different charging and discharging voltages) is needed, as well as power into and out of the battery. That leads to more accurate analyses and better insight into the performance of the battery-bank. Further, it allows the calculation of the non-availability of generated power to the load due to a full battery-bank. These factors are needed to calculate the performance ratio (PR) of the solar PV system, which reveals the interaction between the solar PV system under the prevailing meteorological conditions and the user. This in turn provides feedback about the initial design assumptions and parameters that were originally specified, compared with the real field application and use of the PV system. Since March 2010, the monitored solar PV systems' recorded data, as well as all the other

[40] For more detailed information and technical specifications for the T-Type Thermocouple, see: http://en.wikipedia.org/wiki/Thermocouple#T (accessed July 26, 2012).

monitored systems such as the pico-hydro power plant, greenhouses and solar water heater bathing centres, are accessible via the RIDS-Nepal DataBank[41]:

3.11.1 Pamlatum Village Solar PV Cluster Data Monitoring System

Figure 3.110 shows cluster 4, with 8 homes (Fig. 3.31), solar PV cluster system's performance and use in Pamlatum village.[42] The red curve shows the instant power output of the 75 W_p BP275F solar PV module. The available global solar radiation on the POA is plotted in green. The blue line represents the power going either into the battery-bank, which is mostly during the day when the power input from the solar PV module is greater than the WLED lighting load, or out of the battery, which is after the sunset, during the night and early morning hours.

From 10:00 a.m. onward, the power into the battery-bank drops despite high global solar radiation on the POA. That indicates that the battery-bank has reached its full charge and is in the trickle charging mode. Likewise the solar PV module's power generation drops as the CC short circuits and opens the connection of the

---

[41] The RIDS-Nepal DataBank (http://www.rids-nepal.org/databank/) has been developed by the author and the RIDS-Nepal staff from 2008 onwards, in order to present each monitoring systems' long-term performance in the local geographical, climatic and cultural context. This is important for various reasons: First, to come to a detailed understanding of how the installed solar PV systems perform under this harsh Himalayan climate over the course of the yearly seasons. Second, system downtimes, over- or under-production of electricity as well as how the various PV systems' individual equipment perform under varying conditions, in order to continue to develop them. This information allows a direct evaluation and assessment of the system's design and enables any shortcoming to be recognised in time, enabling the RIDS-Nepal staff to mitigate the shortcomings, or repair the failed equipment. It also provides information for future needed design improvements. Further, as the village based solar PV systems are all "first time" electrification systems for these village communities, it is vital to come to know how the communities learn to live with basic electric indoor lighting. Is their pattern of life changing with the availability of clean, electric light in their homes, and if so in what ways? Is the electricity demand and consumption over the course of the years growing as anticipated, with the systems able to provide the increased demand as predicted in the survey and design, or are adjustments/restrictions to the systems needed? All these issues are of great importance for the design, building and installations of more sustainable, remote rural electrification schemes. The DataBank offers the user the chance to investigate any of the various monitored systems. Within each systems' own data bank the user can select from the different monitored parameters which ones he wants to be plotted, or extracted and downloaded into an Excel sheet for further calculations, assignments or reports etc. for a defined time period. With this tool, RIDS-Nepal offers a wider interest community of like minded community developers, solar PV system developers, scientists and students an important, and thus far missing instrument, to come to a more in-depth understanding of how defined power generation systems, such as solar PV and pico-hydro power systems, work under long-term field conditions.

[42] Pamlatum village in Humla can be seen via Google Earth through the RIDS-Nepal based website link at: http://www.rids-nepal.org/images/stories/explore_nepal/google_earth/Pamlatum.kmz (accessed July 26, 2012).

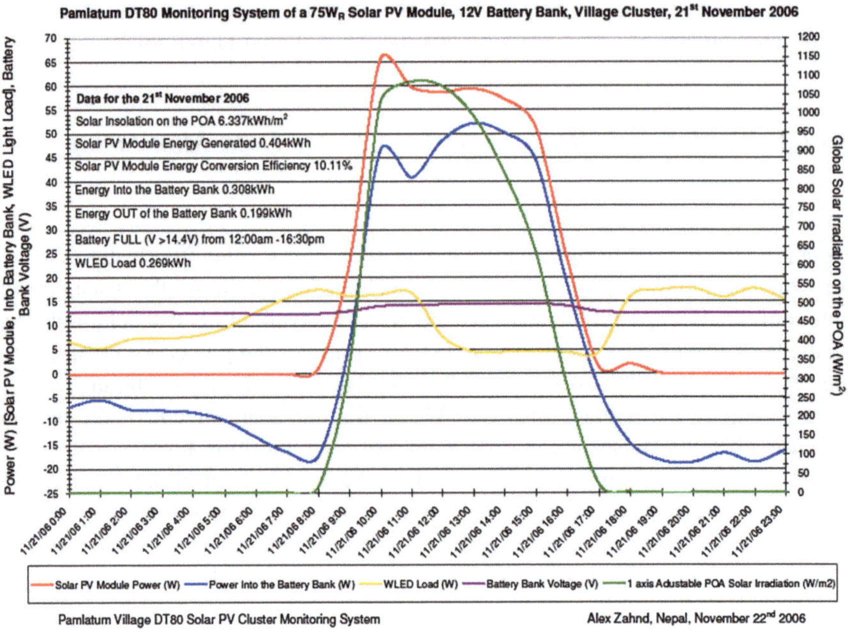

Fig. 3.110 24 hours (21.11.2006 00:00 a.m.–21.11. 24:00 p.m.) solar PV cluster (cluster 4 with 8 homes in Fig. 3.31) power generation and load demand under the given local climatic and geographical conditions of Pamlatum village. Detailed evaluation of the PV system's performance and user demand on the system is possible using these data

solar PV module to the battery, according to the trickle charge mode. Thus it can be concluded that over this 24 hours, the battery-bank was fully charged, and in fact, that there was more energy available to be fed into the battery-bank under the given solar radiation. The yellow line shows the user's load pattern for indoor lighting only. The load consists of a total of 24 WLED lamps, each consuming 1 W, with 3 lamps in each of the 8 homes in the cluster.

The yellow line shows that during the morning hours, from 6:00 a.m. to 11:00 a.m. between 13 and 18 WLED lights were on, or on an average about 2 WLED lamps per household. Likewise in the early evening after sunset, from 17:00 p.m. to 22:00 p.m. 15–18 WLED lamps were in use. As expected, after 22:00 p.m. more and more lamps are switched off, with 5–10 WLED lamps, or an average of one WLED lamp per household, remaining switched on all night long, until the early morning hours at 05:00 a.m.

While all night usage was not included in the original load pattern design, we learned that people like to sleep with one WLED lamp on. The majority of people in this region follow a shamanistic folk religion, and believe that evil spirits roam through the dark nights and do not visit even dimly lit areas. Thus the WLED lamps provide a certain security from fear of being visited and possibly even harmed by an evil spirit.

Humli houses are very dark inside, and the yellow line shows that an average of 4 lamps were in use in various homes throughout the daytime during the period measured. Other important results, such as the efficiency of the solar PV equipment, can be calculated from the recorded data. The performance ratio (PR) of the whole solar PV cluster system can be calculated as follows (IEA 2000, 2007; Mayer and Heidenreich 2003):

$$PR = \frac{E_{Load}(Ah) \times V_{mean\,charge}(V)}{H_{ref}(Wh/m^2/day) \times \eta_{STC}(\%) \times A_{Module}(m^2)} = \% \tag{3.1}$$

This calculation takes the daily load current E_{Load} for the WLED lamps (21 Ah/day) into account at a defined $V_{meancharge}$ (mean value of charge voltage of 13.5 VDC). This value is divided by the product of the available daily global solar radiation H_{ref} on the POA (6,337 Wh/m^2), multiplied by the Standard Test Conditions (STC) BP275F PV module efficiency (12.04 %) and the PV module's area A_{Module} (0.630 m^2). Thus the PR for the solar PV cluster system for the 21st November 2006 was 59 %, which we consider to be satisfactory under the recorded circumstances (full battery-bank and thus trickle charging). Because the system is designed to provide for 95 % of user's load demand even during the poorer solar radiation days of the monsoon season, it is rather oversized for the bright and sunny November days, resulting in a slightly lower PR factor.

In addition to the graphs shown above, each of the parameters we monitor can be plotted against one another in relevant and indicative ways. In this way one can understand in great detail how the solar PV system and the user interact with each other and how the system performs under the given local climatic and geographical circumstances. That provides insight into how the system is actually functioning under real field conditions, which is usually quite different from ideal lab conditions or STC under which each type of solar PV module is tested. Monitoring the equipment this carefully also enables us to understand the system losses incurred through non STC conditions, non ideal MPP power generation, PV module power generation decrease due to increased cell temperatures, battery storage losses and the influence of dirt or shading on the PV array.

Further, monitoring the system allows us to record the load demand for the WLED indoor lighting of the consumers and will allow us to predict the load demand growth pattern in the years to come, as the users experience electric lighting for the first time in their lives. Thus we are able to study how people learn for the first time to use electric light in their homes and to analyse important anthropological issues for a deeper understanding of the "software" issues of a culture as it adopts and adjusts to a new, life altering technology.

The monitoring system of all three different solar PV system approaches will continue to provide very valuable insight, not only into the technical performance of the equipment we install, but also how people learn to live with electric light either as an individual family, a group of families or as a whole village. As monitoring continues, we expect that there will be important lessons to learn about changing social patterns, which will have an impact on the long-term sustainability and suitability of each of the three systems described above.

Can a village community get organised in order to maintain their communally owned solar PV system properly, or is it easier for groups of people to maintain a cluster system? How much does the personal pride of a single SHS system owner drive him to carefully maintain the system so that he does not lose face with a neighbour whose system is running well/better? These and other "software" issues of a project with critical impact on its long-term sustainability are likely to be revealed as we continue to monitor.

Undoubtedly, monitoring PV systems costs time and money. The initial monitoring system design and equipment cost is very high, and the ongoing time demand to support the systems, analyse and present the data is significant. Every month the three monitoring systems are visited, data are downloaded and the systems are checked. Frequently, our staff have to trouble shoot problems that arise. Data have to be carefully checked and cross checked by field staff for inconsistencies and errors.

Time spent on these tasks is saved for our field staff since realistic benchmarks were defined for each parameter, in order to help staff immediately identify erroneous readings. There is no short cut in learning to understand how field based solar PV systems perform and interact with their users. But the final outcome is worth the effort in regard to new scientific and applied knowledge and it is also invaluable for understanding how PV systems are accepted in different cultural settings, and how our services and collaborative projects with local people can become more contextualised, appropriate and sustainable.

3.11.2 Tulin Village Central Solar PV System Data Monitoring

The second solar PV data monitoring example is in Tulin which is located at latitude 29°59′23.48″ North and a longitude 81°46′57.05″ East, 2,377 m.a.s.l. It is a 4 h walk along the trekking route to Tibet from Simikot. Twenty-nine households of the village are electrified with a 300 W_p, 2-axis, self tracking, solar PV system powering a 300 Ah, 24 VDC deep-cycle battery-bank with an autonomy period (days without sunshine) of 3 days. Each house is connected through underground cabling to the battery-bank and has 3 WLED lamps installed, consuming 1 W each (Fig. 3.111).

The eleven main parameters of the 22 measured and recorded parameters to determine the status and the performance of the solar PV village system and its equipment are presented in Table 3.3. Again, a DT80 data logger is used to record and store the data (Figs. 3.112, 3.113). The measured parameters are identified in Table 3.3.

Fig. 3.111 Tulin village with 29 households and with the 2-axis tracking central PV system

Table 3.3 Main parameters measured and equipment/type of sensors used

S. no	Parameters	Sensor type	Units
1.	Ambient temperature	T-type thermocouple	°C
2.	Battery-bank temperature	T-type thermocouple	°C
3.	PV Cell temperature	T-type thermocouple	°C
4.	Horizontal global solar radiation	80SPC, SolData Pyranometer	W/m^2
5.	Global solar radiation on the plane of array (POA)	80SPC, SolData Pyranometer	W/m^2
6.	Solar PV array current	HXS 20-NP[a] Current Transducers	Amperes
7.	Solar PV array voltage	Direct to data logger DT80 connected	DC volts
8.	Battery-bank voltage	Direct to data logger DT80 connected	DC volts
9.	Battery-bank current in/out	HXS 20-NP Current Transducers	Amperes
10.	WLED load voltage	Direct to data logger DT80 connected	DC volts
11.	WLED load current	HXS 20-NP Current Transducers	Amperes

[a] HXS 20-NP Current Transducers, http://www.lem.com/hq/en/content/view/148/126/ (accessed July 26, 2012)

The DT80 data logger system is installed in the power house, along with the battery-bank and the main analogue volt and ampere indicators (Fig. 3.113, top left corner).

The dataTaker DT80 samples each of the parameters every 10 s and averages them over 1 min. The averaged values are recorded in the DT80's internal memory under three different time schedules:

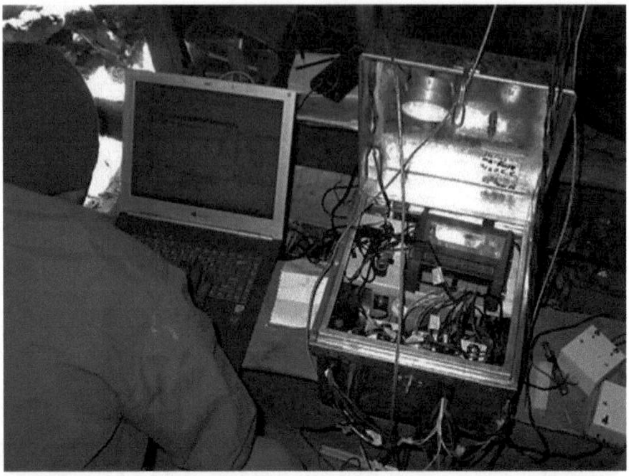

Fig. 3.112 Sensors connected to the DT80 data logger. A special protective casing was fabricated for the monitoring equipment. The DT80 is separately powered with two 20 W_p Solar PV modules and a 30 Ah sealed deep-cycle battery. Care needs to be taken that the mice can't easily get access to the wires

Fig. 3.113 Installation of the data logging system in Tulin village. The program, specifically written for the data logger DT80, is loaded and tested. The sensors' cabling is carefully connected to the data logger

- 10 min (averaged)
- 1 hour (averaged)
- 1 day (24 hours averaged)

Monthly, the data is downloaded from the DT80 to a USB memory stick (Figs. 2.19, 3.109).

3.11.3 Analysis

3.11.3.1 Locally Available Solar Energy Resource

An important parameter for the design of a solar PV system is the peak sunshine hours per day (PSH) over the course of each month of the year. The global solar

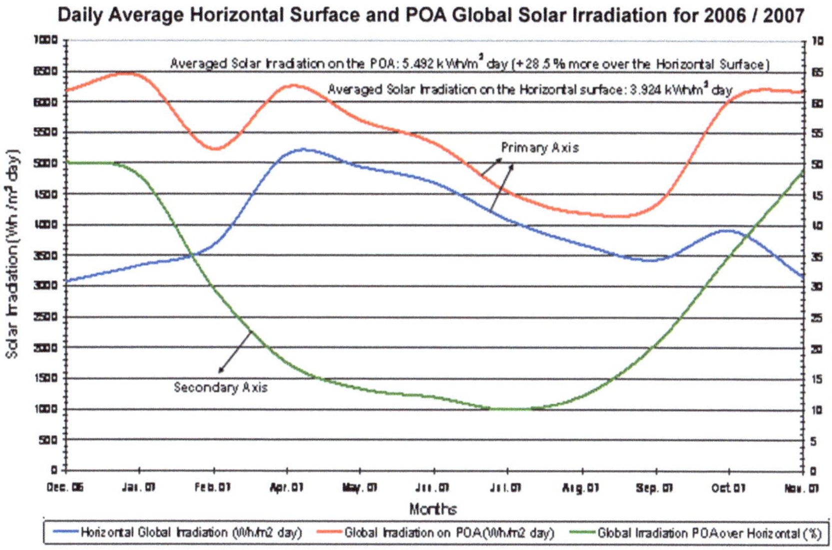

Fig. 3.114 Daily average horizontal and POA global solar irradiation per month for the year 2006/2007

radiation in Tulin is measured on two surfaces, the POA, representing the received global solar radiation on the solar PV modules in the array, and the horizontal global solar radiation, which is the international standard, allowing a comparison with other international data. These two parameters are measured each with a pyranometer, a calibrated silicon solar cell, SolData 80SPC,[43] with an accuracy of ±3 % against a thermopile Kipp&Zonen CM-21 pyranometer.[44]

Figure 3.114 shows the averaged daily global solar irradiation (Wh/m^2 per day) received in Tulin village over the period of a year, from December 2006 to November 2007. The important features identified from the data analyses are:

1. The average solar irradiation on the 2-axis tracked surface over the year was 5.492 kWh/m^2 per day.
2. Annual solar irradiation on the tracked surface was 28.5 % more than on the horizontal surface.
3. The maximum solar irradiation on the 2-axis tracked surface over the year was 6.412 kWh/m^2 per day, while the minimum was 4.194 kWh/m^2 per day.

Figure 3.115 shows that Tulin receives fewer hours of sunlight due to the high surrounding mountains. Sunlight is especially limited during the 4 winter months

[43] SolData Pyranometer 80SPC, http://www.soldata.dk/PDF/pyrano%2080spc%20A4%20-%20UK.pdf (accessed July 26, 2012).

[44] Kipp&Zonen CM 21, http://www.kippzonen.com/?product/1491/CMP+21.aspx (accessed July 26, 2012).

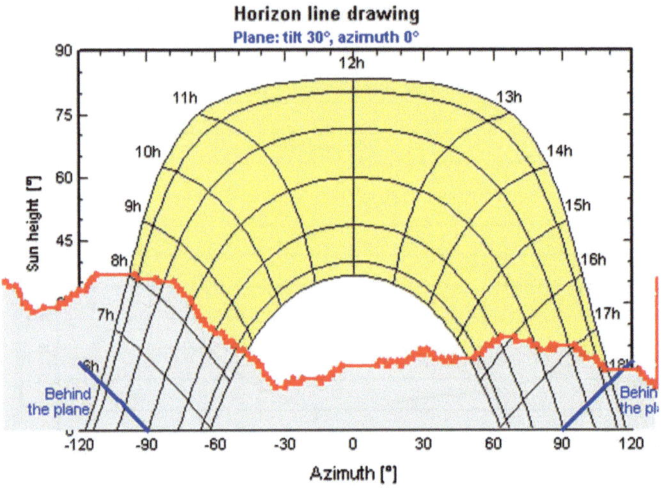

Fig. 3.115 Tulin's high horizon (360° horizon around Tulin village)

from November to February. Due to the clustered arrangement of the houses (Fig. 3.50) and limited peak sun hours, a 2-axis central tracking PV system was chosen.

Tracking from East–West is achieved with a bi-directional 1.5 VDC motor, powered by the relative voltage difference between two 2.5 W_p solar modules positioned at 30° to each other (Figs. 3.41, 3.49, 3.116, 3.118, 3.124). Back tracking to East in the morning is done by a 2.5 W_p module positioned at the East side of the PV array frame (Figs. 3.40, 3.117). The second axis, the North–South axis, is manually adjusted from 5° to 60° (Table 3.4), depending on the seasonal variation of the sun-path angle (Figs. 3.40, 3.117).

The increase in solar radiation in the POA shows the benefit of our locally developed and manufactured 2-axis tracking system (Figs. 3.38, 3.39, 3.40, 3.41, 3.49, 3.65, 3.111, 3.116, 3.117, 3.118, 3.124). Periodic follow-up visits during the period under study revealed, that the monthly adjustment of the North–South tracker was not routinely carried out by the locally trained users. If adjustment had been carried out as instructed (Table 3.5), it could have increased the system's annual solar reception by an additional 8–12 % (Sect. 3.12.2.2). This demonstrates the necessity for routine staff training as well as support to the end-users in the villages.

In Fig. 3.116 the various components of the Tulin 300 W_p 2-axis Solar PV Tracking System are indicated. Also it is important to notice that the central tracking system is installed without any cement due to the prohibitive transport cost (13 fold price compared to the purchase price) and thus the inability of the local people to maintain it if in the future additional cement is needed.

Instead the frame is developed in such a way that it can be filled with stones, which are widely and easily available in the village. Further, the PV system does

Fig. 3.116 Tulin 300 W_p central PV System. Two 2.5 W_p solar modules positioned at 30° to each other power the 1.5 VDC motor which turns the tracker daily from East–West according to the sun's daily path

Fig. 3.117 While the sun's daily East–West path is automatically tracked, the tracker's seasonal elevation is manually adjusted fortnightly. This allows a much simpler technology to be applied for the remote place, increasing the sustainability while minimising the maintenance requirements

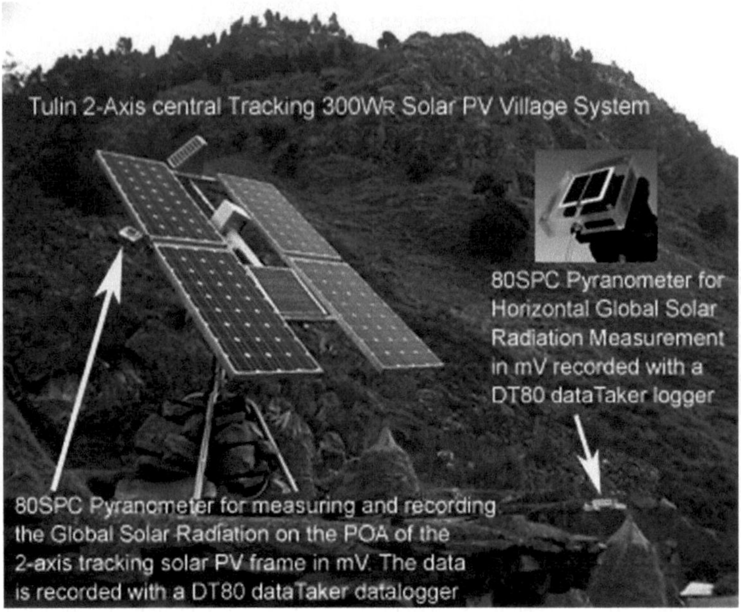

Fig. 3.118 Tulin's 300 W_p solar PV village system with a 2-axis tracker

not take much space away from the already utilised and limited roof top. Thus the tracker system integrates seamlessly with the daily tasks such as drying grass, vegetables and fruits as well as the children's games. This is part of the development of contextualised technologies.

3.11.3.2 Power Generation

Tulin's solar PV village system has 4 × 75 W_p BP275F Si mono-crystalline solar PV modules installed. The system's voltage is 24VDC, and 300 W_p peak power output.

Figure 3.119 shows the daily average energy generation, Array[45] and Reference Yield[46] over the year. The important features of the Solar PV array performance identified from the data analysis are as following:

---

[45] The Array Yield is defined as the ratio of the daily array energy generation to the peak installed capacity of the system, expressed in kWh/kW_R per day.

[46] The Reference Yield is the total daily solar insolation (kWh/m^2) intercepted in the plane of the array (POA), divided by the array reference irradiance, which is the STC value of 1,000 W/m^2. Thus, it represents the equivalent number of peak sun hours (PSH) a day (number of hours the sun shines per day at 1,000 W/m^2). Therefore, the Reference Yield is equal to the number of peak sun hours. Hence the Reference Yield is calculated from: Reference Yield = Total plane of array insolation (kWh/m^2)/1 kW/m^2

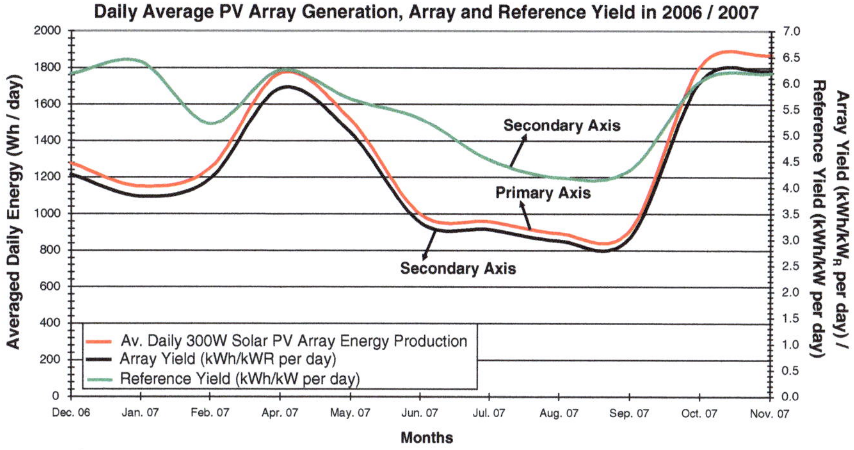

Fig. 3.119 Daily average 300 W_p solar PV array energy generation, array and reference yield for each month from December 2006–November 2007

1. On an average 1.310 kWh of energy per day was generated over the course of a year. During this period the PV system achieved an overall average efficiency[47] of 9.47 %.[48]
2. A minimum of 0.894 kWh of energy per day was generated during August which is the worst month due to the monsoon climate, during which it can rain for days at a time. The average minimum generation is well over the maximum average load described in one of the following sections.
3. A maximum of 1.871 kWh of energy per day was generated during one of the clear, sunny days in November.
4. A clear seasonal energy generation pattern can be seen due to the pertaining seasonal, climatic and meteorological pattern. This is particularly evident when the lower energy generation is observed during the monsoon season and snowfall, from end of May–September and December-January respectively.

[47] The overall average Solar PV Array Efficiency = Power Generated by the PV Array/(Solar Irradiation × Total Area of the Array).

[48] The calculated long-term efficiency of 9.47 % of the 4 serial- and parallel-connected BP275 solar PV modules (to make up the 300 W, 24VDC system) shows, in contrast to a single BP275 solar PV module at STC with a conversion efficiency of 12.04 %, a 2.57 % real efficiency drop, which is a 21.35 % relative drop in the solar PV modules' power generation efficiency due to non-STC conditions and module mismatch losses measured under real field conditions. In addition, the high capacity status of the battery-bank throughout the year increases the potential energy generation losses, thus indicating a lower average solar PV power conversion efficiency, as well. That shows that the local meteorological conditions and the technical performance of the equipment need to be known for the proper design of the whole system. These values are important to know and include in the initial design of new village solar PV systems, in a comparable context, if one aims to install systems with a high availability, minimal LoL (Loss of Load) and high sustainability.

Spring (mid February–May) and early winter (October–November) seasons are sunny with clear skies, consequently generating more energy per day.

3.11.3.3 Energy Storage

A well designed battery-bank energy storage unit is essential in a RAPS PV system which is not connected to the grid, as the power generation is inherently intermittent. Tulin village has a battery-bank with a capacity of 300 Ah @ 24 VDC (6 × 12 VDC, 100Ah deep-cycle batteries). The battery-bank is designed for 3 days of autonomy,[49] which means that the system can meet the average total village load demand without sunshine for up to three consecutive days with a maximum DoD of 30 %.

The battery-bank in Tulin remains on a constant high state of charge (SOC) each day, as the average daily DoD measured is low, between 3.29 and 8.85 %. The charge-controller switches to trickle charge by open circuiting the PV array, as the battery-bank capacity reaches >90–95 %. This keeps the battery-bank protected from being overcharged. Thus the PV array power output is not fully utilised in relationship to what it could produce under the prevailing meteorological conditions.

Over the year the array efficiency was recorded to be 9.47 % with a maximum of 11.89 % in October. A lot of energy is being lost due to the high capacity status of the battery-bank. This is expected for the first few years of the Tulin PV system as the system was designed for a 10 years maximum battery-bank life-expectancy. Thus the system and battery-bank is designed to provide the full energy demand by the local community after 5–6 years of population growth and increase in energy requirement.

Figure 3.120 shows the average battery-bank voltage and the battery-bank's average DoD per month over the year. The important features of the battery-bank performance identified from the data analysis are as follows:

1. The battery-bank's voltage never dropped below 25.84 VDC, with a yearly average of 26.27 VDC.
2. Over the course of the year on average 0.939 kWh energy per day, generated by the PV array is fed into the battery-bank.

---

[49] The Australian Standard AS 4509.2-2002 Stand-alone power systems—System design guidelines, Chapter 3.4.7.8 PV, wind or hybrid systems with no genset, page 35, recommends that stand-alone solar PV systems are designed with a typical autonomy of 5–10 days IF 100 % of energy is generated by solar energy. This, of course is dependent on the climatic conditions at the site. However, for the remote, high-altitude places in the Himalayas, an autonomy of 3 days has been chosen due to the fact that very low power consuming loads, only WLED lamps, are used, and the batteries have to be air lifted into the area (which makes them very expensive) and then carried by porters to the actual project site. Also, in the high-altitude areas of the Himalayas it is very rare that the sun is not shining for more than 2–3 days, due to the prevailing high-altitude climate. These factors, as well as considering the economic situation of these impoverished communities, justify a 3 day autonomy assumption for the battery-bank design.

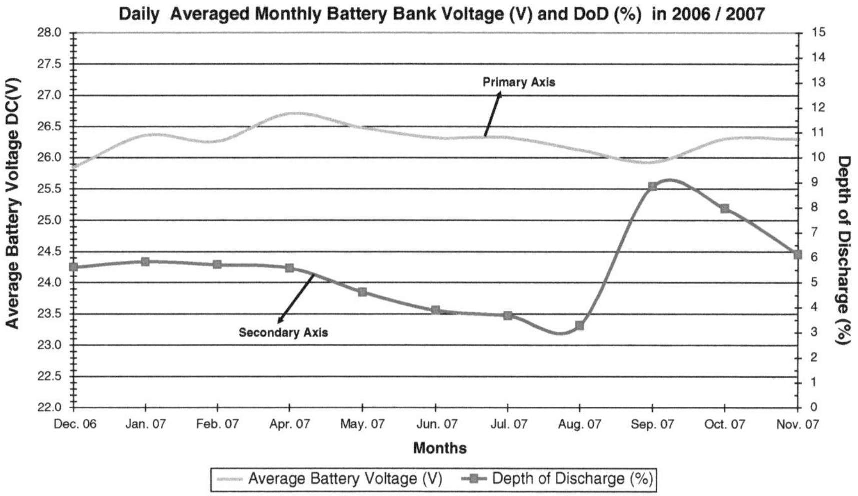

Fig. 3.120 Daily average battery-bank voltage and depth of discharge per month for 2006/2007

3. The maximum energy input into the battery-bank recorded was 1.510 kWh/day in November, whereas the minimum was 0.613 kWh/day during the monsoon month of August.
4. The daily DoD, on average, was 5.57 % over the year. The maximum daily DoD was 8.85 % in September, with a minimum of 3.29 % in August.
5. The DoD trend reflects the living pattern of the villagers. During the summer months (June–August) some of the villagers move to the higher altitude (3,000–4,500 m.a.s.l) meadows with their animals. This reduces the village population and hence results in lower daily DoD during this period. Once they returned with their livestock in early September, more people were in the village, and the longer winter nights slowly set in. Both of these factors increase the demand for the lights, resulting in higher daily DoD of the battery-bank.

The constant high battery-bank voltage level indicates a good, high SOC of the battery-bank throughout the year. Thus, over the year all the load demands for the designed indoor lighting could be provided without interruption. That means there was 0 % loss of load (LoL) due to insufficient power generation by the solar PV system or insufficient energy storage capacity of the battery-bank.

The low DoD in the initial design has certainly added to the overall system cost, but factors such as expected population growth, days of autonomy period, potential of unauthorised connection of additional loads (or misuse of the PV system) and extended battery-bank life-expectancy were accounted for in the design.

3.11.3.4 Load Demand

Each of the 29 houses is connected through armoured underground cabling (Fig. 3.135) to the battery-bank and has three, 1 W WLED lamps. Thus a total theoretical maximum load of ~100 W (29 households × 3 W plus ~15 % system losses) at any given time is expected.

Figure 3.121 shows the average daily load consumption per month over the year. The important features of the end-usage pattern identified from the data analysis are as follows:

1. 0.562 kWh was calculated as the average daily consumption for indoor lighting over the period of a year.
2. Maximum energy demand of 0.707 kWh/day was recorded in November (during the longer winter months when the lights are on for longer time periods) and a minimum of 0.419 kWh/day in August (during the shorter summer months).
3. While a maximum load demand of 101 W was recorded in August, an even higher load of 225.5 W was once recorded in September. This suggests that an additional, unauthorised (or misuse of the PV system) external load/appliance was connected to the system. An average base load of ~20 W was recorded over the year, suggesting that even during the day people are using some lights. Though daytime usage was not initially anticipated during the project system design phase, it is understandable, since local homes are built with only very small windows with no glass to keep out the cold and wind. Thus people learned quickly the benefits WLED lamps provide during the day time as well as at night. In addition to the darkness of the home interiors leading to increasing usage during the day, the recorded data leads us to believe that some people may leave the lights on all night long to drive away evil sprits.
4. The graph in Fig. 3.121 reflects a clear pattern of the seasonal load demand as previously discussed in Sect. 3.11.3.3, point 5.
5. The data shows that the average daily load profile differs from what was expected. Initially, the solar PV system's load profile was defined to be ~7–8 hours a day, with two peak energy demands, one in the morning hours and one in the evening hours. However, our data suggest that for a few short time periods, additional loads were connected to the system. This could have been for charging additional external batteries for flashlights or for other equipment such as radios. These additional, unauthorised loads can lead to faster discharging of the battery-bank and lower average battery-bank voltage, and thus faster ageing of the battery-bank. Also the PV system is exposed to a higher risk of short circuits or the destruction of system parts. This information was shared with the villagers, and they agreed to look out for any misuse and to put an end to such activities. For future system designers this experience serves as an indication of the necessity to include unforeseen loads in the system design and to include necessary precautions to protect the system from unanticipated high loads. This could be achieved by including electronic power limiting circuit

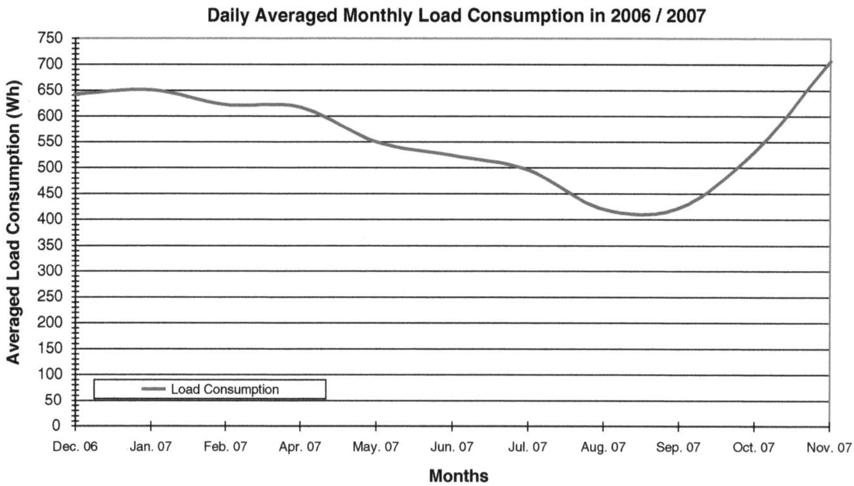

Fig. 3.121 Daily average Tulin village electricity load (Wh) per month for the year 2006/2007

breakers or relevant low voltage recognition equipment which, once a set threshold is reached, cut the village power supply.

3.11.3.5 System Performance Ratio (PR)

The PR is used to indicate the overall effect of losses on the solar PV array's rated output under defined meteorological conditions. Losses are induced due to the solar PV modules' increased (>25 °C) temperature, only partial utilisation of the available solar irradiation (due to a full battery-bank, inaccurate PV module angle or shading etc.), system component inefficiencies and cable losses or system equipment failures (IEA 2000).

The Performance Ratio is the ratio of Final Yield[50] to the Reference Yield i.e.

$$PR = \frac{Final\,Yield\left(kWh\,per\,Day/_{kW_P}\right)}{Reference\,Yield\left(kWh/m^2 per\,Day/_{1kW/m^2}\right)} \tag{3.2}$$

Figure 3.122 shows the daily average system PR per month from December 2006 to November 2007. Some key performance features of the system are:

[50] The Final Yield is defined as the ratio of the daily load to the peak installed capacity of the system, expressed in $kWh/day/kW_R$

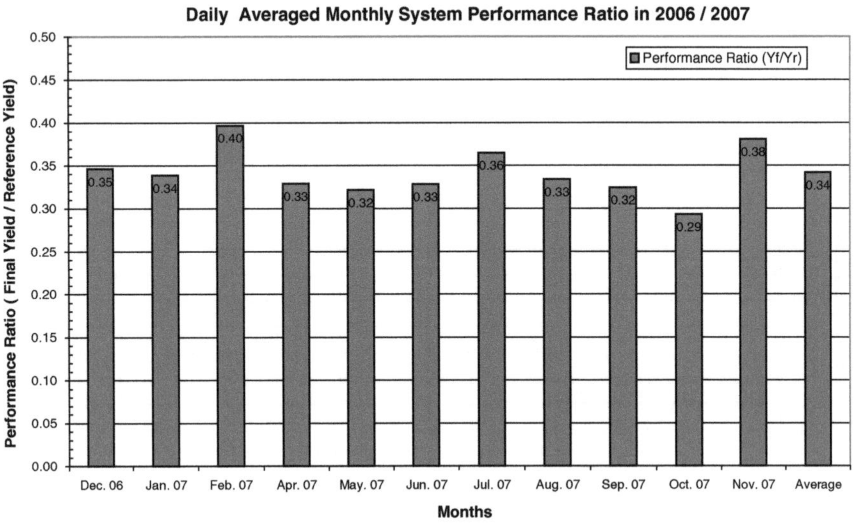

Fig. 3.122 Daily average Performance Ratio per month for 2006/2007

1. A maximum PR value of 0.40 was recorded for the month of February. This value shows lower capture losses[51] compared to other months, as the sun is not yet as strong and the sun hours not as long as during the later spring months. Further, the February winter month nights are still long, corresponding with higher load.
2. A minimum PR value of 0.29 was recorded on October. This is a result of higher energy capture losses as higher daily PSH values are recorded in October, resulting in surplus energy generation and the battery-bank being often >90 % full. The DoD graph in Fig. 3.120 is not directly indicative of this finding, since we suspect that in the month of September extra loads were plugged into the system, discharging the battery-bank more than otherwise would have been the case. As the PR graph presents the monthly average values, the October value is not necessarily indicative of the lowest PR value over the year.
3. On a yearly average the PR is 0.34. From an analysis of 260 systems, the IEA-PVPS T2-01 report (IEA 2000) mentions that the annual PR of stand-alone (no backup) solar PV systems ranges from 0.1 to 0.6. The Tulin village solar PV system is a prototype and was designed for minimum LoL in order to have continuous performance data and uninterrupted indoor lighting for the local users. Due to these demands set for the system at the design stage, we expected a slightly decreased average PR value and an increased total system cost.

[51] Array capture losses incur mainly when the battery-bank is nearly full (90–95 % SOC) or full, as then the charge controller open circuits the PV array power flow to the battery-bank. Thus the available excess power cannot be utilised and thus is considered as loss. Other capture losses include PV module temperature incurred losses and shading losses.

The PR value is dependent on the load usage. An oversized RAPS Solar PV System as in this case has higher system losses which result from a more frequent disconnection of the PV array from the fully charged battery-bank. Correspondingly, it is not entirely correct to say that a system with a higher PR value indicates a "better" system. The Tulin village solar PV system was intentionally oversized to allow higher reliability, lower LoL, and a population and load demand growth over the course of its first 5–6 years of operation. Further, it is a first time experience for this community to have basic electric indoor lighting. With these parameters under consideration, it is expected that the performance ratio of the Tulin solar PV system will improve over time.

3.11.3.6 System Efficiency

The PR value in itself is not sufficient to give the overall picture of the system's operation. Hence additional parameters have to be monitored for a more complete understanding of the system. One such additional parameter is the system efficiency, which can be calculated by specifying the ratio of solar PV array power generation and load at standard test conditions (STC).

$$\text{SystemEfficiency} = \text{FinalYield}/\text{ArrayYield} \tag{3.3}$$

Accordingly the average system efficiency of the Tulin solar PV system over the year was 78.8 %. On an analysis based on more than 30 stand-alone systems installed worldwide, with peak power varying from 450 to 5,000 Wp, Mayer and Heidenreich conclude that "good" systems are those with efficiencies ranging from 75 to 95 % and with a PR greater than 0.3. A high efficiency value indicates few or no technical problems in a system (Mayer and Heidenreich 2003). Computing the system efficiency factor along with the PR value provides a fuller picture of the system's operation.

3.11.3.7 Summary of the Tulin and Pamlatum Village Power Projects

Comprehensive monitoring is essential in order to determine a system's performance under real-field conditions. For instance the Tulin solar PV array under study in this project was found to be operating at an overall efficiency (generated electrical power by the solar PV array surface per m^2 divided by the received global solar radiation per m^2) of 9.47 % over the year. Slightly higher future efficiency values are expected with the increase of the village load demand compared to the current average load.

The local geographical and meteorological conditions justify the use of a solar PV array tracking system with the POA results showing that 28.5 % more energy is captured with the POA than the horizontal array. Furthermore, the monitoring system provides a detailed understanding of the varying village load patterns over the course of each day, month and season of the year. From these findings, important lessons can be learnt.

The monitored data also indicate that the village solar PV systems may occasionally have been misused, which otherwise would have not been possible to know. One of the lessons learned is the importance of incorporating a load disconnection unit to protect the system from higher than designed currents being drawn. Additionally, the data revealed that the battery-bank needs to be designed with appropriate energy storage capacity, in order to provide the villagers with the energy they need during the predicted number of days without sunshine.

Overall, the data collected by the monitoring systems show that the Tulin and Pamlatum solar PV systems are performing well in general, as designed and as expected. Crucial parameters such as the locally available solar resource, the solar PV array's power generation, the battery-bank's daily voltage, SOC and energy storage levels, as well as the village load consumption, were monitored and recorded. These data give an excellent overall understanding of the system's performance, providing a detailed record of each individual parameter and the operation of the system overall. This would not be possible without a data logging system, which is lacking in solar PV systems installed in similar contexts throughout the region.

Monitoring of a solar PV system under various meteorological conditions over the course of several years allows for its improvement and optimisation through adjustments made on the initial design calculations. These adjustments then help to avoid unnecessary system failures, and improve economic efficiency of new PV systems based on real-field experience.

Careful monitoring and system adjustment in this fashion helps to improve the delivery of quality energy service to the consumers by minimising any system downtime. Further, the data and knowledge gained from these experiences, appropriately contextualised, can provide the development community with important tools to improve the design of solar PV systems for particular contexts. This is beneficial for end-users, for equipment manufacturers, and for researchers interested in solar PV RAPS systems.

A professional and contextualised solar PV system design for a defined local context and application is essential for long-term reliability and sustainability. This information can only be acquired through detailed and long-term data collection and analysis for whole solar PV systems in the installed context. To date, the long-term performance of the small scale solar PV systems installed elsewhere in Nepal has been inconsistent. The approach described above offers a solution to this situation. Eventually the carefully monitored and optimised project approach described is able to transform the performance and sustainability of other systems in regions throughout Nepal.

3.12 Identification and Evaluation of the Losses Occurring in a Solar PV System under Field Conditions in Humla, Nepal

The DC rated output of each solar PV module is provided by the PV manufacturer in the form of the values measured under STC. While this defined environment allows the direct comparison of different solar PV modules, it hardly ever reflects the field environment a solar PV module is exposed to.

Further, based on the local solar radiation received, installation issues such as the correct array orientation, tilt angle and partial shading considerations play a crucial role in the overall power output of a solar PV system.

In most cases, the understanding of the real field conditions in a given environment, as well as the main losses occurring in a solar PV system (e.g. due to the solar PV cell temperatures, PV module mismatches, wrong battery-bank usage etc.) are lacking, or not discussed or considered in the design of a solar PV system prior to its installation.

In the present study, in particular, the losses occurred due to non-standard temperature conditions, non-ideal solar PV module angle position, increased solar PV cell temperature, solar PV module mismatch, and battery-bank and wire losses are considered. These are presented and discussed with recorded field data, graphs and photos. This will help us to achieve a more responsible, professional and contextualised design of a solar PV system, with an increased assurance of it performing to the end-users' expectations and satisfaction. It assists the design and choice of technologies for a village solar PV system in a defined context.

3.12.1 RIDS-Nepal's Simikot Office Solar PV System Definition

The solar PV system investigated is installed at RIDS-Nepal's Simikot office in Humla, as shown in Fig. 3.123 (Latitude: 29°58′22.07″ North, Longitude: 81°49′05.63″ East, Altitude 3,000 m.a.s.l.).[52] The office is powered by a 900 W_p PV system, consisting of three 300 W_p, 2-axis Solar PV tracking systems.

[52] The dataTaker DT605 was one of the main data logger models that dataTaker marketed until 2006, when the DT80 was introduced, replacing the DT605. Thus there is no web site anymore available for the DT605 model. However, a technical specification sheet for the DT605 can still be downloaded from ITS (Industrial Temperature Sensors), a distributor of dataTaker products. The web site can be visited at: http://www.itsirl.com/Product_Data_Sheets/DT500600_TS0056C1.pdf (accessed July 26, 2012)

Table 3.4 Main technical specifications for one BP275F Solar PV Module

Nominal peak power (Pmax)	75.00 W
Voltage @ Maximum Power (Vmpp)	17.00 VDC
Open Circuit Voltage (Voc)	21.40 VDC
Current @ Maximum Power (Impp)	4.45 A
Short Circuit Current (Isc)	4.75 A

Fig. 3.123 RIDS-Nepal Simikot office in Humla Nepal at 3,000 m altitude, with a 900 W_p monitored Solar PV system. The system includes a DT605 dataTaker and a CEM20 extension module, monitoring in detail its performance throughout the years since 2004

Fig. 3.124 A 2-Axis Solar Tracker as used in the RIDS-Nepal office in Simikot, Humla, each with four BP275F solar PV modules

While the sun's daily east–west path is tracked automatically, the sun's seasonal angle at the horizon is adjusted bi-weekly manually. In this way the PV modules are exposed almost all the time perpendicular to the sun's incoming radiation. The PV system provides AC output from an inverter.

Each tracker is composed of 4 solar PV modules. Two modules are connected in series and two in parallel so that the maximum power point current of the array (I_{mpp}) is 8.90 A and the maximum power point voltage (V_{mpp}) is 34.00 VDC at STC, generating 300 W_p per tracker. The three trackers are connected in parallel, thus generating at the maximum power point (MPP) 26.70 A at 34.00 VDC, delivering around 900 W_p under STC.

In total, 12 × 75 W_p BP275F PV modules (Fig. 3.124) are used to power the battery-bank which consists of 16 × 100 Ah @ 12 VDC flooded lead-acid deep-cycle batteries, connected as a 24 VDC system. Thus a total energy capacity of 19.2 kWh is available (Table 3.4 and Fig. 3.124).

3.12.2 A Solar PV System's Key System Losses

The following analysis is based on averaged hourly data recorded in 2006/2007 for all the monitored parameters including soil temperature data for 2009. The system losses were calculated with reference to the PV module's rating under STC.

3.12.2.1 Solar PV Module Production Losses

The BP275F solar PV module manufacturer guarantees that each module generates a minimum of 70 W_p under STC. This amounts to 5 W or 6.7 % less than its intended peak power. Consequently, for the RIDS-Nepal Simikot Office Solar PV system, the manufacturing tolerance related loss of the whole PV system can amount up to 60 W.

3.12.2.2 Tilt Angle and Sun Tracking

The global solar radiation incident angle on the PV module should be as close to 90° under the sun's daily path as possible, throughout the day. Thus manual or automatic tilt angle adjustment can increase the total incoming solar radiation to electricity conversion efficiency considerably. In Humla, for small, fixed angle solar home systems (16 W_p), the yearly average maximum power output is achieved with a PV module tilt angle of 30–35° south, equalling the location's northern latitude. In order to maximise the power output of the RIDS-Nepal Simikot office PV system, three 2-axis tracking frame systems were installed.

The Sun's daily East–West path is automatically tracked and the seasonal North–South angle manually adjusted on a fortnightly basis. The trackers' manual

Table 3.5 Manually adjusted North–South angle for the RIDS-Nepal Simikot Office PV system trackers according to the month

Month	Degrees south inclined	Remarks
January	50–55°	Latitude +20° to +25°
February	40–45°	Latitude +10° to +15°
March	30–35°	Latitude +0° to +5°
April	20–25°	Latitude −5° to −10°
May	10–15°	Latitude −20° to −15°
June	5–10°	Latitude −25° to −20°
July	10–15°	Latitude −20° to −15°
August	15–20°	Latitude −15° to −10°
September	25–30°	Latitude −5° to +0°
October	35–40°	Latitude +5° to +10°
November	45–50°	Latitude +25° to +30°
December	55–60°	Latitude +25° to +30°

North–South angle adjustments, to comply with the Sun's changing altitude angle (declination) over the year between the horizontal and the axis through the Sun, varies for Humla between 5° and 60°, the minimum and maximum values in June and December respectively. For the manual North–South tracker axis adjustment, which takes less than a minute, the author derived the following values for each month of the year (Table 3.5).

Thus, alongside the daily East to West automatic tracking for the maximum power output throughout the year, the tracking frames' North–South angle can be manually adjusted within the above indicated angles. This allows the PV modules to be exposed daily perpendicular to the sun's varying path over the course of the year. The following two graphs in Figs. 3.125 and 3.126 show the difference of the intercepted global solar radiation measured at the horizontal position (international standard); 30° south inclined (standard angle for fixed installed PV systems in Nepal); and the 2-axis tracking frame, alongside the averaged ambient daily temperature for the years 2006/2007.

The main findings can be summarised in the following points:

1. Over the course of these 2 years, the maximum daily averaged horizontal global solar radiation was 692 W/m^2, the maximum averaged 30° south inclined was 815 W/m^2 and the maximum averaged tracked was 881 W/m^2. The maximum and minimum ambient temperatures recorded were 20.5 and 8.4 °C respectively. This shows clearly, that the available PV module power output is considerably less under real field conditions compared with the STC under which the PV modules are rated.
2. The averaged yearly (over both years) intercepted global solar radiation for the 2-axis tracking system was 7.225 kWh/m^2/day, which is 56.9 % more than received on the horizontal (4.607 kWh/m^2/day) and 31.5 % more than received on the 30° south inclined surface (5.496 kWh/m^2/day).

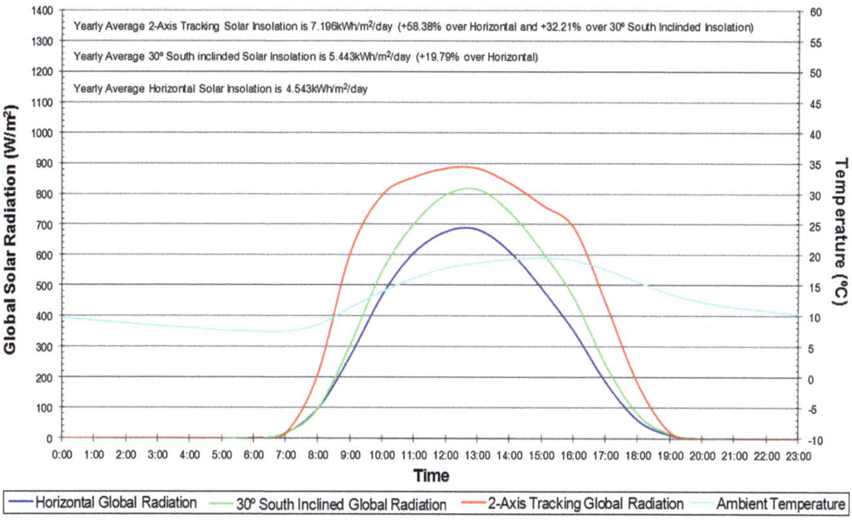

Fig. 3.125 Daily hourly averaged global solar radiation and ambient temperature for the year 2006

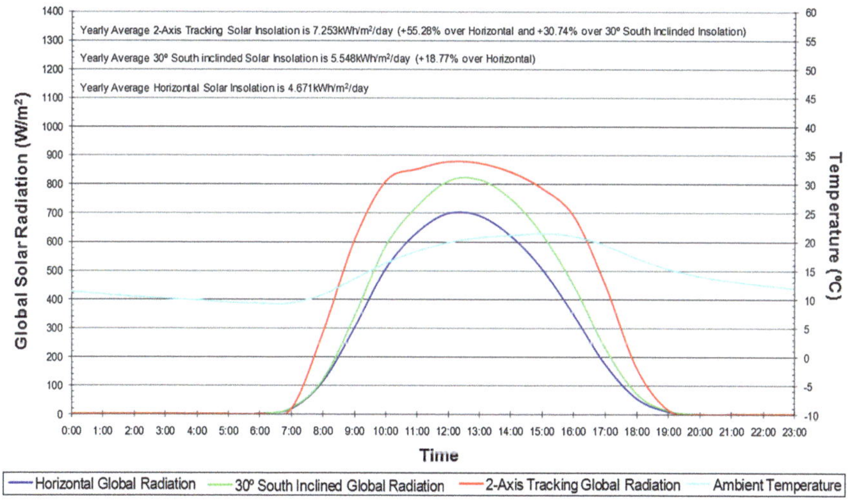

Fig. 3.126 Daily hourly averaged global solar radiation and ambient temperature for the year 2007

From the above data, it can be seen that through the easy periodic manual adjustment of the North–South tilt angle and the automatic East–West tracking system the average daily increase in energy generation over the course of the year is around one third more compared to the present Nepali standard of 30° south inclined, fixed position, solar PV installations.

3.12.2.3 Solar PV Module Temperature related Losses

The cell temperature of the PV array will vary drastically due to ambient conditions such as sun intensity, air temperature, wind speed and other external factors.[53] Most Si crystalline PV modules have a temperature related power output reduction coefficient of around −0.40 %/ °C to −0.45 %/ °C of their rated power output >25 °C (STC).

From the graphs in Fig. 3.127 the following results emerge:

- The PV solar module temperature exceeds the STC temperature of 25 °C on an average from 10:00 a.m. to 4:00 p.m., or for about 6 h per day. Thus during the main sunlight hours each day, some small losses, which are induced through increased PV module temperatures have to be considered in the daily energy generation calculation.

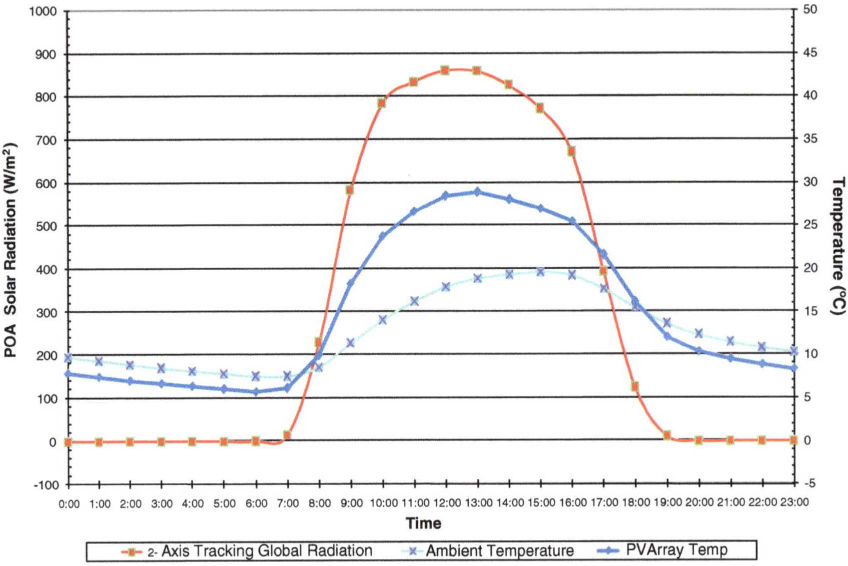

Fig. 3.127 Averaged daily global solar radiation versus ambient and solar PV module temperatures for the combined years 2006/2007

[53] Other external factors which influence the PV module cell temperatures include the soiling (dust, dirt, snow, ice) of the cell, the age of the cell and the serial cell connection wiring, which, based on the PV module's history and exposure can have increased resistances. This can often be identified through a discoloring (turning more dark yellowish) of the cells serial string connections.

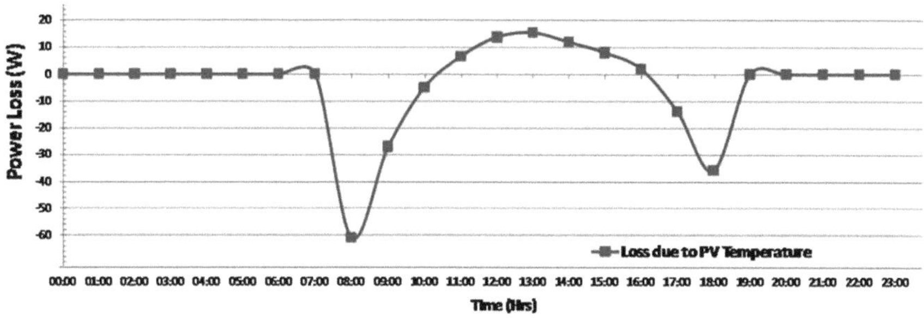

Fig. 3.128 Averaged daily solar PV module temperature induced power generation variations (losses and gains) for the combined years 2006/2007

- As the solar insolation level increases, the ambient temperature rises and so proportionally does the PV module temperature. But in the early and later hours of the day, the PV modules' temperatures are even lower than the ambient temperature. This is because the ambient temperature thermocouples are shielded from the direct sun and wind, while the wind can easily flow through and around the 2-axis frame to keep the PV modules as cool as possible (Figs. 3.116, 3.117, 3.124, 3.118).

Considering that the average daily sunshine hours are from 8:00 a.m. to 6:00 p.m. and the rated output power of the system is 900 W_p, the following findings are apparent from the graph in Fig. 3.128:

1. The temperature related PV losses are negative (power gain) for 2 hours in the morning from 8:00 a.m. to 10:00 a.m. and then losses are positive until 4:00 p.m. as the PV module temperature exceeds 25 °C. Again after 4:00 p.m., the losses are negative until 6:00 p.m. As the graph clearly shows, for most of

[54] In the southern, flat parts of Nepal, with its tropical climate, the ambient temperature frequently reaches 40–45 °C during the 3 months of summer. It is realistic to say that the solar PV module temperature during a sunny day has about a 20–25 °C higher temperature compared to the ambient temperature, of 60–70 °C. With a power reduction temperature coefficient of 0.44 % per °C (as per the manufacturer's specification) above 25 °C (as defined in the STC), e.g. the BP 75 W_p monocrystalline solar PV module (as used by RIDS-Nepal for all its cluster and central tracking village systems) loses 15.4–19.8 %, or 11.55–14.9 W of its 75 W_R power output at STC, due to the increased PV module/cell temperature. These are not insignificant values and thus have to be considered and included in the power/energy generation calculation of the solar PV system. On the contrary, in cold, high-altitude climate areas, such as Humla, where RIDS-Nepal is working (2,500–4,000 m.a.s.l.), during the 4–5 winter months at middays, high solar radiation values up to 1,150 W/m^2, are intercepted by the solar PV module/cell, at PV module/cell temperature not higher than 10 °C. That means, that e.g. the BP 75 W_p monocrystalline solar PV module generates at these conditions 6.6 %, or 4.95 W more power due to the lower than STC module temperature condition, and 15 %, or 11.25 W, more power due to higher than STC intercepted global solar radiation condition. That amounts together to 16.2 W per PV module, or 21.6 %, more power output than its rated power output at STC, due to the local field conditions. Again, these conditions have to be known and need to be taken into consideration during the solar PV system design and calculation as

the time during the day the temperature induced losses are very low (PV module >25 °C) and in the morning and evening even negative (PV module <25 °C). This is because the PV system is installed in the cold, high-altitude location of Simikot. In more temperate or even tropical climates, the power losses due to increased PV module temperature would be much greater and thus are often one of the main loss contributors to a solar PV system's non-standard condition performance.[54] In cold areas such as Humla, where the PV module temperatures are often <25 °C throughout the day, especially through the four cold winter months, Si poly/mono-crystalline PV modules generate overall more energy compared to Si amorphous PV modules. This, because Si poly/mono-crystalline PV modules have in general a greater temperature related power output reduction coefficient compared to Si amorphous PV modules (often in the range of −0.15 %/ °C to −0.25 %/ °C), thus generating more power at <25 °C. Thus, during the shorter winter days, with longer nights, when more energy for lighting is needed, the PV array can generate more power. This important finding, based on the recorded data, shows that it is essential to know and understand the local context for any solar PV system, in order to provide the most appropriate, reliable and cost effective solar PV technology and system design.

2. Si poly/mono-crystalline PV modules are about twice as efficient in converting the received solar radiation into electricity compared to Si amorphous PV modules. That is important as all equipment has to be air transported and carried by porters in Humla. Thus weight, which is therefore significantly lower for Si poly/mono-crystalline PV modules per W_p, plays an important role.

3.12.2.4 Battery Losses

The RIDS-Nepal Simikot Office PV system has a battery-bank with a total energy storage capacity of 19.2 kWh. It consists of 16 flooded lead-acid, deep-cycle batteries, manufactured in Bangladesh by the company Volta.

Each battery produces 12 VDC and has a capacity of 100 Ah (C_{20}).[55] Being a 24VDC system, the battery-bank consists of 2 battery strings of 8 parallel and 2 series connected batteries.

(Footnote 54 continued)
the system cabling sizes and fuses have to be increased accordingly (at least 1.2 times the maximum current) in order to carry the increased currents without jeopardising the system's security.

[55] The "C" stands for the Capacity in Ah (ampere-hours) of a battery. The "20" indicates the amount of hours the battery has been discharged with a constant load (ampere) at an ambient temperature of 20 °C, so that the battery is considered flat or empty (according to the manufacturer's specification, which is most commonly for flooded lead-acid batteries 1.75 V per cell) after 20 h. The total amount of Ah discharged specifies the battery's capacity (Ah) under these defined circumstances.

Table 3.6 Technical specification of the Volta, solar deep-cycle batteries used in the RIDS-Nepal Simikot Office

Manufacturer	Volta
Country	Bangladesh
Model	6SB90
Type	Flooded lead-acid
Rated voltage (VDC)	12
Capacity (Ah)	100
C value	C_{20}
Temperature	20 °C

Fig. 3.129 Similar battery-bank

The detailed specifications are given in Table 3.6 and Fig. 3.129.

From Fig. 3.130, averaged hourly graphs, the following findings can be extracted:

[56] The daily 15 % energy loss of the battery-bank can be explained as follows. The higher voltage during the battery-bank's charging process (on a daily average ~13–13.5 V) compared to the lower discharge voltage (on a daily average ~12–12.5 V), the battery-bank's internal losses due to the chemical reactions, the increased temperatures occurring during the battery-bank's charging and discharging processes, and the losses occurring due to the resistances of the batteries' terminals and wires connected to the battery-bank.

[57] The Coulombic battery efficiency calculation takes only the Ah (ampere-hour) or capacity into and out of the battery-bank into consideration, avoiding the voltage difference losses of the battery-bank's higher charging voltage and lower discharging voltage. Thus the Coloumbic battery efficiency value presents the battery-bank's technical performance efficiency, which includes the battery-bank's internal losses due to the chemical reactions, the increased temperatures occurring during the battery-bank's charging and discharging processes, and the losses occurring due to the resistances of the batteries' terminals and wires connected to the battery-bank.

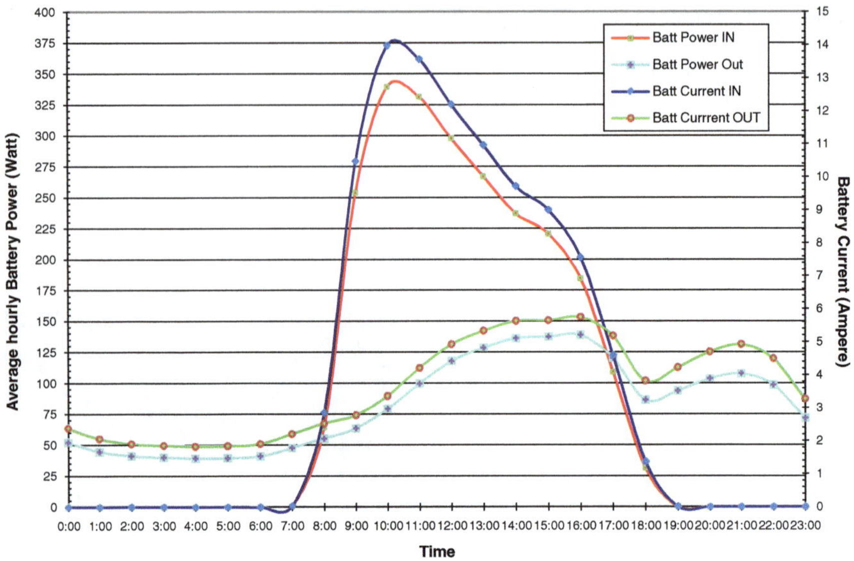

Fig. 3.130 Power in and out of the battery-bank for the combined years 2006/2007

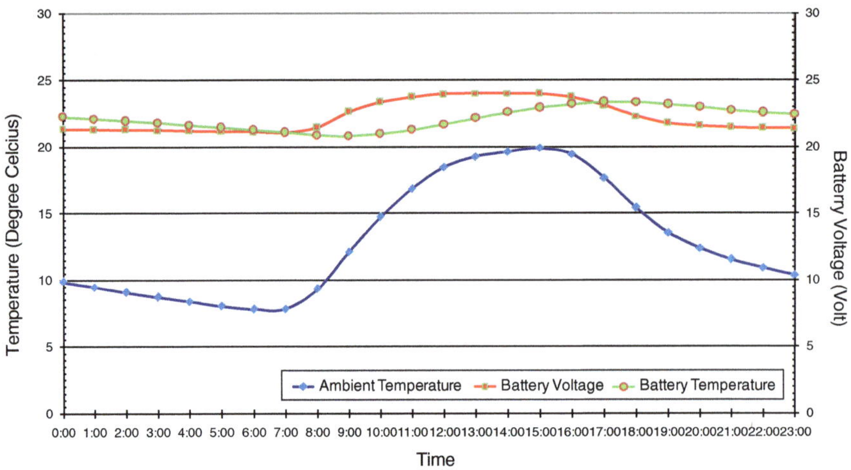

Fig. 3.131 Daily averaged hourly battery-bank Voltage vs. Temperature for 2006/2007 combined

1. An average daily energy of 2,344 Wh is fed into the battery-bank while an average daily energy output from the battery-bank is recorded as 1,992 Wh. Hence the battery-bank's overall energy input to output efficiency is 85 %, with an average daily energy loss[56] of 15 %.

2. An average daily current flow of 96.5 Ah goes into the battery-bank while an average daily current output of 87.05 Ah per day is withdrawn from the battery-bank. Hence the Coulombic battery efficiency[57] is 90 %.
3. The battery power and current graphs show that the evening and night time power demand is significant with over 100 W between 8:00 p.m. and 9:00 p.m. Thus in order to provide the peak evening demand as well as the overnight energy demand with an acceptable average daily maximum battery-bank DoD ≤10 %, the battery-bank needs to be of large capacity.

From the graphs in Fig. 3.131 the following can be observed:

1. On average the battery-bank's voltage over the year was between 22.28 VDC and 24.06 VDC, which is not ideal, as the battery-bank's average voltage rating should not be below 24.00 VDC.
2. A clear daily pattern of the battery-bank's voltage can be seen, which is directly related to the daily solar PV array energy generated input and the daily battery-bank output, defined by the daily office load demand.
3. An unambiguous trend is seen in the battery-bank's temperature variations, which are directly related to the chemical process of charging/discharging the battery-bank. Higher current input during the sunshine hours, demand a faster chemical reaction, increasing the battery-bank's temperature, while lower current output during the evening and night times, lowers the battery-bank's temperature. The gradual rise and decline of the battery-bank's daily temperature curve indicates that no significant, prolonged daily peak load occurs.
4. The average ambient temperature was 12.97 °C with a minimum of 7.83 °C and a maximum of 19.93 °C, the average daily battery-bank temperature over the year was 22.2 °C, with a minimum of 20.86 °C and a maximum of 23.4 °C. This is within a battery's ideal "comfort" zone, allowing it to perform at its best in regard to energy intake, energy storage and energy output. Further, in order to have a low self-discharge rate, it is crucial for a flooded lead-acid battery-bank to be kept within the battery's "comfort" temperature range of 15–25 °C. Thus the battery-bank achieved good performance values with overall energy and Coulombic efficiencies of 85 and 90 % respectively.

---

[57] The Coulombic battery efficiency calculation takes only the Ah (ampere-hour) or capacity into and out of the battery-bank into consideration, avoiding the voltage difference losses of the battery-bank's higher charging voltage and lower discharging voltage. Thus the Coloumbic battery efficiency value presents the battery-bank's technical performance efficiency, which includes the battery-bank's internal losses due to the chemical reactions, the increased temperatures occurring during the battery-bank's charging and discharging processes, and the losses occurring due to the resistances of the batteries' terminals and wires connected to the battery-bank.

Fig. 3.132 The combination of Si poly-crystalline PV modules (PhotoWatt PW750) with Si amorphous PV modules (Free Energy Europe FEE-20-12) in one array results in increased energy generation losses due to solar PV technology, size and brand mismatches

Fig. 3.133 The Si amorphous PV module (three Free Energy Europe FEE-20-12 modules) array's daily energy generation output is strongly affected by daily, long-term partial shading and insufficient free air flow around and through the array

Table 3.7 Technical specification of the a PhotoWatt PW 750 and a FEE-20-12 Solar PV module

Parameter	PhotoWatt PW 750	FEE-20-12[a]
Nominal Peak Power (Pmax)	80.00 W	19 W
Voltage @ maximum power (Vmpp)	17.30 VDC	16 VDC
Current @ maximum power (Impp)	4.60 A	1.18 A
Short circuit current (Isc)	4.75 A	1.45 A
Open circuit Voltage (Voc)	21.40 VDC	22.8 VDC

[a] Free Energy Europe (FEE) Fee-20-12 specification sheet: http://www.freeenergyeurope.com/pdf/FEE-20-12.pdf (accessed July 26, 2012)

3.12.2.5 Solar PV Array Mismatch Losses

Mismatch losses are caused by the interconnection of solar cells or modules. They are induced by connecting several PV modules from the same brand and type and even more losses occur if different PV types, brands or sizes are interconnected with each other (Fig. 3.132). This is because none of the PV modules, not even from the same brand, type or lot has identical performance properties (Fig. 3.133).

Thus, the output of the entire solar PV array is determined by the solar cell/module with the lowest output under worst case conditions. The common professional practice is not to combine different PV technologies, brands and sizes of modules in one array. We include in the design a cell/module mismatch value of 5 % for the same PV modules interconnected in an array (Table 3.7).

3.12.2.6 Dirt, Snow and Dust related Losses

Over time, a PV array will be affected with fine dirt and dust particles from the prevailing winds or even covered with snow during the winter months in high-altitude places such as in Humla. All these have an effect on the interception of the global solar radiation and the albedo[58] value on the surface of the PV modules and thus influence the array's power generation. The amount of power-loss due to these

Fig. 3.134 In Humla, it is not uncommon that during the 4–5 winter months (November–March) snow covers the solar PV array fully

[58] The Albedo value, also called surface reflectivity, is the fraction (or ratio) of incident solar radiation reflected by the earth's, or a body's, surface to the amount incident upon it (commonly expressed as a percentage) back into space or an interrupted object/surface, such as a solar PV module. http://en.wikipedia.org/wiki/Albedo (accessed July 26, 2012), and http://nsidc.org/arcticmet/glossary/albedo.html (accessed July 26, 2012). A snow covered environment increases the ratio of reflected solar radiation (higher Albedo value) and thus solar PV modules, if they are free of snow, can, if their tilted angle is favorable, receive significant higher solar radiation values and thus generate more power.

Fig. 3.135 The three different sized, 4, 2.5, 1.5 mm^2, armoured cables used in the Humla solar PV systems

factors depends on the location, season, and types of dust and particulates as well as their frequencies and time of occurrence. Thus in any solar PV system design such losses have to be considered and kept to a minimum. That is done by surveying the local situation and conditions and bearing in mind that periodical cleaning/maintenance is needed to keep the losses to a minimum (Fig. 3.134).

However, a realistic value for the power/energy loss calculation varies between 2 and 5 % depending on the local circumstances and occurrences, which have to be identified in the initial survey. In Humla, it is not uncommon during the 4–5 winter months from November to March that snow covers the solar PV array fully. So the dirt/dust/snow derating factor of 2–5 % is a conservative estimate.

3.12.2.7 Cable Sizing

Proper wire sizing is another essential aspect of solar PV system design. It is important to choose the proper sized wires in PV systems to ensure safe operation and to minimise voltage and therefore power losses due to increased resistance and heating of the cables in the system. There are detailed wire size standards for most countries which have to be strictly adhered to in order to fulfil the local requirements.

Three different sized armoured cables are used in the Humla Solar PV system (Fig. 3.135). They are reinforced with galvanised steel wires and placed underground for additional protection and to minimise the losses due to increased temperatures. Their conductor cross section sizes are 4, 2.5, 1.5 mm^2, respectively. Each of the copper wires, inside and outside, is covered with a PVC layer, providing additional protection. The cables are manufactured in Nepal. According to the manufacturer, the 4 mm^2, 2 strings with 7 copper strands, can carry up to 41 amps and 1,100 V, at up to 70 °C conducting temperature. It has a maximum resistance R of 1.23Ω/km at 20 °C. The total outer diameter of the wire is 7 mm including 1 mm thick insulation. The cables are buried 200–400 mm deep underground.

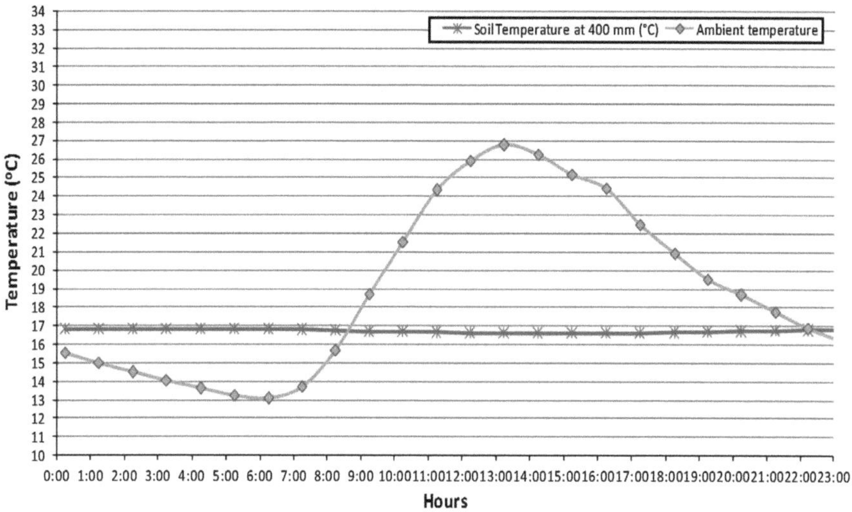

Fig. 3.136 Daily hourly average soil and ambient temperatures for March–May 2009

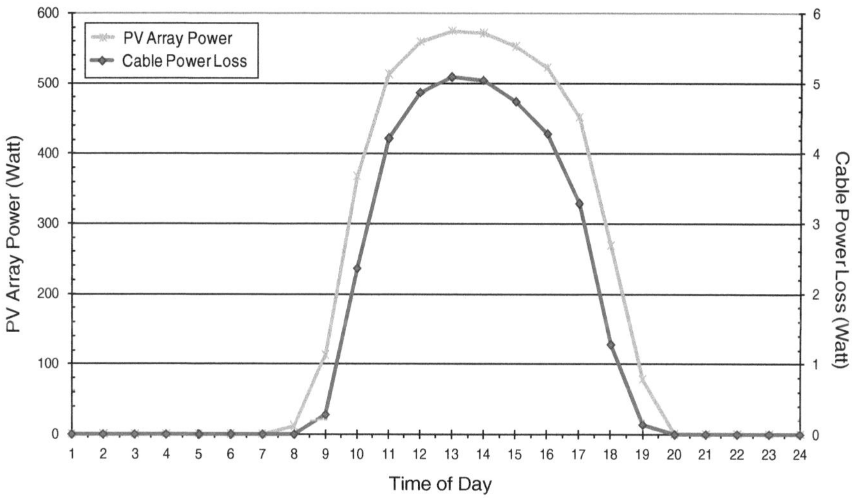

Fig. 3.137 Daily average solar PV array power versus cable induced power loss for the combined years 2006/2007

The length of the armoured cables from the PV arrays to the battery-bank in the Simikot solar PV system is 10 m. From Fig. 3.136, the following can be observed:

1. On average the soil temperature at 400 mm depth remains almost constant at 16.7 °C throughout the three months (from March to May 2009), whereas the ambient temperature has a clear daily pattern with a minimum of 13.6 °C and a maximum of 26.75 °C.

2. A constant, low temperature of the transmission cables keeps the ampere flow losses at a minimum. Transmission losses are further minimised with an increased system voltage, which is 24 VDC in the Humla PV system.
3. Additionally, the underground buried cables provide high protection and security for the solar PV system and for people, which is important in a village setting.

From the graphs in Fig. 3.137, the following can be observed:

1. The total daily energy loss due to the cables resistance from the PV array to the battery-bank, the inverter and the dump load is on an average 35.6 Wh, which is 0.77 % of the total daily average generated PV array energy of 4,581 Wh. The cable resistance induced power loss is calculated for 10 m length with the formula: where R is the resistance and I is the current flowing in the cable

$$\text{Power Loss in Cable} = I^2 \times \text{Resistance}(R) \tag{3.4}$$

2. As the cable power loss increases with the square of the current, the cable resistance induced power losses are higher around midday.
3. Considering the average daily energy of 2,344 Wh fed only into the battery-bank, the cable losses account for 12.6 Wh, which is 0.54 % of the total energy fed into the battery-bank.

3.12.2.8 Inverter Losses

With the RIDS-Nepal Simikot office being a project office, the needed load is mainly AC. Thus, an inverter (Fig. 3.138) is used to convert the DC power drawn from the battery-bank into AC power for all office equipment. The installed sinusoidal inverter, from the company Studer in Switzerland, has an AC power capacity of 800 W_p.

For any solar PV system with an inverter, DC–AC conversion losses have to be included. In this case, an efficient, imported inverter is used. While this minimises the losses, it adds to the risk of not being able to repair it in case of failure or damage. Thus, it is preferable to choose locally built inverters. In Nepal the most efficient locally built sinusoidal inverters have an efficiency of around 80–85 %. This shows that significant conversion losses occur, which can not be neglected in the initial design stage. Thus a choice must be made in the initial stage of a project's design, between sustainability and local availability. Reducing efficiency values and price, but increasing local availability and maintainability, or highly efficient, and often more expensive, imported equipment.

Type	Joker 802-S[a] (with solar charger)
Serial number	MJ02279
Pnom [W]	800W constant
Ubatt [VDC]	21-32VDC
Uout [VAC]	230VAC/50Hz
Power 30 min @ 25°C	1,300VAC
Power 5sec @ 25°C	2,800VAC
Maximum Efficiency	94%

Fig. 3.138 Technical specification of the Joker sinusoidal inverter

3.12.2.9 Summary

This section provides a series of important conclusions regarding the analysis of losses in PV systems:

- The use of a 2-axis PV tracking system results in capturing more energy compared to a fixed angle PV array. The results show a yearly energy generation increase of 56.4 % compared to horizontal and 31.5 % compared to 30° south inclined systems.
- The solar PV arrays operate more efficiently in cold areas such as Humla, where the solar modules can operate at <25 °C. The lower average ambient temperatures and higher average solar insolation help to generate more power/energy in the shorter winter months. Si mono- and Si poly-crystalline PV modules generate more energy in the cold season when compared to Si amorphous PV modules. Additionally the 2-axis tracker design enables air to flow freely through air flow gaps around the PV modules to keep them as cool as possible.
- The data reveal that the battery bank's energy (Wh In/Out) efficiency is 85 % and the Coulombic efficiency (Ah In/Out) is 90 %. It shows that maintaining the battery-bank temperature through appropriate insulation and air ventilation between 20 and 25 °C provides satisfactory results. Further, periodic maintenance (adding distilled/rain water, cleaning of contacts etc.) and a low daily DoD of $\leq$10 %, will maximise the battery-bank's life-expectancy.
- Solar PV arrays suffer from increased mismatch losses due to the use of different solar modules (technologies, brands and sizes), different illumination, shading and dust/dirt/snow. Thus all these factors have to be considered during the survey and design stage, and need to be minimised for any time of the year.
- Small losses, induced through cable resistance and increased temperature losses have to be calculated in the range of up to 1 % of the daily PV array generated energy. Cables buried in the ground minimise the temperature induced effect, though the cables and installation are more costly.

- The losses in the inverter depend on the DC-AC conversion technology used (transformers or transformer-less), its conversion efficiency (the quality of the equipment and materials used), the current and voltage at the terminals (the higher the current the higher the losses) and the temperature at which the inverter runs (the higher the temperature the higher the losses).

Providing electric power to remote, cold regions at high-altitude can be an expensive and a technically challenging task. Solar PV systems provide a reliable and cost effective solution yet their potential is underdeveloped, in part because of a lack of knowledge about their system performance in such regions. This chapter tried to identify the performance of a solar PV system, categorising and evaluating the losses of the systems operating in a cold, high-altitude climate.

The main lessons learnt from several years of data recording of the solar PV system at the high-altitude RIDS-Nepal office in Humla are:

1. Selecting the most appropriate PV module technology for a defined climate and average level of insolation and using only the same kind of PV modules in an array maximises the annual energy yield.
2. To insulate the battery-bank so that it can be kept within its ideal temperature "comfort" zone of 15–25 °C, minimises the losses and increases the battery-bank's efficiency and life-expectancy.
3. To limit the daily DoD of the battery-bank to $\leq$10 % increases its life-expectancy.
4. Unexpected high losses occur due to dirt/dust/snow covering of the PV modules.
5. Inappropriately sized/poorly chosen inverters increase system losses significantly.
6. A professional system design demands that all factors which induce losses, and thus reduce a solar PV system's performance, must be identified through an initial survey of the local context. Considering and integrating these data and information into the initial system design, enables a solar PV system to perform more reliably and satisfactorily under real conditions.

3.13 Possible Human Development Index (HDI) Improvement Through Elementary WLED Indoor Lighting

The comparison of different aspects of WLED lighting with traditional fuel based lighting shows that it not only improves the quality of lighting, but also the general living standard of the people, resulting in a notable improvement of the United Nation's Human Development Index (HDI) factor for the community. Replacing fuel based lighting by WLEDs can also contribute to the overall development of the underprivileged and underdeveloped communities by helping to improve the health, education and life-expectancy of the people as well as their income generation.

3.13.1 HDI for the Humla District in Nepal

The United Nations HDI is a widely accepted tool to rank a country's development status in regard to poverty, literacy, education and life-expectancy. Thus, it is good professional practice to apply such methodologies to determine, track and compare a defined group of people's living standard/conditions within their transforming context.

The United Nations Development Programme's annual *Human Development Reports,*[59] since the first in 1990, have published HDI as a composite measure of human development. HDI measures the average achievements in a country in three basic dimensions of human development: *long and healthy life*, as measured by the life-expectancy at birth; *knowledge*, as measured by the adult literacy rate and the combined gross enrolment ratio for primary, secondary, and tertiary schools; and *a decent standard of living*, as measured by Gross Domestic Product (GDP) per capita at Purchasing Power Parity (PPP) in US dollars (UNDP 2005).

Each of the three basic dimensions counts a third towards the HDI of a particular group of people, community or nation. As the HDI is identified for each country, one has to be careful in interpreting the given value, as the local context of a community or group of people can be very different, especially in certain fast growing urban areas in developing countries. The national HDI of Nepal in 2011 was 0.458[60,61] (UN 2011) though the HDI for Humla was only 0.244 (KIRDRC 2002). In the following, the Humla district is reviewed at its present development stage.

The first, "long and healthy life" indicator, identifies a life-expectancy of 85 years as highest (factor 1), and 25 years as lowest (factor 0). The average life-expectancy of Humla women and men is 54 years (KIRDRC 2002), resulting in a life-expectancy index of:

[59] The United Nations' *Human Development Reports*, from the first in 1990 to the most recent, are all freely available for download at: http://hdr.undp.org/en/reports/(accessed July 26, 2012).

[60] While the UN calculated the HDI for Nepal to be 0.428 and 0.458 in 2010 and 2011 respectively, the HDI value for the years 2003-2004 was 0.499-0.504 (NPC/UN 2005). That shows that either the UN had to review their values calculated in previous years, or Nepal's actual HDI values have indeed decreased over the course of the last 5 years. The author is of the opinion that the previous HDI values were too optimistic, with the most recent HDI value for 2011 being more realistic.

[61] The sinusoidal inverter Joker 802-S, as used in the RIDS-Nepal HASR office since 2004, from Studer in Switzerland is since 2009 new identified as AJ-1300-24, http://studer-inno.com/index.php?cat=sine_wave_inverters/id=430/pId=1197/tab=1#ul, and http://www.studer-innotec.com/upload/temp/Datasheet%20AJ%20series.pdf (both accessed July 26, 2012)

$$Life\,Expectancy\,Index = \frac{54\,years - 25\,years}{85\,years - 25\,years} = 0.483 \tag{3.5}$$

The second, "knowledge" indicator measures adult literacy and combined gross enrolment ratio for primary, secondary and tertiary schools. With an average adult literacy rate of 18 % (women 4.8 %, men 31.2 %) and an estimated gross enrolment of 33 %, the combined education index is calculated as:

$$Adult\,Literacy\,Index = \frac{18 - 0}{100 - 0} = 0.180 \tag{3.6}$$

$$Gross\,Enrollment\,Index = \frac{33 - 0}{100 - 0} = 0.330 \tag{3.7}$$

$$\begin{aligned} Education\,Index &= {}^{1}\!/_{3}\,Adult\,Literacy\,Index \\ &\quad + {}^{1}\!/_{3}\,Gross\,Enrollment\,Index = {}^{2}\!/_{3}(0.18) + {}^{1}\!/_{3}(0.33) = 0.23 \end{aligned} \tag{3.8}$$

The third indicator, the "GDP per capita", measures the annual income of the people. A logarithmic curve is assumed, as achieving a decent living standard does not need unlimited financial resources. In calculating the GDP index, US$40,000 represents factor 1. In Humla the average per capita income is US$72 (KIRDRC 2002), derived mainly from selling firewood and trading goods between China, Nepal and India. As more than 95 % of families in Humla are farmers, an additional estimated annual US$40 value from their annual crops can be added. Thus the GDP index factor is calculated as following:

$$GDP\,Index = \frac{\log(112) - \log(100)}{\log(40000) - \log(100)} = 0.019 \tag{3.9}$$

The final HDI is the average of the individual three indexes:

$$\begin{aligned} gatheredHDI &= {}^{1}\!/_{3}(Life\,Expec\tan cy\,Index) + {}^{1}\!/_{3}(Education\,Index) \\ &\quad + {}^{1}\!/_{3}(GDP\,Index) \\ &= {}^{1}\!/_{3}(0.483) + {}^{1}\!/_{3}(0.230) + {}^{1}\!/_{3}(0.019) = 0.244gathered \end{aligned} \tag{3.10}$$

3.13.2 Possible HDI Improvement Through Elementary WLED Indoor Lighting in Humla Homes

In order to utilise the synergistic benefits that elementary indoor electric lighting can provide, each electrification project RIDS-Nepal is implementing in Humla is part of a long-term HCD project. The main factors of our HCD projects include the building of a pit latrine for each family, the installation of an efficient smokeless metal cooking and heating stove, and a commonly owned village drinking water system. It has to be pointed out that due to the long time frame of each HCD project (~20 years), with increasing individual and even more synergistic benefits expected through each project over time, it is physically impossible to measure and express the changes in living conditions in numerical terms. Consequently, the following HDI calculation parameters used to determine the possible HDI improvements through a HCD project including WLED indoor lighting are assumptions made by the author based on his living and working experience in Nepal since 1983. The numerical assumptions derived from the expected changes an implemented HCD project can bring forth after a ten years time span, compared to the unchanged neighbouring villages.

3.13.3 Healthy Long Life

The installed WLED lighting instead of the "*jharro*", in addition to the installed smokeless metal stove for cooking and heating, enable clean indoor air. PM_{10} (Particulate Matter less than 10 μm) measurements in traditional houses in Humla since 2006 showed that over a 24 h time span, the average indoor air pollution was ~10–50 times less with WLED lighting and a smokeless metal stove compared to the use of "*jharro*" for lighting and cooking on an open-fire place. These improvements over 10 years time will incur considerable positive health improvements, especially in regard to chronic respiratory chest diseases. It is assumed that this will enable a prolonged life-expectancy of about 10 years, from the present 54–64 years. Thus, the "healthy and long life" HDI factor increases from 0.483 to 0.650.

3.13.4 Education

Non-formal education classes for adults and out of school children are part of the RIDS-Nepal HCD projects in Humla. The installed WLED lighting is providing clean and smoke free light to enable people to read. That means that one of the synergistic benefits from the WLED lighting is that the educational level is foreseen to significantly increase over the next 10 years. Based on our NFE program's mothers and children enrolments since 2005, it is realistic to say, that also the number of children joining school after our NFE program will increase substantially, because also many mothers became aware of the changes education brings to their village, life and the future of their children. It is therefore realistic to assume a doubling of the literacy rate and school enrolment over a ten year time span. Thus, the combined "education" HDI factor increases from 0.230 to 0.460.

3.13.5 GDP Per Capita

Adequate lighting inside the home provided by WLEDs, clean indoor air due to the use of a smokeless metal stove, a pit latrine, access to clean drinking water, enable all synergistic benefits if implemented alongside each other in a village. The increased social gatherings in a more comfortable environment bring forth new ideas for community and income generation activities such as knitting, bamboo basket weaving and other simple handicrafts.

A basic village electrification system using WLEDs is often the very first step of a family to start learning and experiencing electric light inside their home. Such an electrification system does not aim to generate enough power for tools or electrical equipment for the production of home based commodities. While this may be the ultimate aim in the years to come, a community has first to learn and experience the benefits of elementary electric lighting. Undoubtedly, WLED based lighting allows an increase in home based income generation abilities, supported by better health and educational status. But, compared to the first two HDI indexes, it may not be the major factor. Thus it is reasonable to assume a 40 % increase of the annual income can be achieved over 10 years, improving the GDP HDI factor from 0.019 to 0.075.

Adding up the re-calculated HDI factors increases the overall HDI index from 0.244 to 0.395, an increase of 60 % over the assumed time span of 10 years. This expected increase in the HDI index shows that replacing fuel based lighting with WLEDs in combination with other related community development projects can substantially improve the socio economic development of the communities.

3.13.6 Summary

While electric lighting is taken for granted in the industrialised world, about one quarter of the world's population is still using fuel based lighting for their basic lighting needs. The pine stick based lighting used in rural Humla in Nepal was found to be more than 300 times less efficient than the new WLED based lighting (Sect. 3.8.2/Appendix 13.2). Still, the WLEDs currently used in the villages are not the most efficient as the manufacturer has already announced the introduction of WLEDs with 150 lm/W luminous efficacy compared to the 29 lm/W WLEDs used in the measured installations (Nichia 2006). Although the present lighting levels provided by the WLED based lighting are minimal, they can be considered adequate for a first time indoor elementary lighting installation. A whole village of up to 30 homes, each with 3 WLED lamps, can be lit with the comparative power of one 100 W incandescent lamp. With the continuous advancement in the efficiency of WLED technology, more light in the future will be available using the same amount of energy.

References

Australian Standard, 1990. *AS 3595-1990 Energy Management Programs—Guidelines for Financial Evaluation of a Project*, available for purchase at: http://www.saiglobal.com/PDFTemp/Previews/OSH/As/as3000/3500/3595.pdf (accessed July 28, 2012).

Basnyat, N. M. S. (2006). Nepal's Umbilical Cord. *Nepalnews*, August 2006. http://samudaya.org/200608/nepals-umbilical-cord/ (accessed March 29, 2010, on July 20, 2012 not available anymore).

Bhusal, P., Zahnd, A., Eloholma, M., & Halonen, L. (2007). Replacing fuel based lighting with light emitting diodes in developing countries: Energy and lighting in rural Nepali homes. *LEUKOS, The Journal of the Illuminating Engineering Society of North America,3*(4), 277–291. http://www.iesna.org/leukos/Volume3/number4.cfm (accessed July 22, 2012).

Blumenthaler, M., Ambach, W., & Ellinger, R. (1997). Increase in solar UV radiation with altitude. Elsevier, Journal of *Photochemistry and Photobiology, Biology, 39*, 130–134, Abstract available and paper to purchase at. doi:10.1016/S1011-1344(96)00018-8 (accessed July 28, 2012).

CEN (Clean Energy Nepal). (2003). *Kathmandu's air quality*, Kathmandu, Nepal.

Choe, M. K., Thapa, S., & Irawati, S. A. (2001). *Surveys show persistence of teenage marriage and childbearing in Indonesia and Nepal*, Asia-Pacific Population & Policy, No. 58. http://scholarspace.manoa.hawaii.edu/bitstream/10125/3893/1/p%26p058.pdf (accessed November 1, 2011).

Craine, S. (2004). *Technical and economic optimization of villageelectrification systems*, MSc Eng. Dissertation, Curtin University WA, pages 9, 11, 14, 36.

Energy Programs and Fisheries Division Department for Primary Industries and Energy. (1994). Saving Energy through Lighting Measurement. Energy Management Advisory Booklet, Australia Government Publishing Service, Canberra, pp. 4–6.

Foster, R. E., Orozco, R. C., & Romero-Paredes Rubia, A. (1999). Lessons Learned from the Xcalak Village Hybrid System: A Seven Year Retrospective. *ISES Solar World Congress*, Jerusalem, Israel. http://solar.nmsu.edu/publications/lessons.pdf (accessed July 22, 2012).

Goldemberg, J. (2000). *World Energy Assessment: Energy and the Challenge of Sustainability*, chapter 10: Rural Energy in Developing Countries, UNDP 2000, p. 380. http://www.undp.org/content/dam/undp/library/Environment%20and%20Energy/Sustainable%20Energy/wea%202000/chapter10.pdf (accessed July 23, 2012).

Haddix, M. K., & Zahnd, A. (2007). Responses to innovation in an insecure environment: Participation rates in holistic community development schemes in the midst of the Maoist revolution in Nepal. *Mountain Research & Development (MRD),27*(4), 302–307, November 2007. http://www.mrd-journal.org (accessed March 3, 2010), also available at: http://www.rids-nepal.org/index.php/Download_document/120_Responses_to_Innovation_in_an_Insecure_Environment_in_Rural_Nepal.html (accessed July 24, 2012).

IEA (International Energy Association). (1999). *Stand-Alone Photovoltaic Applications: Lessons Learned*. IEA PVPS Programme, Utrecht, The Netherlands, published by James & James, ISBN 1-873936-91-5, overview available at: http://openlibrary.org/books/OL8654509M/Stand-Alone_Photovoltaic_Applications (accessed July 20, 2012).

IEA (2000). *Analysis of Photovoltaic System*: Report IEA-PVPS T2-01:2000. http://www.iea-pvps.org/products/download/rep2_01.pdf (accessed March 27, 2010, on November 1, 2011 not available any more).

IEA (2002). *Energy and poverty*. Chapter13, World Energy Outlook.

IEA (2007). *Cost and Performance Trends in Grid-Connected Photovoltaic Systems and case Studies*. Report IEA-PVPS T2-06:2007. http://www.iea-pvps.org/products/download/rep2_07.pdf (accessed March 25, 2010, on November 1, 2011 not available any more).

IEC (International Electrotechnical Commission). (1998). *Photovoltaic System Performance Monitoring—Guidelines for Measurement, Data exchange and Analyses*, IEC 61724.

ITDG (n.d.). *Reducing exposure to indoor air pollution*, ITDG publication, p. 14, http://www.itdg.org/docs/smoke/itdg%20smoke%20-%20part%203.pdf (accessed July 20, 2012).

KIRDRC (Karnali Integrated Rural Development and Research Centre) (2002). *Governance in the Karnali, an Exploratory Study*. August 2002, Jumla, Nepal.

Mayer, D., & Heidenrich, M. (2003). Performance analysis of stand-alone pv systems from a rational use of energy point of view. In *Proceedings of the 3rdWorld Conference on Photovoltaic Energy Conversion, Osaka, Japan, 2003*.

Mills, E. (2005). The specter of fuel based lighting. *Science,308*(5726), 123–126.

NEA (Nepal Electricity Authority). (2009). *A Year in Review—Fiscal Year 2008/2009*, Annual Report, pages 9, 83–84, Kathmandu, Nepal. http://www.nea.org.np/reports/annualReports/YjqRDwQ9E3anrep09.pdf (accessed March 23, 2010, on November 1, 2011 not available any more).

Nichia (2006). previously available at: http://www.nichia.com/about_nichia/2006/2006_122001.html.

Reddy, A. K. N. (2000). *Energy and Social Issues—World Energy Assessment: Energy and the Challenge of Sustainability*, chapter 2, New York, United Nations Development Program (UNDP), pp. 39–60.

Reinders, A. et al. (1999). *Sukatani revisited: on the performance of nine-year-old solar home systems and street lighting systems in Indonesia*. Renewable & Sustainable Energy Reviews, Pergamon, pages 15, 25, 27, 28.

Saghir, J. (2005). *Energy and poverty: Myths, links, and policy issues*. Energy Working Notes, World Bank: Energy and Mining Sector Board, No.4 May, 2005.

Shakya, P. (2003). *Air Pollution in Kathmandu Valley*, Kathmandu Metropolitan City Office, pp. 4–5. http://www.unescap.org/esd/environment/kitakyushu/urban_air/city_report/kathmandu.pdf (accessed July 20, 2012).

Smith, Nigel. (1997). *Motors as generators for micro-hydro power*. London: Intermediate Technology Publications.

SolRC (Solar Radiation Calculator) (2001). SolRC version 1.3. (since end of 2010 access is only for registered users, accessed March 1, 2011, on November 1, 2011 not available any more).

Tiwari, A. (2006). Recognise Remittances—Overseas earnings have saved us from collapsing altogether. Nepali Times, Issue #306, 14—20th July 2006. http://www.nepalitimes.com/issue/306/StrictlyBusiness/12156 (accessed March 29, 2010).

UN (2011). *Human Development Report 2011—Sustainability and Equity: A Better Future for All*, p. 129. http://hdr.undp.org/en/media/HDR_2011_EN_Complete.pdf (accessed July 24, 2012).

UNDP (2005). *Human Development Report 2005*. United Nations Development Programme, New York: Oxford University Press. Technical notes 1.

WEA (World Energy Assessment) (2000). *Energy and the challenge of sustainability*. New York: UNDP. http://www.undp.org/content/undp/en/home/librarypage/environment-energy/sustainable_energy/world_energy_assessmentenergyandthechallengeofsustainability.html (accessed July 24, 2012).

WECS (2003). *Rural Energy-Pressure, State, Impacts and Responses*. WECS Data-2, Chapter 3, Kathmandu, Nepal.

World Bank (2010). *The little Green Data Book*. International Bank for Reconstruction and Development/The World Bank, page 153, ISBN: 978-0-8213-8245-5, 1818 H Street, N.W., Washington, D.C. 20433, USA, http://www.semide.net/media_server/files/semide/topics/water-data/little-green-data-book-2010-world-bank/GreenDataBook2010_WorldBank.pdf (accessed July 24, 2012).

Zahnd, A. (2004a). *Case Study of a Solar Photovoltaic Elementary Lighting System for a Poor and Remote Mountain Village in Nepal*, MSc RETs Dissertation, Murdoch University, Perth, Western Australia. http://www.rids-nepal.org/index.php?option=com_docman&task=doc_download&gid=2&Itemid=104 (accessed July 24, 2012) and http://www.rids-nepal.org/index.php?option=com_docman&task=doc_download&gid=3&Itemid=104 (accessed July 24, 2012).

Zahnd, A. (2004b). Handling (Managing, Administrating, Governing) Available Resources Responsibly—Case Study of a Renewable Energy Project with a Remote and Poor Mountain Village in the Nepal Himalayas. *World Renewable Energy Congress VIII*, Denver Colorado, USA, 28.8.–3.9. 2004, available at: http://www.rids-nepal.org/index.php?option=com_docman&task=doc_download&gid=9&Itemid=104 (accessed July 24, 2012).

Zahnd, A. (2005a). A holistic approach to community development. *World Rivers Review, 20*(1), 8–9,15 http://www.internationalrivers.org/files/WRR.V20.N1.pdf (accessed July 24, 2012).

Zahnd, A. (2005b). Three research project reports published on their NEWS web site about the application of datataker's DT80 and DT605 data loggers in the RIDS-Nepal Humla and Kathmandu University based research projects. *May 2007 News dataTaker, PD189, PD190 News 2007 and AN34 January 2005 News.* Melbourne, Australia. http://www.datataker.com/index.html (accessed November 1, 2011, July 24, 2012 not available any more). http://www.datataker.com/newsletters/sitemap_u1.html (accessed November 1, 2011, July 24, 2012 not available any more).

Zahnd, A., & Kimber, H. M. (2005). Design of an Optimised PV System for a Remote Himalayan Village. *ANZSES 2005,* Otago University, Dunedin, New Zealand, 28.11.–30.11.2005. http://www.rids-nepal.org/index.php?option=com_docman&task=doc_download&gid=14&Itemid=104 (accessed July 24, 2012).

Žukauskas, A. (2007). *ENLIGHTING Project Lecture Series on Semiconductors and White LED.* University of Vilnius, Lithuania, p. 86, 107, Kathmandu University. Nepal.

Chapter 4
Thermal and Processing Technologies

"Poverty alleviation and development depend on the access to energy services that are affordable, reliable and of good quality" (Reddy 2002; Saghir 2005), and as the WEA/UNDP (2000) report confirms, there are "clear linkages between improved access to energy services and reduced infant mortality and fertility rates and increased literacy and life-expectancy". But too often one is inclined to limit energy services to the availability of electricity, leaving out other important energy services such as cooking, room heating, and water heating for improved hygiene. These thermal services are all equally important for sustainable and holistic community development among the poor and marginalised communities. Thus this chapter addresses some key thermal process technologies for improved thermal energy services. These have been developed and tested in executed projects, as part of the long-term HCD concepts of the "Family of 4" and the "Family of 4 PLUS" projects, through RIDS-Nepal and the local communities in Humla, Nepal.

4.1 High-Altitude Smokeless Metal Stove: Research, Development and Implementation Project

This section aims to highlight the importance and need for a Smokeless Metal Stove (SMS) for high-altitude rural communities, in order to improve their health conditions through improved indoor air conditions, and to help in active ways to slow down the increasingly widespread deforestation. The health impairing indoor air conditions of the Humla people are highlighted, as well as the improved living conditions and benefits people themselves have identified, through the use of a SMS in their home. Anecdotal and practical experiences from families who use a "Jumla Design " SMS in their homes in Humla, as well as Kathmandu University (KU) laboratory data and results of the first generation SMS, the "Jumla Design" and the new design (secondary combustion stove) prototypes KU-1 and KU-2 are discussed and presented through pictures, data and graphs.

A. Zahnd, *The Role of Renewable Energy Technology in Holistic Community Development*, Springer Theses, DOI: 10.1007/978-3-319-03989-3_4,

Fig. 4.1 Open-fire places for cooking, room heating and light create heavy smoke in the homes in Humla. Especially the mothers and children are exposed to the smoky indoor fires, causing severe health hazards. Thus asthma, respiratory chest diseases, heart diseases, eye infections etc. are very common

The communities living in those rural areas are economically so weak that the energy from refined fuels such as Liquefied Petroleum Gas (LPG) and kerosene is not an option. Therefore the families have no alternative to meet their daily energy service demands other than to rely on their locally available firewood resources. This heavy dependence on fuel wood for cooking and space heating on open-fire places (called "*odhan*", Fig. 4.1) poses immense health hazards for the people, especially to women and children.

According to the Nepal Water and Energy Commission Secretariat (WECS), about 30 %, or 42,240 km^2 of Nepal's total landmass of 147,181 km^2 is covered with forest. Monitored data shows that the forest areas are annually reduced by 1.7–1.8 % (Koirala n.d., World Bank 2010). This leads to the scarcity of local firewood, forcing women and children to spend increasingly longer hours collecting their daily firewood needs for cooking, heating and lighting. Thus to provide the energy services for cooking and space heating within a clean indoor environment through a SMS, with less firewood consumption, is of high value and interest to the local people.

4.1.1 Firewood Consumption, Cooking Method and Health Impact in Rural Areas

Fuel wood, used for cooking, space heating and lighting, provides $\geq$95 % (Fig. 4.2) of the total energy consumption in high-altitude remote areas such as Humla, with some consuming 20–40 kg of firewood per family per day (Zahnd 1998). According to the survey carried out by WECS in rural villages, 98 % of the total energy consumption is met by biomass (Fig. 4.2), of which about 90 % is met

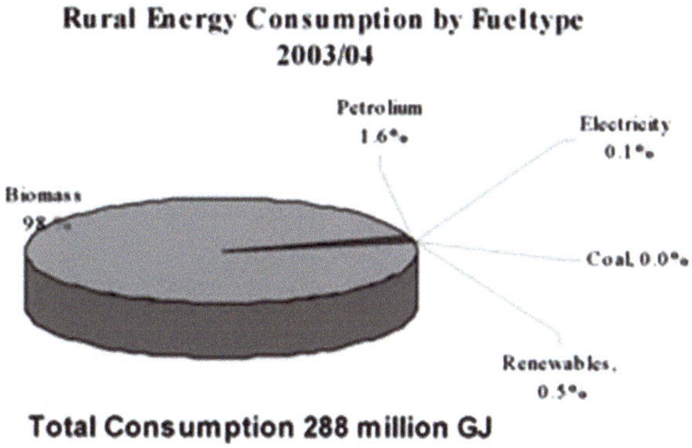

Fig. 4.2 Average rural household energy consumption in Nepal by energy type for the years 2003–2004 (WECS 2006)

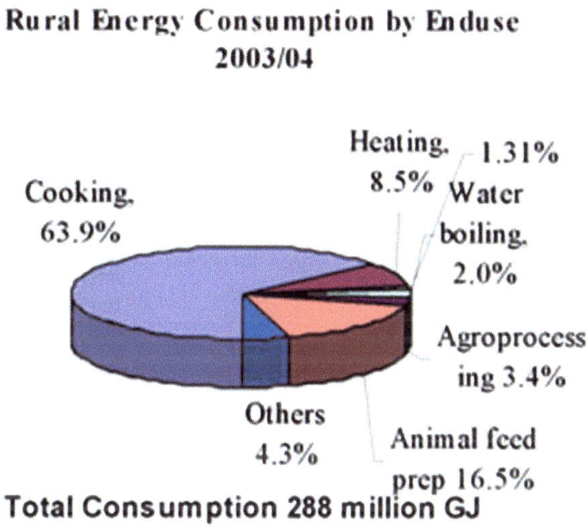

Fig. 4.3 Average rural household energy end uses in Nepal for the years 2003–2004 (WECS 2006)

by fuel wood, followed by animal dung and agricultural residues (Zahnd 1998). Figure 4.3 shows that 63.9 % is used for cooking, 19.9 % for animal feed preparation and agro-processing, 8.5 % for space heating, 2 % for water boiling and 4.3 % for other uses such as indoor lighting (Winrock 2004).

In contrast, Rai and Zahnd's research shows, that in the high-altitude mountain areas 32 % of household energy is used for cooking and 56 % is used for space heating, while in the mid-hill areas families use 40 % for cooking and 36 % for heating. The remaining 12 and 24 % respectively, is used for lighting, boiling water and other agro processing activities (Rai et al. 2005). That indicates that the fuel wood demand for space heating is greater than the demand for cooking. This high dependence on fuel wood to meet the local population's daily energy service demands had already caused alarming conditions in regard to the country's deforestation (Figs. 4.4, 4.5).

Fig. 4.4 The widespread deforestation leads to an increased scarcity of local firewood resources, forcing people, mostly women and children, to spend hours collecting firewood for their daily energy needs

Fig. 4.5 Every other day the women have to spend 7–8 hours for firewood collection

In rural and high-altitude mountain areas, women are responsible for cooking and other household work and the increasing deforestation forces these women and children to spend 7–8 hours every second day, to gather fuel wood further a field (EIA 2002, Haddix and Zahnd 2005). Therefore women, in particular, place a high value on a smokeless metal stove which consumes up to 40 % less firewood and provides a clean indoor air environment for the whole family.

Smoke from open-fireplaces is among the four greatest risks of diseases and injuries leading to death, identified by WHO. Smoke from burning of solid fuel kills 1.5–1.6 million people each year, of which 66 % are children aged under five (WHO 2006; Winrock 2004). Children, if exposed to indoor air pollution, are 2–3

Fig. 4.6 "Odhan" (open-fire place) is the traditional way of cooking and room heating. The people are exposed to very high, health impairing indoor air pollution

Fig. 4.7 The "Jumla Design " smokeless metal stove enables clean indoor air, cooking the traditional food in one go and, if used properly, consumes up to 40 % less firewood

times more likely to contract acute respiratory infection, which is the main cause of children's death from indoor air pollution. RIDS-Nepal data show that women in the remote areas who use firewood for cooking, room heating and lighting spend about 6 hours per day cooking food on a traditional open-fireplace, called an "*odhan*" (Figs. 4.1, 4.6, 4.9). The smoke and soot from the open-fire are highly toxic pollutants, resulting in eye infections, heart diseases and irreversible impacts on the respiratory system. These chronic, long-term health impairments are, beside the harsh living conditions, a major reason for the low life-expectancy for women, who are four times more likely to suffer chronic bronchitis, and the high death rate of children <5 years of age. Illness caused by smoke kills more children annually than malaria or HIV/AIDS (Warwick and Doig 2004).

4.1.2 High-Altitude "Jumla Design" Smokeless Metal Stove

The "Jumla Design" SMS (Figs. 4.7, 4.8, 4.10 and 4.11), developed by the author, is designed for cooking and space heating for communities within their cultural and environmental context of Nepal's remote, high-altitude mountain areas. It has three burners for cooking people's traditional dish, "*dal, bhat and tarkari*" (rice, lentils and vegetables) simultaneously and a stainless steel hot water tank attached to provide continuous hot water for other household purposes. In addition to this, a slot is made in which "roti" (the traditional unleavened bread) can be baked in a traditional way directly on embers (Fig. 4.8, 4.10, 4.12).

Fig. 4.8 Teaching the owner how to use the new SMS in her restaurant. The SMS is in use from 05:00 to 22:00, or 17 h a day. By May 2012 it passed 40,000 working hours, more than fulfilling its designed life-expectancy. However, a visual, detailed check by the SMS designer confirmed that it is still in very good condition

Fig. 4.9 A personal air sampler is used to measure and record the smoke and air pollution exposure from the open-fire cooking and heating. While the restaurant used to use up to 120 kg of firewood a day with the open stove, they need now only ~40 kg/day with the new SMS (Fig. 4.8), and no IAP

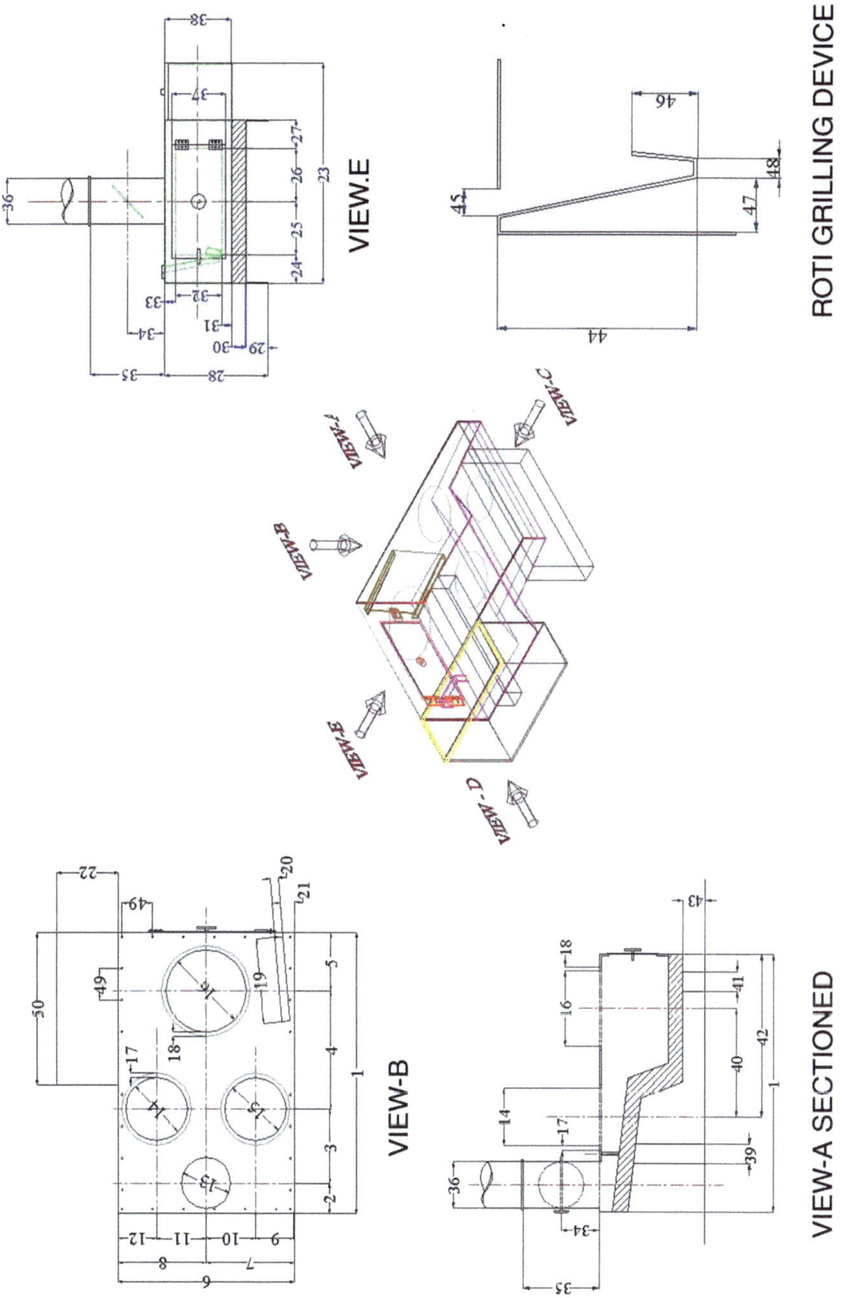

Fig. 4.10 "Jumla Design" smokeless metal stove drawing

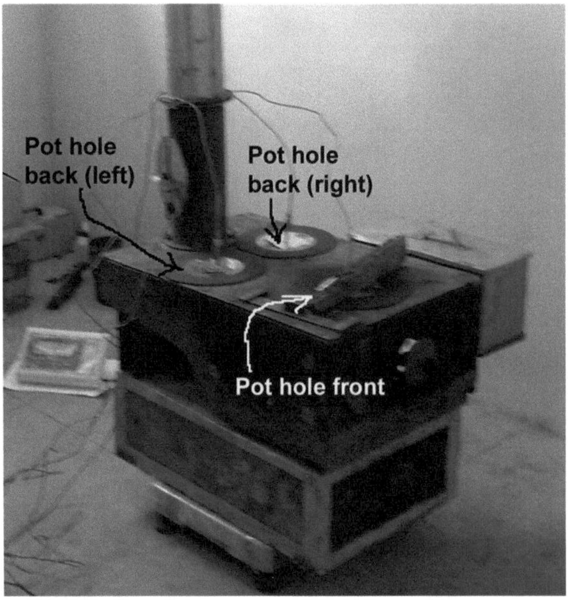

Fig. 4.11 "Jumla Design" smokeless metal stove being tested in the Kathmandu University (*KU*) stove laboratory. Cooking (on the three pot holes), water heating (in tank) and room heating efficiencies are tested

Fig. 4.12 The KU stove laboratory and equipment. Thermocouples ≤1,200 °C, digital weighing scales (measuring the initial amount and ongoing burning rate of firewood) and a gas analyser measuring O_2, CO_2, CO, NOx and flue gas temperatures are the main equipment

In order to minimise the downside heat loss, the stove has a mud filled double bottom. To transfer the heat more efficiently towards the cooking pots, an air draught allows the regulation of the combustion air through an adjustable air vent in the main door and a damper in the flue pipe. With 1.5 mm thick steel walls and a 4 mm thick cooking surface plate and three 3 foot flue pipes, this stove weighs 38 kg.

Experience shows that when properly installed and used, the "Jumla Design" SMS consumes ~40 % less firewood compared to cooking on open-fire places. The ongoing monitoring of the firewood consumption in RIDS-Nepal's Simikot office and HARS shows that the average firewood consumption per person per day for all food and heating services is 1.8–2.6 kg, compared to 3.3–6.6 kg in local homes without SMS.

Field tests with the SMS show that 14–22 % of the calorific value of the used firewood was utilised for cooking and water heating, with the rest heating the room (~20 %) and escaping as hot flue gases through the exhaust pipe. Considering wood energy as 100 % energy input, 15 % is distributed towards the three burners (pot holes) and 4 % towards the 9 liter stainless steel water tank. Laboratory tests showed, that from the remaining 81 % an estimated 15–28 % of the energy input can be utilised for space heating (Rai 2005).

4.1.3 KU-1 Secondary Combustion Smokeless Metal Stove Prototype

The "Jumla Design" SMS has three burners, a "roti" grilling slot, the ability to generate hot water, as well as providing appropriate room heating. These energy services were identified by the local communities as their main energy service demands to be met. Thus this SMS is an example of contextualised technology as it took, besides the environmental constraints, also the social and cultural context into consideration for its design. While this makes this SMS highly acceptable, field experience indicates that the SMS's overall efficiency can be improved. This can be achieved if a secondary combustion[1] inside the SMS takes place, which requires a combustion air temperature of ≥630 °C to burn/ignite the volatile unburned gases. The KU lab tests have shown that this cannot be achieved with the "Jumla Design" SMS. This stove has no secondary combustion chamber, nor can

[1] There are two sources of heat generation in a biomass/firewood, stove. The first source of heat is generated in the first/primary combustion chamber, where you can see the wood being consumed by the burning fire. In the primary combustion, moisture is driven off, and the firewood is ignited and volatilized, releasing unburned particulates. The secondary heat, which can be significantly greater than the heat generated from the first combustion process, is generated in the secondary combustion chamber, sometimes also called "afterburner", usually positioned above/after the first combustion chamber. In this, secondary combustion chamber, the unburned gasses which are released during the burning of the firewood in the first combustion chamber, are instantaneously ignited, increasing the heat output of the firewood substantially (e.g. 50–200 % in modern, airtight stoves for individual household heating systems in Europe), if the main condition, the ignition temperature of the volatile gasses, is met. Through laboratory testing we evaluated the temperature range for a successful ignition of these gasses. For the local Humla firewood this lies between 630–650 °C. That means that the oxygen, which needs to ignite the volatile flue gasses in the secondary combustion chamber, has to be at least 630 °C. This has to be reached by pre-heating the air for the secondary combustion through the first combustion chamber. While for bigger, modern airtight stoves, as used in Europe, this is possible, for our local Humla context, with a much smaller stove which is frequently opened and closed, this is a significantly greater challenge and the main focus of the stove research project described in this chapter.

Fig. 4.13 Manufactured KU-1 secondary combustion stove prototype

the needed combustion air reach ≥630 °C due to the SMS' uninsulated stove walls and single combustion chamber. This led the Kathmandu University (KU) in 2002 to start a research project, under the author's leadership, for a secondary combustion process Smokeless Metal Stove for high-altitude areas.

The first prototype, secondary combustion SMS, was developed and built in 2002. It was named KU-1 (Figs. 4.13, 4.14) and has two separate combustion chambers, one primary and one secondary. The primary combustion chamber is provided with two burners and a separate air vent/channel to control the needed combustion air flow. The unburned and volatile gases from the first combustion chamber are directed into the secondary combustion chamber. The supplied air is intended to be first preheated to ~630–650 °C by flowing around the first combustion chamber and is then led via a separate channel into the bottom opening of the secondary combustion chamber.

An ashtray placed underneath the primary combustion chamber serves also as an air intake passage for both primary and secondary combustion chambers. A stainless steel water tank, holding 9 liters, is attached to the primary combustion chamber. It provides hot water for cooking, drinking, washing and other household purposes. The KU-1 stove weighs 37 kg.

Tests on the KU-1 stove showed that the required secondary air temperature of ≥630 °C to burn the volatile and unburned gases in the secondary combustion chamber could not be reached. Too much energy is conducted through the uninsulated metal stove walls into the water tank and the environment. This led to the development of the KU-2 secondary combustion SMS.

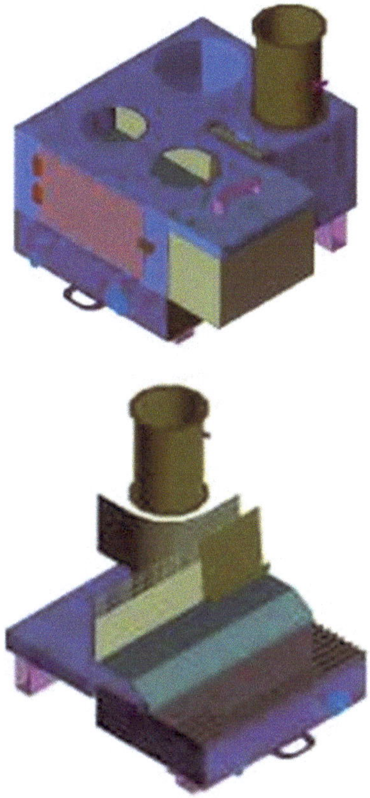

Fig. 4.14 CAD drawings of the KU-1 secondary combustion stove

4.1.4 KU-2 Secondary Combustion Smokeless Metal Stove Prototype

Starting in November 2002, with the sponsorship of the Norwegian research institute SINTEF, the KU-2 prototype was designed and tested in the KU stove laboratory. This design includes a well insulated primary combustion chamber, a secondary combustion chamber, a better heated secondary air route and an uninsulated heating tank. Each of the three major parts of the prototype is manufactured as a module, in order to change each part separately for the various laboratory tests and for ease of carrying and transporting.

Figures 4.15, 4.16, 4.17, 4.18, 4.19, 4.20 show the KU-2 secondary combustion stove. The primary and the secondary combustion chamber air flows in separate channels below the floor of the first combustion chamber to get preheated. The preheated air enters the first combustion chamber at the back of the chamber (Figs. 4.17, 4.18), while the preheated air for the secondary combustion chamber enters much more preheated through two layers of air nozzle inlets underneath and at the beginning of, the secondary combustion chamber (Figs. 4.12, 4.16, 4.18).

Fig. 4.15 Overall picture of the fitted KU-2 secondary combustion smokeless metal stove with all three chambers

Fig. 4.16 Mud filled double walls and the two layers of air nozzle inlets for the secondary combustion chamber

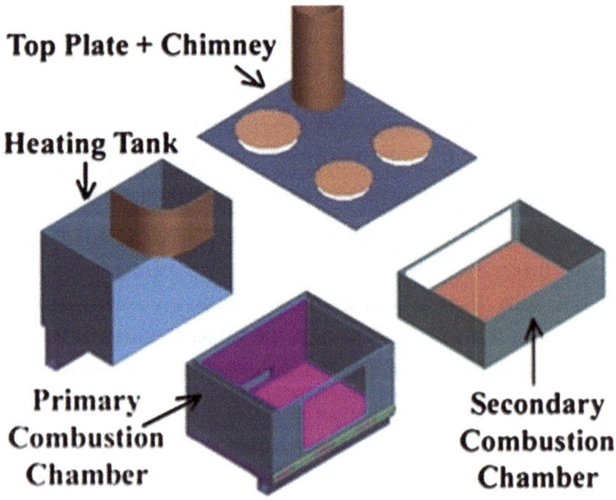

Fig. 4.17 KU-2 secondary combustion stove CAD explosion drawing with each main module

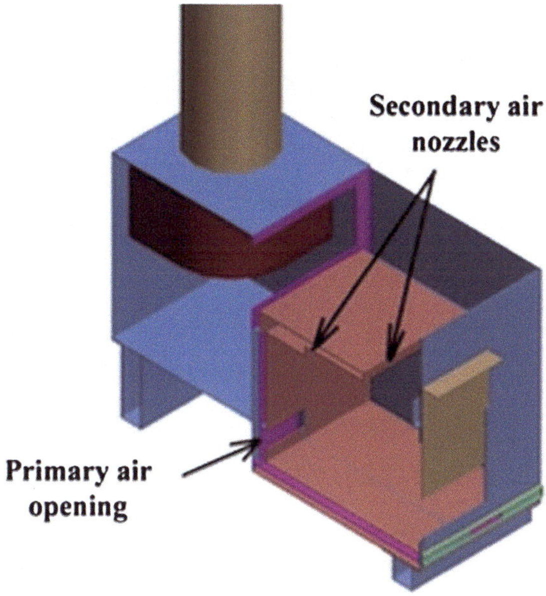

Fig. 4.18 KU-2 secondary combustion stove CAD cut through of the combustion chambers

Fig. 4.19 KU-2 primary combustion zone with the open firewood intake door

Both, primary and secondary combustion chambers are insulated through mud filled double walls (Fig. 4.16).

A baffle shield below the chimney (Figs. 4.17, 4.18) reflects the flames and hot flue gases toward the third burner. That increases their resident time inside the stove, improving the convective heat transfer. The air flow for the primary combustion is controlled by one air sliding vent (Fig. 4.15 left bottom sliding vent), while air flow for the secondary combustion is controlled by two sliding air vents (Fig. 4.15 middle and right bottom sliding vents). A "butterfly valve" damper allows the regulation of the flue gases passing through the exhaust pipe.

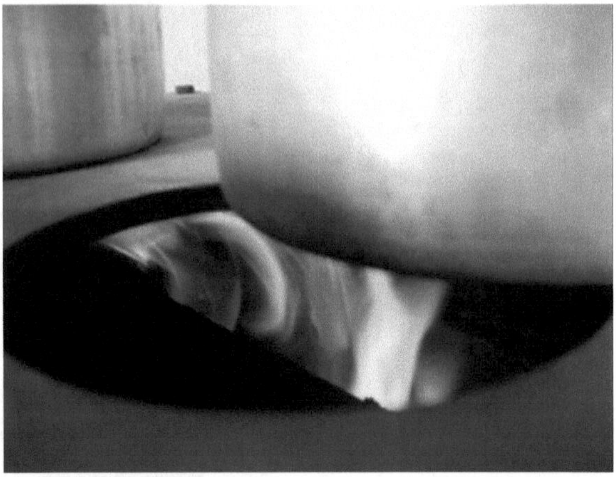

Fig. 4.20 KU-2 secondary combustion zone with sporadic secondary combustion

4.1.5 Comparison "Jumla Design" and the KU-2 Secondary Combustion SMS

4.1.5.1 Firewood Consumption and Burner Temperatures

The main objective is to develop a stove with secondary combustion, leading to a measurably higher efficiency. Thus the KU-2 stove was compared with the "Jumla Design" stove. The experiments were carried out after an initial ~20 min warming up period of the stoves using 1 kg of firewood for each. The same experimental conditions, equipment (including the same exhaust pipe length), arrangements and measuring equipment were used. The same regular firewood inputs were maintained throughout the tests.

It can be seen in Fig. 4.21 that the "Jumla Design" SMS burner (pot hole) temperatures vary with the burning rate, which further depends upon the amount of air supplied for the combustion. With a properly adjusted air vent and chimney damper, it took ~21 min. to burn 1 kg of pinewood. The burner temperatures attained were 250–350 °C for most of the test period and the maximum temperature recorded was 380 °C. With the right air/fuel mixture ratio, it took 30–40 min to cook a traditional meal for an average family of 5 people (Figs. 4.22, 4.23).

Figure 4.24 shows that for the KU-2 prototype SMS the front burner temperatures are between 400–550 °C for most of the test period, with a maximum of 580 °C. The KU-2 secondary combustion stove took 31 minutes to burn 1 kg of the same firewood. Lower firewood consumption and increased burner temperatures indicate that the KU-2 stove transfers the combustion energy more efficiently (by radiation and convection (Fig. 4.22)) to the cooking utensils, and over a longer period, compared to the "Jumla Design" stove (Fig. 4.23).

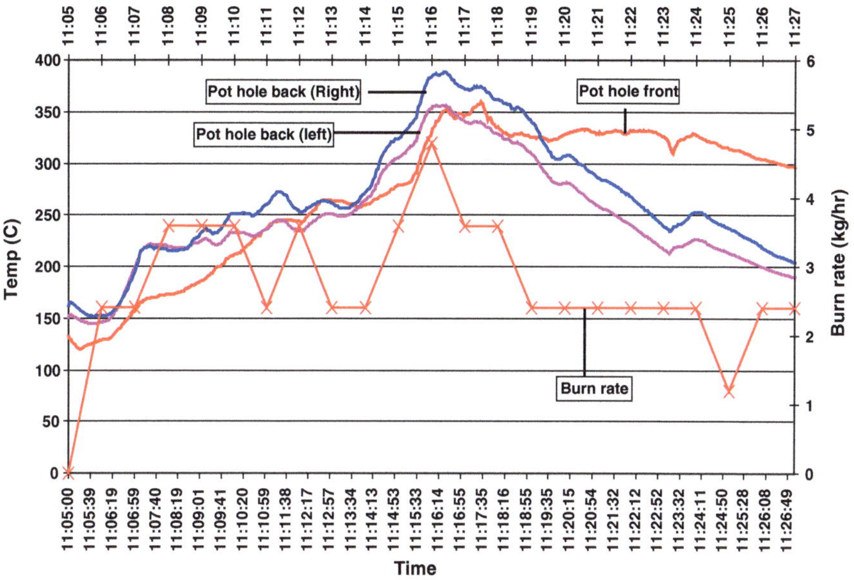

Fig. 4.21 Firewood consumption and pot hole temperatures for the "Jumla Design" SMS

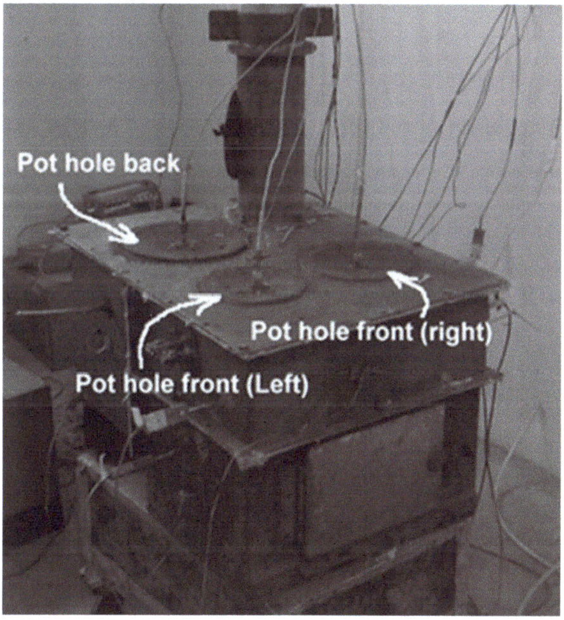

Fig. 4.22 KU-2 secondary combustion SMS

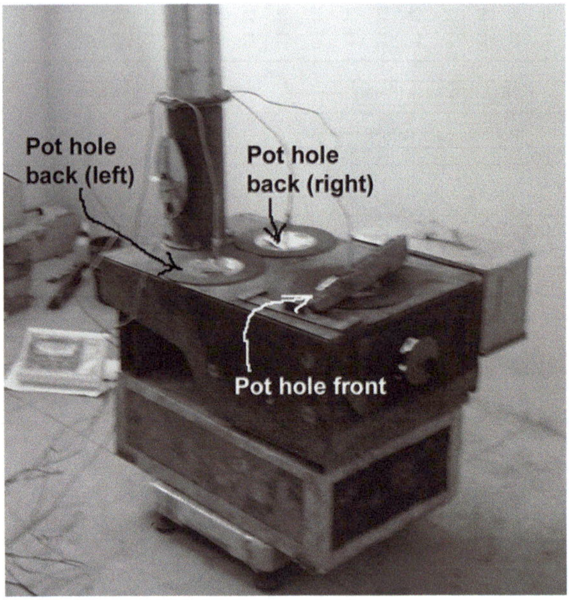

Fig. 4.23 "Jumla Design" SMS

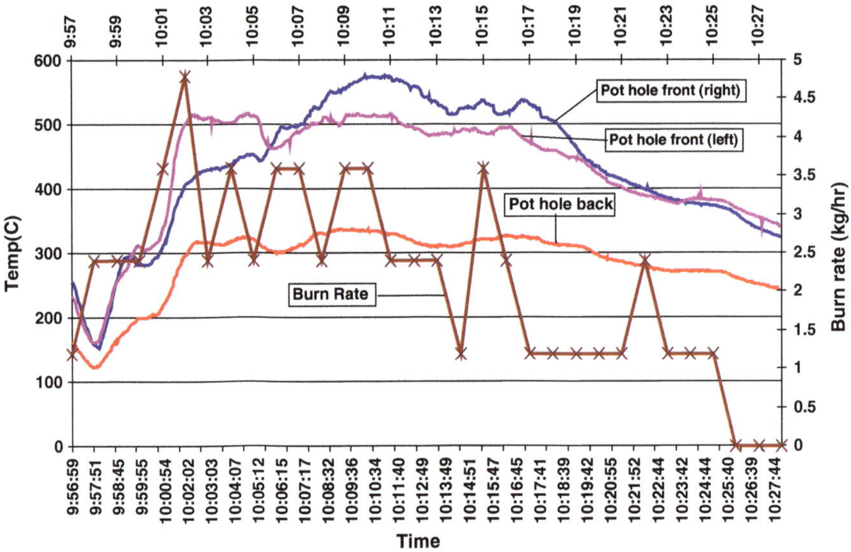

Fig. 4.24 Firewood consumption and pot hole temperatures for the KU-2 prototype SMS

4.1.5.2 Gas Flue and Chimney Temperature

A gas analyser (Figs. 4.25, 4.26), measuring the O_2, CO_2, CO and NO_x content of the flue gases in percentage in the chimney at two temperature points, was used. These data allow us to test experimentally the stove's best air/fuel ratio in order to minimise the emissions and maximise the combustion efficiency. The higher the chimney temperature the more hot flue gases escape without being used for cooking, water heating and space heating. That results in overall lower stove efficiency.

Figures 4.27, 4.28 show the flue gas temperatures measured over the course of the laboratory tests for both, the "Jumla Design" and KU-2 secondary combustion SMSs.

With a Chimney Temp.1 of ~400 °C and a Chimney Temp.2 of ~250 °C the "Jumla Design" SMS loses considerable energy through the hot flue gases without converting them into the expected energy services, resulting in an overall lower stove efficiency.

The KU-2 stove's chimney Temp.1 is around 250 °C throughout the test, while its chimney Temp.2 is between 150–200 °C. These lower temperatures compared to the

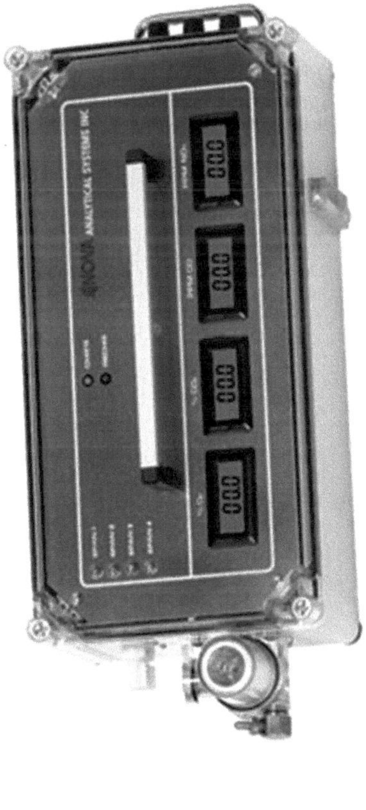

Fig. 4.25 Gas analyser, measuring flue gases' O_2, CO_2, CO and NO_x contents (%) in the chimney

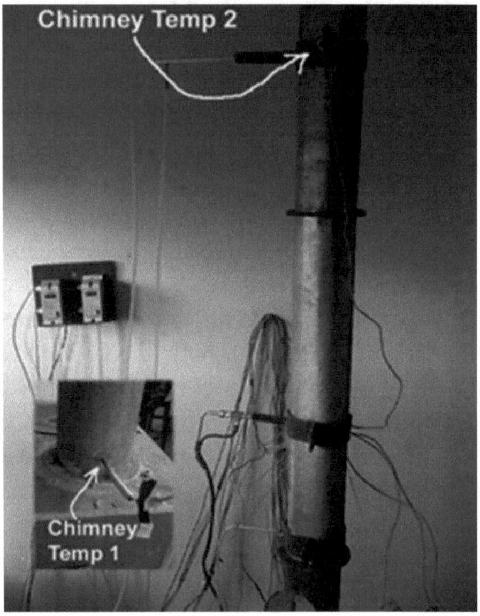

Fig. 4.26 Chimney temperature measurements: at the bottom (Chimney Temp.1) and at the middle of the nine ft long chimney (Chimney Temp.2)

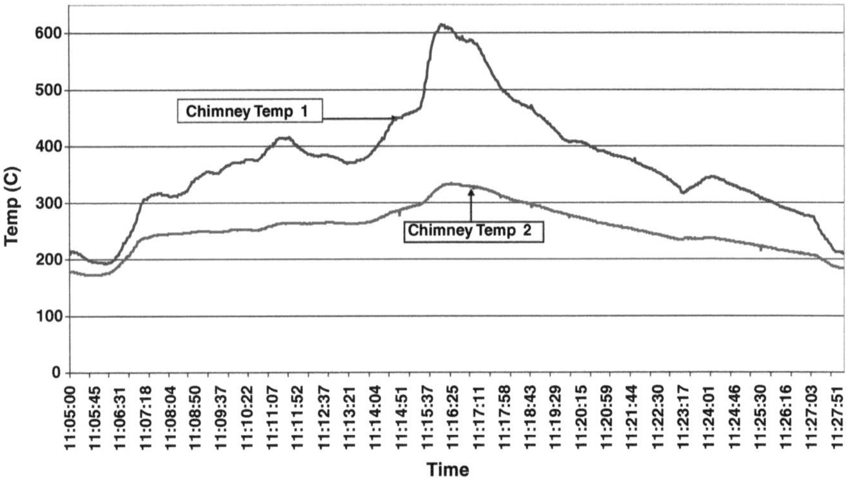

Fig. 4.27 Flue gas temperatures "Jumla Design" SMS

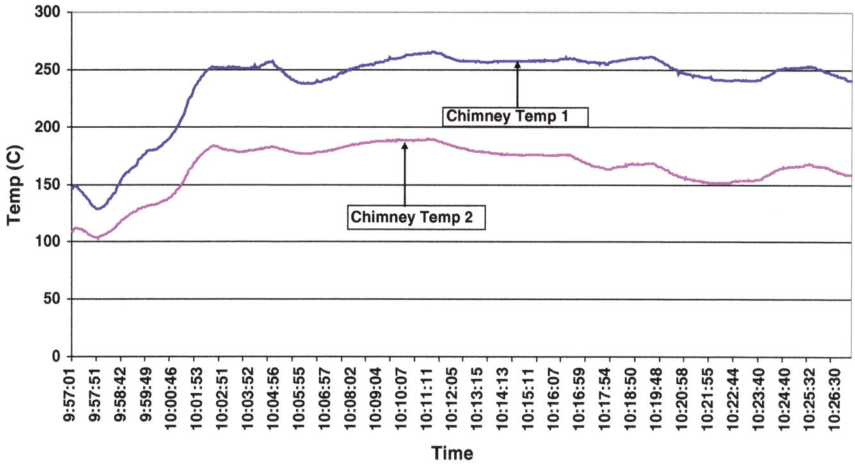

Fig. 4.28 Flue gas temperatures KU-2 SMS

"Jumla Design" stove indicate that more converted energy is utilised for cooking, heating water and the kitchen. This in turn results in overall higher stove efficiency.

4.1.5.3 Combustion Chamber Temperature

Figure 4.29 shows that the Chamber Temp.1 is around 600 °C for the "Jumla Design", with maximum values of up to 700 °C over a short time period. Chamber Temp.2 is lower for the first half of the test period and then reaches similar temperatures during the second half of the test period.

Comparing the graphs and values it can be concluded that the combustion process in the "Jumla Design" stove is less efficient compared to that in the KU-2 stove. The burning of volatile gases in the secondary combustion chamber of the KU-2 stove should generate higher temperatures than in the primary combustion chamber. But the primary combustion chamber in Fig. 4.30 produces temperatures between 600 and 700 °C, while the secondary combustion chamber produces 450–600 °C. Thus, it can be concluded that no steady secondary combustion process has taken place. However, the sudden increases in secondary combustion and burner temperatures between 10:07–10:10 (Fig. 4.22 pot hole front (right) graph), suggest that secondary combustion probably occurred, though only sporadically and over short time periods.

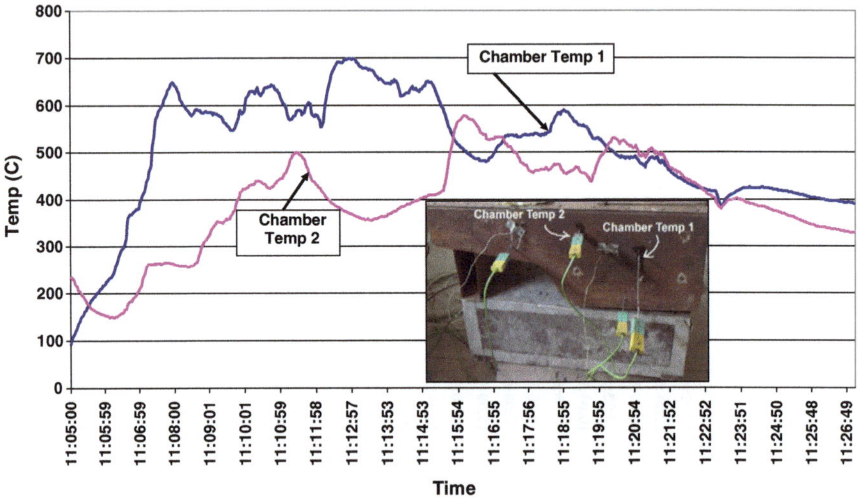

Fig. 4.29 Combustion chamber temperature of the "Jumla Design" SMS

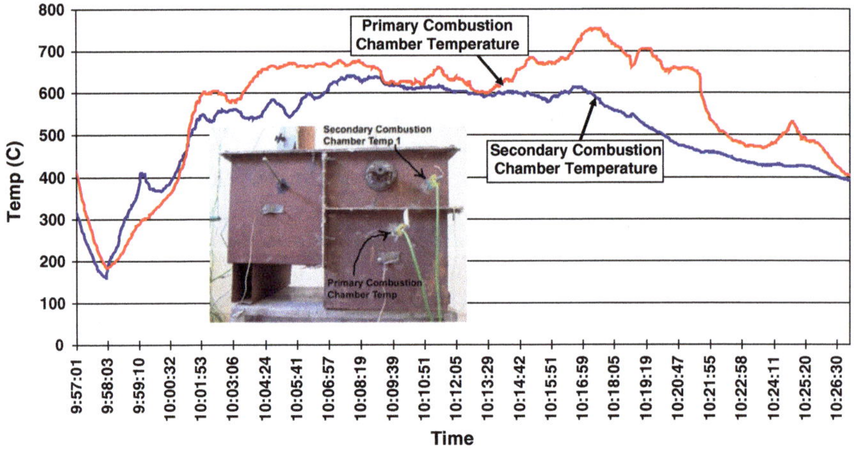

Fig. 4.30 Combustion chamber temperature of the KU-2 Prototype SMS

4.1.6 Secondary Combustion Air Temperature for KU-2 Prototype

The products of primary combustion (aromatic compounds, methane etc.) can ignite when they are mixed with adequate hot air. As the tests showed, the KU-2 stove's hottest zone is within the primary combustion zone (600–700 °C) rather than in the secondary combustion zone.

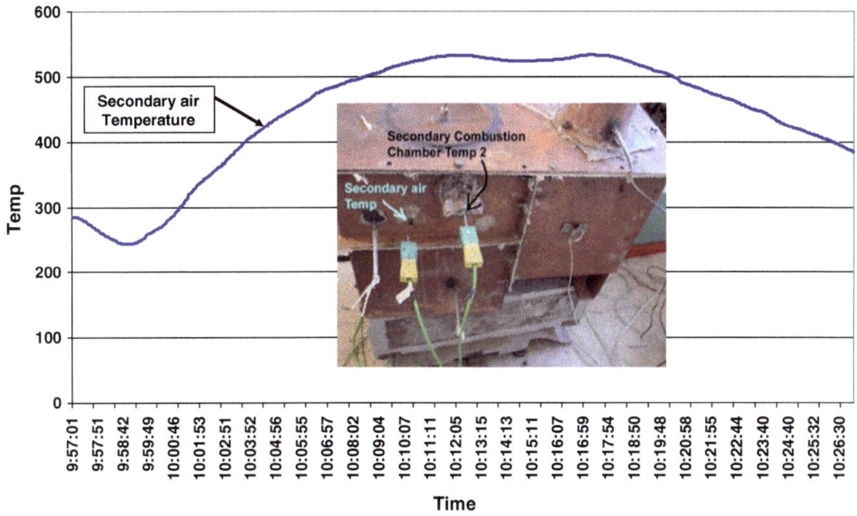

Fig. 4.31 Secondary air temperature in KU-2 Prototype

Ideally the secondary zone should reach even higher temperatures, but with the present KU-2 design the secondary air can not be preheated up to ≥630 °C, needed to ignite the volatile gases for a secondary combustion process. As Fig. 4.31 shows, the secondary air temperatures were recorded in the range of 450–530 °C with a maximum of 540 °C.

The test results and data support the conclusion that the combustion process is not only taking place more efficiently in the KU-2 stove primary combustion chamber compared to the "Jumla Design" stove, but that the volatile gases may be ignited before reaching the secondary combustion chamber, thus providing the high primary combustion zone temperatures of 600–700 °C.

However, the KU-2 stove's overall performance and test results show that higher combustion temperatures are achieved in both chambers and thus also at the burners. Further, the flue gas temperatures in the KU-2 stove are clearly lower throughout the test phase, pointing to the higher utilisation of the converted energy for the needed energy services of cooking, water and room heating. Thus the test results justify stating that the KU-2 stove achieves measurably better results compared to the "Jumla Design" stove.

4.1.7 KU-3 Secondary Combustion Stove Prototype

The years of experience in designing, manufacturing, lab testing and data monitoring of all the "Jumla Design", KU-1 and KU-two stoves, as well as the extended field experience with the "Jumla Design" SMSs, provide enough data and information to continue the research and development of the next model of secondary combustion stove, the KU-3. The main aim is to achieve a more stable secondary combustion process and thus improve the combustion efficiency, while keeping the cultural appropriateness of the stove. The modifications for the KU-3 stove, based on the lessons learned thus far, include:

- The chimney position at the centre of the back end of the stove to achieve a more uniform temperature/energy distribution field for the two back burners' cooking surface.
- Redesigning the baffle sheet before the exhaust pipe, to ensure more heat transfer toward the cooking surface.
- Redesigning of the cooking surface to ensure better radiation and convective heat transfer to the cooking pots.
- Improve the secondary air passage to prevent hot air leakage between adjacent joints.
- Reduce the water heating tank area to increase cooking and room heating capacity and to minimise the manufacturing cost and weight of the stove.

4.1.8 Conclusions

Due to the low income and limited resources of the people living in the harsh environments of Humla's high-altitude mountain areas, firewood will continue to be their main energy source for decades to come. But their fragile environment, including their sustainable timber supplies and their vulnerable hillside slopes without trees, as well as their health, must be protected and sustained. Therefore they must have the means to burn firewood far more efficiently at affordable cost. Further, solar energy and pico-hydro power services need to be made available to reduce the local people's dependence on firewood. But, first and foremost, an improved smokeless metal stove is needed to reduce the hazardous indoor air pollution, to improve the overall health conditions of the people, and to reduce the drudgery for the women and children of collecting firewood each day. Further, lower firewood consumption will help to reduce deforestation.

Finally, it should be noted that a SMS is just one of the projects, which has to be implemented as part of a sustainable, long-term holistic community development project with the local community as project partner. The "Jumla Design" SMS has always been installed by RIDS-Nepal as one part of an integrated HCD project such as the "Family of 4", in all the working areas and communities. Other projects such as a pit latrine for each family, basic indoor lighting and access to clean drinking water, are the other key pillars for a long-term HCD project. Together, they bring synergistic benefits beyond those achievable with individually implemented projects.

4.2 Thermal Evaluation of a Greenhouse in the Remote High-Altitude Area of Humla, Nepal

4.2.1 Introduction

Infant mortality in Nepal, compared to developed nations, is high ranging from 53 to 86 per 1,000 live births for the richest and poorest 20 % of the population respectively. The UNDP report (UNDP 2006) estimates that in 2001–2003, 17 % of the nation's population was estimated to be malnourished. In children <5 years, this was reflected in 48 and 51 % being underweight and underheight for age respectively. Over 20 % of babies were born underweight. These average statistics, alarming as they are, mask a more serious situation for rural and remote people in Humla, where malnutrition and its effects are much worse, as previously discussed.

Food security is defined as "a situation that exists when all people, at all times, have physical, social and economic access to sufficient, safe and nutritious food that meets their dietary needs and food preferences for an active and healthy life" (FAO 2003). As previously mentioned, out the 75 districts in Nepal, Humla belongs to the poorest. Using a ranking of 1 (best) to 75 (worst), Humla was ranked in the overall index at 74th position in terms of poverty, deprivation, socioeconomic and infrastructural development, and women's empowerment (NIDS 2003).

Due to a shortage of arable land and a harsh climate, high-altitude areas of Nepal suffer both chronic and acute malnutrition. Appropriate greenhouse technology can extend the growing season considerably. Experience in the Ladakh region of India indicates that year round cropping is possible in greenhouses in cold mountainous areas. A simple 50 m^2 greenhouse has been constructed in the RIDS-Nepal office in Simikot, Humla at 3,000 m altitude. This section describes the evaluation of the thermal performance of that greenhouse. Both, measurement during the cold winter months and simulations were used in the evaluation. Measurements indicate that the existing design is capable of producing adequate growing conditions for some vegetable crops, but that improvements are required if crops like tomatoes are to be grown successfully for longer time periods. Options to improve the thermal performance of the greenhouse building envelope, such as wall insulation, double glazing and using a thermal screen, have been investigated through simulation with a validated TRNSYS[2] (2005) model, a modular simulation program for transient systems. The impact of the addition of night time heat from internal passive solar water collectors was also simulated.

[2] TRNSYS is a program designed to simulate transient systems. It is the international benchmark software for predicting the performance of solar energy systems, but is also used extensively by researchers interested in building thermal performance. It is modular in nature and components of engineering systems are described in general terms as subroutines. A model of a particular "system" is created by linking relevant subroutines together and by supplying operational parameters specific to that system. TRNSYS web site, with the latest version (version 17 March 2011) is available at: http://sel.me.wisc.edu/trnsys/features/ (accessed July 26, 2012).

Table 4.1 Averaged climatic summary of data collected at Simikot, Humla from May 2004 to March 2007. Figures in brackets indicate number of days of complete data

Month	Average monthly horizontal solar radiation ($MJ/m^2/d$)	Average monthly ambient temperature (°C)	Average monthly maximum ambient temperature (°C)	Average monthly minimum ambient temperature (°C)
January (82)	12.6	6.6	13.2	−1.4
February (55)	14.0	6.6	14.5	1.2
March (79)	17.4	8.1	15.4	2.5
April (44)	20.1	11.5	17.8	6.0
May (93)	19.0	14.5	20.0	9.4
June (90)	17.2	17.1	22.6	12.3
July (93)	14.9	17.9	22.5	14.9
August (77)	14.7	17.2	22.2	14.0
September (57)	17.7	16.8	22.6	12.4
October (39)	16.7	12.4	19.5	6.4
November (42)	14.2	8.5	17.4	1.1
December (93)	12.1	7.4	16.0	1.1

4.2.2 Humla Valley

Less than 1 % of the land in the Humla Valley is arable due to steep slopes, rocks and forest cover (DDC 2004). In addition, snow cover for 4–5 months of the year further reduces the opportunity for year round fresh food production. As a result, and unsurprisingly, there is chronic malnutrition among the local population. Low average daily ambient temperatures (Table 4.1), accompanied by even lower average overnight minimum temperatures and snow, means that little can be grown outside from November to March in the Humla Valley. Planting outside begins in early April but little fresh produce is harvested until the end of May resulting in almost seven months of fresh food deficit, particularly vegetables. Despite the low monthly winter temperatures, the solar radiation levels over the winter months (Table 4.1) suggest that an appropriate greenhouse technology could be used to increase food production. Greenhouses built in other high-altitude areas have demonstrated that growing seasons can be greatly extended, even achieving year round production in some cases.

4.2.3 High-Altitude Greenhouses

While protected horticulture is increasingly practiced in many developing countries, the particular characteristics of Nepal pose special problems. The country is among the poorest in South Asia and its mountainous terrain has severely impeded

Fig. 4.32 Example of a simple polyethylene growing structure in Simikot

infrastructure development. In Humla, as noted earlier, there are no vehicles or electricity and only a limited number of commercial enterprises. Thus solutions which might have been used elsewhere are often simply not feasible. Some success with high-altitude greenhouses for vegetable production in the Ladakh region of India has been reported (Cooper and Fuller 1983). The walls of these 50 m^2 solar greenhouses were constructed from sun dried bricks, insulated with straw and UV-stabilised polyethylene was used as the glazing. More than 100 greenhouses were built within three years from 1998. It is estimated that it is possible to earn US$50 per month for about 2 hours of work inside a greenhouse each day.

Experience with greenhouses for food production at high-altitudes in Nepal, is more limited, although the idea of protected cropping is not unknown. Some small, low polyethylene covered structures can be seen around Simikot (Fig. 4.32). Introduced in 2004 by a Nepali NGO (Appropriate Technology Asia), they are used only for frost protection. Other than these, there are no greenhouses designed for year round intensive food production in the area, except the RIDS-Nepal greenhouse in Simikot and this is the focus of this evaluation.

4.2.4 The RIDS-Nepal Greenhouse in Simikot

As part of their food security and nutrition programme, RIDS-Nepal constructed a greenhouse at HARS and RIDS-Nepal office at Simikot in September/October 2004 (Fig. 4.33). The HARS greenhouse design is simple and it is built largely from locally available materials. The overall inside dimensions of the greenhouse are 10 m by 5 m. The south facing roof is inclined at 15° to the horizontal. This slope was dictated by site considerations and is far from ideal in terms of snow shedding or maximising penetration of winter solar radiation. There was also some

Fig. 4.33 External view of the HARS greenhouse at Simikot

Fig. 4.34 Internal view of HARS greenhouse at Simikot

shading at certain times of the year due to an adjacent building. The low profile of the building was also not ideal ergonomically, but the impact is minimised by short working periods.

The walls are made of stone and mud, and the roofing frame is constructed from 75 mm by 50 mm timber. The roof is supported internally by vertical posts located in the ground. The glazing material is UV-stabilised polyethylene imported from India. Until late December 2006, a single glazing layer was used. Subsequently, a second layer of polyethylene, with a 10 cm gap, was installed. The crop, which to date has largely been leaf vegetables such as spinach, is grown directly in the ground (Fig. 4.34).

The local RIDS-Nepal staff use the HARS greenhouse to grow their out of season vegetables and to raise more awareness in the local communities. The thermal performance of the greenhouse, however, was not evaluated, and therefore possible improvements have not been identified. Until 2011, 35 similar, though often smaller, greenhouses have been built in various villages in the upper Humla Valley in response to villagers' requests. It is important that the best design, consistent with local needs and infrastructure, is promoted. Therefore the thermal performance of the HARS greenhouse has now been evaluated using a combination of direct measurement and simulation.

4.2.5 Measurement

Various thermal performance data was collected from the greenhouse in the winter months November 2006–March 2007. Total solar radiation on the horizontal plane outside the greenhouse was measured using an unshielded photoelectric pyranometer (Pacific Systems Pty. Ltd., SolData SPC80-SN243, spectral range 300–2,800 nm). Ambient temperature and relative humidity were also measured by a weather station (Spectrum Technologies, Inc., Watch Dog Model 2900ET), which is part of the long-term data collection system at the HARS. The weather station is located adjacent to the greenhouse.

Battery powered data loggers (Onset Corp., Hobo H8 and U12-006 Series and TMC HD temperature sensors) were used to measure internal greenhouse air and humidity at four locations within the greenhouse. Soil temperatures at the surface (50 mm), 200, 600 and 2,000 mm depth inside and outside the greenhouse were also measured with this type of sensor. These devices were cross calibrated against each other at the start of the experiments. A hand held hot wire anemometer (TSI Inc., VelociCalc Model No: 8350-1) was used to measure air movement at various points in the greenhouse.

The data collected over the winter months of 2006–2007 have been analysed to understand the greenhouse performance. Figure 4.35 illustrates the typical daily behaviour of the greenhouse in terms of temperature and relative humidity in mid January 2007. The outside ambient overnight minimum temperature of −5.8 °C occurs at approximately 6.30 a.m. On average, between midnight and sunrise, the greenhouse air temperature is approximately 7.5 °C above the outside ambient temperature. Sunrise occurs at approximately 7 a.m and the outside ambient and greenhouse air temperatures start to rise soon afterwards. During the daytime the average greenhouse air temperature is approximately 12 °C above the outside air temperature, with a peak of 34 °C at approximately 1 p.m. The average greenhouse relative humidity is 65 %, approximately 25 % above the ambient level. Horizontal solar radiation on the day was 13.8 MJ/m^2.

Figure 4.36 illustrates the typical daily behaviour of the soil inside and outside the greenhouse in terms of temperature in mid January 2007.

The temperature just below the surface of the soil inside the greenhouse follows approximately the same pattern as that outside the greenhouse but with a reduced swing and higher peak. During the night time, the sub-surface soil in the

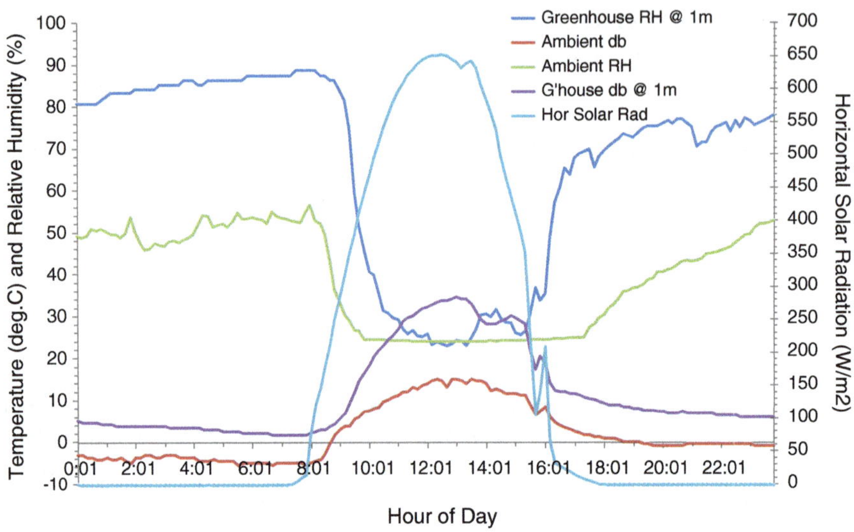

Fig. 4.35 Typical daily greenhouse air and relative humidity levels in mid January

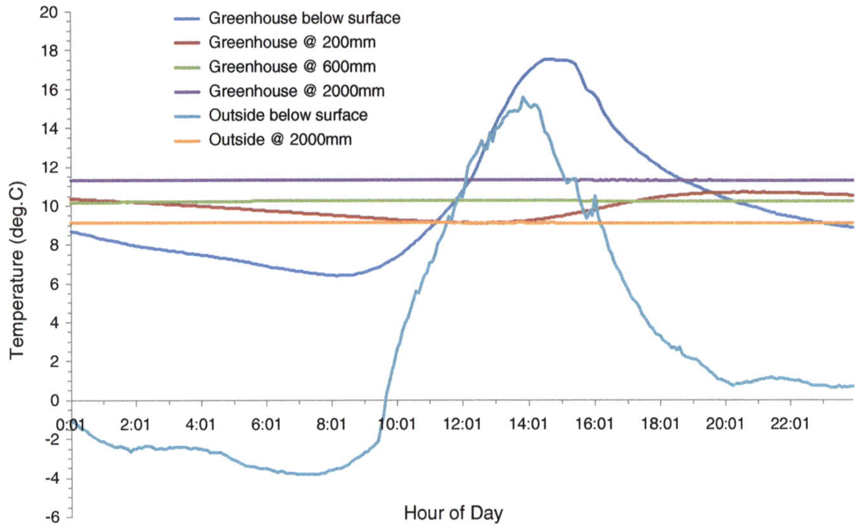

Fig. 4.36 Typical daily greenhouse and outside soil temperatures in mid January

greenhouse is approximately 7–9 °C warmer than the outside sub-surface soil. At 200 mm, the temperature of the soil in the greenhouse does not fall below 9 °C. At 600 and 2,000 mm depths, the soil temperature in the greenhouse is relatively constant at 10.3 °C and 11.3 °C respectively.

Figures 4.35, 4.36 indicate that, while not ideal, growing conditions in the existing greenhouse are certainly suitable for some vegetable crops such as spinach, carrots and beans. Improvements in the thermal performance of the greenhouse such as increasing the overnight minimum air and soil temperatures will be beneficial to crop growth and productivity. The options for performance improvement can be best identified by simulation.

4.2.6 Simulation Model

Numerous dynamic thermal models have been developed to investigate the thermal performance of greenhouses (Cooper and Fuller 1983). Most of these are relevant to the advanced type of protected horticulture practiced in industrialised countries. These models have been developed for structures where the glazed to opaque surface ratio is very high. In addition, cropping levels and therefore latent loads are also high. Thermal mass levels (apart from the floor) are very low. In these respects, the HARS greenhouse differs substantially from the modern horticultural greenhouse. The roof and less than half of the side walls are the only glazed surfaces, and even here the ratio of structural members to glazing is high because the use of locally available materials and construction skills has been maximised. All four walls are made of local stone, which provides substantial amounts of thermal mass capacitance. With respect to these two factors, the structure has more similarities to traditional building rather than a greenhouse. Cropping levels at this stage in the development of this technology in Nepal are also low compared to modern greenhouse practice. For the above reasons, a more traditional building model has been used to investigate the thermal performance of the greenhouse, rather than a specialist greenhouse model. The standard building sub-routine within TRNSYS (Type 56) has been used. This model allows the user to define a building of any size or shape. User supplied parameters define material properties and dimensions. Schedules can be used to determine inputs at a specified time. A radiation processor within TRNSYS uses horizontal radiation data to calculate hourly values of solar radiation on all of the building's surfaces.

4.2.6.1 Model Description

The Type 56 sub-routine requires various fixed parameters to be supplied and assumptions to be made by the user. Table 4.2 indicates the material properties used.

Several assumptions have been made and included in the model. An infiltration rate of one volume air change per hour is reasonable for a polyethylene covered greenhouse. Until April 2007, there were no vents installed. The doors are small and often in winter these are not opened. Although a ground reflectance value of 0.7 is normally used, when the ground is covered by snow, a value of 0.2 was used

Table 4.2 Material properties used in TRNSYS simulations

Material	Density (kg/m^3)	Conductivity (W/mK)	Specific heat capacity (kJ/kg°C)	References
Granite	2,640	3.00	0.82	Cornwell (1977)
Earth	1,200	0.37	0.88	Rozis (1996) and Garzoli (1988)
Timber	700	1.40	2.60	Cornwell (1977)

in these simulations because in mid November this had not yet occurred. The north wall of the greenhouse is almost entirely buried by soil. Both this wall and the earth floor are assumed to be bounded by a thermal layer at a constant temperature of 12 °C. This temperature was determined from soil temperature measurements made in November. Sky temperature was assumed to be constant throughout the simulations and set at ambient air temperature less 12 °C. Although lower levels will occur when there are clear night skies, no cloud cover observations were available and an average value was therefore assumed. The TRNSYS window library does not contain data for polyethylene and so a single layer of glass was assumed. While the solar and thermal properties of thin film polyethylene and glass are different, several factors reduced the impact of these differences. The polyethylene used was canvas grade and therefore significantly thicker than the normal 150 μ horticultural grade, there was some dirt and moisture on both sides of the glazing, reducing transmittance in all wavelengths, and finally an opaque canvas was pulled over the glazing at night.

4.2.6.2 Model Validation

The TRNSYS model of the HARS greenhouse was validated using one week of data collected in November 2006. A climatic data file of hourly values of global horizontal solar radiation, and ambient dry bulb temperature and relative humidity was constructed from the weather station data. Hourly predictions of greenhouse air temperature, relative humidity and soil surface temperature from the model were made and compared against actual measurements made at the same period (Figs. 4.37, 4.38, 4.39). Predicted air temperatures follow the same diurnal pattern as the average of the measured values, with the predicted rates of rise and fall in temperature following actual measurements well (Fig. 4.37). Although the model tends to under predict the daily maximum, the prediction of daily minimum temperatures shows better agreement. The under prediction may be due to the higher solar transmittance of polyethylene film compared to glass.

The model also calculated the temperature of the soil at the surface. Since actual measurement of this temperature is not available, the prediction has been compared with the temperature of the soil 5 mm below the surface (Fig. 4.38). Some difference between the prediction and the measurement is therefore understandable. The surface responds instantaneously to thermal inputs, whereas there is some delay, shown by the offset, in the response of sub-surface layers. Figure 4.36

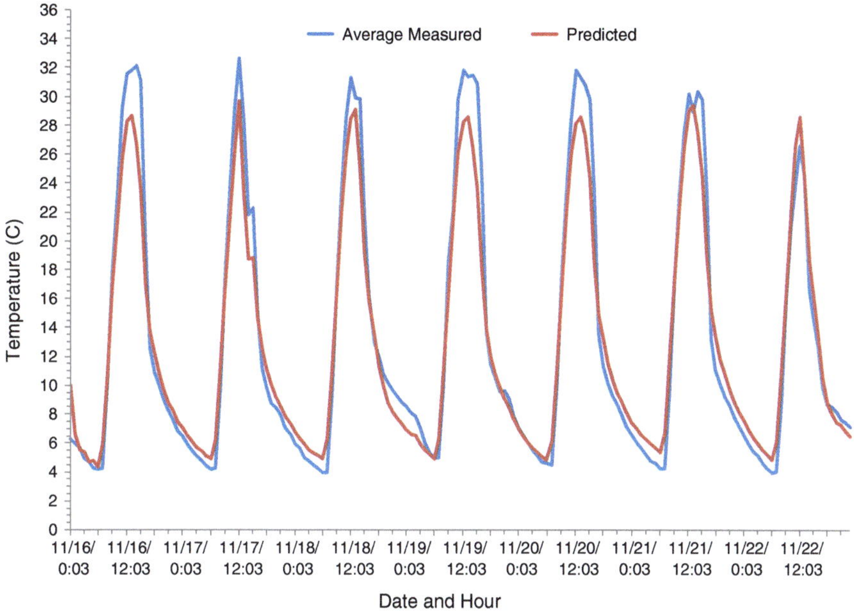

Fig. 4.37 Comparison of measured and predicted greenhouse air temperatures during November 2006

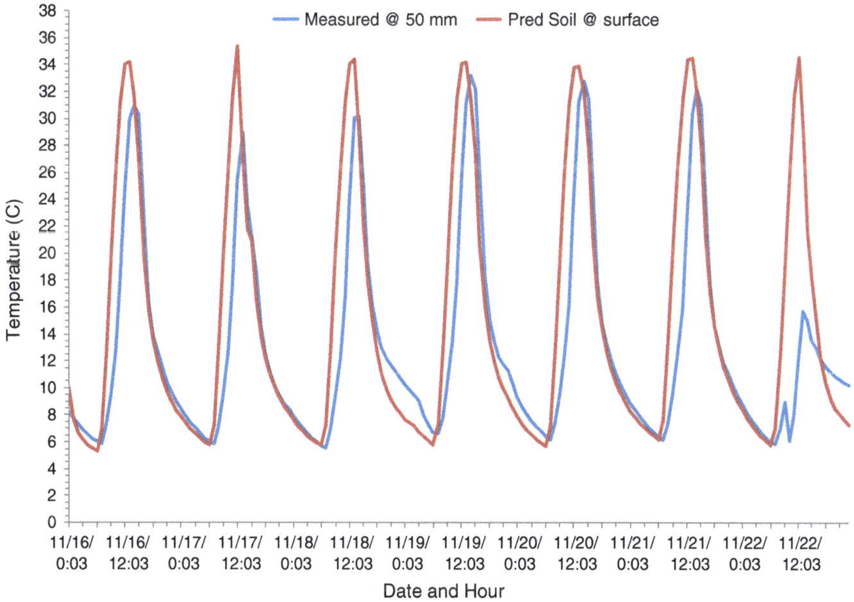

Fig. 4.38 Comparison of measured and predicted greenhouse top soil temperatures during November 2006

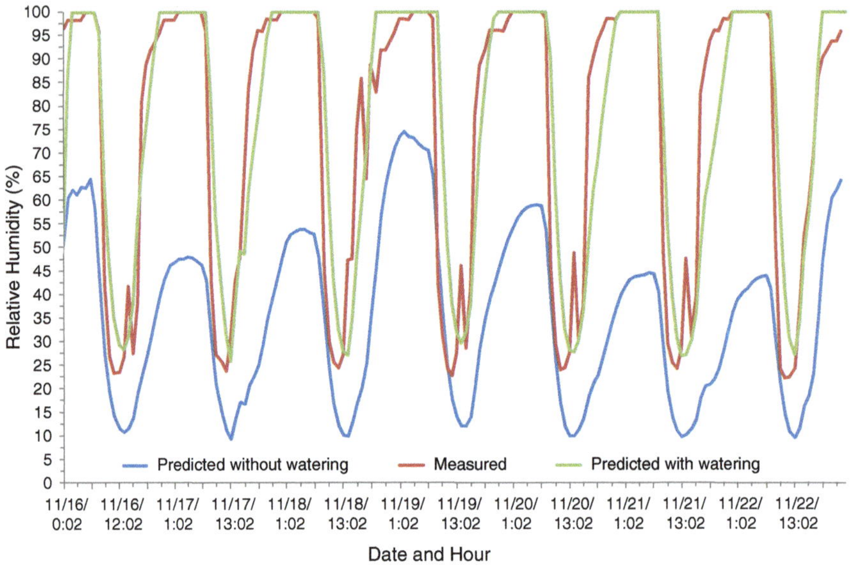

Fig. 4.39 Comparison of measured and predicted greenhouse relative humidity—with and without watering during November 2006

also illustrates this behaviour in both outside and inside greenhouse soil temperatures.

Type 56 is a TRNSYS sub-routine designed to simulate the performance of buildings. Although the HARS greenhouse has many similarities to a conventional building, it also contains plants and soil, which add moisture to the surrounding air. Initial simulations did not include any humidity gain and the poor agreement between predictions of relative humidity and the measured values at one metre is evident in Fig. 4.39. Unfortunately there are no records of water use in November 2006. However, in the first three months of 2007, daily water use varied from 11 to 24 liters per day. Photographic records also indicate only a relatively low cover of young spinach plants in November 2006. Assuming that only ~5 liters of water were used a day and that evaporation occurred evenly each day, the model was programmed to assume an hourly addition of 0.23 kg of moisture. Figure 4.39 shows that this assumption produced a good representation of reality, as the predictions of greenhouse relative humidity rises and falls in line with measurements.

4.2.7 Performance Improvements

The measured and predicted results over a one week period in November 2006 were considered to be accurate enough to have confidence in the use of the model for long-term predictions. The validated model was therefore used to investigate various strategies to improve the performance of the HARS greenhouse. These include the use of a thermal screen, wall insulation and "solar sleeves". The following sections describe how these strategies were simulated and their impact.

4.2.7.1 Thermal Screen

Thermal screens or night blankets made from multi-layers of advanced synthetic materials have been used in greenhouses in industrialised countries for approximately 30 years. Simpler versions of night blankets are in use in developing countries. Screens made of various materials are drawn across the crop at night and reduce night time heat losses from the greenhouse glazing caused by convection and radiation. A screen also reduces the volume of air in the greenhouse at night and this subsequently reduces heating requirements. Depending on the quality and complexity of the screen, fuel savings of up to 60 % have been reported when screened and non-screened greenhouses have been compared.

It was not possible to simulate directly the addition of an insulating layer within the greenhouse because any layer must be withdrawn during the day because the Wall Layer defined in the Type 56 subroutine is fixed throughout the simulation. An alternative approach was therefore used to simulate the effect of a thermal screen. The energy "saved" by the change in the overall heat loss of the greenhouse was calculated at each time step. The energy saving is a function of the reduction in the overall heat loss for the greenhouse due to the thermal screen, the glazing area and the temperature difference between the greenhouse and outside air. This amount of energy was then introduced into the greenhouse as an energy "gain" during the hours of thermal screen use. A reduction in overall heat loss of 1.8 W/m^2/°C was considered a realistic saving for the simple sort of thermal screen envisaged for the HARS greenhouse (FAO 1988).

4.2.7.2 Wall Insulation

The present walls of the HARS greenhouse are 450 mm thick and made of granite. Some mud is used as mortar. Granite has a higher conductivity than concrete and is a poor insulator. An alternative construction technique could be envisaged in which a layer of insulating material is placed between two 200 mm thick walls. The most available insulating material in Humla is dried pine needles, which are already used as insulation in the roofs of the traditional dwellings in the area. Values of thermal conductivity (0.09 W/m^2 °C), specific heat (0.28 kJ/kg °C) and density (0.27 kg/m^3) for thatch (Garzoli 1988), were used in the simulations in lieu of more precise data.

4.2.7.3 Solar Sleeves

Passive solar water heaters or "solar sleeves" have reportedly been used successfully by researchers and growers in at least nine countries in southern Europe (FAO 1988). Water filled transparent polyethylene film sleeves (Ø200 mm) are placed on strips of black plastic on top of 50 mm of insulation material between the crop rows (Fig. 4.40). The ends of the sleeves are tied to stakes driven into the soil at the ends of each tube.

During the day, solar radiation passes through the transparent sleeves, and is absorbed by the black plastic sheet. The heat generated is transferred to the water in the sleeves. At night, the heat in the water is transferred to the greenhouse air by convection and to the adjacent plants by radiation. As the plants grow, they increasingly shade the sleeves and reduce their effectiveness. Their main benefit is therefore in establishing and promoting early growth. The system is both simple and inexpensive and therefore is potentially feasible for use in the Humla greenhouses.

A four days experiment to determine the effectiveness of solar sleeves was conducted in April 2007 in the HARS greenhouse. Figure 4.41 shows the temperature rise and fall in the water contained in an Ø200 mm tube, placed on flat black plastic polyethylene socks filled with dry pine needles. On the basis of these experiments, it was calculated that on average 94.0 MJ could be captured in the greenhouse during the day and released overnight, assuming that there are seven 5.0 m tubes on the greenhouse floor, containing in total 1,120 liters of water. To determine the potential impact of the solar sleeves on greenhouse thermal performance, this amount of energy was released over a 12 hours period at night in the TRNSYS model.

Fig. 4.40 Solar sleeves inserted between crop rows

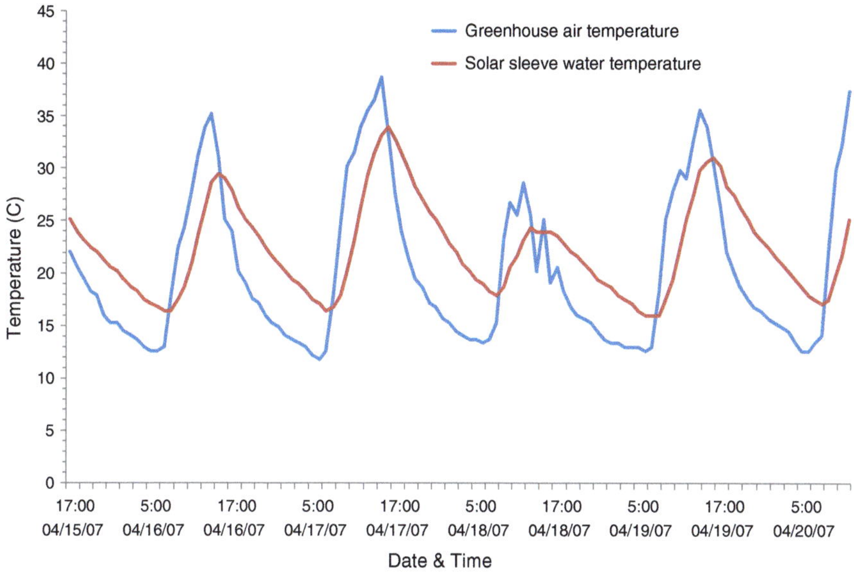

Fig. 4.41 Diurnal temperature of water in clear plastic tube in the HARS greenhouse

4.2.7.4 Impact Strategies

Since the HARS greenhouse is unheated, the impact of the strategies described above was determined by their predicted effect on night time air temperatures, rather than energy consumption. Table 4.3 shows their individual and combined impact during the same week used earlier to validate the model.

Almost all of the strategies impact positively on the indicators chosen. The one exception is that the use of 50 mm of pine needles in a cavity wall, which does not raise the absolute minimum overnight temperature experienced in the period. However, this strategy does raise the average greenhouse air temperature between 6 p.m and 6 a.m. The advantage of this strategy is that no new materials or technologies are required, and no additional user action is needed. This strategy cannot be applied to the current greenhouse, but could be considered for future greenhouses.

The thermal screen increases both the average and overnight minimum temperatures. It also has the potential to provide shade in summer. The disadvantages of a thermal screen in this environment are that the material must be imported and the user must operate the screen daily to achieve its benefits.

The most effective single strategy appears to be the solar sleeves, which raise the absolute minimum by 2.3 °C and the average overnight temperature by 4.2 °C. While they are the most effective in thermal terms, they do have some practical disadvantages. The sleeves reduce the productive growing area and make access in the greenhouse more difficult.

Table 4.3 Predicted TRNSYS simulation impact of individual and combined strategies on various greenhouse air temperatures

Strategy	Average greenhouse air temperature during the week (°C)	Average greenhouse air temperature between 6 p.m–6 a.m (°C)	Minimum overnight greenhouse air temperature during week (°C)
None	13.1	8.2	4.4
Thermal screen	13.8	9.3	5.1
Wall insulation	13.7	8.8	4.4
Solar sleeves	15.7	12.4	6.7
Combined	17.9	15.5	7.1

In combination, the strategies are obviously superior to any single strategy. Overnight temperatures were predicted to be higher by over 7.3 °C, and the minimum experienced during the week was 7.1 °C, compared to 4.4 °C in the current design. The average air temperature in the greenhouse rose by 4.8 °C compared to the current design.

4.2.8 Conclusions

The Humla valley is a remote and mountainous part of North-West Nepal. The area suffers from a permanent food deficit because of the harsh terrain and climate. For the five winter months of the year, it is impossible to grow fresh vegetables outside. Protected horticulture (i.e. greenhouses) is seen as a technology to improve and extend the fresh food supply. A simple greenhouse, constructed from local materials, has been operating at the RIDS-Nepal HARS in Simikot, the main town of the Humla District since October 2004. However, no systematic evaluation of the greenhouse and its thermal performance has been undertaken. The thermal performance of the design needs to be understood so that changes, if required, can be made and so the best and most appropriate technology is offered to potential users.

The greenhouse has therefore been evaluated through measurement and simulation. Measurements indicated that the conditions within the greenhouse are suitable for some vegetables, but need to be improved if a wider variety and greater output is required. In order to evaluate possible design or other improvements, a simulation model using TRNSYS Type 56 was developed and validated against measurements. The predictions of the model were compared with greenhouse air and soil temperatures, and relative humidity measurements collected between 16 and 22 November 2006. The model was considered to be sufficiently accurate to allow various energy saving and collection strategies to be evaluated. These were the provision of wall insulation, a thermal screen and passive solar thermal water storage. These strategies were evaluated separately and in combination. Their effect on average greenhouse air temperature and overnight temperatures was predicted.

Overall, the simulations show that low cost and appropriate changes to increase overnight and daily average temperatures in winter could be made to the existing design. The simulations indicate that the single most promising change is the inclusion of solar passive water storage collectors or "solar sleeves" between the crop rows. The model predicts that these solar sleeves would raise the average greenhouse air temperature by 2.6 °C and the overnight air temperature would be increased by 4.2 °C. However, this benefit would be achieved at the expense of crop production area within the greenhouse.

4.3 Evaluation of a Solar Drier in a High-Altitude Area of Nepal

4.3.1 Introduction

Because Humla is a permanent food deficient area, it is of highest priority to improve general food security. As part of this objective, a solar drier has been introduced into a mountainous region of northwest Nepal with the aim of increasing and improving crop preservation. This section describes the evaluation of this solar drier. The purpose of the evaluation was to determine the effectiveness of the current design. The results of the evaluation indicate that in terms of drying efficiency the solar drier is not superior to sun drying. In large part, this is due to the poor air circulation through the drier. Collector and drying efficiencies of 20–31 and 13 % respectively were calculated. Although theoretically the solar drier could pay for itself in two years of use in the non-winter months, it is hard to justify the additional capital outlay, other than the possibility of more hygienic drying conditions, when sun drying could achieve similar results. As a result of the evaluation, some suggestions are made on an alternative and perhaps more suitable type of solar drier for the Humla District.

According to the local NGO Nepal Trust (2007), "in this food deficient area, a subsistence economy runs on inter-village trade, livestock and the cultivation of food grains". As previously mentioned, less than 1 % of the land is arable and snow cover for 4–5 months of the year reduces fresh food production to just 3–4 months per year. As a result, and unsurprisingly, there is chronic malnutrition among the local population. In a study of children in the districts of Humla and neighbouring Mugu, Emeriau (2006) found that global acute and chronic malnutrition rates were 12.3 and 63.9 % respectively. As Nepal-Trust succinctly states "survival is a serious business for the people of the Hidden Himalayas".

Humla is 430 km air distance northwest of Kathmandu and its main town is Simikot. The author, through Kathmandu University, initially established with RIDS-Nepal the HARS (Fig. 4.42) in Simikot in 2002 for research purposes. Since 2004, the HARS has been run by RIDS-Nepal, an independent Nepali based NGO.

Fig. 4.42 HARS in Simikot, Humla District, North–West Nepal

The station acts as the base from which RIDS-Nepal conducts its long-term community based development programs in the villages north of Simikot.

One of the functions of HARS is to test and evaluate various solar and other technologies prior to their introduction into villages. Currently under evaluation at HARS are a greenhouse and a solar drier, which are seen as integral parts of the current food security and nutrition programme of RIDS-Nepal. Testing and evaluation are vital to ensure the best and most appropriate technologies are introduced to villages.

4.3.2 Solar Drier Description

Between 2001 and 2003, the author and his students of the Department of Mechanical Engineering at Kathmandu University developed a number of small solar driers, one of which is now produced commercially by a Nepali company. Six of these commercial units have been installed in various villages in the Humla Valley. The solar drier can be defined as an "indirect type" of solar drier because the drying cabinet is opaque to solar radiation and heated air is generated in a solar air collector attached to the drying cabinet (Fig. 4.43).

The external collector dimensions are 1.68 $\times$ 1.01 m, giving a collector area of 1.7 m^2. The collector casing is made of galvanised iron (GI) sheet. The glazing is 3 mm thick, toughened window glass. The absorber consists of two layers. The bottom layer is a black painted GI sheet on top of the back insulation. The top layer is a black painted perforated plate, which rest on supports fixed to the bottom layer. Air enters the collector through an insect proof opening on the underside of the bottom of the collector, which is inclined at 40° to the horizontal. A baffle at

Fig. 4.43 External view of the solar drier evaluated

the bottom of the collector directs air in between the two absorbing layers. Some of this air passes through the perforated upper layer. The remainder stays between the two layers. Both air streams meet at the top of the collector, where the heated air enters the drier cabinet through a 7.5 × 95 cm opening covered with insect mesh.

The drying cabinet sits on an angle iron frame. The internal dimensions of the drying cabinet are 55 cm deep by 95 cm wide by 63.5 cm high. The cabinet is constructed from GI sheet on the outside and aluminium sheet on the inside with 25 mm of polystyrene foam between the layers. The cabinet contains six drying trays, each 48 × 93 cm. The useable drying area is 2.39 m^2. The trays are made of timber[3] with insect mesh fixed to the underside. Stops have been welded on the tray runners at various positions so that the drying air is encouraged to flow across the top of each tray. After passing across the trays, air is exhausted from the drying

[3] The first lot of solar driers had timber drying trays, in order to use more locally available material. Experience showed that these timber trays started to warp after some months of use, probably because the timber used was not fully seasoned, and the timber trays were exposed frequently to varying moisture content during the drying process. Besides, the warped trays were more difficult to put into and take out of the solar drier. Further, the solar drier was designed with a defined air flow pattern between the drying trays, like a serpentine path across the surface of the stacked trays, from the bottom tray upwards, to maximize the drying process. The warping of the trays must have caused some short-circuiting of the defined airflow mainly around their edges instead of across the trays as intended. Thus the poor air flow values measured inside the solar drier above each tray. These findings caused the author to change the drier trays from timber frames to aluminium frames.

cabinet through a 16 cm by 16 cm chimney, 84 cm high and insulated with polystyrene foam. A small 12VDC battery powered fan located inside the chimney draws air through the collector and drier.

Overall, the solar drier is well made. Sealing strips have been used around the cabinet doors and under the edges of the glass. Despite these seals, some leakage through the doors is likely to occur because the seals are not compressed enough when the doors are closed.

Some of the drying trays have warped and this is likely to cause some short circuiting of airflow around their edges instead of across the tray as intended.

4.3.3 Evaluation Methodology

4.3.3.1 Flow Rate Measurement

Initially the drier was tested without a crop i.e. in the "no load" condition. Temperature, humidity, solar radiation and flow measurements were taken on each hour between 9 a.m and 5 p.m. The purpose of these measurements was to determine the best achievable collector efficiency and the maximum daily energy that could be collected for the given ambient conditions. These factors are dependent on flow rate through the drier and this was determined using the following procedure:

- The hot wire anemometer was inserted in holes half way up the chimney.
- Three positions: quarter, half way and three quarters across the chimney were chosen as measuring points.
- After inserting the probe at these pre-determined positions, 1 min was allowed for the sensor to equilibrate.
- Four readings were then taken at 15 seconds intervals at each position.
- This procedure was repeated along east–west and north–south axes.
- The average velocity in the chimney was calculated from the 24 readings.

Drying efficiency is determined by how effectively the heated air passes over the crop. Airflow velocities were therefore measured both through and across the centre of each tray by inserting the hot wire anemometer in holes drilled in the side of the cabinet between trays. Flow across and up through the trays was estimated by orientating the tip of the anemometer in the appropriate direction.

Four measurements were made at the centre of each of the six trays. Prior to all flow rate measurements, the gaps around the doors were taped to ensure that there was no entry of air at this point i.e. short circuiting.

4.3.3.2 Efficiencies

There are various efficiencies associated with this type of solar drier and the following equations may be used to evaluate its performance. The overall drying efficiency (E_d) of the drier may be defined simply.

$$E_d = \frac{energy\ content\ of\ water\ evaporated}{energy\ input} \tag{4.1}$$

In general, overall drying efficiencies of 10–15 % might be expected for a solar drier relying purely on natural convection for air movement, while higher efficiencies (20–30 %) might be expected for solar driers using forced convection i.e. using a fan.

Since this solar drier has an attached solar collector, this part of the drier may be evaluated in the conventional way used to evaluate the performance of any solar collector. This procedure involves measurement of inlet and outlet temperatures, mass flow rate and solar radiation on the plane of the collector. Equation 4.2 generally defines solar collector efficiency (E_c).

$$E_c = \frac{energy\ collected}{incident\ solar\ radiation} \tag{4.2}$$

4.3.4 Instrumentation

In order to determine the various efficiencies described above, average global solar radiation on the plane of the collector, and dry bulb air temperatures at the inlet and outlet of solar collector were recorded at hourly intervals. Total solar radiation on the plane of the collector was measured using a photoelectric pyranometer (Pacific Systems Pty. Ltd., SolData SPC80-SN243). Battery powered data loggers (Onset Corp., Hobo-H8-Series) were used for the dry bulb measurements. The loggers were cross calibrated against each other at the start of the experiments. A hand held hot wire anemometer (TSI Inc., VelociCalc Model No-8350-1) was used to measure airflow at various points through the drier.

4.3.5 Results and Discussions

4.3.5.1 No-Load Conditions

Collector inlet and outlet temperatures, as well as those at Tray 2 and Tray 5 (considering Bottom tray = Tray 1; Top tray = Tray 6, i.e. trays numbered in ascending order) and at the entrance to the chimney were measured on 30 March

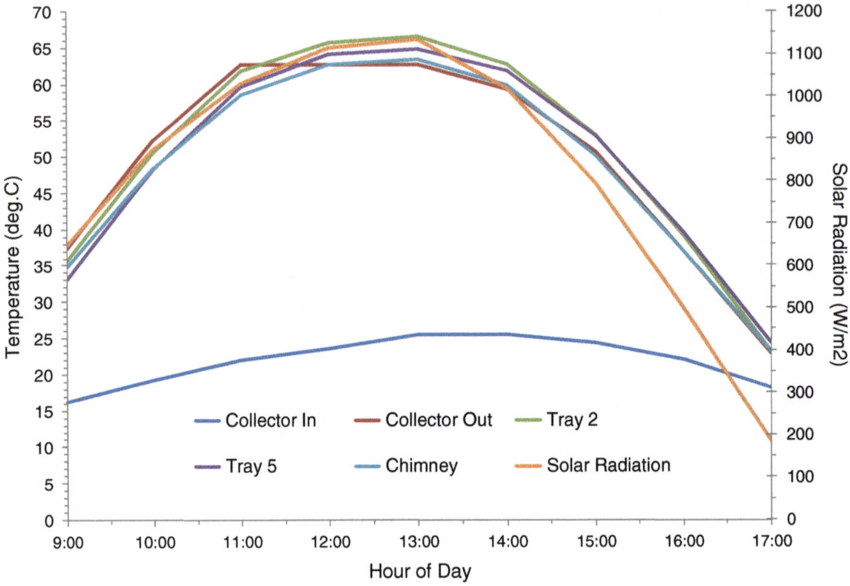

Fig. 4.44 No load collector inlet, outlet, plus other cabinet temperatures, measured on 30/3/2007

Table 4.4 Average velocities of air in chimney along E–W and N–S axes

Time	Average velocity (m/s) along E–W axis	Average velocity (m/s) along N–S axis
9	1.37	1.42
10	1.43	1.51
11	1.37	1.50
12	1.24	1.40
13	1.28	1.45
14	1.39	1.35
15	1.16	1.54
16	1.28	1.41
17	1.33	1.42

2007 (Fig. 4.44). As expected, the conditions inside the cabinet have a similar pattern and value to the collector outlet temperature.

Air velocities were measured at the same intervals in the drier chimney. This location was chosen because it offered the best opportunity to produce repeatable results. Prior to the measurements, the sides of the doors and at the junction of the solar collector and drying cabinet were taped to minimise the ingress of outside air. There was little or no wind during testing.

Table 4.4 shows the average of nine air velocity measurements made across the E–W and N–S axes respectively. The average velocity is 1.38 m/s with a standard deviation of 0.097 m/s. Given the conditions for testing, this was considered acceptable and the average value was used in all subsequent calculations.

The specific air mass flow rate through the solar drier was calculated to be 29.7 kg/h/m^2. This mass flow rate is only 43 % of that suggested by Peck and Proctor (1983) for a forced convection solar air heater. The low flow rate is the main reason for the high temperature of the outlet air (Fig. 4.44). The inlet air temperature is raised more than 40 °C on one occasion.

The relatively low air volume flow rate through the empty drier also indicates that at best, the velocity of air passing the crop cannot exceed 0.03 or 0.5 m/s if the air moves vertically or horizontally respectively. These velocities are significantly lower than the 2–3 m/s used in other forced convection driers. Some horizontal flow (0.21–0.25 m/s) was measured below the bottom tray (Tray 1), but otherwise no flow could be detected. This finding indicates that the strategy to encourage the heated air to follow a serpentine path across the face of the stacked trays is not working.

4.3.5.2 Full-Load Conditions

The solar drier was evaluated under "full-load" conditions by comparing its ability to dry sliced carrots against sun drying the same crop on a traditional bamboo drying tray.

The carrots were first washed and then cut into approximate 5 mm thick slices. A total of 6.738 kg of sliced carrots was spread across the six drying trays and 0.532 kg of slices on the bamboo drying tray. On each solar drier tray, nine slices were used as samples whose weight was measured in the morning and evening of each drying day. The poor airflow identified in the "no-load" condition was evident from these measurements taken at the end of the first day of drying (Fig. 4.45).

The samples on Tray 1 at the bottom of the cabinet experienced the warmest and driest air, and their percentage of original weight lost was the greatest. The adjacent tray, Tray 2, also benefited from being close to the collector outlet. The sample on the bamboo drying tray, however, lost more of its original weight than those on Trays 3–6.

Figure 4.46 demonstrates the temperature variation at various levels in the drying cabinet. Importantly, the temperature of the air at the entrance to the chimney can be seen to be significantly warmer than at Trays 2 and 5, indicating that the heated air is bypassing the trays.

In order to provide the best opportunity for even drying, the tray positions were changed at the start of the second day. All trays were moved one level higher in the cabinet, except Tray 6, which was repositioned at the bottom of the drying cabinet. The effect of the repositioning is evident on Day 2 (hours 23–31) as the samples on Tray 6 now lose the highest percentage of their original weight (Fig. 4.45). Unfortunately there was only one individually weighed sample on the bamboo tray and this sample was somewhat thicker than surrounding slices (Figs. 4.46, 4.47, 4.48).

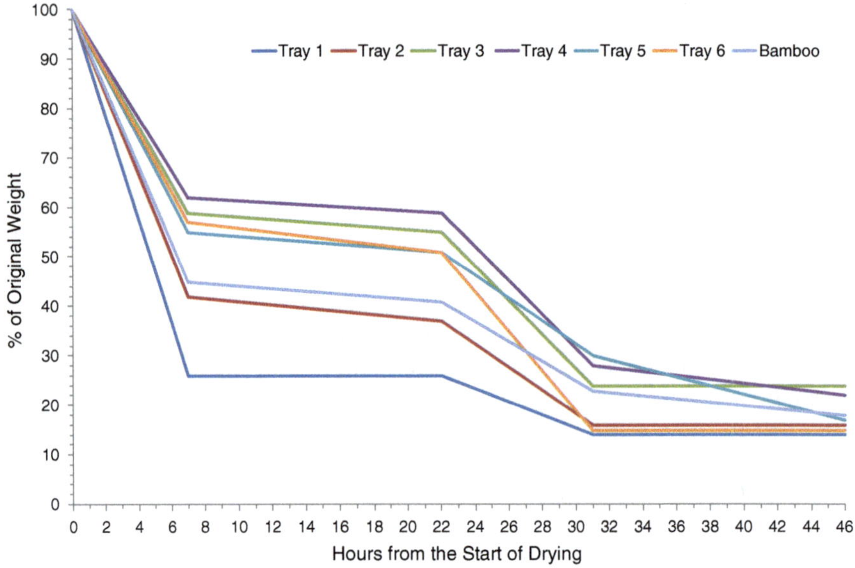

Fig. 4.45 Percentage of original weight of sliced carrot samples during drying trial

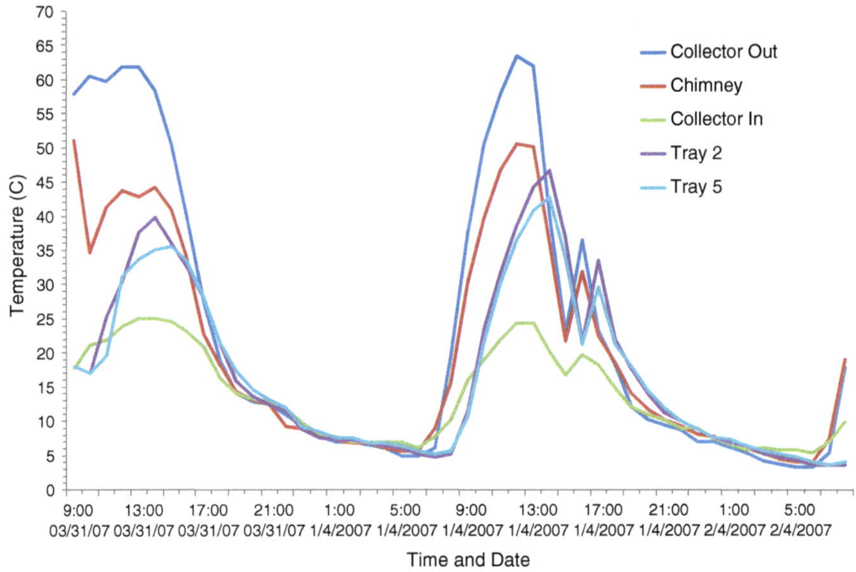

Fig. 4.46 Air dry bulb (db) temperatures at various positions in solar drier

Although not evident by sample weight, the slices on the bamboo tray were overall noticeably drier than the other trays. This observation was confirmed by weighing all of the slices on the bamboo tray at the start and end of Day 2.

Since high resolution weighing scales and a drying oven were not available at the test location, the fresh moisture content of the carrots could not be accurately determined. The moisture content (wet basis (wb)) of fresh carrots is approximately 85 % (Sandhu et al. 1979; Nayak et al. 2006). By assuming a fresh moisture content level and a desired final moisture content level for safe storage, the level of drying that had been achieved by sun drying could be confirmed. Table 4.5 indicates that depending on the initial moisture content, the slices on the bamboo tray reached a 10 % (wb) moisture content sometime on Day 2.

In the light of the above result, indicating that sun drying was as good as the solar drier, the total weight of the slices on each of the six solar drying trays was measured at 8 a.m on Day 3. The slices (5 %) that were judged to require further drying were removed and also weighed. The final desired weight, assuming initial and final moisture contents of 85 and 10 % (wb) respectively, and the actual to desired weight ratios were then calculated (Table 4.6).

Table 4.6 shows that the slices on Tray 1 appear to have been over dried. Tray 1 was close to the outlet of the collector on both drying days. The slices on Tray 6 were at the correct moisture content, while all other slices had not reached their desired final weight.

The actual to desired ratios were confirmed by the percentage of slices on each tray judged to require further drying. Over 70 % of these were on Trays 3 and 4, while Trays 1 and 6 had only 2 % of the total number that needed further drying.

4.3.5.3 Efficiencies

Based on the calculated mass flow rate and hourly temperature and radiation measurements, the efficiency of the collector was calculated in the "no-load" condition. The collector efficiency ranged from 20 to 33 %, which is considered low for this design.

The low mass flow rate is the principal reason for the relatively low range of thermal collector efficiencies. The overall drier efficiency was calculated assuming that 2.5 days were required to dry 6.738 kg of carrots from 85 to 10 % (wb). Since the total solar radiation incident on the collector over the 2.5 days was 63.4 MJ/m^2, then the overall drying efficiency, as defined earlier, is approximately 13 %. This value is considered low for a forced convection drier and is another consequence of the low flow rate through the drier and its poor air distribution.

4.3.5.4 Financial Evaluation

The financial viability of the solar drier for use in the remote villages of Humla province has been determined. The costs and prices used in the evaluation are

Table 4.5 Calculated final weights at 10 % (wb) for various fresh weight moisture content levels

Assumed fresh weight moisture content (% wb)	Dry matter content (%)	Water content at 10 % (wb) moisture content (g)	Calculated final weight of initial load on bamboo tray (g)	Actual weight of crop on bamboo tray (g)	Day and Time
80	20	22	118	119	Day 2, 8 a.m
85	15	17	89	71	Day 2, 5 p.m

Table 4.6 Calculated final weights at 10 % (wb) for each tray plus actual final weights achieved

Tray No.:	Crop fresh weight (g)	Calculated final desired weight of initial load on tray at initial and final moisture contents of 85 and 10 % respectively (g)	Actual weight of crop on tray (g)	Actual: desired weight ratio
1	1,309	218	179.5	0.8
2	1,443	241	258.0	1.1
3	1,185	198	280.0	1.4
4	1,044	174	236.0	1.4
5	912	152	196.0	1.3
6	846	141	134.5	1.0

shown in Table 4.7. Suppose a farmer is able to grow enough surplus vegetables in the months of June–October to use the solar drier continuously. Assuming the drying time for one load is approximately 2–3 days and two loads are dried every week. Assuming there are 21 weeks of drying, then 42 "loads" will be dried. Assuming that the value of the dried crop is equal to the price of fresh produce freighted to Simikot from Nepalgunj and the capacity of the drier is 10 kg, then the value of the crop is nearly 35,000 Nepali Rupees (NRs)—using the cost of carrots as an example.

Every year, the difference between operating costs and value of the dried crop is approximately NRs. 20,000. This means that the drier has paid for itself after nearly two years of operation. Since the drier should last considerably longer than two years, the solar drier appears to be a good financial proposition. However, since the drier does not appear to be superior to sun drying, there appears to be little, if any, justification to use the technology in preference to the traditional drying technique.

4.3.6 Alternative Solar Drier

Apart from the benefits of improved hygiene and rain protection, which can usually be attributed to any solar drier, the unit evaluated in this chapter has no obvious advantages compared to sun drying. In fact, from the perspective of the

Table 4.7 Costs, prices and charges used in financial evaluation of the solar drier in 2007

Item	Cost (NRs)
Cost of solar drier ex Kathmandu	30,000
Air freight charge for transporting solar drier to Simikot, including handling charges	6,500
Cost of transportation of solar drier to remote village	1,200
Total capital costs	*37,700*
Growing costs in village (labour, seed, packaging etc.) for 30 days @NRs 450 per day	13,500
Transportation costs (porter) to Simikot once per month @NR s200 per day	1,000
Total operating costs	*14,500*
Value of 420 kg of fresh carrots in Nepalgunj @NRs 20/kg	8,400
Cost of air freight for 420 kg of fresh carrots to Simikot @NRs 63/kg	26,460
Total value of dried crop	*34,860*

Humla district, there are several important disadvantages. The unit is manufactured in Kathmandu and has no local material or labour content. It is heavy and fragile, and must be transported first by air and then by donkey or porter to the remote villages. It is also expensive and requires electricity for its operation.

An alternative solar drier that minimises these disadvantages and is superior to sun drying, is required. One possible design is the traditional "direct" type or Brace drier. In this design, the crop "sees" the sun directly and the crop temperature is raised by the direct absorption of solar radiation, rather than convection, as in the solar drier described in this chapter. Direct absorption is a more powerful heat transfer mechanism than convection, its main disadvantage in a drier being the possibility of crop discolouration.

Such a drier was tested extensively for drying pineapple slices by Bena and Fuller (2002). Their drier was made of wood and covered with UV stabilised polyethylene. It was mounted on a stone chamber with air inlet holes at ground level. Driven by natural convection, heated air rose through the drying cabinet and was exhausted through adjustable vents at the top and rear of the cabinet. Their solar drier had a 25 % larger drying area than the Humla drier, and a calculated drying efficiency of 22 %, compared to the 13 % efficiency measured in this study.

Importantly, the timber parts of a Brace type drier could be constructed in Simikot, increasing the local labour content. The user could construct the base of the drier with local stones. The only imported parts would be the tray mesh and the polyethylene, both of which are lightweight items. Components could be transported more easily. Capital and transportation costs would be significantly lower. In the light of such a comparison, a trial of such a drier at the HARS site would be instructive.

4.3.7 Conclusions

A solar drier, commercially available in Nepal, has been evaluated in both the "no-load" and "full-load" conditions during the winter "dry season" in a remote mountain area of the northwest of that country. While caution should always be exercised in drawing definitive conclusions from one trial in a particular climatic season, it is believed that the issues discussed above and main conclusions would not change if other trials were conducted at other times of the year.

In the "no-load" condition, the air mass flow rate through the solar collector was found to be low compared to that of another proven and widely used design. The solar drier itself has been designed to encourage the air to flow between the trays in a serpentine path. Measurements above and at the centre of each of the six drying trays failed to record any flow, suggesting that the air was not following this pathway. Some flow could be detected underneath the bottom tray.

In the "full-load" condition, 6.738 kg of sliced carrots were dried in the solar drier. As a control, 0.532 kg of the same crop was sun dried on a bamboo drying tray. The traditional drying method proved to be equal or better than the solar drier in terms of uniformity and time taken to dry. To be competitive with sun drying, the trays in the solar drier had to be repositioned at the start of each day. This action was necessary because the flow of heated air through the drier does not pass over or through each of the drying trays. In addition, some of the crop from the solar drier required additional drying.

The solar collector and drying efficiencies were calculated to be 20–31 and 13 % respectively. These are considered low for this type of drier and reflect the low and poorly distributed airflow through the collector and drying cabinet respectively. A simple financial evaluation of the drier indicated that it could pay for itself in two years, given an ideal set of assumptions. However, given the solar drier's poor performance compared to sun drying, there appears to be no justification for the additional capital expenditure.

4.4 High-Altitude Solar Water Heater

4.4.1 High-Altitude Solar Water Heater Community Bathing Centre in Humla, Nepal

4.4.1.1 Introduction

This section aims to highlight the importance and need for High-Altitude Solar Water Heaters (HASWH) for remote communities in order to improve their hygiene conditions. It draws attention to the utilisation of the abundant, locally available solar energy resource, as a sustainable source of energy for the disadvantaged in the remote Himalayan communities. It emphasises that improved personal hygiene, through appropriate applied renewable energy technologies, is an important and integral part of people's appropriate and sustainable holistic development. In addition, it is an effective preventative health care measure for people in such communities. The section discusses the results of the first two years of service of the 1st generation HASWH, including, its shortcomings and the need for improvements. It discusses the details of the 2nd generation HASWH which has provided, since November 2005, encouraging and satisfactory results.

Scabies, worms and amoebic dysentery (diarrhoea) are among the five most common illnesses in Nepal.[4] These diseases are directly linked with, and are often the major cause of the extremely poor hygienic conditions and the general poor health of the people living in remote mountain communities. Children are particularly vulnerable, as people in these regions are totally unaware that part of their poor living conditions is due to insufficient basic personal hygiene through lack of periodic washing and cleaning. The hardship of the high-altitude mountain regions, with their long, cold winter season from October to March, does not support periodic washing or bathing. Very rarely is sufficient care given to personal hygiene, as for most of the year the rivers' water temperatures are between 4 and 12 °C, with the warmest being during the short summer from June to August, measured as 12–16 °C. Thus people have to heat water on their open-fire places, or on their cooking stoves for bathing and washing their clothes (Figures Appendix 22-Humla_139–145). That puts increased strain on the already scarce firewood and forest resources. Hence people rarely wash themselves during the 5–6 winter months, with some of them rarely bathing at all. Consequently, the incentive to utilise the locally available solar energy resource to provide basic energy services such as hot water generation through solar water heaters (SHWs) (Turkenburg 2000). Solar energy provides an appropriate and reliable alternative to the century old local traditions and pattern of using firewood to meet the daily energy service demands. Further, the lucrative tourist business in many major trekking areas provides a good reason to exploit the local solar energy resource through SWHs, for increased hygiene, comfort and income generation opportunities.

4.4.1.2 Solar Water Heaters in Nepal

A form of solar energy conversion technology that is widely installed in the urban areas of Nepal is the domestic solar water heater (SWH). The thermo-siphon SWH technology was transferred from Switzerland to Nepal in the early 70s and there are now around 220 local, Kathmandu based, SWH manufacturers. Most of the locally manufactured SWHs are still using the same old technology. The main focus of the SWH market in Nepal has been on economic growth and competitiveness rather than the development of appropriate new products. The most

[4] The Nepal Trust, Primary Healthcare Programme report 2011, http://cloud.globalgiving.co.uk/pfil/8179/projdoc.pdf (accessed July 26, 2012).

Fig. 4.47 SWH without pipe insulation, insufficient absorber insulation, and poor, black absorber paint

Fig. 4.48 SWHs are poorly manufactured, with cheap materials, incurring high heat losses

widely available, locally made SWH system consists of a 2 × 2 m^2 absorber with galvanized steel pipes and steel absorber fins, painted with simple black, non-reflective blackboard paint. A SWH unit typically has a 150 l storage tank with 50 mm of glass wool, to provide minimal insulation. The overall efficiency of one of these units (energy gained in the hot water (MJ)/incoming solar irradiation (MJ)) was measured and calculated as part of a Kathmandu University 4th year Mechanical Engineering student research project during 2001–2002. It was found to be rather low, around 22–25 %. It is believed that approximately 80,000 SWH units are operational in urban areas throughout Nepal.

The Nepalese SWHs have single layered ordinary window glass with high, approximately 0.1 %, Fe_2O_3 (iron oxide) content, allowing only 80–82 % of the solar radiation to be transmitted (Duffie and Beckmann 1991; Sorenson 2005). The glass wool insulation is insufficient during the cold winter months (with 0–5 °C night time temperatures in Kathmandu), and they are often very loosely packed due to the high price competitiveness. Further, the lack of any quality control or standards for manufacturers to adhere to, does not allow the customers to make a sound judgment from among the available SWH products on the market. In this context, it is not surprising that the challenge of developing a new SWH for the high-altitude communities, for far more difficult climatic conditions, has not been previously taken up by any manufacturer or institution.

4.4.1.3 Kathmandu University and RIDS-Nepal HASWH R&D

Since 2001 the Kathmandu University (KU) has been involved, through the author's leadership, initially through student projects, in the research and development of Nepalese SWHs, in order to address the urgent need for warm water for personal hygiene, utilising the abundant locally available solar energy. Special emphasis has been given since 2002 to improving personal hygiene of the poor and remote high-altitude communities. Therefore KU's Mechanical Engineering research group focused on a HASWH, which is able to perform satisfactorily throughout the year under the harsh high-altitude climatic conditions. That means it has to be able to generate and store hot water under extreme conditions such as ambient temperatures as low as –20 °C. It also needs to be able to withstand snow storms, hail, high UV radiation and high-altitude intensive solar radiation >1,200 W/m^2, with a guaranteed life-cycle expectancy of 15–20 years. The first small scale HASWH design and prototype was manufactured and installed in RIDS-Nepal's HARS in Simikot Humla, at 3,000 m altitude in early 2004 (Figs. 4.49, 4.50, 4.51). Periodic monitoring and usage revealed its strengths and weaknesses, which provided the basis for the 2nd generation HASWH design. One of the important issues for improvement in the 2nd model was the unsatisfactory hot water storage tank insulation which led to very significant losses during cold winter nights. This, and other factors, were addressed through a number of KU based, research projects. The design changes have been successfully incorporated into the 2nd generation HASWH, which was installed in the HARS in Simikot in November 2005. It has been subjected to detailed data monitoring since then, providing valuable information and field experience for future design improvements and projects.

4.4.1.4 Prototype Testing of the 1st Generation HASWH

Technology is designed and developed to be applied within a defined context and environment. That means that each technology, in order to be appropriate and sustainable, has to be contextualised accordingly. Thus, in order to provide the required hot water services, the environmental, cultural and social conditions of the HASWH end-users, have to be known and defined. They set the peripheral conditions for the design and manufacturing, so that the services can be provided appropriately. Therefore the following criteria have been defined for a HASWH design, based on a survey regarding the needs of the user community within their environment.

- Due to the high-altitude Himalayan mountain climate, considerably increased ambient temperature differences have to be taken into consideration (−20 °C to +30 °C).
- Thicker and different insulation materials have to be used (e.g. polyurethane foam instead of glass wool).
- The HASWH system's main focus groups are whole communities rather than single families (due to the cost, culture and development stage of the remote mountain communities).
- Freeze protection must be in place for the absorbers and piping system as a thermo-siphon system is used. Active systems with a secondary loop and a glycol water mixture with a pump would not be appropriate, as electricity would need to be generated through a solar PV system, which increases the cost and maintenance needs substantially.
- As much as possible, only locally available materials are used to increase sustainability, strengthen the local economy and provide new skills and entrepreneurship training.

With the above criteria in mind the first small size, thermo-siphon system prototype HASWH was designed, locally manufactured, installed and field tested in the HARS in Simikot, Humla, at an altitude of 3,000 m in early 2004. A copper fin absorber, with selective coating, imported from India, with a total absorber surface of 1 m^2 was chosen for this initial trial. The insulated storage hot water tank was designed as an integrated part of the HASWH, with a capacity of 50 l (Fig. 4.49). For all insulations, the locally available glass wool, 65 mm thick for the hot water tank and 50 mm for the absorber box, was chosen.

In order to protect the absorber raiser and header pipes during the 5–6 months of freezing cold, high-altitude winter nights, as well as to increase the interception of the available solar radiation on the absorber during the day, a highly reflective, Styrofoam insulated aluminium lid cover, was added to the system (Fig. 4.49). Beside the insulated reflector lid, which can be manually adjusted according to the seasonal position of the sun and closed before sunset, manual drain down system through valves (Morrison 2003) are installed (one at the bottom of the absorber and one before the hot water storage tank), for additional protection against the extreme ambient temperature conditions during the month of January (<15 °C).

A further challenge was to construct the HASWH in such a way that it is modular, compact and thus easier to transport into the remote areas. Many remote mountain communities can be reached only by airplane and then carried by porters for hours or days. The installation and maintenance need to be easy with simple installation procedures.

In order to understand and analyse the performance of the HASWH prototype, it is regularly monitored through the following nine different parameters which have been recorded with a dataTaker logger DT605,[5] since 1st May 2004:

- Global solar radiation on the plane of the absorber at 30° south inclination.
- Cold inlet and hot outlet water temperatures (hot water storage tank temperature).
- Four absorber temperatures at different raiser heights.
- Ambient temperature.
- Water storage tank insulation temperature at 50 % insulation depth (to measure and calculate the heat losses).

Figures 4.49, 4.50, 4.51 show the 1st HASWH under field test at HARS in Simikot, Humla.

During the month of October the average day time hot water temperature attained is around 30 °C, compared to the cold water intake temperature of around 15 °C. The first HASWH prototype's average system efficiency is only 15–20 %. There is a direct correlation between the incident irradiation and the absorber temperature with prolonged absorption and retention of the absorbed solar heat. The selective coating of the copper absorber supports that due to its high absorption and low re-radiation (emissivity, or high retention of heat) properties. The minimal night time temperature difference between the absorber and the ambient temperature, despite using the insulated cover lid, shows that the lid is not as efficient as anticipated.

The following graph, Fig. 4.52, shows the average daily performance of the 1st HASWH during October 2004. The outlet water temperature (representative of the hot water storage tank temperature) and the 50 % insulation depth (~32.5 mm) temperature lines have a similar daily pattern, indicating that the 65 mm glass wool insulation is not sufficient for high-altitude SWHs, and thus increased heat loss occurs. This will be even more, the higher the ΔT between outlet hot water and ambient temperature. There are a few interesting points to be mentioned:

- The HASWH absorber reacts quickly to the day's first sunshine (7:00 a.m) with increased absorber and outlet water temperature, indicative of a low thermal mass and insufficient insulation. Any energy gained is also quickly lost.
- The average intercepted solar radiation from 15:30 to 16:30 p.m is ~400 W/m^2, indicative of the minimum solar radiation needed to overcome the HASWH's internal heat losses (equilibrium), as the warm water outlet temperature during that time remains constant.

[5] The specifications and details for the DT605 data logger from DataTaker AUS, can be consulted at: http://www.datataker.com/products/dt605.html (accessed March 1, 2011, but from October 2011 onwards not anymore available as the DT605 has been superseded by the DT80 data logger).

Fig. 4.49 The storage tank module is integrated to limit pipe heat losses. The insulated, reflecting aluminium lid acts as a night time cover/day time reflector

Fig. 4.50 Selective coating, copper fin absorber with four thermocouples at different heights to record the temperature profile. The normal window glass cover limits the incoming solar radiation to 80–82 %

Fig. 4.51 Manual water drainage is possible, in addition to the insulated reflector lid. The absorber body is insulated with 50 mm glass wool, while the hot water storage tank is insulated with 65 mm

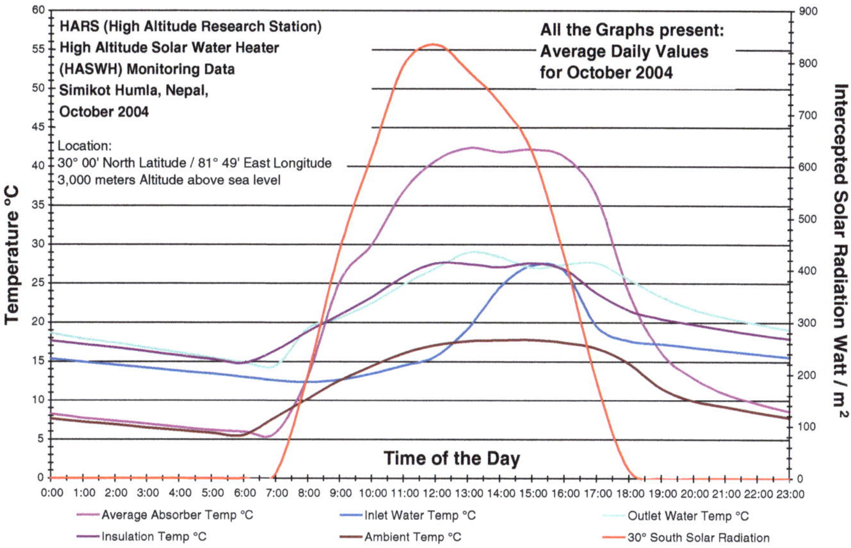

Fig. 4.52 First HASWH average daily performance at HARS at 3,000 m altitude during October 2004

- From 17:00 to 20:00 p.m, and even more from 0:00 to 06:00 a.m the ambient temperature, the tank insulation and storage tank temperature fall at a similar steep rate, indicating the inadequate insulation properties of the loosely packed 65 mm glass wool insulation.

Based on these data and experience, major improvements for the next, 2nd generation HASWH prototype have been identified as:

- Minimising the heat loss in the hot water storage tank and absorber through better insulation.
- For better stratification a 300 l, vertical hot water storage tank, is used.
- Improve cover lid insulation and extend the cover over the absorber's sides.
- Seasonal lid reflector adjustment, to maximise the intercepted solar radiation.
- Increase absorber surface to 1.5 m^2. Join 4 absorbers to a HASWH system unit.

4.4.1.5 2nd Generation HASWH, Prototype Testing and First Results

Based on the first two years of experience with the 1st HASWH prototype and a project proposed for the first community owned high-altitude solar water heater bathing centre for the village of Dharaprori (350 people), a second generation HASWH was designed. The following design drawings show some central parts of the new 2nd generation HASWH, which was manufactured during September–October 2005 (Figs. 4.53, 4.54, 4.55, 4.56).

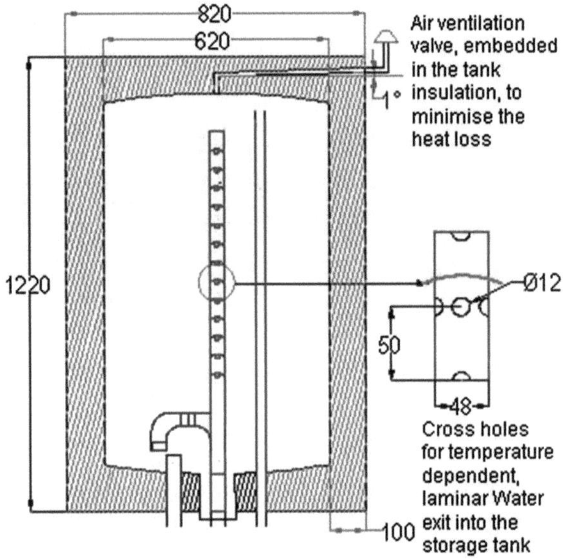

Fig. 4.53 New, 300 liters stainless steel hot water storage tank with 100 mm polyurethane foam insulation

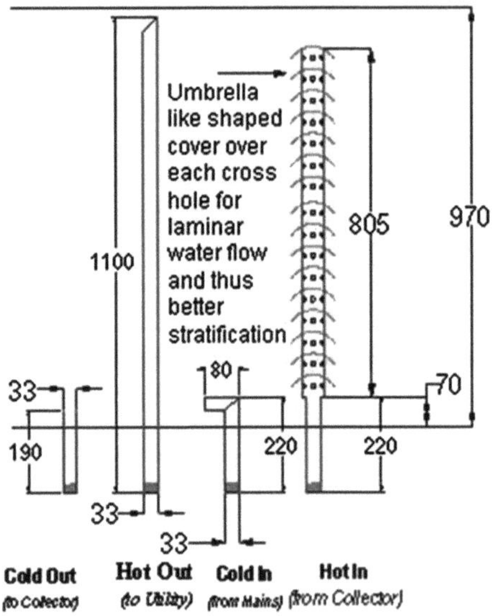

Fig. 4.54 Hot and cold water pipes inside the storage tank. Cross holes and cover minimise turbulence, improving the tank water stratification

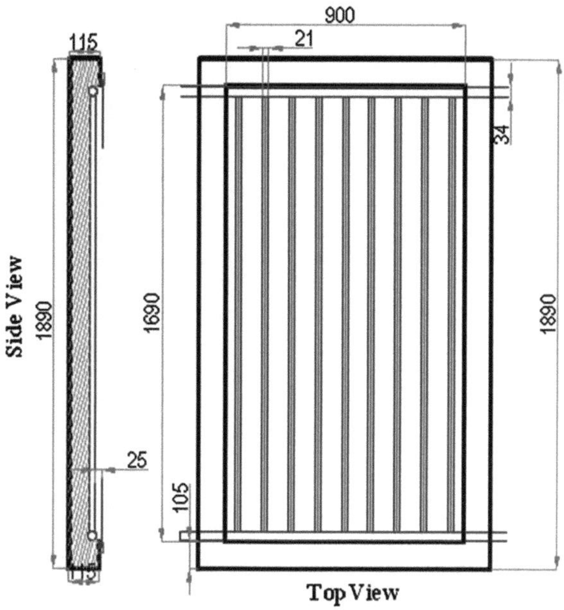

Fig. 4.55 Dimensions of the new absorber, with its nine ½ in. GI raisers, aluminium fins, and one in. headers. Insulation with polyurethane foam

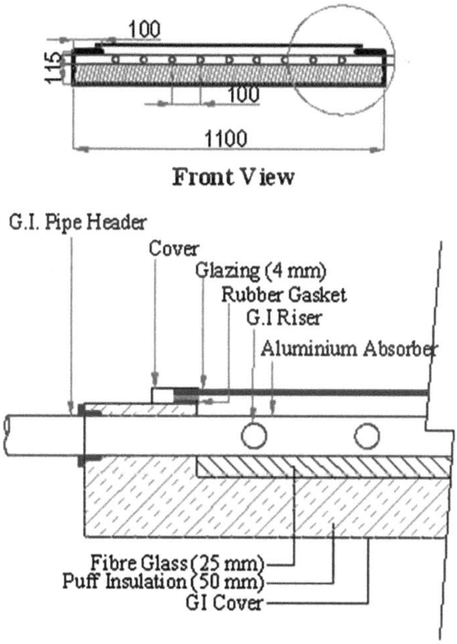

Fig. 4.56 Close up of the new absorber structure with GI pipe raisers, aluminium fins and improved box insulation materials

Fig. 4.57 The 2nd Generation Thermo-Siphon HASWH, with opened, adjustable cover lid reflectors. A horizontal and 40° south angle pyranometer (SolData 80SPC), records the received solar irradiation on the absorber's POA. The absorbers and insulated pipes can be shut off from the storage tank and emptied through hand valves as part of the freezing protection

Fig. 4.58 The HASWH in closed, freezing night position. Absorber boxes and hot water storage tank are insulated with 50 and 100 mm two component polyurethane foam respectively. Four thermocouples on one absorber and four in the hot water tank, each at different heights, record the different water temperatures. All data is recorded with a dataTaker DT605

The first unit (with four absorbers of 1.5 m^2 each and one 300 liters hot water storage tank) was installed in November 2005 in Simikot Humla (Figs. 4.57, 4.58) and has since undergone its long-term, detailed field test and data monitoring.

Figures 4.59, 4.60, 4.61, 4.62, 4.63, 4.64 show some of the changes integrated into the 2nd generation HASWH, based on the 2 year research project experience with the 1st generation HASWHs. These data needed to be evaluated and properly interpreted in order to identify the shortcomings of the 1st HASWH.

Figure 4.65 shows the average daily performance of the 2nd HASWH in March 2006.

The Legend and explanation to the 2nd Generation HASWH graph from the 22nd–23rd March 2006 (Fig. 4.65) is as follows:

1. Sunrise at 07:15 a.m. Before that 100 % diffuse light only.
2. Sunset at 17:15 p.m. After that 100 % diffuse light only.

Fig. 4.59 The 300 l, stainless steel hot water storage tank is insulated with 100 mm of polyurethane foam. The air ventilation valve exits at the top on the side, in order to minimise heat losses. Thermocouples at different heights in the hot water tank measure the attained stratification

Fig. 4.60 Nine ½ in. GI pipe raisers with aluminium fins are used as absorbers, rather than copper, for sustainability reasons, as copper can not yet be welded properly in Nepal. Four thermocouples at different heights measure the temperature profile on the absorber. The top and sides are overlapped with galvanised sheets while at the bottom the rain can run off freely

3. A minimum global solar radiation of 420 W/m^2 is needed to keep the HASWH system in equilibrium (heat gain − heat loss = 0). At this point the HASWH system efficiency is zero.
4. No heat gain into the hot water under these conditions (hot water temperature, solar radiation, heat loss, ambient temperature). Zero system efficiency.

Fig. 4.61 A 50 mm Styrofoam insulated aluminium lid covers the absorber all around tightly, including 50 mm overlapping, in order to minimise the absorber heat losses during the freezing nights. The lid's corners are reinforced to increase its stiffness. The lid acts during the day as a reflector with its high gloss, stainless steel sheet on the inside

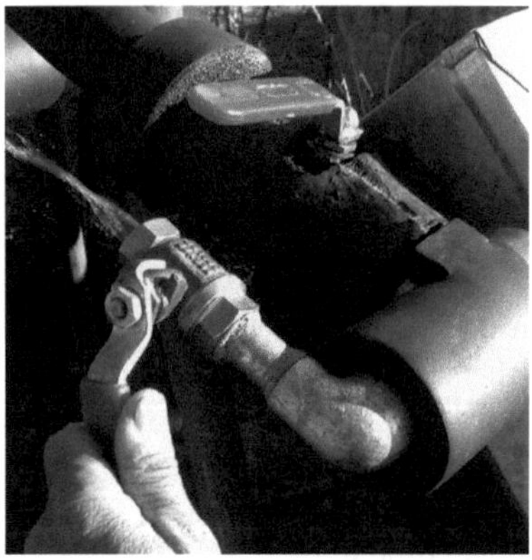

Fig. 4.62 The drain and the hot water storage tank inlet closing valves at the top of the absorber unit. They isolate the storage tank from the absorber unit as part of the freeze protection during the cold, winter nights

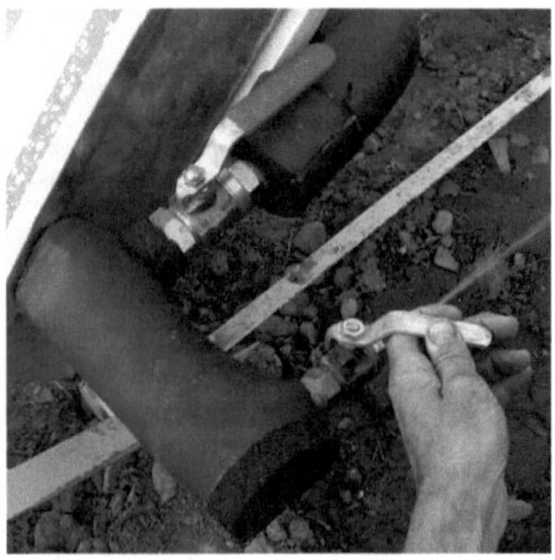

Fig. 4.63 Valves at the bottom of the absorber unit. Once the hot water storage tank inlet valve is closed and the top drain valve is opened, the bottom valve is opened and the absorber water circuit is emptied. In this way only air remains inside the absorber pipes

Fig. 4.64 In order to maximise the winter hot water generation, the absorbers are inclined at 40° South at a latitude of 30° North

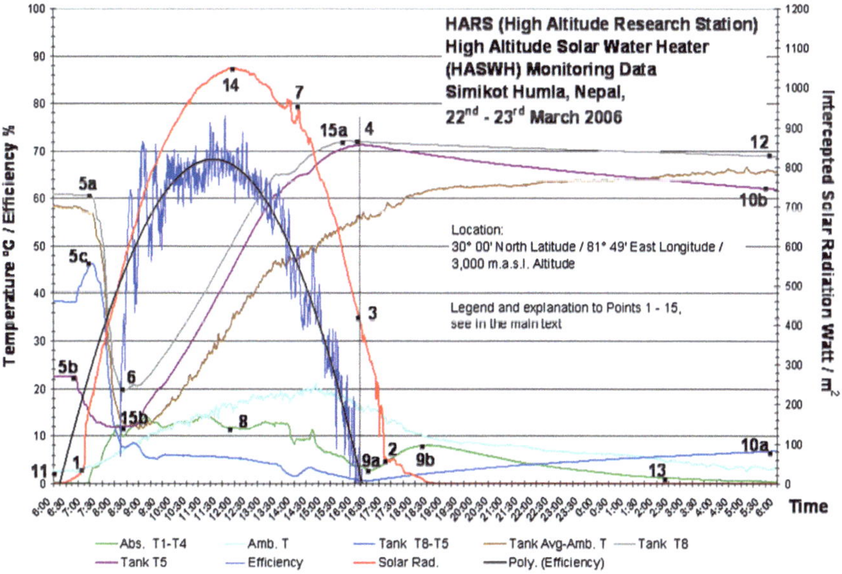

Fig. 4.65 2nd HASWH Performance at HARS at 3,000 m Altitude in Humla from the 22nd–23rd March 2006

5a/5b/5c. A ΔT of ~40–45 °C between 5a (hottest water) and 5b (coldest water) inside the storage tank could be maintained throughout the night with hottest tank water temperature—average ambient temperature of ~60 °C. Good stratification inside the tank. 5c increased initially as hot water was dispatched from the tank and cold water (~10 °C) flows into the bottom of the tank. After 5c the hot water tank is almost emptied and re-filled with cold water.

6. Hot water dispatch stopped, with minimal ΔT (~8 °C) and stratification. Tank is full of fresh, cold water, with an average temperature of ~16 °C.
7. Sudden sharp increase of the beam solar radiation. Possibly caused by the passing by of cumulus humilis clouds. Just as the sun reappears these white sharp edged clouds, can provide a sudden solar radiation increase due to the strong reflection at the white cloud's edges.
8. From about 08:15 a.m–15:00 p.m, T1–T4 (hottest—coldest absorber temperature, Fig. 4.60) is between 10 and 14 °C. A good ΔT to "drive" the received heat energy into the water flowing inside the nine absorber raisers.

9a/9b. 9a. Shortly after the minimal intercepted solar radiation (equilibrium point), the insulated absorber lid was closed, and the freezing protection valves (Figs. 4.62, 4.63) closed and opened respectively, to drain the absorbers. Thus only air remains in the absorber, and thus the ΔT over the absorber increases again slightly till 9b, before the hot air slowly starts to cool as well. Thus the absorber is protected throughout the night from freezing.

10a/10b. 10a. The hot water storage tank temperature ΔT:T8 (hottest)—T5 (coldest) shows increased stratification since the cover lids were closed, as no heat producing solar radiation, nor hot water is dispatched and no cold water enters the tank. The water inside the storage tank remains "calm" and thus can stratify clearly according to the different temperature dependent water densities. 10b (coldest tank temperature T5) shows a steeper drop compared to T8, as over time the cold water descends to the bottom of the tank due to its greater density.

11. "Real life proves theory", as the coldest ambient temperature is normally between 06:00 a.m and 06:30 a.m, just before sunrise (Ahrens 2003, p. 66).
12. The average hot water storage tank heat loss has been measured (over many days) to be 0.3–0.4 °C per hour at an average tank inside—ambient temperature ΔT of 60–70 °C.
13. With the freezing protection lid covered, the air inside the emptied absorber cools down to a more and more equal absorber temperature, just above the ambient temperature in the early morning, thus T1–T4 goes to zero.
14. The total global solar radiation on the plane of the absorbers over the course of the day amounts to: 45.27 kWh/day. (Total measured solar radiation received at 40° South inclination is 7.57 kWh/m^2/day. With a total absorber surface of 5.98 m^2 (one unit with four absorbers, each 1.495 m^2) the total intercepted

global daily solar radiation amounts to: 5.98 m^2 × 7.57 kWh/m^2 = 45.27 kWh/day).

15a/15b. The total HASWH system efficiency is calculated according to the total energy gained by the water, divided by the total energy received by the total absorber unit:

$$\eta_{HASWH\,System} = \frac{V_{Water} \times \rho_{Water} \times \Delta T_{av.\,water\,end\,-\,av.\,water\,beginning} \times C_P}{I_G \times 3.6 \times A_{Total\,Absorber\,Surface}} \times 100 = 45.32\,\% \tag{4.3}$$

with: $V_{Water} = 0.315\,\text{m}^3$ (total water volume of: hot water storage tank and absorber unit circuit)

$\rho_{Water} = 1,000\text{kg}/\text{m}^3$ (density of water)
$\Delta T_{av.\,water\,end\,-\,av.\,water\,beginning} = 56\,^\circ\text{C}$ (Average hot water tank temperature in the evening – average hot water tank temperature in the morning)
$I_G = 7.569\,\text{kWh}/\text{m}^2$ (total intercepted global solar radiation over the day)
$A_{Total\,Absorber\,Surface} = 5.979\,\text{m}^2$
$C_P = 4.186 \times 10^{-3}\,\text{MJ}/\text{kg}/^\circ\text{C}$ (heat capacity factor)

This example illustrates that maximum knowledge of the context in which the HASWH will have to perform for an appropriate life-cycle of 15–20 years, as well as appropriate engineering design, in order to manufacture the HASWH with mostly locally available materials, infrastructure and skills, are crucial to the aim of sustainability.

4.4.1.6 Discussion of 2nd Generation HASWH Prototype Test

Based on the identified major improvements needed (Sect. 4.4.1.4), good performance was achieved during the first 6 years test period of the 2nd generation HASWH, also installed in the HARS in Simikot, Humla. The heat loss reduction due to the improved hot water tank insulation with polyurethane foam has, in particular, been remarkable (see point 12 in the explanation of Fig. 4.65, compared with the heat loss recorded in Fig. 4.52). The vertical, stainless steel hot water storage tank maintained remarkable stratification levels of >40 °C between the top and the bottom water.

The highly reflective, glossy stainless steel, insulated cover lid showed improved performance. The heat loss from the absorber cover glass is less, as the insulated lid also covers 50 mm of the top and the two sides of the absorber (Figs. 4.61, 4.64). The seasonal adjustment is simple and quick to teach and learn, and a single person can open, close or adjust the cover/reflector lid easily (Fig. 4.70). The absorber size of 1.5 m^2 has turned out to be at the upper limit in regard to the ability to transport each absorber by aircraft, and afterwards by porters over the mountains, to the final installation place. The concept of four absorbers joined together on site to form one unit of 6 m^2 total absorber size, seems to be an appropriate system size for community bathing centres. According to the population, hot water demand and affordability, several units can be put together and thus great flexibility for proper sizing for a defined context is possible. The cost per m^2 of the 2nd compared to the 1st generation HASWH is similar as all parts of the new HASWH, except the toughened glass, are manufactured in Nepal.

4.4.1.7 Darapori Village HASWH Community Bathing Centre

After 3 years of testing and data monitoring the 2nd generation HASWH bathing centre at the HARS in Simikot has proven to perform flawlessly, and as expected, no substantial changes are needed. Thus, based on this experience, RIDS-Nepal and the village community of Darapori village decided in 2008 to build the first village based bathing centre in their village in 2009, as part of a long-term HCD, "Family of 4 PLUS" project. Thus the digging, building and installation works started in November 2008, finishing and enabling the first people to have a hot shower after the winter season in May 2009 (Figs. 4.66, 4.67, 4.68, 4.69, 4.70, 4.71, 4.72, 4.73, 4.74, 4.75, 4.76, 4.77, 4.78).

The HASWH community bathing centre in Darapori village in Humla, allows the 350 villagers to take a hot shower once every week, addressing the pressing need to improve their hygiene conditions. The capacity of the HASWH bathing centre was laid out for 50–55 people a day, with each person using ~10 l of 50 °C hot water (or a 35 °C "design temperature" shower as defined in Sect. 4.4.2.3), which can be mixed with cold water at their convenience. As shown in the following, an average daily 5.2 h of full sunshine, or 5.2 PSH (Peak-Sun-Hours) with 1,000 W/m^2 solar radiation, are needed to generate the daily hot water energy.

$$PSH = \frac{N_{people} \times V_{Water} \times \rho_{Water} \times \Delta T_{Water} \times C_P}{3.6 \times \eta_{HASWH\,System} \times A_{Total\,Absorber\,Surface}} \tag{4.4}$$

with: $V_{Water} = 0.01\,\mathrm{m}^3$ (10 l of 50 °C hot water per person per bath/shower)

$\rho_{Water} = 1,000\,\mathrm{kg/m^3}$ (density of water)
$\Delta T_{Water} = 40\,°\mathrm{C}$ (with an assumed minimum water inlet temperature of 10 °C)
$A_{Total\,Absorber\,Surface} = 12\,\mathrm{m}^2$ (for two units with each four absorbers, each with 1.495 m^2 of exposed surface area)
$C_P = 4.186 \times 10^{-3}$ MJ/kg/°C (heat capacity factor)
$\eta_{HASWH\,System} = 40\,\%$ (assumed average)

thus: $PSH = \frac{53_{people} \times 0.01\,\mathrm{m}^3_{hot\,water}\ 1,000\,\mathrm{kg} \times 40\,°\mathrm{C} \times 4.186 \times 10^{-3}}{\mathrm{m}^3 \times 3.6 \times 0.4 \times 12\,\mathrm{m}^2} = 5.2\,\mathrm{PSH}$

The figure of 5.2 PSH per day can be compared with the average daily solar radiation for Nepal of 4.8–6.0 kWh/m^2 per day (Appendix 11.5). The HASWH

Fig. 4.66 Bathing Centre with the two rows of four absorbers each and two hot water storage tanks. Proper distance/angle between collectors is important to avoid shading

Fig. 4.67 While an azimuthal angle of +30° from south to west maximises the wintertime hot water generation, it still provides a very good energy input and hot water production throughout the year

Fig. 4.68 During the day the adjustable and insulated absorber lids act as reflectors, increasing the received global solar radiation

Fig. 4.69 During the night the absorber lids act as insulators for the long and freezing winter nights, besides the available absorber cycle drainage valves

Fig. 4.70 The absorber lids can be easily and quickly opened, closed or adjusted according to the sun's seasonal angle at the horizon

Fig. 4.71 The bathing centre has its own water source and a 750 l cold water tank inside an insulated 1,000 l tank on top of the shower cabins

Fig. 4.72 Six shower cabins allow several local people and tourists to take a shower simultaneously. 50–55 people a day can take a hot shower each day

Fig. 4.73 Inside a shower cabin. Stainless steel shower boxes with hot and cold water mixing station. The walls are 1 m high and surface plastered to keep the showers clean

bathing centre was discussed with the local people and they agreed to provide the locally available materials such as stone, wooden beams, mud and sand, besides all the necessary digging work for the bathing centre. The land was registered under the name of the newly established "Darapori Village Bathing Centre Committee". The committee has been put in charge of the use and maintenance of the bathing centre infrastructure. A caretaker was appointed by the committee for the daily

Fig. 4.74 Two cold water taps and a dedicated, cemented clothes washing and drying area to its south allow people to wash and dry their clothes

Fig. 4.75 Taking showers and washing clothes generates dirty water. The waste water is led into a reed bed, planted with local plants, to be cleaned

operation, maintenance and protection of the bathing centre, earning a small monthly salary, intended to be raised from each family monthly.[6]

[6] While the land is registered under the "Darapori Village Bathing Centre Committee" and the local user committee was formed, after three years of running and using the High-Altitude Solar Water Heater Bathing Centre (HASWHBC), it has been shown that it is very difficult to raise the necessary funds from the local end-users to pay for the caretaker and the needed maintenance. This, as the HASWHBC is a rather new concept within their culture it will take years for the local community to get used to and be able to use it to maximum effect and possible improvement to their personal hygiene. That's why, the topic of washing/showering with hot water is integrated in the NFE classes' workbooks (Appendix 17), with an awareness raising poster (Appendix 22.2) put up in the near by vicinity and a brochure (Appendix 19.6) to teach the local people how to use the showers and clothes washing facility. Thus, till end of 2011 the necessary funds for the caretaker and the maintenance of the HASWHBC have not yet been raised from the local community and therefore RIDS-Nepal's is still required to provide for the needed maintenance funds.

Fig. 4.76 Darapori Village with its 52 families and 350 people. In 2005 a "Family of 4" long-term HCD project started, and each family now has a pit latrine, a smokeless metal stove in the kitchen, WLED electric indoor lighting through a solar PV home system for each home (Sect. 3.3.2.1), and, since June 2012, a new drinking water system. From 2007 onwards additional needs, identified by the local community, and part of the "Family of 4 PLUS", have been addressed. These are: NFE classes for mothers, a nutrition program for malnourished children <5 years of age and their mothers, several greenhouses, a solar drier, several home based slow sand water filters and the HASWH Community Bathing Centre (*bottom left* in the above picture), which was inaugurated in May 2009

Fig. 4.77 A horizontal and a 40°-south inclined (equalling the absorber angle) 80SPC pyranometer, record the global solar radiation

Fig. 4.78 A DT80 data logger measures and records all the various environmental and solar water heater data

4.4.1.8 The HASWH Bathing Centre as an Integrated Part of a Holistic Community Development Project

As already mentioned several times in the previous chapters, the HASWH community bathing centre too, is not an individual project, but one of the "Family of 4 PLUS". It aims to bring long-term synergistic benefits along with the other "Family of 4" and other "Family of 4 PLUS" projects.

4.4.1.9 Lessons Learned

The two years of field experience with the 1st generation HASWH formed the basis for the design and manufacture of the 2nd generation HASWH, which has now been under real field testing since November 2005. Valuable lessons learned, from recorded data and interviews with users, have been evaluated, technically interpreted and realised in the new, 2nd generation HASWH design. Already some new potential improvements for a 3rd generation HASWH have been identified, if in the future a village community requests a HCD project which includes a bathing centre. Some of these improvements are:

- Increased overlapping of the covering galvanized steel sheet for the hot water storage tank and absorbers.
- Better aligned and stronger hinges for the cover lid.
- Thicker valve insulation (with extended valve handles).
- Smaller diameter for the absorber pipe connections, a 450 liters hot water storage tank per HASWH unit with four absorbers.
- Low Fe_2O_3 glass cover (thus far unavailable in Nepal and India).
- Improved manufacturing quality and quality control.

While these improvements will increase the overall efficiency of the HASWH slightly, the cost per m^2 will only change marginally, as the emphasis has to be given to improved manufacturing quality and quality control, rather than to significant technical or material changes.

4.4.2 Prediction of Hot Water Usage in a Solar Heated Community Bathing Centre in Humla, Nepal

4.4.2.1 Introduction

A natural thermo-siphon prototype HASWH community bathing centre has been constructed at HARS in Simikot, in the remote north-western district of Humla, Nepal. This HASWHBC is the prototype for a larger community bathing system proposed for a nearby village. This section describes the community bathing centre and presents the results of a validated TRNSYS model that has been used to predict its performance and to test various operational strategies.

In direct response to the extremely poor hygiene conditions, which many of the high-altitude communities in Nepal have to endure, a community bathing centre was built by the local NGO RIDS-Nepal in the remote village of Darapori, Humla. The bathing centre was designed to have a significant impact on hygiene conditions and hence on the health of the local population. Predictions from the model indicate that the design target of 550 showers can be provided each week if these showers are limited to 10 liters of water at 35 °C and the centre is operated between 9 a.m and 3 p.m. This section describes part of the research undertaken to ensure the successful design and operation of the solar water heating system at the bathing centre.

Figure 4.79 shows a schematic layout of the initial plan for the bathing centre, on which the calculation and simulation in this chapter was based. Later, during the actual village project planning and implementation the bathing centre was reduced to two units each of four absorbers as described in Sect. 4.4.1.7. The solar hot water system for the bathing centre is a thermo-siphon system, developed by the author and his Kathmandu University students. Such a system avoids the additional expense and maintenance incurred if forced circulation is used. The system was built by RIDS-Nepal, which is a local NGO that has been implementing a holistic community based development program in the Humla district since 2002. The community bathing centre is part of that program. The absorbers are made from aluminium sheet with galvanised iron 27 mm internal diameter (ID) headers and 16 mm ID risers. Each absorber has nine risers, which are a "push fit" into the absorber panel. Reflective and insulating covers are placed over the absorbers at night to prevent freezing. The hot water storage tank for the system is made of stainless steel and insulated with 100 mm of two component polyurethane foam (Fig. 4.80).

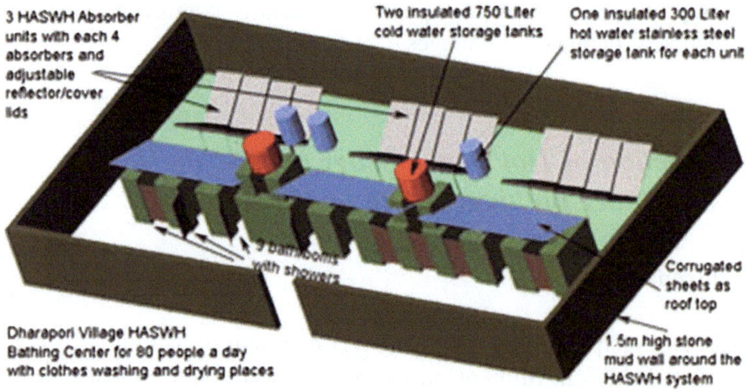

Fig. 4.79 Initial planned Darapori Bathing Centre Layout with three units, each with four absorbers. The actual bathing centre, built in 2009, consists of two units each with four absorbers, two hot water storage tanks and six shower rooms

Fig. 4.80 Absorbers are made from aluminium sheet with galvanised iron headers and nine risers. The stainless steel hot water storage tank is insulated with 100 mm of polyurethane foam

4.4.2.2 TRNSYS Model

The simulation software, TRNSYS (Version 16.0) was used to predict the daily performance of the community bathing centre system. The sub-routine, Type 45, within this program allows the user to describe a thermo-siphon collector storage system. An option in this subroutine is for the internal calculation of internal pressure drop and flow, rather than as user supplied data. Since no data existed for this collector, this option was used in these simulations. The user must supply various system parameters; the key ones are listed in Table 4.8.

Key parameters were determined by calculation and assumptions. The collector efficiency factor (F′) and the collector heat loss factor (U_L) were calculated to be 0.73 and 27.4 kJ/h/m^2/°C respectively. The flow factor (F″) was assumed to be

Table 4.8 Key system parameters

Parameter	Value
Collector area (m^2)	6.0
$F_R\tau\alpha$	0.53
F_RU_L (kJ/h/m^2/°C)	19
Collector slope (°)	40
Riser internal diameter (m)	0.016
Header internal diameter (m)	0.027
Vertical distance between collector outlet and inlet (m)	1.14
Vertical distance between tank outlet and collector inlet (m)	1.55
Diameter of collector inlet pipe (m)	0.027
Length of collector inlet pipe (m)	7.65
Number of right angle bends in inlet pipe	4
Loss coefficient of collector inlet pipe + insulation (kJ/h/m^2/°C)	1.073
Diameter of collector outlet piping (m)	0.03
Length of collector outlet piping (m)	1.58
Number of right angle bends in outlet pipe	3
Loss coefficient of collector outlet pipe + insulation (kJ/h/m^2/°C)	1.07
Tank volume (m^3)	0.3
Tank height (m)	1.2
Height of collector return to tank above bottom of tank (m)	0.875
Effective thermal conductivity of fluid and walls (kJ/h/m^2/°C)	1.5
Overall UA value for tank (kJ/h/°C)	5.76
Ratio of top insulation to side insulation thickness	1.0

0.93 (Whillier and Saluja 1965). Solar transmittance (τ) and absorptance (α) were assumed to be 0.82 (measured) and 0.95 (for black paint) respectively.

Using these parameters, the model was validated against data collected from the HARS thermo-siphon system in 2006. The staff and local residents of Simikot currently use the hot water produced by the HARS system for showering and washing clothes. The hot water is made freely available between the hours of 5 a.m–8 a.m. At all other times, the shower door is normally locked, and neither staff nor residents can access the hot water that is produced on that day. On Saturdays, however, the shower is not locked and hot water can be used at any time.

A sixteen day period in November 2006 was randomly selected and used to validate the model. Water temperatures at four positions in the tank were measured at 10 min intervals and recorded on a data logger. The actual "load" or hot water drawn from the system was calculated by estimating the amount of cold water that must have been introduced into the system to achieve the decline in water temperature at each hour of use. Water temperature changes due to tank losses were ignored because these are very small compared to those caused by the introduction of cold water. Figure 4.81 compares the average measured tank temperature for sixteen days with the model's predictions for the same period.

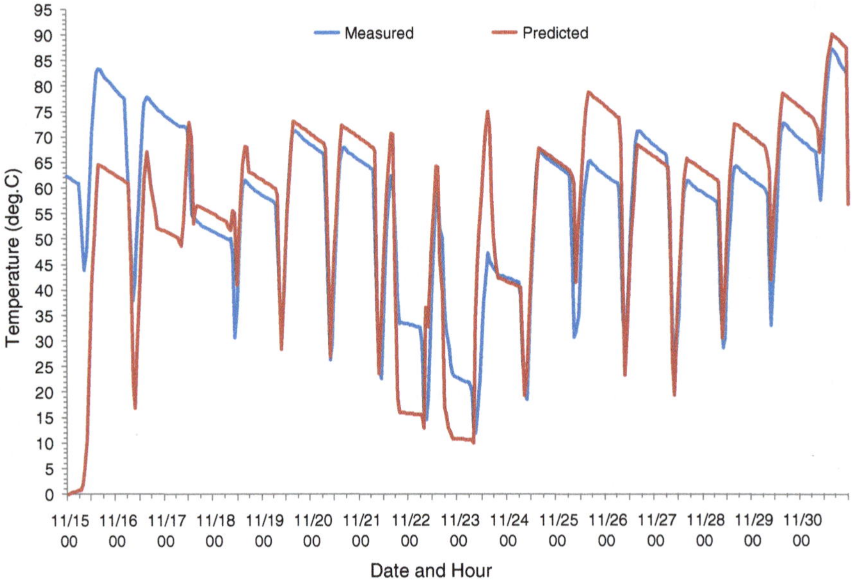

Fig. 4.81 Measured and predicted average hot water storage tank temperatures

One of the deficiencies of the Type 45 subroutine is the inability to set the initial storage temperature and this accounts for the discrepancy between measured and predicted temperatures over the first two days of the period. The model is also prone to instability and "range check" errors. This occurred inexplicably sometimes when small changes in the values of parameters were made. As a result, a higher value of F_RU_L (23 kJ/h/m^2/°C) was used in the simulations than that calculated. However, given the instability of the subroutine, the uncertainties in the hot water load and the assumptions made, agreement between the measured and predicted average tank temperatures was considered to be good enough over the two week period to give confidence that year long predictions of performance could be made.

4.4.2.3 Checking the Design Aim

The size of the SWHs for the bathing centre is based on the aim of providing 10 liters of 50 °C water to 550 people every week (Zahnd and Malla 2006). Water at 45 °C is the limit that can be tolerated for any appreciable time by the human hand (Zahnd, personal observation). If a temperature of 35 °C is used for a "design temperature" shower, it is possible to establish the amount of hot water used from a tank at any particular temperature to provide 10 liters at 35 °C. It is then also possible to readily calculate the number of these 10 liters showers that may be taken from a 300 liters tank at any temperature at or above 35 °C. At that particular temperature, 30 showers are available and no cold water is added. As the tank temperature increases more showers are available because higher temperature water is now mixed with some cold water, assumed to be at 15 °C. It was also assumed, however, that the maximum number of these showers that could be taken in any one hour was ten, based on the allowance of six minutes per shower for undressing, washing ($\sim$2 min), drying and getting dressed again.

These limitations (temperature and time) were initially used to predict the number of "design temperature" showers that could be taken between 5 a.m–8 a.m, because it is the cultural preference to wash early in the morning. A climatic data file (solar radiation and ambient temperature) for one year, constructed using averaged data collected from the HARS site between May 2004 and May 2006, was used in the simulations. The model predicted that the HARS system would allow 3,458 "design temperature" showers to be taken annually. Since the bathing centre system is three times the size of the HARS system, 10,373 early morning showers would be available. The design target is to provide 550 showers for the local people in the village per week or 28,600 annually. The predictions indicate therefore that only 36 % of this target can be reached using this operating strategy, and therefore that an alternative operating strategy is required.

4.4.2.4 Alternative Operating Strategies

Initially there were no time-of-day restrictions on the use of the bathing centre because of the current cultural preference. However, the aim is gradually to introduce a bathing centre operating regime, which will maximise the potential of the HASWH community bathing centre. Ultimately, it is believed that by spreading the hot water demand across the day, say from 9 a.m–3 p.m, this will reduce storage tank heat losses overnight. However, some hot water will remain in the tank to allow showering to take place at 8 a.m. This means that the bathing centre will be open for approximately six hours per day.

As an interim step, however, an operating regime of 7.30 a.m to noon was introduced, but only once the bathing centre was being used regularly by a significant number of villagers. The annual performance of the HASWH community bathing centre with these two alternative operating strategies was therefore predicted with the model. The performance indicator used, as before, is the number of 10 litre "design temperature" showers. The model predicted that with the interim operating strategy (7.30 a.m–noon) the bathing centre would annually be able to supply 18,468 "design temperature" showers i.e. 65 % of the target. The ultimate operating strategy (9 a.m–3 p.m) would be able to supply 31,934 showers, i.e. 614 per week, and therefore exceed the design target by approximately 10 %.

4.4.2.5 Discussion

The previous parts in the chapter have discussed the validation and use of a simulation model to understand the behaviour of a thermo-siphon hot water system. This work is useful because it can guide the designer towards system and operational improvements. The research, however, assumes an ideal world where user behaviour is entirely rational and predictable. Several social and cultural factors make the operation of the proposed bathing centre extremely challenging and need to be continuously considered particularly in operational decisions. The section below discusses some of these considerations.

- This current design is based on providing 10 liters of hot water per person at 50 °C. At best, with a frugal user and a water efficient shower head, this might provide a 2 min shower at 35 °C. However, experience at HARS has shown that many local people (and even HARS staff) do not like the water pattern created by a shower head and prefer to wash under a constant stream of water. Some users even remove the shower rose from its fitting! It is recognised that such cultural behaviour will only change slowly.
- Experience with the HARS system shows that people often bring clothes to wash at the same time as personal showering. This obviously means that more water is used than would be the case if water use was restricted purely to personal washing. Preventing users from washing clothing once they occupy the shower cubicle will be a challenging task for the caretaker. To help prevent such a problem, a cold water clothes washing and drying area is provided adjacent to the showering facilities, as shown in Figs. 4.74, 4.75.
- Controlling the individual allocation of hot water will probably be the most challenging task and would be so in any culture. One of the components of the holistic development program of RIDS-Nepal is non-formal education, which includes a section on how to shower and even how to wash babies. This work will complement specific instructions on the use of the bathing centre.
- One of the main aims of the modelling is to better understand the behaviour of the hot water system and hence try to minimise user disappointment. The community will have invested time and money in the bathing centre and expectations will be high. Experience at HARS indicates that many local people have little or no understanding of the relationship between solar radiation and the availability of hot water. Regulating opening times to when hot water is available will be crucial to the successful introduction of the community bathing centre.
- Once absorbed by the local community's culture, the HASWHBC enables improved personal hygiene for the whole village community. Widely spread skin and other diseases can be addressed preventatively, improving people's health, ability to work and capacity to learn. These are all important factors for improving their overall living conditions. Thus, it is justifiable to say that the proper, long-term use of the HASWHBC will boost the HDI beyond the already

identified HDI increase through a "Family of 4" HCD project, as another synergistic benefit.

4.4.2.6 Conclusions

Hygiene conditions in remote parts of Nepal are poor and this has a high impact on the health of villagers. Scabies ranks high amongst common diseases in rural Nepal and this disease is directly linked to poor hygiene. RIDS-Nepal, a Nepali based NGO has built in 2009 a community bathing centre in the village of Dharaopri in the northwest province of Humla to provide a washing facility for its 350 residents. The prototype of the village system is a 300 liter 6 m^2 thermo-siphon solar water heating system which has been operating at the NGO's HARS in Simikot since November 2005. Based on operational data from that system, the performance of the larger system for the community bathing centre has been predicted.

It was found that in order to meet the design aim of 550 showers per week, the cultural pattern of people's early morning bathing time has to be slowly altered. If a 9 a.m–3 p.m operating strategy is adopted, then over 600 showers can be provided each week, exceeding the design target by over 10 %. While this result is encouraging, it is acknowledged that changing cultural and traditional habits (i.e. the so called non-technical factors or "software issues") will pose additional challenges to the successful operation of the facility.

4.5 Summary

Technology is never developed in a vacuum but always within a defined culture and environment, and thus has to be contextualised to fit the prevailing context in regard to demands and needs identified by the end-users. Skill and experience are needed to bring these various parameters of technology, life and context together under the same denominator. It makes technology tangible, and no longer is the highest efficiency or cheapest cost per energy unit the ultimate goal, but appropriateness and sustainability within a defined environmental and cultural context. With both being dynamic factors of life, technology can never remain at a certain status quo. Rather, the task of contextualising technology is an ongoing, dynamic one. It includes continuing research and development, professional engineering, field data monitoring of the actual performance, environmental impact and cultural changes.

Periodic interviews and impact survey questionnaires for the end-users' feedback are an important part of any project, in particular of technical projects aiming to bring forth changes. These bring to light the more anthropological factors, or "software issues" of a project. They reveal the natural and project implementation related changes perceived and experienced by the users. They are key parameters to make a project more successful and sustainable in the long-term.

Thus it becomes clear that a HCD project approach is necessary in order to be relevant and respectful towards the partnering community's identity and culture. A HCD project is a dynamic process of different disciplines, of which engineering is just one. It pays due respect to the fact that there are crucial social, and technical parameters, within a defined cultural, geographical and climatic context. Also longer time scales (Goldemberg 2000) need to be considered and applied, compared to more traditional development projects which often have initially defined and limited (1–2 years) time frames. Further, they often distinctively dichotomise between project implementer and project end-user/receiver. These matters though are often at the centre of a project's "software issues", and thus need to be considered, as culture and people's ability to adjust to changes and take ownership of projects often takes place at a much slower pace than technological changes.

With this framework in mind the recently developed, contextualised thermal process technologies such as the smokeless metal stove, the greenhouse, the solar drier, the HASWH and the HASWHBC research and implementation projects in local village communities have been launched. The long-term time frame of each project enables these new technologies to become, over time, part of the local community's life and culture, in accordance with their ability to accommodate and integrate changes with their own initiative and ownership. In this way the projects will provide benefits beyond the mere individual project level, as each project and technology is an embedded and integrated part of a community initiated, long-term HCD project, implemented with and owned by, the end-users.

References

Ahrens, C. D. (2003). *Meteorology today* (7th ed., chapter 3, pp. 57–59, 66, Figure 3.14, chapter 11, p. 286). Brookes/Cole.

Bena, B., & Fuller, R. J. (2002). Natural convection solar drier with biomass back-up heater. *Solar Energy, 72*(1), 75–83.

Cooper, P. I., & Fuller, R. J. (1983). A transient model of the interaction between crop, environment and greenhouse structure for predicting crop yield and energy consumption. *Journal of Agricultural Engineering Research, 28*(5), 401–417.

Cornwell, K. (1977). *The Flow of Heat*. Berkshire, U.K: Van Nostrand Reinhold Co. Ltd.

DDC. (2004). *District Profile of Humla (DPH)*. (in Nepali language), Government of Nepal, Simikot, Nepal.

Duffie, J. A., & Beckmann, W. A. (1991). *Solar engineering of thermal processes* (2nd ed., chapter 5.7., p. 231). New York: John Wiley.

EIA. (2002). *World Energy Outlook 2002*. Paris: International Energy Agency, OECD/IEA. Retrieved March 10, 2010, on November 1, 2011 not available any more from http://www.iea.org/Textbase/nppdf/free/2000/weo2002.pdf

Emeriau, S. (2006). *Nutritional exploratory mission—Humla and Mugu Districts, Nepal*. Kathmandu, Nepal: Action Contre le Faim (ACF).

Food and Agriculture Organisation (FAO). (1988). *Utilization of Solar and Geothermal Energy for Heating Greenhouses*, Report and Proceedings of CNRE Workshop, Turkey. CNRE Bulletin No.21, Food and Agriculture Organisation of the United Nations, Rome, Italy

Food and Agriculture Organisation (FAO). (2003). *Trade reforms and food security. Conceptualizing the linkages*. Rome, Italy: Food and Agriculture Organisation of the United Nations.

Garzoli, K. V. (1988). *Greenhouses—handbook for nurserymen, horticulturalists and gardeners*. Canberra, Australia: Australian Government Publishing Service.

Goldemberg, J. (2000). *World energy assessment: Energy and the challenge of sustainability*, chapter 10: *Rural energy in developing countries* (p. 380), UNDP 2000. Retrieved July 23, 2012 from http://www.undp.org/content/dam/undp/library/Environment%20and%20Energy/Sustainable%20Energy/wea%202000/chapter10.pdf

Haddix, M. K., & Zahnd, A. (2005). *A social and health impact assessment in Chauganphaya and Kholsi, phase one*. Kathmandu, Nepal: The ISIS Foundation & RIDS-Nepal.

Koirala, B. S. n.d. *Biomass energy uses: an experience and application of alternative energy technologies in Nepal*, Kathmandu, Nepal: Environment Advisor, Rural Energy Development Program, UNDP. Retrieved July 19, 2012, from http://www.p2pays.org/ref/35/34321.pdf

Morrison, G. (2003). *Solar thermal energy MECH9720, performance of solar water heaters* (p. 272). Sydney: UNSW Australia, School of Mechanical and Manufacturing Engineering.

Nayak, C. A., Suguna, K., & Rastagoni, N. K. (2006). Combined effect of gamma-irradiation and osmotic treatment on mass transfer during rehydration of carrots. *Journal Food Engineering, 74*, 134–142.

Nepal Trust. (2007). *Health—a basic human right for some.....* The Nepal Trust Organisation, Health Projects. Retrieved July 22, 2011, from http://www.nepaltrust.org/

Nepal Institute of Development Studies (NIDS). (2003). *Ranking of districts based on weighted scores*, Nepal Institute of Development Studies. Retrieved November 1, 2011, from http://www.hdihumla.org.np/remote-districts-of-nepal.htm

Peck, M. K., & Proctor, D. (1983). Design and performance of a roof integrated solar air heater. *Solar Energy, 31*(2), 183–189.

Rai, K. (2005). *Development of improved cooking stoves for mountain areas of nepal*. Nepal: Research, Development and Consultancy Unit, School of Engineering, Kathmandu University.

Rai, K., Zahnd, A., & Cannell, J. K. (2005). *High-altitude smokeless metal stove research and development*. Unpublished final Report on High-altitude Smokeless Metal Stove. Dhulikhel, Nepal: Kathmandu University.

Rai, K., & Zahnd, A. (2006). High-altitude smokeless metal stove research and development. *Boiling Point 51*, 37–38. Retrieved July 25, 2012, from http://practicalaction.org/docs/energy/boilingpoint51.pdf

Reddy, A. K. N. (2002). Energy technologies and policies for rural development. In T. B. Johansson, J. Goldemberg (Eds.), *Energy for sustainable development: A policy agenda* (Ch. 4, pp. 115–136). New York: UNDP.

Rozis, J-F., Guinebault, A. (1996). *Solar Heating in Cold Regions*. London, U.K: Intermediate Technology Publications Ltd.

Saghir, J. (2005, May). *Energy and poverty: Myths, links, and policy issues*. Energy Working Notes, World Bank: Energy and Mining Sector Board, No. 4.

Sandhu, B. S., Mannan, K. D., Dhillon, G. S., & Cheema, L. S. (1979). Design, development and performance of multi-rack, natural convection drier. In *Proceedings ISES Congress, Atlanta*, Georgia (pp. 25–28).

Sorenson, B. (2005). *Renewable energy, its physics, engineering, environmental impacts, economics & planning* (3rd ed., p. 359). ISBN 0-12-656153-2, chapter 4, Figure 4.17, Amsterdam: Elsevier.

TRNSYS. (2005). *TRNSYS Version 16.0*, Solar Energy Laboratory, University of Wisconsin, USA, web site with the latest TRNSYS version 17. Retrieved July 23, 2012, from http://sel.me.wisc.edu/trnsys/index.html

Turkenburg, W. C. (2000). *Renewable energy technologies,* chapter 3: *Energy, the environment, and health* (pp. 247–251), World Energy Assessment: Energy and the Challenge of Sustainability, UNDP.
UNDP. (2006). *Human Development Report 2006—Beyond scarcity; Power, poverty and the global water crisis,* p. 294, New York, USA: United Nations Development Program. Retrieved July 24, 2012, from http://hdr.undp.org/en/media/HDR06-complete.pdf
Warwick, H., Doig, A. (2004). *Smoke—the killer in the kitchen.* London: ITDG Publication, ISBN 1 85339 588 9. Retrieved July 24, 2012, from http://practicalaction.org/docs/smoke/itdg%20smoke%20report.pdf
World Energy Assessment (WEA) (2000). *Energy and the Challenge of Sustainability*, UNDP, New York, USA. Retrieved July 24, 2012, from http://www.undp.org/content/undp/en/home/librarypage/environment-energy/sustainable_energy/world_energy_assessmentenergyandthechallengeofsustainability.html
WECS. (2006). *Annual Report 2005.* HMG of Nepal, Kathmandu, Nepal.
Whillier, A., & Saluja, G. (1965). The thermal performance of solar water heaters. *Solar Energy, 9*(1), 21–26.
WHO. (2006). *Fuel for life: Household energy and health.* Switzerland: WHO Geneva.
Winrock. (2004). *Household energy, indoor air pollution and health impacts—status report for Nepal*, Kathmandu, Nepal: Winrock International Nepal. Retrieved July 24, 2012, from http://www.pciaonline.org/files/Nepal_Household_Energy_and_Health_Overview.pdf
World Bank. (2010). *The little Green Data Book.* International Bank for Reconstruction and Development/The World Bank, p. 153, ISBN: 978-0-8213-8245-5, 1818 H Street, N.W., Washington, D.C. 20433, USA. Retrieved July 24, 2012, from http://www.semide.net/media_server/files/semide/topics/water-data/little-green-data-book-2010-world-bank/GreenDataBook2010_WorldBank.pdf
Zahnd, A. (1998). *Firewood consumption survey in Jumla Villages*, Jumla, Nepal. Unpublished data survey.
Zahnd, A., & Malla, A. (2006, September 13–15). *High-altitude solar water heater community bathing centre—renewable energy utilised for a remote and impoverished Himalayan Village in Humla Nepal.* In *ANZSES 2006, Canberra*, Australia. Retrieved July 24, 2012, http://www.rids-nepal.org/index.php?option=com_docman&task=doc_download&gid=11&Itemid=104

Chapter 5
Thermal Comfort in the Home

5.1 Improving Comfort Levels in Traditional High-Altitude Humla Houses

The aim of this chapter is to investigate ways in which the comfort levels in traditional Humli housing might be improved using simple, low cost strategies. As building homes is an integral part of people's lives and culture, they inherently are part of a HCD program. Also, it recognises that the locally available renewable energy resources can be utilised in wise and appropriate ways to improve their housing conditions, while contributing to the effort for more sustainable energy use. This makes thermal comfort an intrinsic part of the key issues for environmental and social sustainability. The research project took place in the remote and harsh, high-altitude climate of Humla (Sect. 1.1.8).

Most Humli people are extremely poor and endure a subsistence culture, living in traditional stone and mud houses. Energy for cooking, heating and lighting comes from fuel wood and "*jharro*", supplies of which are diminishing. Based on anecdotal evidence and unpublished survey data (RIDS-Nepal 2008), it has been found that for approximately 10 months of the year, the open-fire simultaneously provides a source of heat and light, both of which are vital for domestic life. Heat is particularly important in the mountainous areas of Nepal, where low ambient temperatures are common. In an attempt to reduce fuel wood consumption and improve the indoor environment, smokeless metal cooking and heating stoves (SMSs), developed by the author, have been introduced into Humla (Sect. 4.1.2).

From 2002 to 2012, RIDS-Nepal has installed 1,182 SMSs in Humla's villages. Measurements taken by the author indicate that, when properly used, the SMS can reduce the indoor PM_{10} level generated by an open-fire by 90–98 %. Unpublished data measured using a portable air monitor (SKC 2010) shows that over a typical 24 hour period the average indoor PM_{10} in a kitchen with a SMS was 0.056 mg/m^3, compared to 1.28 mg/m^3 in a similar kitchen using an open-fire place. However, there is some concern that comfort levels may not be as acceptable as before because of a reduction in radiant heat. Since there is a need to continue to

A. Zahnd, *The Role of Renewable Energy Technology in Holistic Community Development*, Springer Theses, DOI: 10.1007/978-3-319-03989-3_5,

Table 5.1 HARS Climate data collected in Simikot from May 2004–March 2007

Month	Average monthly horizontal solar radiation (MJ/m^2/d)	Average monthly ambient temperature (°C)	Average monthly maximum ambient temperature (°C)	Average monthly minimum ambient temperature (°C)
January (82)	12.6	6.6	13.2	−1.4
February (55)	14	6.6	14.5	1.2
March (79)	17.4	8.1	15.4	2.5
April (44)	20.1	11.5	17.8	6
May (93)	19	14.5	20	9.4
June (90)	17.2	17.1	22.6	12.3
July (93)	14.9	17.9	22.5	14.9
August (77)	14.7	17.2	22.2	14
September (57)	17.7	16.8	22.6	12.4
October (39)	16.7	12.4	19.5	6.4
November (42)	14.2	8.5	17.4	1.1
December (93)	12.1	7.4	16	1.1

introduce improved cooking stoves, because of the health benefits, there is also a need to address comfort levels.

The approach, taken in this research project, is to analyse the thermal performance of the traditional house and investigate simple ways of improving the building envelope to improve comfort levels. A validated dynamic model of a traditional house was developed and current comfort levels were predicted. The model was then used to investigate the effects on those comfort levels from changes to the building envelope. Previous similar research is reviewed and the model used in this research and its validation are described. The impact on comfort levels achieved by implementing various energy conservation measures is then predicted using the validated model. According to Hagen (1980), Nepal can be divided into seven natural topographical "units", which can be clearly distinguished from each other. Humla belongs to the region known as the Inner Himalayas. It is the name given to the valleys, which lie to the north of Nepal's principal and well known chain of mountains, the Himalayas. These inner valleys are described by Hagen (1980) as "the real high mountain valleys of Nepal, surrounded on all sides as they are by ice clad giants". The climate in Humla is challenging. Table 5.1 shows some climatic data collected at RIDS-Nepal's HARS in Simikot from May 2004 to March 2007.

5.2 Previous Research

Although limited, there is some previous research into the thermal performance and associated comfort levels in traditional Nepali houses. A very useful report was prepared by Nienhuys (2003) as part of a project to "develop and apply effective affordable (low cost) thermal insulation solutions for traditional stone dwellings in the Northern Areas of Pakistan". Revised for Nepal, the document contains a list of priorities and suggested improvements to the building envelope. These include:

- closing the open hole in the roof used to vent smoke
- applying internal wall insulation; keeping the roof and walls dry
- installing a suspended ceiling
- installing double-glazing and curtains
- filling cavity walls with insulation
- keeping foundations dry and insulating the floor.

The performance of a two storey, traditional dwelling in an unspecified mountainous area of Nepal was simulated by Rijal and Yoshida (2005). The dwelling had a total floor area of 54.5 m^2, each floor being a single room. A modified Japanese heating and cooling load calculation program was used to perform the simulations. Modifications to the program were reported to be the addition of "the calculation of natural ventilation rates for a building with multiple compartments" and "the calculation of heating and cooling loads, considering simultaneous heat and moisture flow in the building and existing materials in components".

It was found that reducing infiltration by closing of doors and windows, and the addition of roof insulation improved the thermal conditions by between 4.4 and 12.7 K. After reducing fuel wood consumption by 60 %, night time temperatures were still 1.0–4.0 K higher than the unmodified house. This applicability to Humla homes is, however, limited by several factors. Although often multi-storey, the Humli houses are typically much smaller, and unfortunately the climatic data values used by Rijal and Yoshida (2005) are not stated and the hourly variation in measured indoor temperatures is not shown. A 4 day thermal comfort study of 36 residents in the town of Lomantang in the Mustang District of Nepal was also conducted by Rijal and Yoshida (2006). This area is mountainous, and located at 3,705 m, which is nearly 700 m higher than Simikot, the location of this research. Although both locations experience severe winters, there are some important differences. While the mean monthly outside ambient temperature for the two locations in May is very similar, Simikot is significantly warmer in January. Relative humidity levels for the two locations are also quite different. In Mustang, mean monthly relative humidity levels are reported to be 97 and 71 % in January and May respectively, while in Simikot 2006 data indicate average values of 29 and 62 % respectively.

The researchers found that the residents of three houses investigated were "highly satisfied with the thermal conditions of their houses". Rijal and Yoshida (2006) also established that the mean temperature for the surveyed residents was only 10.7 °C, although there was considerable variation in the temperatures established in the three houses measured. The detailed analysis (Rijal 2006) used a 9 point scale, in preference to the 7 point scale used by ASHRAE. Because there was no electricity in the houses investigated and residents went to bed at 8 p.m., the study was only conducted in the daytime. The researchers established high levels of personal insulation, measured in CLO,[1] by weighing clothing. These values were 2.87 for men and 5.96 for women. Body surface areas of 1.57 and 1.44 m^2 were calculated for men and women respectively.

In comparison with most traditional Humli house standards, however, the houses investigated were quite different. Firstly, they were 2 storey courtyard houses, where the first storey is used for animals and storage. Secondly, the Mustang houses are constructed from sun dried 450 mm thick bricks, as opposed to granite rock[2] in Humla. These factors, together with climatic differences, limit the value of the study by Rijal and Yoshida (2006) for the present research.

5.3 Typical Humli House

A typical Humli house is usually multi-storey. The ground level rooms are used to stable animals, while the middle and upper storeys often consist of a living room and a store room. Such a house was identified in Simikot, the main town of Humla district, and was used for this study.

Two adults and four small children permanently occupy the house. The normal routine of this working family is to awaken between 5 and 6 a.m. and to go to bed sometime after 9 p.m. Main meals are eaten twice a day. Breakfast cooking begins at 7 a.m. and it is eaten at about 9 a.m. Cooking dinner begins at about 6 p.m. or after sunset and takes 2–3 hours. This meal is eaten between 8 and 9 p.m. Some snacks might be cooked on the stove during the day between 1 and 2 p.m. The four principal rooms in the house were identified as follows:

- Living Room 1 is on the middle level. It is the main room of the house, containing a SMS and is where the family of six sleep at night.
- Store Room 1 is also on the middle level and is adjacent to Living Room 1.
- Living Room 2 is a spare room on the upper level, used by visiting relatives and guests. Two adults occupied this room at the time of this study.
- Store Room 2 is also on the upper level, adjacent to Living Room 2.

The four rooms are small and their principal internal dimensions, together with those of the animal stables, are given in Table 5.2 and the floor plans of the three levels are shown in Fig. 5.1.

---

[1] CLO: Clothing Level. ASHRAE provides a clo calculator to measure people's clothing level (by weight) to determine their living "comfort" level.

[2] Some granites have a very high uranium and thorium content and give off large amounts of radon gas. Radon is a serious carcinogen and when combined with house dust or smoke it will adhere to the lungs and respiratory tract and produce tumours (personal communication with Prof. Dr. Philip Jennings). As far as the author knows this has never been investigated in Humla, as it has e.g. in the USA, Sweden and the Netherlands. If it is a problem, there are ventilation and hygiene techniques that can reduce the risk. This is a field which needs additional research and investigation.

Table 5.2 Principal dimensions of the rooms in the Humli house

Room	Area (m^2)	Height (m)
Living room 1	15.8	1.81
Store room 1	5.8	1.81
Living room 2	3.4	2
Store room 2	4.2	2
Animal stable	27.9	1.9

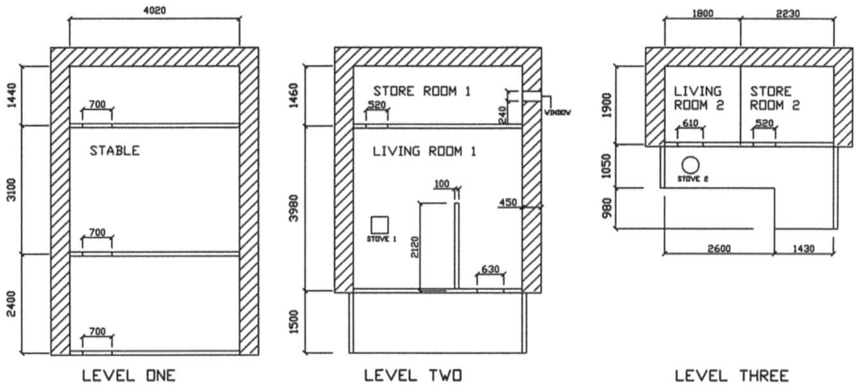

Fig. 5.1 Floor plans of 3 level traditional Humli house

Fig. 5.2 Loose fitting timber is used to create a semi-enclosed area in front of the rooms on top level of the Humla house

In front of the rooms on the top and middle levels, a semi-enclosed area has been built with loose fitting timber (Fig. 5.2). This construction offers some protection from cold southerly winds off the mountains, but also shades the main wall from solar radiation. These semi-enclosed areas measure 6.0 and 5.6 m^2 for the middle

Fig. 5.3 Humli houses are built attached to each other horizontally and on top of each other

and top levels respectively. There are also 0.5 m overhangs projecting from roofs of the semi-enclosed areas. The roofs of level 1 and level 2 are made of various layers of timber, bracken, plastic sheet, slate, finished on top with earth/mud. They serve as the access walkways for the levels above. In this Humli house, there was (from outside to inside) 40 mm of mud, 30 mm of slate, a plastic sheet, 20 mm of bracken and 25 mm of timber, supported on a network of beams.

Figure 5.3 shows the middle and top storeys of the houses adjacent to that used for this study. Inaccessibility prevented the actual house from being photographed effectively. Access to other levels is often by log ladders[3] and several of these can be seen in Figs. 5.3, 5.4. The constructional walls are stone, usually 450 mm thick. At ground level, the constructional wall is exposed (Fig. 5.4), as are the end walls of any particular row of houses. In this instance the studied house is an "end" house (Fig. 5.5).

5.4 Indoor Thermal Environment

Ambient conditions inside the house were measured at hourly intervals over a 12 day period in early April 2007. Although the last snow had fallen in early March, ambient temperatures at night still fell below 10 °C every night (Table 5.3). Battery powered data loggers (Hobo-H8) were installed in each of the two living rooms and storerooms, in each case at a height of approximately 1.5 m. These devices were cross calibrated against each other at the start of the experiments.

[3] "Log ladders" are called "lisnu" in the local language, made out of a round tree, into which "foot steps" have been cut into in the vertical axis, to gain height as in a ladder. They are the traditional and most commonly used means to climb from one floor to the next of one house, or the next house, which is often built on top of another house as can be seen in Figs. 5.3, 5.4.

Fig. 5.4 View of Level 1, showing mud roof and exposed wall

Fig. 5.5 East side of studied house

Table 5.3 Summary of temperature and relative humidity in two living rooms

Conditions	Living room 1 Temperature (°C)	Living room 2 RH (%)	Outside ambient Temperature (°C)
Average	21.2	33	16.2
Maximum	31.9	47	21.7
Minimum	14.9	23	10.6

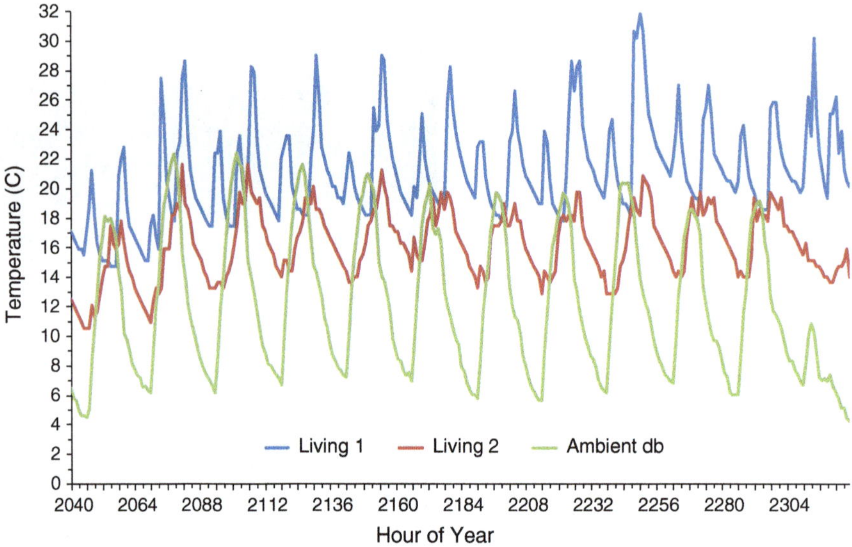

Fig. 5.6 Dry bulb (*db*) temperatures inside and outside living rooms of Humli house in early April

Figures 5.6, 5.7 show the hourly dry bulb (db) temperature and relative humidity measured in the two living rooms and outside the house over the period and Table 5.3 provides a summary of those conditions, in terms of an average, maximum and minimum, together with outside ambient conditions.

High temperatures in Living Room 1 are due to the operation of a SMS in that room. The logger was approximately 1.8 m from the stove. The smaller and larger peaks on Fig. 5.6 show the impact of stove operation in the mornings and evenings respectively. The mean minimum temperatures are higher in Living Room 1, not only because of the use of the cooking stove, but also because the whole family sleeps in this room at night. High humidity levels at the end of the evaluation period were caused by rain.

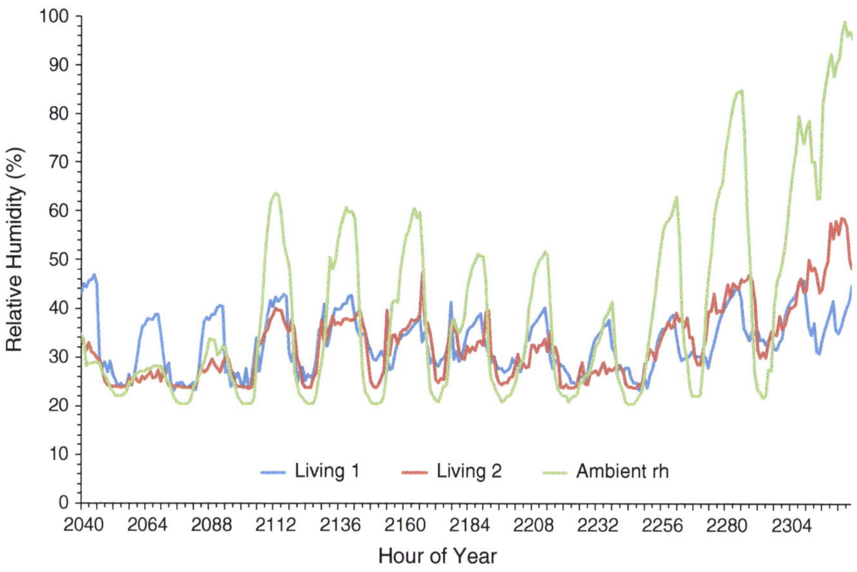

Fig. 5.7 Relative humidity levels inside and outside living rooms of Humli house in early April

5.5 Comfort Assessment

There are two basic approaches to the assessment of thermal comfort (Nicol 2002). The "rational index" approach is based on the response of subjects placed in climate controlled chambers and is the basis of the ISO 7730[4] and ASHRAE 55[5] Standards. Some studies (Brager 1998) have found that this approach does not describe comfortable conditions adequately, particularly in naturally ventilated buildings. This inadequacy has led to the other "adaptive" approach, which is based on the premise that a subject is an active agent in any given thermal environment and that there is some thermal adaptation. Thermal comfort has been shown to be a function of the mean monthly outdoor temperature for the location and various algorithms have been suggested for calculating the "comfort" or neutral temperature (Humphreys 1978).

4 ISO 7330:2005, Ergonomics of the thermal environment—Analytical determination and interpretation of thermal comfort using calculation of the PMV and PPD indices and local thermal comfort criteria, http://www.iso.org/iso/catalogue_detail.htm?csnumber=39155 (accessed July 27, 2012).

5 ASHRAE 55. 1995. *Thermal Environmental Conditions for Human Occupancy*. American Society of Heating, Refrigerating and Air-Conditioning Engineers. Atlanta Georgia, USA, ASHRAE 55-2010 for purchase is available from: http://www.techstreet.com/ashrae/standards/ashrae/55_2010?product_id=1741646 (accessed July 27, 2012)

Perceptions of comfort vary. The longitudinal study by Nicol (1996) of subjects in five climatic zones of Pakistan demonstrates this difference in perception. The winter comfort temperatures vary by 5.3 °C between the coldest and warmest zones. This study was conducted using city dwellers. Although Humla has a similar mean winter temperature to the two coldest Pakistani cities, the perceptions of comfort of Humli residents may not be the same.

People in such isolated locations simply do not have a way of evaluating "comfort" in the same way as a person living in a city and this may lead to surprising results. As described earlier, previous researchers (Rijal and Yoshida) studying other traditional people living in mountains of Nepal, calculated a mean temperature of 10.7 °C, which is over 9° lower than the coldest winter value determined by Nicol (1996). Despite the difficulty of comparing peoples' perceptions of comfort, the author knows from his own experience of living and working in Humla that people do feel cold and that this affects their health and well being.

The prime objective of this research was to obtain some understanding of the thermal behaviour of a typical Humli house in order to improve the internal conditions. Such improvements could be embraced by RIDS-Nepal's "Family of 4 PLUS" HCD program. No such study has been conducted previously. The purpose of this work was not to assess people's feelings with respect to comfort or to derive a particular comfort temperature, but rather to use calculated comfort levels as an indicator of the effect of building envelope modifications. As cited earlier, some low-cost methods of improving the thermal performance of traditional Nepali houses have been suggested by Nienhuys (2003). In order to assess building envelope modifications, some readily derived indicator was required to assess their individual or collective impact on comfort.

An indication of the comfort levels experienced during the measurement period was determined using a commercially available thermal comfort prediction tool prepared for ASHRAE (Fountain 1996). Although the tool contains eight thermal comfort models and calculates a number of indicators of thermal comfort, the "PMV-PPD" model was chosen to evaluate the conditions in the Humli house because its indicators are widely used and easily understood. The PMV (Predicted Mean Vote) index represents the sensation of a large population to a particular thermal environment. A PMV value represents the predicted vote on a seven point thermal sensation scale.

If the calculated value is within the range of −0.5 and +0.5, then the conditions are within the ISO (International Standards Organisation) comfort zone. The PPD (Predicted Percentage of Dissatisfaction) is a function of PMV and indicates the number of individuals likely to be unhappy with the conditions. This value never drops below 5 % because of the assumption that there will always be a certain level of dissatisfaction, even if the PMV index is zero.

To use this tool, some assumptions were made because more input data was required than was measured in the Humli house. The mean radiant temperature (MRT) was assumed to be, on average, 0.2 °C higher than the air temperature. This assumption was based on the temperatures measured over a 10 day period in May 2007 at the HARS of RIDS-Nepal. The internal surface temperature of a rendered

Table 5.4 RH (relative humidity), PMV (predicted mean vote) and PPD (predicted percentage of dissatisfaction) values in two living rooms at occurrence of average minimum dry bulb (db) temperatures

Conditions	Living room 1				Living room 2			
	Temperature (°C)	*RH* (%)	*PMV*	*PPD* (%)	*Temperature* (°C)	*RH* (%)	*PMV*	*PPD* (%)
Mean minimum	17.3	31	−1.68	61	12.9	33	−2.88	98

external stone wall and the room air temperature were measured on an hourly basis. The air velocity in Living Room 1 was measured using a hand held hot wire anemometer[6] in various positions and found to be low, i.e. in the range 0.0–0.1 m/s.

The residents were assumed to be seated most of the time because the rooms are small and there is little room to carry out any other kind of activity. A MRT value of 1.0, recommended for persons sitting quietly, was therefore used. Residents were asked about their clothing levels during the period of the study and the CLO calculator provided in the ASHRAE tool was used to calculate a value of 1.0. This was also assumed to be identical for all occupants, regardless of age and gender. Using the mean minimum dry bulb (db) temperatures and their corresponding relative humidity (rh) values, PMV and PPD values were calculated with the comfort tool (Table 5.4).

Both PMV values were found to be outside the ISO comfort zone. At present, increasing the minimum temperature inside the house involves using the stove more often and this means using more fuel. For the poor in Simikot, obtaining fuel wood already involves significant labour and drudgery. For a 25 kg load, a family member must walk for 2–3 hours, and then spend 2 hours collecting the fuel. The return home walk with the fuel wood on their back takes 2.5 hours, meaning around 7 hours has been spent on this one task.

If the family is lucky enough to own a SMS, one 25 kg load will last ~2 days. If they are still using an open-fire, one load will only last ~1 day. In general, it is the women and children of the family who perform the arduous task of fuel wood collection and they must do it every other day. Purchasing fuel wood is simply not an option for the poor since it costs 250–300NRs or US$3–4 per 25 kg load.

Any appropriate strategy that might improve the comfort levels in a traditional Humli house therefore will have significant impacts. At an individual level, apart from a general improvement in wellbeing and health, less time would be spent on fuel wood collection. At a societal level, a reduced demand for fuel wood would ease pressure on dwindling supplies.

In order to investigate possible options for comfort improvement, a TRNSYS thermal model of the Humli house was developed and validated, and that process is described in the sections below.

[6] TSI Inc., VelociCalc Model No 8350-1

Table 5.5 Material properties used in TRNSYS simulations

Material	Density (kg/m^3)	Conductivity (W/mK)	Specific heat (kJ/kg/ °C)	References
Granite	2640	3	0.82	Cornwell (1977)
Earth	1900	3.6	0.23	Rozis (1996) and Garzoli (1988)
Timber	700	1.4	2.6	Cornwell (1977)

5.6 TRNSYS Model

In the case of the Humli house, the standard building subroutine within TRNSYS (Type 56) has been used. Predictions of thermal performance were made on an hourly basis and measurements of air temperature and relative humidity within the house have been used to validate the model. This approach has allowed various strategies to improve the thermal performance of the house to be investigated.

The Type 56 subroutine requires various fixed parameters to be supplied and assumptions to be made by the user. Table 5.5 indicates the material properties used. Other assumptions made in the model include: a fixed sky temperature = ambient air temperature less 12 °C, a ground reflectance = 0.2 and a density of air[7] = 0.837 kg/m^3. It was assumed that the house was occupied by the family and guests between 5 p.m. and 10 a.m. and that the four animals were housed in the stable between 7 p.m. and 7 a.m. Human activity levels in the house were assumed to be low, primarily for eating and sleeping, and therefore equivalent to "seated, at rest", as defined in ISO 7730 within TRNSYS. The metabolic heat generation of an animal was assumed to be twice that of a human.

5.6.1 Model Validation

In order to validate the TRNSYS Humla house model, a climatic data file of total solar radiation, outside ambient temperature and relative humidity was constructed. These data were measured at a weather station,[8] which is part of the long-term data collection system at the HARS. The mass of fuel wood used daily by the family was measured on a weighing scale at 6 a.m. every morning over 8 days (Table 5.6). The validation period was therefore from 27th March to 3rd April inclusive.

---

7 Air density calculation: http://en.wikipedia.org/wiki/Density_of_air, and http://wahiduddin.net/calc/density_altitude.htm (both accessed July 27, 2012)

8 Spectrum Technologies, Inc., WatchDog Model 2900ET, installed at HARS in Simikot, Humla. http://www.specmeters.com/WatchDog_Weather_Stations/WatchDog_Model_2900ET_Weather_Station.html (accessed July 27, 2012)

Table 5.6 Daily firewood use (kg) in Humli house over 8 day period

Date	Morning	Evening
27th March 2007	4	4.5
28th March 2007	4	6
29th March 2007	4	4
30th March 2007	5	4
31st March 2007	4.5	5
1st April 2007	4	5
2nd April 2007	6	4
3rd April 2007	5	4
Average	***4.6***	***4.6***[a]

[a] An ~10 kg firewood consumption a day for a family of 2 adults and 4 small children is very low compared to the average village family (20–40 kg/day). Possible reasons: This family lives in the main upper Simikot village and thus firewood is very expensive or time demanding to collect. It is a small family with only small children and they are probably very poor, as many low caste families live in this village, thus minimizing their firewood consumption to the bare minimum

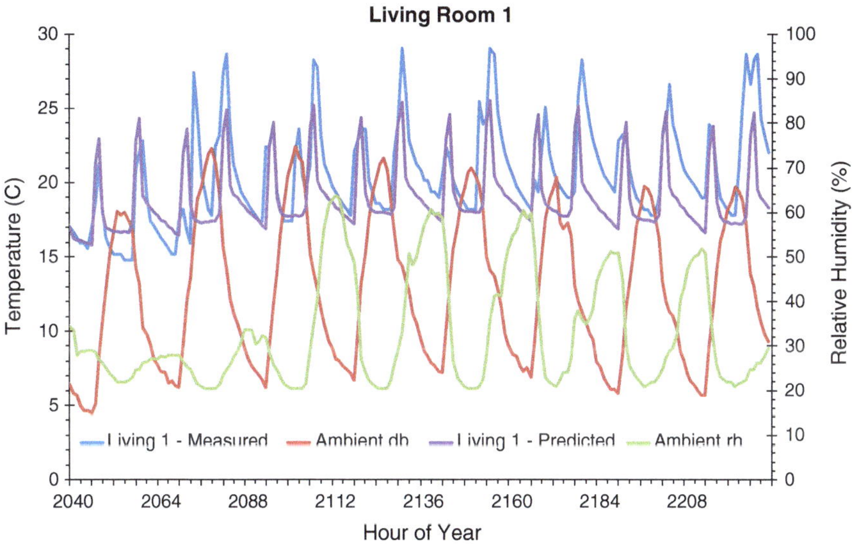

Fig. 5.8 Measured and predicted temperatures in Living Room 1 during 8 day period

Indoor temperatures in the four rooms of the selected house were predicted on an hourly basis over the 8 day period and compared with measured temperatures. Figures 5.8, 5.9, 5.10, 5.11 show these comparisons for Living Room 1, Store Room 1, Living Room 2 and Store Room 2 respectively. Outside conditions, db temperature and relative humidity are also shown.

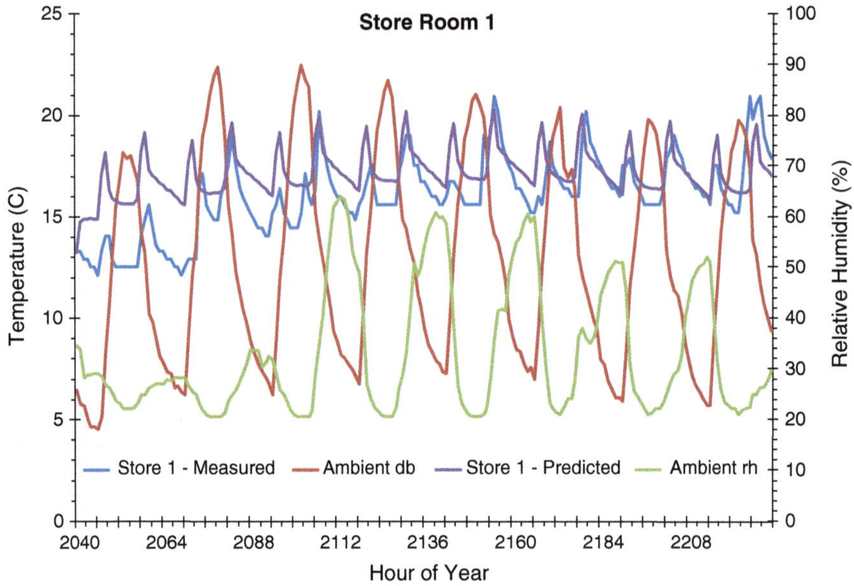

Fig. 5.9 Measured and predicted temperatures in Store Room 1 during an 8 day period

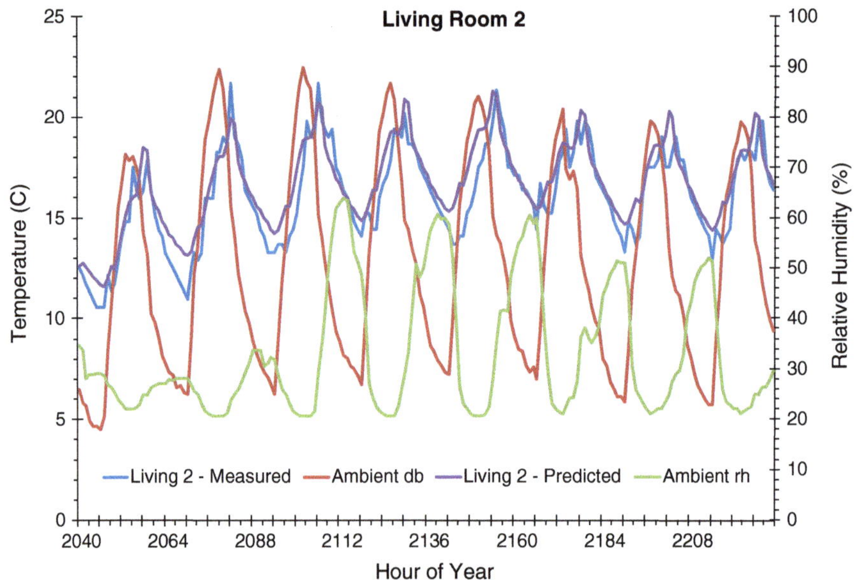

Fig. 5.10 Measured and predicted temperatures in Living Room 2 during 8 day period

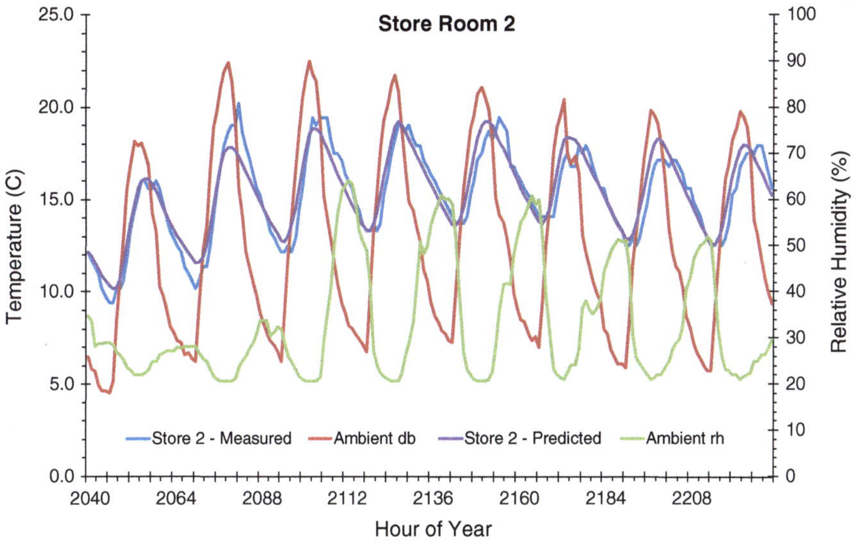

Fig. 5.11 Measured and predicted temperatures in Store Room 2 during an 8 day period

The main stove (Stove 1, Fig. 5.1) used by the family is located in Living Room 1 (Fig. 5.12). In the model, cooking was assumed to take place from 07:00–09:00 a.m. to 18:30–20:30 p.m. The average value of fuel wood mass (Table 5.6) was assumed together with a fuel calorific value and stove efficiency of 15 MJ/kg (Mande et al. 1999) and 30 % respectively (Thakuri et al. 2006). Heat output from the stove into the room was assumed to be equally convective and radiative, based on average stove and wall temperatures of 100 °C and 15 °C respectively.

An air infiltration of three air changes per hour was assumed. The validity of this assumption was confirmed by measurements of airflow through a permanently open ceiling vent above the stove. The predicted maximum indoor air temperatures, as a result of stove use, generally occur at the same time as the measured peaks (Fig. 5.8). In ~50 % of occasions, the predicted peak temperature is similar to the measurement, while peak temperatures are under predicted at the remaining times.

The temperatures in Store Room 1 were found to be closely coupled to those in Living Room 1. Peak air temperatures in Store Room 1 were coincidental, but reduced in amplitude, compared with those in Living Room 1. Air exchange between these rooms was assumed to be 102 kg/h. A low airflow velocity of 0.1 m/s through a 0.34 m^2 gap between the two rooms would produce such a coupling. These assumptions produced acceptable predictions, compared to measured temperatures (Fig. 5.9).

There is a small inefficient fire place (Stove 2, Fig. 5.1) in the "verandah" area outside Living Room 2, which was evidently being used by visitors during the time

Fig. 5.12 Smokeless Metal Stove (SMS) installed in Humli houses

of this research. No fuel wood consumption data was available, and therefore an estimate of 500 kJ/h was assumed to be the heat input by convection only. The stove was assumed to operate between the hours 05:00–05:30 a.m. and 18:30–20:00 p.m.

Air infiltration was assumed to be high in this room with 6 air changes/per hour (AC/H) due to exposure and poor sealing. These assumptions produced good agreement between measured and predicted air temperatures in Living Room 2 (Fig. 5.10).

The air exchange rate in Store Room 2 was assumed to be only 1.0AC/H. Temperatures in this room appear to be largely determined by ambient conditions. Measured and predicted air temperatures compare well in terms of amplitude and time of occurrence (Fig. 5.11).

The air exchange rates used in the simulations of the four rooms are based on a combination of experimental measurements, and trial and error. These air exchange rates were fixed for the entire 8 day simulation period.

In such traditional homes which are poorly sealed, one might expect that variations in local winds would be very influential in determining internal thermal conditions. In reality, the winds are quite predictable in this location. In the mornings, there is little or no wind, but in the afternoon the wind velocity increases to such an extent that it is considered too dangerous for aircraft to land or take off after midday. Despite this daily variation, acceptable agreement was achieved between the measured and predicted temperatures in each room using fixed values.

5.7 Performance Improvements

The validated model has been used to investigate three strategies to improve the performance of the Humli house. These strategies were:

- reducing infiltration
- improving ceiling insulation
- the creation of a "sunspace" in front of each living room.

These strategies and their impact on comfort levels are described in more detail below.

5.7.1 Reduced Infiltration

The validation modelling and some measurements with a hot wire anemometer indicated that air infiltration rates into the house are high. In Living Rooms 1 and 2, it was necessary to assume infiltration rates of 6 and 3AC/H to produce reasonable agreement between measured and predicted internal dry bulb temperatures. Reducing these high rates of infiltration to the levels more commonly experienced in modern housing could be achieved by better construction and sealing methods. The effect of reduced infiltration rates was therefore investigated by reducing AC/H in both rooms to 1.5.

5.7.2 Improved Insulation

The existing structure has a minimal level of insulation which has been achieved using between 20 and 50 mm of bracken, sandwiched between various roof layers. An improvement in comfort levels should occur if insulation levels are increased. These have been predicted by adding a 50 mm layer of insulating material on the inside surface of the ceilings in Living Rooms 1 and 2. The R-value of the insulation material was assumed to be 0.62 m^2K/W, similar to that for 50 mm thick compressed straw (AIRAH 1989).

5.7.3 Sunspaces

The semi-enclosed area in front of the two living rooms has been created using loose fitting heavy timber planks (Fig. 5.2). The space could be converted into a sunspace by replacing the timber planks with a semi-glazed wall and door. A heavy duty flexible ultraviolet (UV)-stabilised plastic sheeting would be used for the glazing to ensure longevity and to permit easy opening (by rolling up) in the event of overheating. This conversion would allow heat generated by absorbed solar radiation to be captured and transferred to the living rooms. Infiltration rates were reduced to 1.0 AC/H in the two living rooms because of the sunspace. Infiltration rates in the two sunspaces were set at 1.5 AC/H because it was assumed that improved sealing would be used compared to the original construction.

5.7.4 Sunspaces Plus Improved Insulation

The combined effect of simultaneously improving insulation and creating two sunspaces has been investigated using the same parameters described in Sects. 5.7.2, 5.7.3, but in combination.

5.7.5 Impact of Improvement Strategies on Comfort Levels

The model was used to predict the average minimum temperature and corresponding relative humidity levels in Living Rooms 1 and 2 in the existing condition and using the four strategies outlined above. These values, as well as the predicted PMV and PPD indicators, are compared with the same comfort indicators, calculated using the predictions of the validated model in the existing state (Tables 5.7, 5.8).

In both living rooms, reducing the (assumed) high infiltration levels is more effective than increasing the ceiling insulation levels. The model predicts that average temperatures in Living Room 1 increase by 0.8 °C with the former strategy, but only 0.3 °C with the latter, compared to predictions for the existing state. Changes in the PMV and PPD reflect the effect of these improvements.

Conversion of the semi-enclosed area in front of each of the living rooms into a sunspace is the most effective single strategy to raise comfort levels. A sunspace not only transfers heat to the living rooms but also reduces the outside air infiltration rate into these rooms. Model predictions of the average temperatures in Living Rooms 1 and 2 are increased by 1.7 and 2.3 °C respectively with this strategy, compared to predictions for the existing state. As expected, increasing the amount of insulation in the ceilings at all three levels in combination with a

Table 5.7 Predicted impact of strategies to improve comfort levels in living room 1

Strategy	Average temperature (°C)	Average RH (%)	PMV	PPD (%)
Existing (measured)	17.3	31	−1.68	62
Existing (predicted)	16.9	54	−1.68	61
Reduced infiltration	17.7	67	−1.43	47
Improved insulation	17.2	53	−1.64	58
Sunspace	18.6	73	−1.26	38
Sunspace plus improved insulation	19.1	72	−1.01	27

Table 5.8 Predicted impact of strategies to improve comfort levels in living room 2

Strategy	Average temperature (°C)	Average RH (%)	PMV	PPD (%)
Existing (measured)	12.9	33	−2.88	98
Existing (predicted)	13.9	65	−2.48	93
Reduced infiltration	15.1	100	−2	77
Improved insulation	14.1	64	−2.43	92
Sunspace	16.2	100	−1.91	72
Sunspace plus improved insulation	16.4	100	−1.62	58

sunspace produces the greatest improvement in overall thermal conditions. The average temperature rises between 2.2 and 2.5 °C with the combined strategy.

Reducing the rate of infiltration of outside air, however, increases the inside relative humidity level. In the case of Living Room 2, the model predicts that the relative humidity reaches the saturation level. In practice, this would be unacceptable. However, Tables 5.7, 5.8 indicate that the model over predicts the average relative humidity levels in both living rooms, so in practice the saturation levels predicted are also likely to be an over prediction. The reason for the over prediction of relative humidity by the model is not clear and requires further investigation. Although more accurate (lower) relative humidity predictions will result in a rise in the values of PPD at the average dry bulb temperatures measured in this study, the relativity of the results is unlikely to change.

5.8 Conclusion

The thermal conditions in a traditional house in Simikot, the main town of Humla District in northwest Nepal, have been investigated. Minimum ambient conditions inside the house in early April indicate that comfort levels are far below what is internationally recognized as acceptable. In the main winter months (November–February), average indoor temperatures will be significantly lower because of colder outside ambient temperatures, so corresponding comfort levels will be much worse in winter. Any strategy that raises comfort levels is likely to be beneficial in terms of general health and well being, and could possibly even result in reduced fuel wood consumption.

The thermal performance of the house has been predicted using the thermal simulation program TRNSYS. The model has been validated using temperature and fuel wood consumption data gathered over a 12 and 8 day period respectively. An acceptable level of agreement between measured and predicted temperatures in four rooms of the house was achieved. The model indicated that high infiltration rates of outside air were likely to occur in the two principal living rooms. This finding was consistent with the poor level of construction, particularly sealing between adjacent building elements.

The validated model was used to investigate three separate strategies to reduce energy losses and thus improve thermal comfort conditions. Two of the strategies (reduced infiltration and additional ceiling insulation) are relatively low-cost. Of these, reducing the level of infiltration of outside air was found to be more effective. A third strategy, which involved converting the existing semi-enclosed areas in front of each living room into a sunspace, would be more costly and time consuming. However, it is considered to be practically and financially possible, particularly if low cost UV-stabilized plastic film were used as the glazing. To confirm that the idea would be culturally acceptable would require further investigation. The benefits of sunspaces in terms of increased average air temperatures were significantly higher than the other two strategies. As expected, a combination of all three strategies achieved the best results, but at a relatively higher cost.

References

AIRAH (The Australian Institute of Refrigeration Air Conditioning and Heating). (1989). *Handbook*. Australia: Adelaide.

Brager, G. A., & de Dear, R. J. (1998). Thermal adaption in the built environment: A literature review. *Energy and Buildings,27*, 83–96.

Cornwell, K. (1977). *The Flow of Heat*. Berkshire: Van Nostrand Reinhold Co. Ltd.

Fountain, M. E. & Huizenga, C. (1996). A thermal comfort prediction tool. *ASHRAE Journal, 38*, 39–42.

Hagen, T. (1980). *Nepal-the kingdom in the Himalayas*. Berne: Kummerly + Frey, Geographical Publishers.

Garzoli, K.V. (1988). *Greenhouses - Handbook for Nurserymen, Horticulturalists and Gardeners*. Canberra, Australia: Australian Government Publishing Service.

Humphreys, M. A. (1978). Outdoor temperatures and comfort indoors. *Building Research and Practice,6*(2), 92–105.

Mande, S., Kumar, A., & Kishore, V. V. N. (1999). A study of large cardamom curing chambers in Sikkim. *Biomass and Bioenergy,16*(6), 463–473.

Nicol, J. F., & Roaf, S. (1996). Pioneering new indoor temperature standards: The Pakistan project. *Energy and Buildings,23*, 169–174.

Nicol, J. F., & Humphreys, M. A. (2002). Adaptive thermal comfort and sustainable thermal standards for buildings. *Energy and Buildings,34*, 563–572.

Nienhuys, S. (2003). Thermal insulation for houses in high-altitudes. *Renewable energy technology for rural development (RETRUD-03) conference* (pp. 12–14). Kathmandu: Tribhuvan University.

RIDS-Nepal. (2008). *Evaluation & follow-up base-line-survey & re-survey*. Rural integrated development service–Nepal.

Rijal, H. B., & Yoshida, H. (2005). Winter thermal improvement of a traditional house in Nepal. *Proceeding 3:1035-1042 of the 9th International IBPSA Conference Montréal*, Canada.

Rijal, H. B., & Yoshida, H. (2006). Winter thermal comfort of residents in the Himalaya region of Nepal. *Proceeding of International Conference on Comfort and Energy Use in Buildings-Getting them Right (Windsor)*. Organised by the network for comfort and energy use in buildings.

Rozis, J. F., & Guinebault, A. (1996). *Solar Heating in Cold Regions*. London: Intermediate Technology Publications Ltd.

SKC. (2010). Haz-dust environmental particulate air monitor, SKC Inc., Model EPAM-5000. Retrieved July 23, 2012, from http://www.hazdust.com/epam5000.php

Sujit, T., & Zahnd, A. (2006). High-altitude smokeless metal stove–a research and development and implementation project through the Kathmandu University. *RETRUD 2006 Conference in Kathmandu, Nepal*, Kathmandu, Nepal, 12.10.–14.10.2006. Retrieved July 23, 2012, from http://www.rids-nepal.org/index.php?option=com_docman&task=doc_download&gid=24&Itemid=104

Chapter 6
Social and Environmental Issues

This chapter highlights social and environmental issues encountered by the local communities we live and work with. In their quest for better living conditions they are always confronted with issues such as tackling the need for improved energy services, through projects that involve social changes. Often technical projects don't have the mandate and scope to address social and environmental issues, even though they are equally important for sustainable community development. Thus, the sustainable development and use of the local renewable energy resources, within a defined environmental and societal context, are an important part of the wider framework of a HCD project.

6.1 Solar Battery Recycling in Rural Electrification

6.1.1 Introduction

Ever since "biodiversity hotspots" have been introduced into the literature (Myers 1988, 2000) much research has been conducted on the options to conserve these hotspots and the causes underlying the threats to their survival. Many scholars have investigated the relationship between poverty and biodiversity (Fisher 2007; Adams 2004; Brockington 2004; Cincotta 2000). Although we acknowledge that poverty alleviation may impose a threat to biodiversity, due to the development/ depletion of natural resources, we argue that it is really the poor physical and institutional infrastructure associated with poverty which poses the greatest threat to the natural environment. For people living in lightly populated areas, the absence of paved roads hinders the recycling of inorganic solid waste (Van Beukering 2001). That notion does not necessarily relate to the economic aspects of poverty for an individual. Poverty imposes a cap on the buying power of individuals and increases the individual's utility discount rate (Becker 1997; Ogaki 1997; Lawrance 1991). As a consequence, short run economic value may supersede long-term value, and people may, for example, (be forced) to buy products and services with a short expected

A. Zahnd, *The Role of Renewable Energy Technology in Holistic Community Development*, Springer Theses, DOI: 10.1007/978-3-319-03989-3_6,

useful life, or they may not internalise the long run economic costs associated with the externalities of the depletion of natural resources. Particularly for poor people, recycling waste or reusing (parts of) goods is commonplace, though often driven by a necessity. After all, for many millennia, indigenous people have protected their natural habitats through minimising waste disposal.

With the introduction of modern industrial products into remote rural areas, however, people are no longer able to recycle or reuse their solid waste. Here, the economic status of these individuals is not necessarily the cause of an environmental problem, but rather the lack of infrastructure (Satterthwaite 2003). That infrastructure may be both institutional or physical (Barrett 2006). In rural areas in developing countries, many roads are unpaved. In mountainous areas, transport and trade are often facilitated by porters. In this chapter we argue that the pricing of the transportation medium cannot be made incentive compatible. We illustrate the nature of the problem by means of a case study, for which we sketch three potential policy scenarios. First, however, we illustrate the severity of the problem by introducing the theme.

Solar systems are commonly used for rural electrification in developing countries. These target areas often lack any sort of physical infrastructure such as fresh water, sanitation, energy grids, or paved roads. In these settings, solar home systems (SHSs) are becoming increasingly popular as an affordable means of electrification. The delivery of SHSs to households either occurs in a self-organised fashion (a household buys a SHS at a retailer), or occurs through a project organisation (a utility company arranges the delivery, after sales support, and often the financing).

After approximately 2–4 years, the batteries of the SHS have to be replaced. In most cases, a complete business infrastructure is lacking since both project organisations and retailers target higher end systems with minimum quality standards and with large volume transactions (Van der Vleuten 2007). As a consequence of failing reverse logistics, the net environmental benefits of renewable energy for rural electrification through SHSs may be nil. As an illustration, in January 2007, the Dutch Minister for Development Aid submitted a report to Parliament in which a specific development aid program was evaluated. Of all projects evaluated, the Bolivia rural electrification project stood out in terms of environmental impact as mentioned in the ORET/MILIEV report (2006, 52):

> The high expectations of the environmental benefits of solar energy in Bolivia were never met. From an environmental perspective, solar energy has advantages if and when the solar system batteries are being recycled (or the energy systems installed do not require batteries). Although in Bolivia the use of NiCd batteries was reduced by 10 %, the solar batteries were never recycled, and the overall environmental gains were negligible.

Humla, our target area for this research project, readily overlaps with the biodiversity hotspots introduced by Myers (2000). For these areas, Fisher and Christopher (2007) argue poverty can be a significant constraint to conservation. We acknowledge poverty may form a major restriction in preventing large scale damage to delicate ecosystems indeed, but this is an untouched theme in the literature so far.

6.1.2 Analysis

We use one case study to illustrate the problem. Following an introduction into the product market and into the geographic setting, we sketch three logical policy scenarios a government can pursue. For each of these scenarios we describe how infrastructure may be a bottleneck, the expected environmental impact and economic cost, and the barriers towards designing an incentive compatible contract with end-users.

Though we focus on the Himalayas as the geographic setting and on porters as the medium of transportation, the message may easily be extended to alternative settings. For example, consider a remote archipelago where ferries occasionally bring new equipment to the islands, but where litter is not returned to the mainland. Entire cars may be dumped on a coral reef, which in the absence of internalising environmental costs, probably becomes the most economic way to get rid of it. In case of an archipelago, the transportation medium is the ferry, or the skipper of a small private boat, instead of the porter that we use. The mechanisms, however, remain the same.

6.1.2.1 The Nepali Market for Solar Home Systems

Since the 1980s, solar PV panels have been considered a highly promising renewable energy source. With soaring prices of fossil energy, SHSs are becoming increasingly economically attractive, particularly in remote areas where no electricity grid is available. SHSs are mostly produced and sold by multinational companies. When these companies target developing countries, they first tend to focus on the more wealthy consumers (London 2004). When an institutional infrastructure for collecting the waste is built, the focus is put on large scale facilities (Hart 1997). In case of solar PV systems, this is no different. Van der Vleuten (2007), for example, criticises the reluctance to build an organisational infrastructure for SHSs in developing countries: "Solar companies 'prefer to be sitting in the capital city, waiting for the next tender' rather than developing the networks to serve the needy clients in the distant rural areas". This is in line with the conclusions of Van Beukering and Bouman (2001) who find lead scrap is hardly recycled in developing countries because of a lack of infrastructure. Hence, if SHSs are either sold in developing countries, through retailers or through subsidised large scale development projects, we may at best expect the latter to develop some organisational infrastructure for recycling the batteries. In a survey completed in the 1990s in Kenya, it appeared that the expected lifespan of a system's battery was some 2–3 years (Van der Plas 1998). These data referred to locally produced batteries. Though battery technology has improved over time, we may readily assume a lifespan of up to some 4–5 years as being realistic. In Nepal, there is neither a domestic market for producing batteries nor for processing lead. Most batteries are imported from India, Bangladesh and Indonesia. The absence of an aftermarket for the waste products takes away most of the economic incentives to care for the used batteries. For other products, people may come up with innovative practices to reuse or recycle waste, but for toxic solid waste such behaviour is definitely undesirable (Goldman 2002).

In Nepal, the Alternative Energy Promotion Centre (AEPC) of the Ministry of Science and Technology keeps track of the dissemination of SHSs. Through email communication with AEPC, the author found out that from 1999 until 26 April 2010, a total of 182,648 SHSs, with a capacity of 4,873.6 kW, have been installed, while another 37,170 SHSs were installed by the end of 2011, amounting to a total of 219,818 SHSs.[1] In their survey on the status of SHSs in Nepal, the dispersion of SHSs across the country correlates with population density. Though the Nepalese government has kept track of most SHSs since the late 1980s, the surveys only occur at very discrete time intervals (AEPC 2006).

While the solar PV technology has become known as a sustainable and appropriate technology for basic rural village electrification schemes to reach the countless remote communities too far away for the grid, the solar PV market is often strongly centralized in the main capital. This is also the case in Nepal. The now 56 registered, and 42 AEPC approved,[2] solar PV companies have their main offices in Kathmandu, Nepal's capital. Some of the major PV suppliers have a loose network to a few other urban centres around the country in the lower altitude flat parts, mainly in the form of sales offices. This kind of network is suitable for serving large scale customers and projects, usually driven by large international donor organizations, but not for serving individual households. This is true in particular for SHSs that aim at providing improved elementary (initial) electrical services (usually limited to basic indoor lighting) for a particular context and demand, which require often a "customized" kind of PV system. They need to be designed and installed according to the user's/owner's initially defined energy services and affordability. This approach is more time consuming as specialists are required to interview and interpret the user's defined needs. Such a "customized" SHS is usually also more expensive, but typically this will distinguish it in regard to used components, size and installation from a SHS "mass product" for large scale solar PV home system projects with a standard design, predefined system components and often lower quality benchmarks.

---

[1] Through personal email communication between the author and AEPC in March 2011, AEPC provided the author with the detailed figures for all the SHS installed under their subsidy programs from March 1999–December 2011. More background information about the various AEPC subsidy programs and donors, but without detailed data of the installed SHSs per time period is available at: http://www.aepc.gov.np/index.php?option=com_content&view=category&layout=blog&id=82&Itemid=132 (accessed July 27, 2012)

[2] In October 2011 there were a total of 56 Solar PV companies registered with the Nepal government and out of them 42 were approved by AEPC as so called "pre-qualified" solar PV companies, thus able to sell SHS under the AEPC subsidy program. Under the 2007–2011 APEC solar PV subsidy program, the pre-qualified solar PV companies are able to cash in 8,000–12,000 Nepali Rupees per SHS sold into rural areas from AEPC.

While the main parameters of a large SHS project are minimal cost, high turnover and short installation time, the main focus of subsidised, customized SHSs is on sustainability, appropriateness of the system for the user's context and identified needs, strong feeling of ownership, training in operation and maintenance, follow-up visits, as well as the establishment of a relationship between the user and the project implementer for long-term access to spare parts and eventual repair services.[3] A SHS cannot be compared with a DIY tool as the users cannot be expected to understand the detail of a SHS and its need for proper professional installation and maintenance, however minimal they may be. Thus, independent of the financing of the SHS (e.g., directly purchased from a solar PV firm, or acquired under a subsidised program), proper installation and user training are crucial success factors. The power generation of a poorly installed SHS can be significantly lower, thus not providing the expected daily energy services. Further, under such conditions it is likely that the batteries are not able to be charged properly, are discharged too deeply on a daily basis and may even suffer from too low or too high temperatures as they have not been positioned and/or insulated according to the local climatic conditions. Such conditions can considerably reduce the life-expectancy of the battery-bank, as these are crucial parameters besides the more commonly known needs for periodical topping up of flooded lead-acid batteries with distilled/rain water.

6.1.2.2 A Description of the Humla Region

Chapter 1 has provided a general and more detailed overview of Nepal and the remotely located, impoverished and underdeveloped district of Humla, in the far North-West of Nepal, bordering Tibet. Lack of roads and extremely difficult air access are perhaps the most important constraints to tourism development in Humla[4] which is elevated between 1,524–7,031 m. Humla is not only rich in species biodiversity but also represents a broad range of habitat types. Subedi (1998) identifies among others a number of wild fauna including the following eight protected and endangered species in the Humla region: the snow leopard (Panthera uncia), the clouded leopard (Neofelis nebulosa), the musk deer (Moschus moschiferous), the gray wolf (Canis lupus), the leopard cat (Felis bengalensis), the wild yak (Bos mutus), as well as the impeyan pheasant (Lophophorus impejanus), and satyr pheasant (Tragopan satyra). In addition to the fauna, Humla

[3] One of the main differences in a subsidized SHS project, such as we run through RIDS-Nepal, is that an initial project budget is developed which will be funded by donors and only in part, usually ~10–20%, by the local people in financial terms. Thus, there is no "pressure" to make maximum profit through a SHS project as there is for the commercial solar PV companies.

[4] See for a generic description of the biodiversity hotspot that applies to the Eastern Himalaya. http://www.conservation.org/where/priority_areas/hotspots/asia-pacific/Himalaya/Pages/biodiversity.aspx (accessed July 27, 2012)

alone has about 1,500 species of high-altitude plants (Burch 2003). The protection of medicinal plants is an issue of particular concern in Humla (Kunwar 2006).

6.1.2.3 Scenario 1: Laissez-faire

In the base case, households freely buy solar systems in the market place. These may be directly purchased in a shop, or be subsidised by a government agency. At present, there is no battery recycling system in Nepal, and government does not make any attempts to collect these batteries, store or recycle them. Government monitoring takes place at very discrete time intervals only (in which an inventory of SHSs is made, but batteries are not labelled). Hence, even if the government were to set up a battery collection and recycling system, it would not know whether a battery comes from distant rural areas, or from areas near to the collection facilities. Despite the serious environmental implications, the potential for a (currently non-existent) battery recycling program is low. Batteries from remote areas are unlikely to be collected, since the premium for handing them in is likely to be lower than the opportunity cost of the trip. Bearing these limitations in mind, we analyse how the environmental impact may still be minimised.

RIDS-Nepal, has investigated SHSs in Humla and found that SHSs have been installed either by their owners, if they purchased it themselves, or more often by an installer who was contracted under a subsidised SHS program. Often the SHS batteries (Sect. 3.7.5) are found not to perform to the users' satisfaction within a matter of a few months, not providing the promised energy services, which is in most cases limited to basic indoor lighting. One of the most commonly observed shortcomings for batteries was the inappropriate place of installation and lack of needed thermal insulation due to the cold climate (Figs. 3.73, 3.74, 3.75, 3.76, 3.77 and 3.78). These can easily be avoided with locally available materials (Figs. 3.91, 3.92, 3.104 and 3.107) if the users have a minimal basic knowledge about SHSs and if they have undergone training in operation and maintenance. Once the batteries have been found to be "dead", they are usually disposed of either inside or outside the homes (Figs. 3.73 and 3.78). Some "dead" batteries are emptied, with the sulphuric battery acid being poured on the ground, polluting the surrounding area and possibly even the ground water sources nearby. The discarded batteries can easily be accessed by children and other people. That brings potential hazards to people and the environment. A possible mitigation is to neutralise ($pH = 7$) the battery-acid. The neutralisation of sulphuric battery acid (H_2SO_4) on site is possible by using diluted sodium hydroxide (NaOH). The reaction product is aqueous sodium sulphate, which can be discarded into the septic tanks of the existing village pit latrines, without causing environmental harm. This disposal process demands a high level of precision which must be performed by an experienced professional. Furthermore these jobs are dangerous and risky and can not be left to the local users.

In this base case scenario, the following conditions for setting up an efficient recycling contract are violated.

- Firstly of all, the limited monitoring (in terms of data richness) hampers any discrimination between remote batteries and batteries installed close to (yet to be launched) collection facilities. In addition, the low frequency of the monitoring (once in a couple of years) may overlook batteries that have had a short life.
- Secondly, the support by NGOs (such as RIDS-Nepal) has not been institutionalised, and hence provides a very fragile solution for damage control. That is, if such NGOs face income cuts, or if they become understaffed, they may decide to lower the quality or quantity of monitoring and assistance.

Given the above restrictions, we foresee that this base case scenario is most likely to be adopted by governments that are on a tight budget, and where the impact of dead battery disposal in the environment is roughly equal throughout the country. The voluntary support by NGOs definitely provides a major improvement, but their effort is unpredictable. In this case, the lack of institutional infrastructure probably has a larger impact on the environmental problem than the physical infrastructure.

6.1.2.4 Scenario 2: Monitoring Through NGO but No Recycling

Alternatively to letting households go unassisted, some third party might assist the households with the proper installation of the solar PV systems. Apart from the benefit that the systems will most likely last much longer (due to professional installation and training and damage is less likely or postponed) compared to the base case, a major advantage of third party assistance is that it provides the opportunity to exactly locate all systems, and to set up a registration system with identified geographical location, battery type and capacity, date of instalment, etc.

In order to address and improve the present SHS scenario, in particular in regard to guaranteeing the SHSs' intended energy generation and energy services, we can consider the experiences of the local NGO RIDS-Nepal. In 2002, RIDS-Nepal set up a long-term HCD project, which included the design, training, installation and maintenance of solar PV systems in partnership with the users. The project only runs in Upper-Humla, and has not been institutionalized by the national government. Therefore the sketch below of the program is only an example of what NGO assistance might look like in this "Scenario 2".

In the RIDS-Nepal project, three different solar PV system approaches have been developed, each depending on the locality of the user(s) homes, the geography, and on the meteorological conditions (Sect. 3.3.2). For example, for individual homes, stand-alone SHSs are designed, whereas in the case where 4–12 homes are located in close proximity to each other (Fig. 3.31), cluster solar PV systems are designed. Lastly, central solar PV systems are set up for a whole village in which the homes are very close to each other (which is the main traditional building style in Humla, (Figs. 3.38 and 3.50). All the village based solar PV systems are designed to provide initial elementary indoor lighting services.

Solar PV systems are never installed by themselves, but as part of the NGO's HCD project approach, the "Family of 4" (Sect. 2.10). That means that every household in a village agrees to participate in a project which includes a pit latrine, a smokeless metal stove, indoor lighting (through an appropriate renewable energy technology such as a solar PV system, a pico-hydro power plant, or a small wind generator, utilising the locally available energy sources) and a village drinking water system. In this way each project can bring forth synergistic benefits through the other, jointly implemented, projects, multiplying the final impact of each project.

As an initial step before a HCD project is considered in a village, the local people must show their keen interest and willingness to participate in the project. That is mostly done by writing a request letter to the NGO to partner with them in a long-term HCD project in their village. Thus in the case of the "Family of 4" project, the local people are involved and advised which of the various solar PV system approaches (SHS, cluster or central PV system) is the most appropriate solution in regard to their geographical, meteorological and climatic conditions. Further, prior to the installation, the locally established "solar PV committee" decides who is to be sent to the 10 days solar PV basics and operation and maintenance training the NGO has developed[5] as an integrated part of each HCD project.

Once the needed skills are taught, the local trainees can apply their newly learned skills and earn their respect among the village community by participating in the solar PV system installation(s) in the village. Periodic follow-up visits by the NGO staff, who live and work in the area, help to support the operation and maintenance and provide an ongoing source of real data and facts about each solar PV systems' performance and condition. Thus any flaw or shortcoming in power generation, battery energy storage or energy service provision is recognized early on and timely appropriate mitigations, maintenance or repair can take place. That helps to keep the PV systems up and running and the users' confidence and pride in their system. Undoubtedly this procedure results in more expensive solar PV systems. However, the NGO's experience since 2002 have clearly shown that the installed PV systems (now in 10 villages with a total of 420 different solar PV systems for 820 homes (June 2012), one school and one health post), have performed significantly better with minimal battery failures. The first battery sets had to be exchanged after 4 years in 6 systems after they had been heavily misused by the local users. Still, there is no battery recycling process in place and the "dead" batteries from the last 10 years have been collected and stored in the NGO's main

[5] RIDS-Nepal has designed and written its own solar PV system maintenance training manual (Appendix 18), which is also available for download from the RIDS-Nepal web site http://www.rids-nepal.org/index.php?option=com_docman&task=cat_view&gid=183&Itemid=32 (accessed August 10–27, 2012)

office in Simikot Humla, which is a 1 day trek away from the villages where the "dead" batteries were located.

In this second scenario, setting up an efficient contract for recycling batteries is greatly enhanced through improved monitoring. Also, assistance in installation, and training in operation and maintenance has extended the lifespan of batteries significantly. Nevertheless, there is still no official database in which key battery data[6] are stored. As a consequence, all improvements relative to the base case depend upon the NGO. That is, the NGO has all relevant information on the SHSs, but since this information is not shared with other parties, it is the same NGO that has to act as the manager if a battery collection program were to be set up. Since battery collection need not be the prime task for the NGO (and since the NGO need not be compensated financially, or supported institutionally), there is still an institutional problem to overcome.

6.1.2.5 Scenario 3: Monitoring and Collection Through NGO

In addition to scenario 2, the third party could also collect the batteries. Since the same NGO also keeps track of the batteries from the installation onwards, it is not necessary for the NGO to take care of the transportation from the households to the NGO's office. Given full information symmetry, the only task is to set up an incentive compatible contract.

The large numbers (estimated to be >220,000 in December 2011)[7] of SHSs and batteries installed in Nepal since 1999 under the current solar PV government subsidy programs, point to the need for a recycling program. It may not be a popular topic with the solar PV companies, as that could force them to put an additional price tag on new batteries to pay for the recycling process. It would also demand institutional and physical infrastructure to collect and manage the old batteries. Neither will it be well liked by the government, as this would demand a fairly complex project with huge costs for such new infrastructure, recycling processes and logistics to get the old batteries in and the recycled batteries back onto the market. That requires a country wide network of collection places with trained staff. The shear costs involved in such an endeavour, considering the majority of the batteries are in very remote places, is mind boggling. Thus this topic has been avoided and has in fact not had a chance to be publicly raised or discussed with the necessary seriousness and urgency. To tackle the issue of identifying, collecting and recycling old solar PV system batteries is also not a

---

[6] Key battery data for a database will include: Type (technology) of battery, brand, capacity, specific gravity and amount of electrolyte used, when and to whom sold and where installed (system size and use, coordinates, climate context).

[7] The estimated number of >220,000 installed SHS (December 2011) includes only the AEPC (government) subsidized SHS, amounting to 219,818, and not the thousands of SHSs which have been installed by many NGOs/INGOs over the course of the years since the mid 90s. Thus the "dead" battery problem is rather more serious than anyone imagines at the present stage.

project that will reap big applause from the public or donor organisations, as it just aims to manage "waste", and that is difficult to "sell" and fund. However, in order to be responsible towards the solar PV system end-users and the environment, we can no longer ignore that issue and thus need to start finding appropriate solutions, small and insignificant though they may be for the first few years, while considering the scope of the problem at hand.

6.1.2.6 Setting-up a Recycling Programme in Practice

NGOs, such as RIDS-Nepal, who are involved in the design and implementation of solar PV system projects, could take a lead in tackling the issue and showing the way ahead. The problem could be tackled from two sides. On the one hand through a more reliable installation and operation of the SHSs. That is, if SHSs are both designed and installed in accordance with the needs of the local context of the end-user(s), batteries will last longer. On the other hand, a battery recycling program has to be put in place in order to minimise the possible toxic waste damage to the environment and health risks to the local people.

The following shows a step by step process of such a possible trial "battery program" project, paying due respect to all parties. All the steps are done either through professional staff of the implementing NGO or by subcontracting the task to an external professional solar PV designer.

(1) Define with the end-user(s) the needed and affordable energy services.
(2) Collect, monitor and record or simulate (e.g. with METEONORM) the defined location's average daily solar energy resource for an average year.
(3) Design the solar PV system based on the needs, financial affordability and locally identified minimal solar resource. That can be done by experienced solar PV designers through hand calculations or with the help of professional software such as PVSyst (http://www.pvsyst.com/) or RETScreen (http://www.retscreen.net/). Design the battery-bank (BB) for 3–5 days independence with a maximum depth of discharge (DoD) of 35 % (i.e., the BB can provide the same daily energy services without sunshine and with a maximum DoD of the BB of 30–35 % of its full capacity after 3–5 days).
(4) Define the project cost and include an "old battery collection and recycle" budget according to the project's location and size.
(5) Negotiate the end-user(s)' financial and voluntary contribution (in the form of labour and local materials) to the project and start raising funds for the remaining cost from external donors and sponsors.
(6) Create a database, identifying and numbering each solar PV system's equipment. Each battery is listed according to its technology, brand, manufacturer, price, purchase date and company, capacity, installation date, location and designed use (energy in/out per day, DoD, temperature, and sulphuric acid density in the case of liquid acid batteries).

(7) Conduct a basic solar PV training for installation, operation and maintenance for the chosen, village based operators, including hands on training and participation in the system(s) installation(s).
(8) The trainees participate in the installation of the solar PV systems in the village (providing them with practical experience and allowing them to earn the respect of the village community).
(9) Develop a follow-up program to check each installed solar PV system and record the performance and status of the system and battery-bank.
(10) Set up a dry and enclosed "old battery" collection storage place.
(11) Sign a "battery recycle" contract with an appropriate company[8] (with the necessary professional skills and infrastructure) to recycle and refurbish the old batteries.
(12) Organise the transport (porters, air plane/helicopter, road transport) of the old batteries to the recycling company.
(13) Develop a test procedure with benchmarks for the "refurbished" batteries and test each one before it leaves the company. Identify each "refurbished" battery once it has passed the test successfully and add it to the database.
(14) Develop a "re-allocation" program for refurbished batteries in the project area.
(15) Include the refurbished battery again in the follow-up program.

This proposed "battery program" project, of which points 1–10 are already in operation, is a first attempt of a trial project in a limited working area of the NGO RIDS-Nepal, which has been engaged in solar PV system projects with the local communities since 2002. It aims to focus on the "life-cycle" of a solar PV system with particular emphasis on the life-cycle of the battery-bank. The project involves detailed monitoring with sophisticated data loggers, of four chosen solar PV systems, using different approaches (recording data since 2006 and 2007) and anecdotal monitoring and data recording of the other installed battery-banks. The aim is to understand the most appropriate working conditions for batteries in a defined user context, the recycling and refurbishing of the old batteries and re-integrating them into solar PV projects, or local manufacturing of batteries (if suitable skills and companies can be contracted). To understand the life-cycle of batteries, enables designers and project implementers to reduce the potential harm of toxic waste in endangered and fragile ecosystems. That allows for a smaller footprint of solar PV systems, in the quest for sustainable HCD in partnership with remotely located people/communities of Nepal.

[8] Till July 2012 there was no company in Nepal able to recycle and refurbish old batteries in a professional way, so that they could be used again e.g. in a solar PV home system.

6.1.3 Conclusions

This section has discussed an incentive incompatibility problem that causes dramatic damage to delicate ecosystems through unsafe disposal of "dead" batteries of SHSs. The aim is to rethink policies that facilitate the introduction of toxic solid waste into markets with an imperfect infrastructure and in the absence of an aftermarket for the solid waste. A case study in Nepal is sketched, where rural electrification through solar systems leads to the dumping of used batteries. Porters are the most popular transportation medium in the Himalayas. Independent of whether the owners of the solar home systems act as porters or hire porters, it is the opportunity cost or the expense for the travel that forms the weak link in the reverse logistics.

Among the three outlined institutional scenarios to respond to the ecological problem, the third, "monitoring and collection through NGO", is considered to be the most responsible solution under the given context, as the toxic waste is collected and disposed of through neutralisation of the sulphuric battery acid on site. It demands that the NGO keeps track of the batteries from the installation of the SHS onwards. With this information the NGO does not need to take care for the transportation of the "dead batteries" from the end-users' household to the NGO's office. The only task is to set up an incentive compatible contract that matches the data. Further, the necessary conditions for a possible, incentive compatible battery recycling program are sketched out.

6.2 Responses to Innovation in Rural Nepal

6.2.1 Introduction

Humla District in Nepal is a very remote area, suffering from constant food shortages and is characterized by a harsh environment. The livelihoods of agro-pastoralists in this district became much more vulnerable during the Maoist insurgency from 1996–2006, and this vulnerability was particularly acute in some areas. As a result, people in different villages responded quite differently to an externally funded RIDS-Nepal HCD project—one of the only projects the Maoists allowed to proceed in Humla during the height of the unrest.

Villagers' responses to this health and conservation oriented development project seem to correlate most closely with their socio-economic status and ability to extract resources from the local environment, as well with the nature of their relationship with the local Maoist cadres. Villagers' perceptions of the risks of becoming involved in the HCD projects in this area, and their ability or willingness to take part in them, are analysed here, drawing on anthropological analyses of perceptions of risk and the diffusion of innovations. The chapter concludes with brief recommendations based on this evidence and our experience in the field.

Fig. 6.1 Simikot, Humla district headquarters. The project site is within a 1–2 days walk

6.2.2 Agro-Pastoralists: "Conservative"?

Development workers' understanding of subsistence farmers' and herders' reluctance to adopt new technologies has been informed for several decades by a focus on the perception of risks and on patterns of diffusion of innovations.

People living on the produce of their own land and herds have been described as "risk averse" and culturally "conservative," and as preferring not to adopt new technologies and farming or husbandry methods (Wharton 1971) that might negatively impact their ability to make it through the next dry or cold season. We have been living and working with agro-pastoralists in the Humla District in northwest Nepal since 2002 (Fig. 6.1), running HCD projects with them that aim to conserve and utilise their local resources and improve human health and sanitation. Our experience has made us aware that farmers' perceptions of the risks involved in change are far from uniform, even in apparently similar contexts, and has led us to reflect on why this is so.

6.2.3 Vulnerable Livelihoods in Humla District

Nepal's poverty is well known, but even in this context Humla District stands out. As discussed in Sects. 1.1.6 and 1.1.7, with a per capita GDP of US$72 (KIRDRC 2002) and one of the highest infant mortality rates in the country, Humlis live in dire straits. Further, identified contributing factors are Humla's permanent food shortage, with little trade, next to no cash economy, high elevation, and a short growing season.

Fig. 6.2 Open-fire cooking and heating. Indoor air pollution in a typical Humli home

6.2.4 Health and Environmental Concerns

Our surveys show that some 65 % of Humla's children $\leq$5 years of age are seriously malnourished (Sect. 1.1.8). Nearly all inhabitants are subsistence farmers or herders, using husbandry and farming methods that are ancient and adapted to the local climate and elevation (2,700–3,500 m). Homes are small and unhygienic, and settlements are extremely dense. Water is drawn from streams or stream fed unfiltered tap stands, and there is no local custom of human waste disposal, so trails in and around the villages are typically polluted by human waste.

Indoor air pollution is extreme; in combination with malnutrition and the lack of hygiene and sanitation described above, this plays a large role in the high mortality rate. Open-fires burn in fire pits in each home (Fig. 6.2), for cooking and heating, and house interiors are lit by the traditional, resin rich, smoky wooden stick known locally as "*jharro*" (Figs. 2.3, 2.15, 3.3, 3.81, 3.82 and 3.101).

All the energy use in remote mountain areas such as Humla is based on firewood. The local high-altitude pine trees (*Pinus wallichiana*[9]) serve this purpose (Fig. 6.3). This means that households in our project area devote up to 40 hours per week to collecting firewood. When they are not collecting firewood, villagers are engaged year round in tasks associated with crop production and livestock husbandry—some at home while others practice transhumant pastoralism, moving animals between pastures at lower and higher elevations depending on the season.

[9] See note 31 "*Pinus wallichiana*" for more details about the "*Pinus wallichiana*" in Chap. 3.

Fig. 6.3 Collection site for *jharro*, the resin rich wood traditionally burned to provide indoor light

Nearly everyone winters at home in the village, with the animals packed into the livestock area on the first level of the home and humans above, usually gathered around the hearth in the common room. Women are required by customary Hindu beliefs about ritual and bodily pollution to leave the common area and refrain from using tap stands during menstruation. In some households, traders leave Humla in winter and head south to trade in Kathmandu or elsewhere.

Most households do not produce enough food for the year; on average, households can feed members from the fruits of their own labour for only 3–6 months in our project area. When they run out of food, they borrow from wealthier members of the community, or head south to the district headquarters (Simikot) to get subsidized rice at the government rice depot. Sometimes the pass to Simikot is closed by snow, or villagers arrive only to find that rice supplies are exhausted, and they return home empty handed.

6.2.5 The Maoist Insurgency

This was the situation into which the cadres of the Maoist insurgency entered. They were engaged in a struggle against the ruling government and ancient systems of land ownership, debt inheritance, caste, and gender discrimination. The Maoists added to the situation of extreme poverty and constant uncertainty about crop failure, wintertime food shortages and mortality by bunking with villagers and consuming resources that were always in short supply.

They demanded time from already pressed villagers during the short growing season for educational "camps", requiring villagers to have written permission to travel outside the village boundaries, and they engaged in violence. Sometimes violence was perpetrated by individuals who would enter villages and homes in search of insurgents or people thought to be aiding and abetting the cadres. Due to the unrest and the demands imposed on development by the circumstances relating to the insurgency, most development efforts in Humla were brought to a standstill during the insurgency. RIDS-Nepal continued to function, bringing a multi-pronged, HCD approach to the region, designed to address some of the environmental and human health challenges described above.

6.2.6 The Comprehensive "Family of 4" Concept

While living and working with local communities, we found that villagers defined their most urgent household and health related needs as a toilet located near the home, a smokeless stove for cooking and heating, minimal electric indoor lighting, and clean drinking water. This is how the "Family of 4" concept used in the projects came into being.

The "Family of 4" is a set of innovations that are installed, as a group, in each home in a target village. It includes:

(1) A pit latrine;
(2) A smokeless metal stove;
(3) Elementary lighting (tapping local solar or water energy resources); and
(4) Access to a safe drinking water system.

Villagers participate in the details of project planning and determine the pace and nature of the project, attend non-formal education (NFE) classes about the basic thinking behind each part of the project, and are involved in all follow-up and trouble shooting activities, and in ongoing technical support. Individual contributions to each project component include labour and materials, as well as a small payment in cash for the subsidized stove and solar PV system.

By and large, although considerable behavioural change is required for villagers to successfully integrate each of the new technologies into their lives, this approach has met with considerable success and popularity in the region. Measurements of stove efficiency show that the stoves use ~40 % less firewood than the traditional open-fires; moreover, in combination with solar lighting, stoves decrease PM_{10} (particulate matter $Ø \leq 10$ μm) levels very substantially (peak PM_{10} measurement with open-fire: 20 mg/m^3, with smokeless metal stove and solar lighting: 1.4 mg/m^3) (Figs. 6.4 and 6.5).

These benefits to environmental and human health are the most obvious ones and are easily appreciated, contributing, we believe, to the popularity of the project approach. People from villages where we have not worked sometimes walk all the way down the valley to the project office to ask that we work in their village next.

Fig. 6.4 A SMS (smokeless metal stove) and WLED lights in the main living room/kitchen enable a clean, hygienic and healthy indoor air environment for the whole family

Fig. 6.5 The SMS provides cooking, heating and hot water services, while the WLED lights provide the minimum needed brightness to learn and socialize

6.2.7 *Unexpected Differences in Responses to Innovation*

We were interested to find considerable non-uniformity in participation rates in the projects. The villages in the region, though culturally and linguistically similar and largely lacking in socio-economic status differentiation, were by no means a homogenous group and did not respond in the same manner to the idea of the innovations being introduced.

This variation is most obviously captured by comparing one village, Bhajgaon (a pseudonym) with our experience elsewhere. In this village, chosen due to its proximity to other target villages, we found very substantial reluctance about the idea of participating in the projects. We were surprised to find this, since our initial view was that this village differed very little from the others—on the surface, caste, direct perceived poverty level,[10] subsistence system, and other indicators did not distinguish this village in obvious ways from the rest.

6.2.8 *The Influence of the Degree of Poverty*

Based on our reading of the literature on the diffusion of innovation and the perception of risk (Wharton 1971), we decided to analyse baseline data from Bhajgaon to help us understand the resistance we were encountering. We decided to test the hypothesis that people living in "peasant" societies cannot be lumped together as a uniform whole, despite past theorizing that all people living in subsistence farming communities are reluctant to change and therefore resist innovation.

Instead, analysts have shown (Wharton 1971) that it is often the case that poorer families are later adopters, waiting instead to follow the lead—and successful experience with a new technology—of a wealthier household. Additionally, we

[10] A direct perceived poverty level can be understood in "seeing" the poverty of a family or community through an initial visit. Thus, things like people wearing torn, or inappropriate clothing, the lack of a stove for indoor cooking and heating, inability to read and write, the lack of access to clean drinking water, etc. are marks of "poverty" which can be perceived quickly as "signs" of being "poor". However, there are many other parameters, such as having less arable land to cultivate and grow their food, fewer animals, limited or no access to health services, higher debt rates per family to land owners, less access to cash/wealth, stronger dependence on shamanistic rituals to address physical ailments and sicknesses, child marriages and child labor, mental stressors etc., all issues which strongly influence the "level" of poverty of a family or society. These can only be recognized once a detailed base-line survey (such as RIDS-Nepal carries out in each village prior to the start of any long-term HCD project, see Appendix 14) is carried out and evaluated. Thus, in the case of the villagers in Bhajgaon, while they "initially" did not look to be "poorer" than their neighboring villagers, after the base-line survey it became clear that there are substantial and real differences in regard to living conditions and level, which distinguished the Bhajgaon villagers as being substantially "poorer" than their neighboring villagers.

wished to determine if the social perception of risk would influence people's willingness to adopt change, and if people living in environments that are inherently riskier will be less likely to adopt a change than people living in less risky places.

In order to carry out these investigations, we analysed our cultural, socioeconomic, and attitudinal data from the baseline studies we conducted in Bhajgaon village prior to entering the project planning stage. The results showed a remarkable resemblance to what might have been predicted by the theorists of the diffusion of innovation and perception of risk: villagers in Bhajgaon did differ in some very substantial ways from other villages, despite bearing a superficial resemblance to others in the area.

Bhajgaon was substantially poorer than other villages, with fewer herd animals and less land under cultivation per household. There was also less variability in wealth—the vast majority of households were in the poorest wealth category, whereas in other villages there were many more households in a higher wealth category (i.e. households that could be appealed to for help in times of scarcity).

Households in Bhajgaon carried a much higher than average debt load, too. Some of this was generations old debt that people had inherited from their forefathers. Food security was also slightly worse than in the rest of the region. Households were producing less of the food they needed to feed their families than in the other villages where we had worked, and were more reliant on loans and the government subsidized rice available in winter.

Anecdotally, we also knew—though we did not collect any data on this topic—that the villagers of Bhajgaon had an antagonistic relationship with the local Maoist cadres, more so than in other villages in the region, and had experienced many anxiety provoking interactions with them.

Drawing from the attitudinal data we had collected, we analysed the ways in which the villagers of Bhajgaon responded to questions about the expected impacts of specific project components. We found that they were much less likely to imagine any impact at all than villagers in other villages, and more likely to imagine negative impacts when they answered specifically. Furthermore, when queried about the long-term changes to the village as a result of the "Family of 4" approach, they were much more likely to give neutral answers, whereas in other villages, people answered specifically and concretely about the positive changes they anticipated. For instance, in other villages, a typical non Bhajgaon male farmer's response is:

> The new stove does not produce smoke [in the house] and uses less firewood.

Thus, non Bhajgaon people clearly identify and confirm that the new smokeless metal stove would save wood, save time, and cook faster (Figs. 6.4 and 6.5) at a much higher rate than villagers in Bhajgaon. In contrast, a more typical, Bhajgaon male farmer's answer to the question about the impact of a smokeless stove is:

> It is difficult to warm ourselves in the winter with the new stove.

In other villages, people named benefits to safe drinking water and latrines such as increased cleanliness and overall health outcomes, while in Bhajgaon, people were more likely not to be able to identify any benefit. In describing the sources of stress that they have felt most keenly over the last decade, people in Bhajgaon were much more likely to cite the lack of resources, while villagers in other places answered that poor health and educational opportunities were problematic.

Given that our results showed substantial differences between Bhajgaon and the other villages, we found the resistance of villagers there quite understandable. If they felt they were being harassed by the Maoists, it made sense that they would want to bunker down and avoid the assumption of any additional risk through associating themselves with outsiders, and with an internationally funded development project, and with change itself.

Given the more severe poverty in Bhajgaon, it was also understandable that people might fear the addition of practices that could strain a situation that was already very difficult. For instance, if the new stove stopped working; used more rather than less firewood; set the house on fire; or otherwise malfunctioned; villagers who were already pressed to the limit and had little hope that other villagers might assist them, would naturally be reluctant to adopt change.

6.2.9 Advantages of a Holistic and Iterative Approach

These differences highlighted 2 lessons for us: the first was the importance of doing baseline studies of communities targeted for development projects, in order to appreciate the existing parallels between and points of divergence from other communities. This helps us to understand the ways in which our work will be received by the people with whom we are hoping to collaborate, and anticipate areas needing extra care or consideration.

The second lesson was that even in a seemingly homogeneous group of villages—where homogeneity is perceived by local people, not just the project staff—important differences are likely to exist, and responses to the innovations to be introduced in the project framework will differ accordingly.

Based on our experience, we would offer the following recommendations to community development project designers and implementers:

- Holistic, long-term, and culturally nuanced projects that honour local priorities and social hierarchies and prioritize local participation have a good chance of long-term success.
- Projects that address only one part of a complex problem and that do not take into account local socio-economic variations, different understandings of development, and people's own development goals and expectations will not succeed.
- Target "easier" villages first and develop model households/wards/villages; then move on to poorer, more stressed villages and apply awareness training, support, education, and implementation, with commitment to a period of intensive follow-up support.

- Project planners and implementers must have up to date information about sources of stress and perceptions of risk within the target community, and deal with these factors in a culturally appropriate fashion.

References

Adams, W. M., Aveling, R., Brockington, D., Dickson, B., Elliott, J., Hutton, J., et al. (2004). Biodiversity conservation and the eradication of poverty. *Science, 306*(5699), 1146–1149.

Alternative Energy Promotion Centre (AEPC). (2006). *Energy Sector Assistance Programme–Phase II (ESAP-II)*, Programme document Volume 1, Nepal: Government of Nepal–AEPC/ESAP.

Barrett, C. B., & Swallow, B. M. (2006). Fractal poverty traps. *World Development, 34*(1), 1–15.

Becker, G. S., & Mulligan, C. B. (1997). The endogenous determination of time preference. *Quarterly Journal of Economics, 112*(3), 729–758.

Brockington, D., & Schmidt-Soltau, K. (2004). The social and environmental impacts of wilderness and development. *Oryx, 38*(2), 140–142.

Burch, W. R., Singh, S, Kanel, K. R. (2003). Mid term evaluation report of enterprise based biodiversity conservation—Nepal project. Implemented by enterprise works worldwide (eww) and asia network for sustainable agriculture and bioresources (ansab), *USAID Global Conservation Project* (GCP).

Cincotta, R. P., Wisnewski, J., & Engelman, R. (2000). Human population in the biodiversity hotspots. *Nature, 404*(6781), 990–992.

Fisher, B., & Christopher, T. (2007). Poverty and biodiversity: Measuring the overlap of human poverty and the biodiversity hotspots. *Ecological Economics, 62*(1), 93–101.

Goldman, L., Tran, N. (2002). Toxics and poverty: The impact of toxic substances on the poor in developing countries. *Technical Report*, World Bank.

Hart, S. L. (1997). Beyond greening: Strategies for a sustainable world. *Harvard Business Review* (January-February), 66–76.

KIRDRC (Karnali Integrated Rural Development and Research Centre), 2002. *Governance in the Karnali, an Exploratory Study*. August 2002, Jumla, Nepal.

Kunwar, R. M., Nepal, B. K., Kshhetri, H. B., Rai, S. K., & Bussmann, R. W. (2006). Ethnomedicine in Himalaya: A case study from Dolpa, Humla, Jumla and Mustang districts of Nepal. *Journal of Ethnobiology and Ethnomedicine, 2*, 27.

Lawrance, E. C. (1991). Poverty and the rate of time preference: Evidence from panel data. *Journal of Political Economy, 99*(1), 54–77.

London, T., & Hart, S. L. (2004). Reinventing strategies for emerging markets: Beyond the transnational model. *Journal of International Business Studies, 35*(5), 350–370.

Myers, N. (1988). Threatened biotas: Hot spots in tropical forests. *The Environmentalist, 8*(3), 187–208.

Myers, N., Mittermeier, R. A., da Fonseca, G. A., & Kent, J. (2000). Biodiversity hotspots for conservation policies. *Nature, 403*(6772), 853–858.

Ogaki, M., & Atkeson, A. (1997). Rate of time preference, intertemporal elasticity of substitution, and level of wealth. *Review of Economics and Statistics, 79*(4), 564–572.

ORET/MILIEV. (2006). *Evaluation 1999–2004: Final Report*, November, p. 52.

Satterthwaite, D. (2003). The links between poverty and the environment in urban areas of Africa, Asia, and Latin America. *Annals of the American Academy of Political and Social Science, 590*(1), 73–92.

Subedi, B. P. (1998). Participatory utilization and conservation of medicinal and aromatic plants: A case from Western Nepal Himalaya. *Proceedings of the International conference on medicinal plants*, 16–19 February. 1998. Bangalore, India: Asia Network for Small Scale Resources.

Van Beukering, P. J., & Bouman, M. N. (2001). Empirical evidence on recycling and trade of paper and lead in developed and developing countries. *World Development, 29*(10), 1717–1737.

Van der Plas, R. J., & Hankins, M. (1998). Solar electricity in Africa: A reality. *Energy Policy, 26*(4), 295–305.

Van der Vleuten, F., Stam, N., & Van der Plas, R. (2007). Putting solar home system programmes into perspective: What lessons are relevant? *Energy Policy, 35*(3), 1439–1451.

Wharton, C. R. (1971). Risk, uncertainty, and the subsistence farmer: Technological innovation and resistance to change in the context of survival. In G. Dalton (Ed.), *Studies in Economic Anthropology*. Washington: American Anthropological Association.

Chapter 7
Discussion

7.1 Achievements

The thesis' title is "The Role of Renewable Energy Technology in Holistic Community Development". In order to discuss achievements, there need to be tangible, measurable results, which require that projects have been designed and implemented, so that they can be monitored and evaluated.

Further, it should be noted that the thesis is the outworking of developed and ongoing projects, which initially started in 1996. Since 2002, this work has been carried out through the non-profit NGO, RIDS-Nepal, registered in 2005, of which the author is a co-founder and the project director. Thus, rather than the thesis triggering the development and testing of HCD projects, based on the development of a theoretical model, the urge to write the thesis emerged out of the desire to document the achievements and impacts of implemented projects with the local communities in the remote high-altitude mountain areas of north-west Nepal, since 1996.

Table 7.1 provides a summary of the main HCD project components, all part of the long-term "Family of 4" and the "Family of 4 PLUS" projects, carried out through RIDS-Nepal and the local communities in the district of Humla since 2002.

Notes for Table 7.1:

- *#*: Number of items/projects installed
- *PL*: Pit Latrines (one per family, Sect. 2.10.1)
- *SMS*: Smokeless Metal Stove (one per family, Sects. 2.10.2, 4.1)
- *Solar PV System* (Sect. 3.3.2): This number includes SHS (Solar PV Home Systems, Sects. 2.10.3, 3.3.2.1) for individual homes (378), Solar PV Cluster Systems for clusters (Sects. 3.3.2.2, 3.5.5) of households up to 12 homes (29), and Solar PV Central Tracking Systems (Sect. 3.3.2.3), for whole villages, Health Posts, Local Schools or NGO offices (13)
- *Pico-Hydro Power Plant*: Sect. 3.3.1.1
- *GH*: Greenhouses (to grow vegetables, herbs and fruits, Sects. 2.11.1, 4.2)
- *NFE*: Non-Formal Education Classes (Sect. 2.11.2) for out-of-school children (5) and women (12)
- *SSWF*: Slow Sand Water Filter to purify and store drinking water (Sect. 2.11.5)

A. Zahnd, *The Role of Renewable Energy Technology in Holistic Community Development*, Springer Theses, DOI: 10.1007/978-3-319-03989-3_7,

Table 7.1 RIDS-Nepal implemented programs/projects from 2002–2012

RIDS-Nepal Humla HCD Programs/Projects 2002–2012 (June)			
"Family of 4" programs/projects	#	"Family of 4 PLUS" programs/projects	#
# of PLs constructed	1,198	# of greenhouses constructed	35
# of SMSs installed	1,182	# of solar driers installed	18
# of solar PV systems	420	# of NFE classes	17
# of pico-hydro power plant	1	# of SSWFs installed	75
# Village drinking water systems	6	# of mal-nourished children <5 years	70
		# of village solar hot water bathing centres	2
		# of scholarships	8
		# of data monitoring systems	11

- *Malnourished Children <5 years*: Children and their mothers who are enrolled in a RIDS-Nepal long-term mal-nourished children program (Sect. 2.11.3).
- *Village Solar Hot Water Bathing Centre*: As discussed in Sects. 2.11.4, 4.4.1.5, 4.4.1.7, 4.4.2
- *Scholarship*: Young Humla women and men, supported by the RIDS-Nepal scholarship scheme for 2½ years to learn a professional skill at the KTS (Karnali Technical School) in Jumla.
- *Monitoring System*: Performance monitoring of Solar PV systems (Sects. 3.10.5, 3.11), Greenhouses (Sect. 4.2), Hot Water Bathing Centres (Sects. 4.4.1.5, 4.4.1.7), e-coliform water testing and the enrolled mal-nourished children.

Table 7.1 includes the major physical and education programs, which are implemented with, and have changed the living conditions of, over 1,000 families. The table does not include all of the operation and maintenance training conducted for the end-users of all the power generation projects, greenhouses and village drinking water systems. Neither does it include all of the village Base-Line and Follow-Up surveys, nor the drinking water quality testing and the Indoor Air Pollution (IAP) measurements, even though they are all an integral part of the HCD programs.

While numbers are indicative and thus important, they are not able to provide the most essential information, which is the impacts and changes the implemented projects and programs have brought over the years for the end-users in their wider community and environment. Therefore it is essential to highlight and comment on some of the monitored and recorded impacts and results that the running HCD projects thus far have achieved.

7.2 Findings

As explained in detail in Sect. 1.3.3, RIDS-Nepal conducts a Base-Line Survey in each new village, where it has been invited to consider a "Family of 4" HCD project (Sects. 2.8, 2.10) in joint partnership with the village community. This aims to inform RIDS-Nepal about the village's and the people's initial living conditions and context. About 1–2 years after the implementation of the "Family of 4" HCD projects a second, or Follow-Up Survey, is carried out, in order to understand the initial impacts and changes that the "Family of 4" HCD projects brought forth. Further it aims to determine what further, new needs the local people may have identified. These further needs have been instrumental for the development of the "Family of 4 PLUS" program, aiming to build upon the impact and synergistic benefits of the "Family of 4" projects.

To evaluate, assess or measure the impacts on, and changes to the living conditions of the people through implemented projects, be it in quantitative or qualitative terms, is inherently difficult. This is because mere numerical values cannot comprehensively explain the effects a project has upon a person's health, living condition and environment. The situation becomes even more complex if a holistic community development approach is followed such the "Family of 4" HCD concept, as each individual project has a direct and indirect influence and impact on each other, as well as enhancing the overall outcome and results through synergistic benefits. Further, it is important to be aware that each person of a family, and each family in a village, understands and uses the various project components of a HCD project differently. Some experience significantly more benefits in their living conditions, health and development than others, who have not yet been able to fully utilise the potential benefits of a HCD project.

With these difficulties in mind the Base-Line and Follow-Up survey questionnaires were developed (Appendix 14) so that various numerical data, as well as anecdotal information, can be recorded. They enable RIDS-Nepal to identify the changes and perceived impacts the HCD projects brought over time. In the following sections some examples of recorded changes between the initial Base-Line Survey and the Follow-Up Survey, conducted 1–2 years later, are presented and discussed.

7.2.1 Impacts of the "Family of 4" HCD Projects

Tables 7.2, 7.3, 7.4, 7.5 present some health conditions and changes identified by households in the same context in the Base-Line Survey (in 5 villages and, 315 HH (Households)) and the Follow-Up Survey conducted (in 3 villages and 112 HH) 1–2 years after the "Family of 4" HCD projects were implemented. Table 7.2 shows the data for firewood usage and collection time before and after the commencement of the project.

Through the installation of the SMS (Smokeless Metal Stove) the average family has reduced their weekly firewood consumption by ~40 %. This is significant and shows that the new stove technology provides substantial firewood savings, while providing the same, or even improved, energy services (cooking, room heating and water heating all with the same SMS).

An important secondary impact, which will be recognised only after 10–20 years, is that it helps to actively reduce the extensive deforestation which is

Table 7.2 Data for the firewood usage before (base-line survey) and after (follow-up survey) in the homes surveyed

Firewood usage in the home	Base-Line 315 HH in 5 villages	Follow-Up 112 HH in 3 villages	Change (± %)	Remarks
# of loads of firewood used per week	5	3	–40	One load is 20–30 kg firewood. Firewood is used for cooking, room and water heating and before the implementation of the "Family of 4" also for indoor lighting
# Hours collecting wood per week	21.2	19.8	–7	

the consequence of the need to collect the firewood increasingly further away from their villages. This is most likely also the reason that the firewood collection time has only dropped by 7 %. It can be argued that it is not the actual collection of firewood that is the time consuming task, but the necessity to get to the places where firewood is still available. Thus, with the use of the SMS, in the long-term, the forests have a greater chance to recover from the decades of over usage, which in time should also increasingly reduce the firewood collection time.

Table 7.3 shows the data for intestinal infection before and after the implementation of the "Family of 4" project.

This table shows that through the use of the newly built PL (Pit Latrine) for each household, people experience better hygienic conditions around their houses as well as inside their homes. The impact of the overall cleaner fields and walking paths around and in the village, which were previously used by the adults and children respectively to relieve themselves, shows that fewer people in general suffered from diarrhoea during the Follow-Up Survey.

Table 7.3 Data for intestinal infection during the surveys

Diarrhoea reported at the time of the survey	Base-Line 315 HH in 5 villages	Follow-Up 112 HH in 3 villages	Change (± %)	Remarks
Female 0–4	17	6	–65	% of households in sample with one or more symptomatic individuals per household
Female 5–14	21	21	0	
Female 15–49	8.5	7	–18	
Female 50+	1	1	0	
Male 0–4	17	14	–18	
Male 5–14	20	21	+5	
Male 15–49	5	3	–40	
Male 50+	1	3	+300	This result could be season dependent

Table 7.4 Data for intestinal infection within a time period of 3 months prior to the survey

Diarrhoea experienced in 3 months prior to the survey	Base-Line 315 HH in 5 villages	Follow-Up 112 HH in 3 villages	Change (± %)	Remarks
Female 0–4	9	7	−12	% of households in sample with one or more symptomatic individuals per household
Female 5–14	16	14	−13	
Female 15–49	13	8	−48	
Female 50+	2	1	-50	
Male 0–4	14	9	−36	
Male 5–14	14	15	+7	
Male 15–49	19	5	−74	
Male 50+	2	1	-50	

However, some of the results demand further investigations as they don't fall in line with the expectations. The increase in cases with diarrhoea could mean that the Follow-Up Survey was carried out during the rainy season, during which a much higher incidence of cases of diarrhoea occurs. The sample sizes for some age groups are also too small to draw definitive conclusions. In order to minimize these random errors in the data, the survey includes as well data from the project for a longer, 3 month time span (Table 7.4).

The longer, 3 month time period, presents a more realistic and consistent picture of the pit latrine's impact in regard to the incidence of diarrhoea. It enables us to check the consistency of the data on the potential health improvements people can experience through the building and using of a pit latrine.

Table 7.5 shows the data for intestinal worm infection, over a 3 month time span, before and after the implementation of the project. This is another indicator of the effectiveness of the PL.

Table 7.5 Data for intestinal worms within a time period of 3 months prior to the survey

Intestinal worms experienced in 3 months prior to the survey	Base-Line 315 HH in 5 villages	Follow-Up 112 HH in 3 villages	Change (± %)	Remarks
Female 0–4	5	3	−40	% of households in sample with one or more symptomatic individuals per household
Female 5–14	18	11	−39	
Female 15–49	22	9	−59	
Female 50+	2	1	-50	
Male 0–4	9	11	+22	
Male 5–14	19	5	−74	
Male 15–49	15	5	−67	
Male 50+	2	0		

These numbers clearly indicate that people suffer less from intestinal worms since they have begun to use a pit latrine. The increase of intestinal worms among the 0–4 year old male children could be season dependent, or due to the fact that the small children often play and crawl on the mud floor at home, with the farmer's animals. This improved health condition can also be due to the access to pure drinking water, which was not available previously. Thus, the pit latrine and the access to clean drinking water both contribute to the improved health status of the villagers.

Thus it is reasonable to say that if only one of either, the pit latrine or the access to clean drinking water had been implemented, the improved results may have been less significant. That means that together, these projects enable synergistic benefits, resulting in better health outcomes. The difficulty in this case, as in any other case where several, mutually supportive projects are implemented alongside each other, is to identify which project, in quantitative and qualitative terms, enabled how much improvement.

To define this accurately is not possible, as too many immeasurable factors and behavioural patterns of the users play a part. But what is certain is, that while each project, if individually implemented, will improve people's health condition, both projects implemented together, as in the case of the "Family of 4", bring forth synergistic benefits which are likely to be greater than the sum of the benefits of each individual project.

Tables 7.6, 7.7, 7.8 and 7.9 present some of the attitudinal factors initially identified by the local people during the Base-Line Survey and the differences in

Table 7.6 Percent of households identifying primary sources of stress/anxiety

Primary sources of stress/anxiety	Base-Line 315-HH in 5 villages	Follow-Up 112-HH in 3 villages	Change (± %)	Remarks
Resources	86	88	+2	% of households mentioning this factor
Education	44	57	+30	
Health	23	23	0	
Conflict	3	0	−100	During the base-line survey the maoist conflict was still going on in Nepal

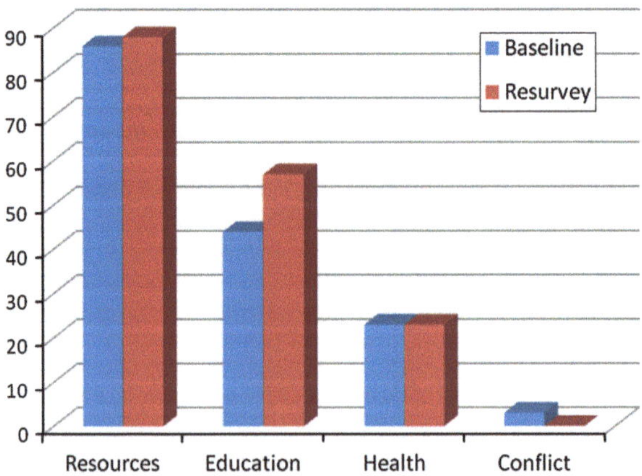

Table 7.7 Percent of households identifying positive changes over the past 10 years (1998–2008)

Positive changes over the past 10 years	Base-Line 315 HH in 5 villages	Follow-Up 112 HH in 3 villages	Change (± %)	Remarks
Resources	27	45	+67	% of households mentioning this factor
General "things seem better"	5	21	+420	
Health	4	15	+375	
Education	2	13	+650	

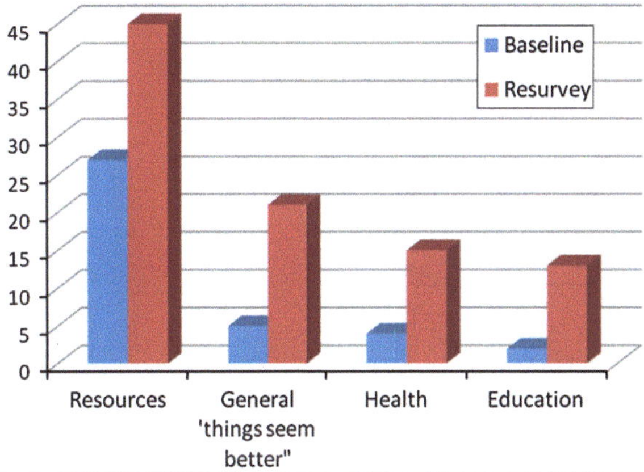

life perception and attitude 1–2 years after the "Family of 4" HCD projects have been implemented. During the Follow-Up Survey each family was interviewed about the changes experienced and how their attitude towards their life circumstances and environment have altered.

Table 7.6 shows the primary sources of stress/anxiety for the villagers before and after the project was implemented.

The Follow-Up Survey in Table 7.6 shows, that the "Family of 4" HCD program has stimulated also attitudinal changes in regard to the beneficiaries' living circumstances and environment. We know, these are all very impoverished, food stressed households, with household members who are malnourished and poorly educated. Through the SMS inside their home, which consumes on an average 40 % less firewood, and the access to clean drinking water, they have started to become more aware of the value, and thus the need to take care and protect their natural resources such as their forests and water resources.

Interestingly, we can see a nearly 30 % increase in the number of villagers mentioning education as a concern in the Follow-Up Survey. The operation and maintenance training, provided by RIDS-Nepal to the people, on how to use and

Table 7.8 Percent of households citing types of changes anticipated (base-line) and observed (follow-up) in their villages since the implementation of the "Family of 4"/"Family of 4 PLUS"

Changes anticipated (baseline) or observed (resurvey) in the village since implementation of the "family of 4"	Base-Line 315 HH in 5 villages	Follow-Up 112 HH in 3 villages	Change (± %)	Remarks
Increased cleanliness	85	91	+7	% of households mentioning this factor
Increased awareness	23	48	+209	
More resources	17	43	+253	
Increased happiness	5	1	–500	
Fewer resources	2	0		

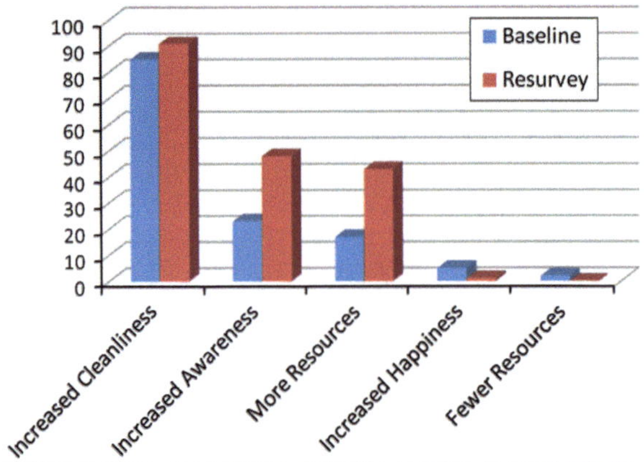

maintain their PL, SMS, the basic indoor lighting for reading and social gatherings and their village drinking water scheme, probably contributed to this trend. People have become more aware of the benefits of education and thus the need for more education. This is further supported through the NFE classes, running in several of the villages that were resurveyed, which highlight the need for more knowledge and understanding due to the introduction of novel technologies.

Health, a permanent issue in such an underdeveloped, harsh and food deprived area, is still identified by almost a quarter of the people as a source of stress. It is the author's and RIDS-Nepal's practical experience that the general health conditions of the people with running HCD programs do improve over the course of time. But, rather than expecting significant health improvements within 1–2 years, monitoring needs to take place over 10–20 years, to allow the HCD projects to become a more integrated part of people's daily life.

Surprisingly, very few villagers mentioned "Conflict" as a source of stress, either during the Base-Line, or during the Follow-Up Survey. This may be because the stress of the daily "fight to survive" is already so severe in these households that the addition of the civil unrest (the civil unrest and violence between the

Maoist and the Nepali government from 1996–2006) was not a significant additional stressor, even during the Maoist insurgency. Since 2006, after the initial, national peace agreement between the Nepali government and the Maoists was signed, it was clearly less of a concern.

The overall picture in Table 7.7 is clear. Life in the villages is believed to have changed for the better as a result of HCD. The number of villagers who feel life in general has improved within the last 10 years, increased dramatically between the Base-Line and the Follow-Up surveys.

These results are a clear indication that the HCD approach, as applied through the "Family of 4" and the ongoing "Family of 4 PLUS" projects, is heading in the right direction.

Villagers had a variety of interesting responses, as shown in Table 7.8, to questions about the changes they either anticipated (Base-Line Survey), or observed (Follow-Up Survey) in their villages since the beginning of the "Family of 4" HCD projects. Thus, 91 % of villagers listed increased cleanliness, 48 % increased awareness of how to achieve better hygiene, and 43 % having more resources, during the Follow-Up Survey.

Interestingly, 5 % of households listed "Increased Happiness" as a possible outcome of the arrival of the "Family of 4", but only 1 % of households listed this at the Follow-Up Survey. This may be the result of a heightened awareness of other, or new, unmet needs. However, this is a very small number and could be just a random fluctuation.

As seen in Table 7.2, villagers used ~40 % less firewood at the Follow-Up Survey, although only about 2 h a week in time saved in collecting it. Further, we were interested, as presented in Table 7.9, in understanding how the villagers would use this time saved, plus the time saved through the new devices, such as the pit latrine, the smokeless metal stove, the new lighting schemes, which all cut down on time and effort for the basic tasks and energy services.

We found that the villagers reported that the saved time overwhelmingly went toward work, with childcare being listed a bit less than half as frequently as work as an activity taking up "saved time". Noteworthy though, is the immense increase between the surveys of the increased awareness that more time should be spent in childcare and teaching children.

This, we believe, is another result and synergistic benefit of the NFE women's classes and the program for malnourished children <5 years of age and their mothers, running alongside the more technical "Family of 4" project components. Thus, the NFE program includes themes not only about the "Family of 4", but as well topics and approaches toward family planning, food preparation and preservation and the value of hygiene. These attitudinal changes are expected to become even more obvious in the years to come, when the HCD programs become more indigenous and more absorbed within the local culture.

Table 7.9 Percent of households citing activities they would do with saved time (Base-Line) and Observed (Follow-Up) with "Family of 4" and "Family of 4 PLUS" projects

Activities respondent would do with saved time	Base-Line 315 HH in 5 villages	Follow-Up 112 HH in 3 villages	Change (± %)	Remarks
Work	51	97	+90	% of households mentioning this factor
No saved time	6	2	–67	
Childcare	3	38	+1267	
Leisure	3	4	+33	
Teach children/studies	2	4	+200	

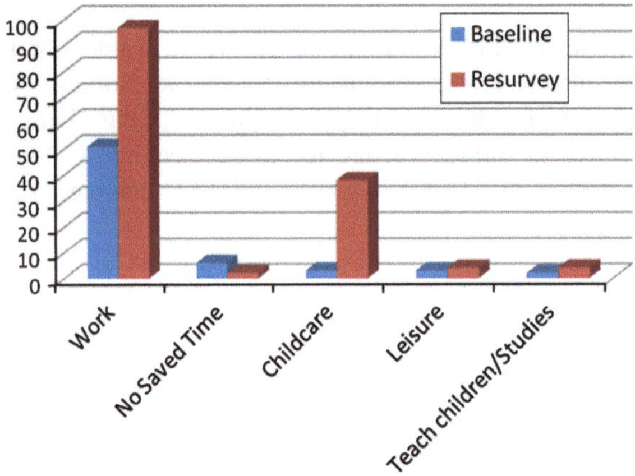

7.2.2 *Impact on Respiratory Ailments Through Reduction of Indoor Air Pollution by the Smokeless Metal Stove and Indoor Lighting*

Health data revealed that villagers suffer from the usual developing world complaints: gastro-enteritis, respiratory ailments, heart disease, and eye conditions. Fever and miscellaneous skin complaints were also reported and the list has not changed between the Base-Line Survey and the Follow-Up Survey.

However, when data between the surveys was statistically analysed, it was found that the number of villagers with symptoms of respiratory infections had significantly dropped. These changes we attribute to the improved indoor air quality after they have installed the smokeless metal stoves and the WLED lamps as part of the "Family of 4" HCD project.

We know from photographs, measured data and experience that the amount of smoke in homes decreases very significantly after the family makes the switch from cooking/heating on an open-fire place in the middle of their main room, to cooking/heating with a SMS (compare Figs. 2.1, 3.4, 3.81, 3.82, 3.101, 4.1, 4.6 and 6.2, Appendix 12.3 with Figs. 2.2, 3.44, 3.63, 3.100, 3.102, 4.7, 6.4 and 6.5).

We also know that the introduction of basic electric indoor lighting with WLED lamps, powered either by a solar PV system or a pico-hydro power plant, removes the need to light the house by burning "*jharro*", which generates extreme amounts of smoke and black soot (compare Figs. 2.3, 2.15, 3.3, 3.81, 3.82 and 3.86, Appendix 12.2 with Figs. 2.4, 2.5, 2.6, 2.8, 2.9, 3.14, 3.15, 3.45, 3.46, 3.62 and 3. 97, Appendix 12.4). So it is not surprising to find that when these two technologies are combined, the amount of IAP decreases very substantially.

The use of solid fuel on open-fires and traditional stoves results in high indoor air pollution (IAP), responsible for 1.5–1.6 million deaths a year (WHO 2006a, Winrock 2004). In Nepal, according to a report issued in 2007 by WHO (2007a), approximately 4,820 deaths/year from Acute[1] Lower Respiratory Infection (ALRI) in children <5 years and 2,680 deaths/year from Chronic[2] Obstructive Pulmonary Disease (COPD) in people >30 years, can be attributed to solid fuel use. The burning of solid fuels produces a number of pollutants which are potentially harmful to human health. Particulate Matter (PM) and carbon monoxide (CO) are two of the main components frequently associated with IAP and incomplete combustion of biomass. Due to the traditional social structure in most developing countries, it is primarily women who are responsible for cooking, and subsequently they and their children are exposed to higher levels of these pollutants, in comparison to the men (Figs. 2.1, 3.3, 3.4 and 3.101).

Mann (2009) lists the known health effects of IAP associated with consistent medical evidence as: ALRI (Acute Lower Respiratory Infection) in children, particularly pneumonia; chronic bronchitis and chronic obstructive pulmonary disease; and lung cancer. Other conditions with emerging evidence are: low birth weight, perinatal mortality, asthma, and middle ear infections in children, cardiovascular disease, tuberculosis, nasopharyngeal and laryngeal (noise and throat) cancer, and cataracts in adults (Schirnding 2002).

While the health effects of airborne particulate matter (PM) are broad, they are predominantly associated with respiratory and cardiovascular diseases. Through population studies, the growing amount of evidence suggests that particulate matter adversely affects human health even at relatively low concentrations (Pope 1995). With increased exposure comes increased risk but the exact exposure response relationship is not known.

[1] Acute exposure: decreased lung function measured by Forced Expiratory Volume (FEV) and Peak Expiratory Flow (PEF); increased respiratory symptoms such as cough, wheeze, and combined lower respiratory symptoms; increased respiratory morbidity measured by absenteeism and decreased activity; increased Chronic Obstructive Airway Disease (COPD) mortality and transient COPD exacerbations; pneumonia; and increased cardiovascular disease symptoms such as angina or death (Pope 1995).

[2] Chronic exposure: chronic bronchitis; COPD; chronic cough; Asthma and wheeze; emphysema; a gradual loss of respiratory function (FEV and PEF); and cardiovascular disease (Pope 1995).

7.2.2.1 Indoor Air Pollutants PM_{10}, $PM_{2.5}$ and TSP

IAP is conventionally measured by PM_{10}, $PM_{2.5}$ and TSP. These are suspended particulate matters with diameters of ≤ 10 and ≤ 2.5 μm and the total suspended particulates (TSP) floating in the air. $PM_{2.5}$ is the more dangerous, fine grained particulate that can penetrate deeply into the lungs and bronchia. The WHO has produced air quality guidelines for particles ≤ 2.5 μm ($PM_{2.5}$) and ≤ 10 μm (PM_{10}) based on the United States Environmental Protection Agency (USEPA) figures (USEPA 2009):

- $PM_{2.5}$: ≤ 15 μg/m^3 (0.015 mg/m^3) annual[3] mean
- $PM_{2.5}$: ≤ 35 μg/m^3 (0.035 mg/m^3) any 24 hours mean
- PM_{10}: ≤ 20 μg/m^3 (0.020 mg/m^3) annual mean
- PM_{10}: ≤ 150 μg/m^3 (0.150 mg/m^3) any 24 hours mean

7.2.2.2 Carbon Monoxide (CO)

Carbon monoxide (CO), although not so much of an individual issue in outdoor air pollution due to its diffusion into the surrounding environment, has a much greater role in IAP. Carbon monoxide is a toxic gas, easily absorbed through the lungs, and its effect is dependent upon the duration of exposure,[4] and the relative concentrations of CO and oxygen in the environment.

The majority of absorbed CO (80–90 %) binds to haemoglobin to form carboxyhaemoglobin (COHb), which is used as a biomarker of exposure in the blood (WHO 2000). Carbon monoxide competes directly with oxygen for binding to haemoglobin. The affinity of haemoglobin for CO is 200–250 times greater than its affinity for oxygen (WHO 2000). The consequences of this competitive binding are reduced oxygen carrying capacity of the blood and therefore impaired oxygenation of extra vascular[5] tissues. The toxic effects of CO become evident in organs and tissues that consume large amounts of oxygen such as the brain, heart, skeletal muscles and the developing foetus (WHO 2000). When exposed to a constant CO level, the COHb concentration initially increases rapidly, begins to level off after 3 hours and after 6–8 hours reaches a steady state.

Animal studies showed that there is a significant lag time between mother and foetus with regard to the CO uptake, occurring up to 40 h after exposure creating an environment where the mean carboxyhaemoglobin levels in the foetus significantly exceed the mother's (Longo 1977). Data on chronic CO-exposure during pregnancy has also demonstrated its association with low birth weight and increased neonatal mortality rates (Longo 1977).

---

[3] 1 mg = 1000 μg, thus 15 μg/m^3 = 0.015 mg/m^3

[4] A high concentration of CO can reach the same blood level rate in 1 h as low concentrations can reach in 6 hours (Forbes 1945).

[5] Located or occurring outside a blood or lymph vessel, http://www.answers.com/topic/extravascular (accessed July 27, 2012)

Thus, it comes not as a surprise, that CO has a particularly harmful effect on pregnant women because the foetus is especially sensitive to the gas (WHO 2000). Additionally, in the foetus, the half-life for elimination of CO is significantly longer than in the pregnant mother (WHO 2000).

Studies have also shown noticeable neuropsychological impairments from an acute low level exposure of CO in the form of "dysfunctions in memory, new learning ability, attention and concentration, tracking skills, visuomotor skills, abstract thinking, and visuospatial planning and processing" (Amitai 1998). As with PM, there is little known about the exposure response relationship. By using an equation which takes into account all of the known physiological variables affecting CO uptake a set of minimal guidelines for CO exposure have been established by the WHO (2000) and USEPA (2009):

- 100 mg/m^3 (90 ppm) for 15 min
- 60 mg/m^3 (50 ppm) for 30 min
- 40 mg/m^3 (35 ppm) for 1 hours
- 10 mg/m^3 (9 ppm) for 8 hours

These values are recommended in order not to exceed a COHb concentration of 2.5 %, which corresponds to the onset of the hypoxic[6] effects of CO upon vulnerable groups such as the elderly, people with latent coronary artery[7] disease and the foetuses of pregnant women (WHO 2000 and Lawther 1975). In contrast, the author has frequently measured COHb contents between 10–17 % in Humla women, cooking on open-fire places.

7.2.2.3 Respiratory Symptoms Before and After the Introduction of the Smokeless Metal Stove and Indoor Lighting

The study in Sect. 7.2.1 looked at some of the more general impacts of the "Family of 4" for 315 households during the Base-Line Survey and for 112 households during the Follow-Up Survey. The present study, highlights and summarises some of the main findings of a Masters dissertation on "The Impact on Respiratory Ailments through Reduction of Indoor Air Pollution by Smokeless Metal Stove and Solar Lighting: Results of a Study in Rural Nepal" (Mann 2009). These are RIDS-Nepal's Humla survey data and the author was the Nepal based supervisor, focusing on respiratory ailments through IAP.

[6] Hypoxia generally refers to a lack of oxygen in any part of the body. It is the condition in which tissues are starved of oxygen, or in a low oxygen state. For the people in the Himalayas, such as in high-altitude Humla, that can happen from low blood oxygen content if one suffers from altitude sickness, or from low blood oxygen carrying capacity, such as happens through carbon monoxide poisoning through the high Indoor Air Pollution caused by open-fire cooking and heating in the homes of the Humla people. (extract and summarised from: Britannica Concise Encyclopaedia: hypoxia, http://www.answers.com/topic/hypoxia (accessed July 27, 2012))

[7] Either of two arteries that originate in the aorta and supply blood to the muscular tissue of the heart http://www.answers.com/topic/coronary-circulation (accessed July 27, 2012)

For this study, out of 404 Base-Line and 265 Follow-Up surveys, the data from 105 households from both surveys was used, because these were the only households that were surveyed during the same season (summer) before and after the implementation of the smokeless metal stove and basic solar powered indoor lighting. Consequently, a higher accuracy of the association between respiratory symptoms and the interventions is expected.

We examined the association of respiratory symptoms and the presence of the SMS and WLED indoor lighting, that is, the incidence rates of people's self-reported respiratory symptoms before and after the intervention. In the surveys, respiratory symptoms were assessed by the reported presence or absence of symptoms related to AURI (Acute Upper Respiratory Infection) and ALRI among household members. Consequently, each person was categorised into either AURI or ALRI during the surveys. For the analysis though, these categories were combined in a single category of respiratory symptoms. The household data was collated to provide numbers of people affected by respiratory symptoms according to age group (0–4 years, 5–14 years, 15–49 years and 50+ years) and gender. While the number of people per house according to age group, sex and smoking status could be determined, no direct relationship between the prevalence of smoking[8] and respiratory symptoms could be drawn.

Mann (2009) analysed the data of the 105 households chosen for both sexes and found that there was a statistically significant difference between the frequencies of respiratory symptoms at the time of the Base-Line Survey compared to the time of the Follow-Up Survey.

Table 7.10 breaks down the respiratory symptom results into gender and Fig. 7.1 provides a graphical representation of the 95 % CI[9] (confidence intervals) in relation to the prevalence rates for each gender individually and collectively.

---

[8] Smoking also has the potential to influence the data due to the well known link between smoking, passive smoking and lung disease (WHO 2000). The surveys revealed a significant difference of people smoking in winter (24.61 %) compared to summer (35.15 %) before the installation of the SMS and indoor lighting, and after the installations, with 40.5 % in winter and 37.8 % in summer. Thus, a significant increase in smoking from before and after has been recognized, for which there is no clear and easy explanation other than once people had less IAP due to the SMS, they felt the need to smoke more in order to "satisfy" their bodies' previous long-term exposure to smoke and CO. Thus all household interviews conducted in winter have been excluded from the present study, and only summer data in matching households have been considered.

[9] 95 % CI (Confidence Index) means that approximately 95 % of all observations fall within 2 σ (standard deviation) of μ (mean value) on a Normal (or Gaussian) distribution curve, which is symmetric, single-peaked and bell-shaped. All Normal distributions have the same overall bell shape with one defined peak. The exact shape of the distribution curve for a particular distribution is described by its mean μ and its standard deviation σ. Together, μ and σ determine the Normal curve's shape. The mean is located at the centre of the symmetric curve and the deviation is the measure of spread for the Normal distribution. For more information: "Confidence Interval", http://en.wikipedia.org/wiki/Confidence_interval (accessed July 27, 2012), and "Normal Distribution", http://en.wikipedia.org/wiki/Normal_distribution (accessed July 27, 2012)

Table 7.10 Respiratory symptoms reported during the base-line and the follow-up surveys

	Base-Line Survey	Follow-Up Survey	Total	Chi Square[a]
Males				
# with respiratory symptoms	73	45	118	9.799
% Males with symptoms	21.86 %	12.82 %		[b]p-0.0018
Total # males	334	351	685	
Females				
# With respiratory symptoms	55	45	100	2.686
% Females with symptoms	18.64 %	13.80 %		p-0.1021
Total # Females	295	326	621	
Males and females				
# with respiratory symptoms	128	90	218	11.6
% of people with symptoms	20.35 %	13.29 %		p-0.0006
Total # of people	629	677	1306	

[a] The Chi Square test (X^2) compares the observed and expected counts. It tells whether the observed differences are statistically significant. The X^2 is a measure of the distance of the observed counts from the expected counts. As a distance, it is always zero or positive, and it is zero only when the observed counts are exactly equal to the expected counts. Large values of X^2 are evidence against the null hypothesis (often expressed as "there is no relationship between two quantities") because they say that the observed counts are far from what we would expect if the null hypothesis were true. Small values of X^2 are not evidence against the null hypothesis. (Moore 2007, pages 554–555)

[b] The *p* value is the probability of obtaining a test statistic at least as extreme as the one that was actually observed, assuming that the null hypothesis is true. Often the null hypothesis is expressed as "There is no relationship between two quantities." For example, in market research, the null hypothesis would be "A ten percent increase in price will not adversely affect the sale of this product." Based on survey results, this hypothesis will be either accepted as correct or rejected as incorrect. The lower the *p* value, the less likely the result is if the null hypothesis is true, and consequently the more "significant" the result is, in the sense of statistical significance. One often rejects a null hypothesis if the p-value is less than 0.05 or 0.01, corresponding to a 5 % or 1 % chance respectively of rejecting the null hypothesis when it is true. http://en.wikipedia.org/wiki/P-value and http://www.businessdictionary.com/definition/null-hypothesis.html (both accessed July 27, 2012)

Males experienced a statistically significant reduction in the prevalence of respiratory symptoms from 22 to 13 % (relative risk (RR)[10] 1.71, 95 % CI: 1.21–2.4) between the surveys. Females reported not as much of a reduction, from 19 to 14 % (RR 1.35, 95 % CI: 0.94–1.94). For both genders combined, respiratory symptoms reduced in frequency from 20 to 13 %, which is statistically significant (RR 1.53, 95 % CI: 1.2–1.96).

Another way of expressing the benefit of the SMS and WLED indoor lighting is a 35 % reduction of relative risk of respiratory ailments for the population after the interventions. This means people are 35 % less likely to suffer from respiratory

[10] Relative Risk (RR) is the risk of an event (or of developing a disease) relative to exposure. Relative risk is a ratio of the probability of the event occurring in the exposed group versus a non-exposed group http://en.wikipedia.org/wiki/Relative_risk (accessed July 27, 2012)

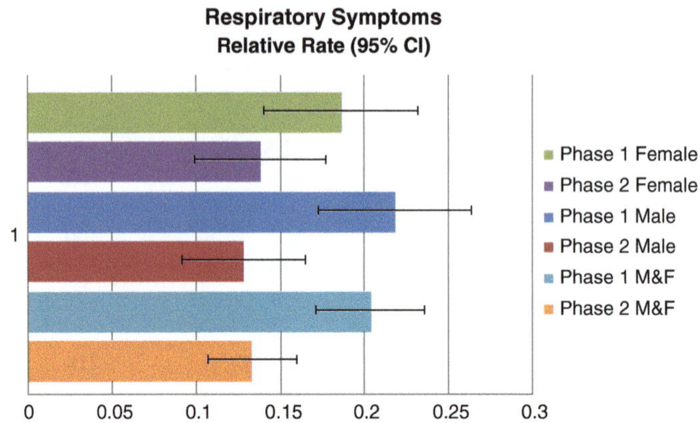

Fig. 7.1 Respiratory symptoms: prevalence rates before (phase 1) and after (phase 2) of the "Family of 4" project, with 95 % confidence intervals

symptoms if they use a SMS and WLED indoor lighting in comparison to the previous traditional methods of providing the same (but in fact considerably improved) energy services for cooking/heating through an "open-fire place" and "*jharro*" for lighting.

When the data is further broken down into age groups, a number of interesting patterns emerge, as presented in Table 7.11. Among females in the age group 0–4 years, the prevalence of respiratory symptoms doubled, while among males aged 0–4 years, the prevalence was unchanged at the Follow-Up Survey.

The highest prevalence of respiratory symptoms was reported by adults aged 50+ years, with 71 % of males and 79 % of females reporting symptoms in the Base-Line Survey. Females in this age group reported a marked reduction in respiratory symptoms during the Follow-Up Survey, with 35 % reporting symptoms, while among males there was no appreciable reduction. Reduction in reported occurrence of respiratory symptoms during the Follow-Up Survey with statistical significance was reported for males in the 5–14 and 15–49 age groups. The other groups reported either no changes or only minor insignificant reductions.

Overall, the age groups 0–4 years and 50+ years together accounted for 70 % of all people with respiratory symptoms, although these two age groups comprised only a quarter of the surveyed population.

While the 0–4 and 50+ age groups had greater percentages of reported respiratory ailments in the Base-Line Survey, the greatest effects on relative risk reduction were seen in the 5–49 years age bracket, particularly among males where there was a significant reduction in prevalence from 12 to 2 % (RR 6.4, 95 % CI: 2.5–16.3). Among females in this age category, the reduction was slightly lower, 8 to 5 % (RR 1.5, 95 % CI: 0.8–3.1).

Table 7.11 Respiratory symptoms according to age group/base-line/follow-up/gender/season (summer) and household

	Base-Line Survey	Follow-Up Survey	Total	Chi Square
Males 0–4				
# Respiratory symptoms	14	15	29	0.0029
% Respiratory symptoms	31.11 %	30.61 %		p-0.9586
Total # Males 0–4	45	49	94	
Males 5–14				
# Respiratory symptoms	8	2	10	4.29
% Respiratory symptoms	8.79 %	2.06 %		(Yates correction[a])
Total # males 5–14	91	97	188	p-0.0440
Males 15–49				
# Respiratory symptoms	22	3	25	17.07
% Respiratory symptoms	14.01 %	1.79 %		(Yates Correction)
Total # males 15–49	157	168	325	p- <0.0001
Males 50+				
# Respiratory symptoms	29	25	54	0.091
% Respiratory symptoms	70.73 %	67.57 %		p-0.7647
Total # Males 50+	41	37	78	
Females 0–4				
# Respiratory symptoms	11	22	33	2.12
% Respiratory symptoms	28.21 %	43.14 %		p-0.1486
Total # females 0–4	39	51	90	
Females 5–14				
# Respiratory symptoms	8	5	13	1.06
% Respiratory symptoms	9.30 %	5.32 %		p-0.3122
Total # females 5–14	86	94	180	
Females 15–49				
# Respiratory symptoms	10	8	18	0.51
% Respiratory symptoms	7.30 %	5.26 %		p-0.4799
Total # females 15–49	137	152	289	
Females 50+				
# Respiratory symptoms	26	10	36	12.45
% Respiratory symptoms	78.79 %	34.48 %		p-0.0005
Total # females 50+	33	29	62	

[a] The English statistician, Frank Yates, suggested a correction for continuity which adjusts the formula for the Chi Square test by subtracting 0.5 from the difference between each observed value and its expected value in a 2 × 2 contingency table. This reduces the Chi Square value obtained and thus increases its p-value. The effect of Yates' correction is to prevent overestimation of statistical significance for small counts, smaller than 5. http://en.wikipedia.org/wiki/Yates%27_Chisquare_test (accessed July 27, 2012)

7.2.2.4 Seasonal Influence

When analysing all the households (404 Base-Line and 265 Follow-Up surveys) we see certain irregularities emerge in relation to season. For conditions known to be related to IAP, such as eye symptoms and respiratory symptoms, different patterns

Table 7.12 Households with eye disease for all 404 base-line and 265 follow-up survey data

Season/house with eye disease	Base-Line survey	Follow-Up survey	Total	Chi square
Winter				
# of households	35	12	47	8.094
% of households	22.20 %	9.50 %		p-0.0047
Total households	158	126	284	
Summer				
# of households	29	20	49	0.396
% of households	12.10 %	14.40 %		p-0.5314
Total households	239	139	378	
Winter and summer				
# of households	64	32	96	2.098
% of households	16.10 %	12.10 %		p-0.1487
Total households	397	265	662	

Table 7.13 Respiratory symptoms between all 404 base-line and 265 follow-up survey data

	Summer	Winter	Total	Chi square
Male and female base-line survey				
# Respiratory symptoms	241	204	445	10.15
% Respiratory symptoms	16.05 %	21.10 %		P = 0.0014
Total # people	1502	967	2469	
Male and female follow-up survey				
# Respiratory symptoms	125	94	219	0.13
% Respiratory symptoms	13.95 %	13.33 %		P = 0.7214
Total # people	896	705	1601	

emerge. Table 7.12 shows the data for eye disease in households before and after the implementation of the HCD project. During the summer, when comparing between the Base-Line Survey and the Follow-Up Survey, there was a small increase in the reported prevalence of eye symptoms, but this did not reach statistical significance. However, during winter (Table 7.12) the incidence of eye symptoms significantly improved after the implementation of the SMS and WLED indoor lighting.

The full set of data, as presented in Table 7.13, shows that there is a significant difference in respiratory symptoms between the Base-Line Survey and the Follow-Up Survey for winter (Chi = 16.776, p <0.0001) compared to summer (Chi = 1.903, p = 0.1682). This significant decrease in reported respiratory symptoms between the surveys has occurred despite a significant increase of 15.91 % in the prevalence rates of smoking (Table 7.14) among the interviewed population for winter (Chi = 69.357, p = <0.0001).

7.2.2.5 Example of 24 h $PM_{2.5}$ and CO IAP Measurements in Humla

The graphs in Fig. 7.2 show a typical 24 h $PM_{2.5}$ IAP measurement from RIDS-Nepal's periodically conducted $PM_{2.5}$, PM_{10}, TSP and CO IAP tests in Humla homes.

Table 7.14 Smoking rates between all 404 base-line and 265 follow-up survey data

	Base-Line survey	Follow-Up survey	Total	Chi square
Summer				
# People who smoke	528	339	867	1.748
% People who smoke	35.15 %	37.83 %		P = 0.1865
Total # people	1502	896	2398	
Winter				
# People who smoke	238	310	548	69.357
% People who smoke	24.61 %	40.52 %		P = <0.0001
Total # people	967	705	1672	

Figure 7.2 shows a 24 h $PM_{2.5}$ and CO IAP measurement test in a traditional Humla home with open-fire place cooking/heating and "*jharro*" lighting (blue and green lines), and in a home with a RIDS-Nepal SMS and indoor lighting installed (red and yellow lines). The PM_{10}, $PM_{2.5}$ and TSP IAP measurements were monitored and recorded with an SKC EPAM-5000 air quality monitor,[11] while the CO values were measured (green and yellow lines) with a PAC-7000 CO[12] monitor.

Figure 7.2 shows data from two, $PM_{2.5}$ and CO, IAP measurement tests. The three spikes correspond with periods of the day when cooking occurs. It can be seen that the IAP is worse and continues for longer in the evening, which is when the whole family is indoors, sitting around the cooking place.

Notice the difference in the scale on the left Y-axis (for the home with an open-fire place) and the right Y-axis (for the home with the SMS). The amount of IAP was more than an order of magnitude less in the room with the SMS, thus the scale had to be adjusted. The peak levels of $PM_{2.5}$ pollution were 124.3 mg/m^3 with an open-fire, compared with 3.9 mg/m^3 with the SMS. The average $PM_{2.5}$ value for the 24 h period with an open-fire place was 3.582 mg/m^3, while for the room with the SMS it was only 0.059 mg/m^3.

The USA Environmental Protection Agency's (EPA) air quality standards are as follows (USEPA 2009): In order to be within the acceptable 24 hours average, the 3 year average of the 98th percentile of 24 hours concentrations at each population oriented monitor within an area must not exceed 0.035 mg/m^3 (35 μg/m^3) for $PM_{2.5}$ values.[13] Thus, the IAP in the home of the family with an open-fire place exceeds the $PM_{2.5}$ limit by 102.34 times, while for the room with the SMS it is 1.68 times over the acceptable limit.

[11] SKC EPAM-5000 Environmental Particulate Air Monitor, http://www.hazdust.com/epam5000.php (accessed July 26, 2012).

[12] Draeger PAC-7000 CO monitor, http://www.cardinalproproducts.com/Misc/Drager/9046061_Pac7000_en.pdf (accessed July 27, 2012)

[13] To attain this standard, the 3 year average of the 98th percentile of 24 h concentrations at each population oriented monitor within an area must not exceed 35 μg/m^3 (effective December 17, 2006). http://www.epa.gov/air/criteria.html (accessed July 27, 2012)

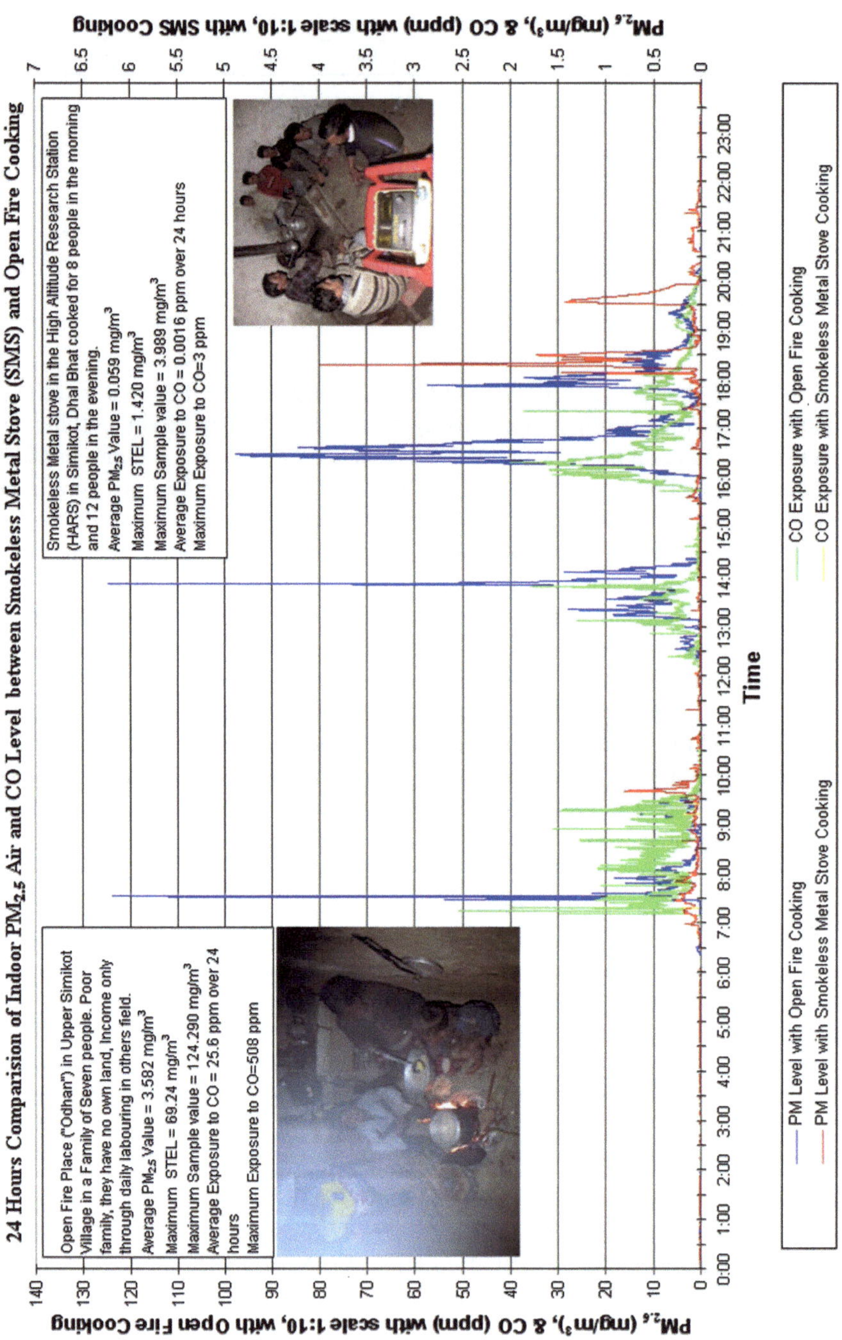

Fig. 7.2 Comparison of 24 h $PM_{2.5}$ (particular matter ≤2.5 μm diameter) IAP measurements in a home with traditional open-fire place and a home with a RIDS-Nepal SMS

In a similar fashion, the CO IAP in the home with an open-fire place is excessive. A maximum value of 508 ppm (parts per million), with a 24 hours average CO level of 25.6 ppm was measured. In comparison, the home with an installed SMS had a maximum CO IAP of 3 ppm with a 24 hours average of 0.0016 ppm.

These values compare to the US-EPA CO air quality standards as follows: 1 hours maximum CO, not to be exceeded more than once a year is 35 ppm, while the 8 hours maximum value not to be exceeded more than once per year is 9 ppm. Considering the 8 hours maximum value, the family cooking/heating with the open-fire place is exposed to 3 times more than the acceptable maximum CO level, and over 24 hours. Further, the family members are exposed to this increased CO level on a daily basis and not just once a year as stipulated in EPA's standards. Thus, the numerical value is multiple imes higher. Considering these high CO levels people (in particular pregnant women and small children) are exposed to all their life, urgent and major improvements are required. In comparison, the home with the installed SMS is well within the US EPA guidelines and thus people are safe from any CO poisoning.

Given this dramatic reduction in IAP, it was reassuring to find a corresponding drop in the percentage of householders with documented symptoms of upper- and lower-respiratory tract infections. We found that for data collected in the winter, when respiratory conditions are likely to be more severe due to people's need to sit for longer times each day around the open indoor fire place to warm themselves, there was a 50 % drop in the average number of people per household with symptoms of AURI (from 14 % with symptoms at the Base-Line Survey, to 7 % at the Follow-Up Survey). This result is statistically significant. ALRI also dropped, though by a smaller percentage, from 9 % of householders per household with documented symptoms to 6 %. The RIDS-Nepal data collectors were trained how to identify the symptoms of upper- and lower-respiratory tract infections, and enumerated by gender and age all household members in all households in the study exhibiting these symptoms during the surveys.

7.2.2.6 Main Study Findings Highlighted

The study conducted demonstrated that the implementation of the SMS and WLED indoor lighting was statistically associated with a significant decrease in reported prevalence of respiratory symptoms among the study population. This section is a summary and interpretation of the chapter "Significance of Main Study Findings" in Mann's (2009) Masters Dissertation (used with permission).

As the data presented indicate, and Fig. 7.2 illustrates for one, 24 hours, $PM_{2.5}$ and CO IAP measurement test, a marked decrease occurs in IAP following the change from the traditional open-fire cooking/heating and "*jharro*" lighting wooden sticks to the SMS and WLED indoor lighting. This reduction in exposure to household pollution is expected to lead to a corresponding fall in occurrence of respiratory ailments. Similar studies into the effects of IAP and flued stoves on respiratory health have shown comparable results. An extensive intervention study conducted in rural Guatemala for example, showed a 40 % reduction in severe cases of Respiratory Syncytial Virus (RSV)[14] following the introduction of a flued stove (WHO 2007b, 5).

The impact of the SMS and indoor lighting had varying effects according to age and gender. The age groups 0–4 and 50+ years produced the majority of the respiratory ailments with the most significant improvements found between the ages of 5–49 years. This group would be the ones most likely to be outdoors most of the time during the day, compared to the very young or elderly, and thus would be less exposed to IAP. However, from the data it appears that males between the ages of 5–49 and females 50+ years benefit the most from the intervention. In local Nepali culture it is mainly the women who cook and thus are exposed to the most severe IAP. Thus, while this fact seems to suggest a greater improvement from health impairments among females, once the SMS and indoor lighting are installed, the recorded data suggests that males receive the most benefit. Thus, while certain roles in relation to gender and age could be hypothesised, e.g. that women spend more time indoors for cooking and other household work, so would have a greater total exposure to indoor air pollution than males, further investigations into specific exposure and people's roles within the family and in turn to the relative exposure to IAP, need to be considered.

The season was shown to have a confounding effect on the prevalence of respiratory symptoms. The analysis indicates marked seasonal differences in the prevalence of respiratory ailments and the relative risk reduction with the use of the SMS and WLED indoor lighting. During winter, people use their stoves more for heating, and as well spend longer periods indoors. In addition, the decreased house ventilation during winter further increases the level of exposure to IAP. Thus, winter with its higher occurrence of respiratory ailments also exhibited a larger reduction in ailments between the surveys when compared to summer. This was also noted in the prevalence of eye disease which had higher initial rates in winter and subsequently had a significantly pronounced drop in rate when compared to summer. The decrease in respiratory ailments during winter occurred even though there was a discernible increase in the rate of tobacco smoking during the winter months. Thus, while seasonality is a recognised determinant of acute respiratory ailments it is a variable that can be controlled in a study by repeating the interviews at the same time each year.

[14] Human Respiratory Syncytial Virus (RSV) is a virus that causes respiratory tract infections (lungs and breathing passages). Bronchiolotis or pneumonia are typical illnesses caused by RSV. It is the major cause of lower respiratory tract infection and hospital visits during infancy and childhood. In adults, it may only produce symptoms of a common cold, such as a stuffy or runny nose, sore throat, mild headache, cough, fever, and a general feeling of being ill. But in premature babies and children with diseases that affect the lungs, heart, or immune system, RSV infections can lead to other more serious illnesses. There is no vaccine. RSV is highly contagious and can be spread through droplets containing the virus when someone coughs or sneezes. Treatment is limited to supportive care, including oxygen. In temperate climates there is an annual epidemic during the winter months. In tropical climates, infection is most common during the rainy season. http://kidshealth.org/parent/infections/lung/rsv.html and http://en.wikipedia.org/wiki/Human_respiratory_syncytial_virus (both accessed July 27, 2012)

Smoking and passive smoking are known causes of morbidity and mortality, with 20 % of all deaths in developed countries attributed to tobacco smoking (Wald 1996). Smoking would undoubtedly have an influence on the occurrence of respiratory ailments in the study population, either directly or indirectly. Smoking during pregnancy has been associated with an increase in foetal and perinatal deaths, low birth weight, preterm delivery and other complications (Edwards 2004). Even passive smoking has been shown to be responsible for respiratory disease and aggravation of allergic responses, particularly in children, with an increase in risk of respiratory infection and childhood asthma (WHO 2000a). The lack of change in the respiratory symptoms' prevalence rates for the 0–5 age group could be partially explained by the high incidence of smoking among the households. Due to the increased vulnerability to passive smoking experienced by small children and the proven relationship between paternal smoking and respiratory symptoms (Cook 1997), the greater the chance that even small doses received by small children through passive smoking, influences the results. In this study we controlled for smoking by having it as a constant and measuring the differences in respiratory symptoms before and after the intervention. This however does not account for the potential bias that may be caused by smoking and passive smoking, especially in relation to children <5 years whose respiratory tract is particularly vulnerable and more susceptible to respiratory disease. However, by reducing IAP through the SMS and indoor lighting intervention, we reduce the dose of overall smoke inhalation and therefore decrease the burden of smoking related disease.

It is well recognised that one's immune response is responsible for the symptomatic expression of disease (e.g. coughing, sneezing or watering of the eyes due to IAP), whether pathogenic (bacterial or viral) or an environmental allergen (CO or particulate matters). Therefore, anything that lowers the body's immune system has a harmful effect on the ability of a person's body to fight ill health. The people in Humla have many poverty related stressors that potentially affect the immune system. The body's ability to respond to its environment can be influenced by physical and/or psychological stressors. Due to their agricultural and labour intensive lifestyle, deprived living conditions, climatic extremes and food security concerns, the issue of lowered immunity of the Humla people is a concern. Prolonged physical exercise (Hoffman-Goetz 1994) and decreased nutritional status (Hamer et al. 2004) can lower the body's immune response and therefore act as confounding factors to the effects of IAP. Smoking also has a negative effect on the immune system (Sopori 2002) and although is equally distributed through the research participants and unlikely to cause internal bias, the overall influence upon the population as a whole could not be controlled for.

Although a reduced immune response would be present throughout the study, its inclusion as a potential confounding factor is because of the potential influence due to other parts of the holistic community development project. Decreased diarrhoea and improved nutritional status, decreased physical exertion due to closer water supply, less firewood gathering due to the more efficient stove and general potential of improved quality of life are all possible outcomes of the project. Therefore, with an improved immunity there is likely to be a change in the severity of the disease process and expression of symptoms.

The significant reported reduction in respiratory ailments following the introduction of the SMS and indoor lighting can be seen as an indicator of how the intervention reduces the effects of IAP. The possible improvement in morbidity and mortality of this impoverished region could advance the overall quality of life experienced by the people. Through a diminished burden of disease a community's potential for improved quality of life and increased life-expectancy can be enhanced. Thus the reduction of IAP, and the subsequent decrease of diseases associated with it, provides better living conditions and health, enabling communities to develop.

By dividing the population into specific groups by age and gender we saw not only how IAP affected specific sections of local society differently but also how the intervention affected these groups in relation to reported respiratory symptoms. The statistically significant improvement in the health of males compared with females was a surprising finding considering that traditionally females spend more of their time indoors and around the open-fire places (previously) and the SMS (now) compared to males. One reason could be that the time span of the interventions (1–2 years) was still too short and thus the previous long-term exposure to high IAP had already caused lasting health impairment/damage to many women's respiratory systems. As expected the very young and very old of both genders had the highest rates of reported respiratory symptoms. However, older females had a statistically significant improvement following the intervention, as would also be expected, in view of the local culture and family roles. The very young had no significant change in reported respiratory symptoms, which was unexpected, considering they are the most vulnerable to IAP. All of these findings lead to the need for closer scrutiny of the behaviours and traditional practices of the family in relation to exposure to pollutants that affect respiratory health.

In a society heavily dependent upon manual labour and lacking in resources and social infrastructure to support disability, the burden of disease has severe ramifications. This, coupled with the community's fuel needs for cooking and heating, places an additional strain on the already diminishing local forests and potentially adds to the cycle of poverty. Due to Humla's geographical isolation, the people are unlikely to be able to change to cleaner fuels such as LPG, so there will be a continued need for biomass fuel. As firewood becomes scarce there is an increased distance to travel, adding time to an already full day. As the need for wood outweighs the rate of natural forest re-growth; there is a need for some sort of intervention. The efficiency of the SMS dramatically reduces the greenhouse gas emissions, wood consumption and cooking time which in turn decreases the time needed for wood gathering and the overall environmental burden of burning biomass fuel. Since it is women who primarily gather the wood, and do the cooking, it is the women who potentially have the most to gain from the SMS. Decreasing the wood gathering time for women can reduce the injuries due to carrying heavy loads, falls, miscarriages, exposure to violence and civil unrest. This is an aspect that can be studied in future evaluations of SMS.

Table 7.15 Number of households identifying what other changes, beside the "Family of 4", are needed?

What other changes are needed	Base-Line 315 HH in 5 villages	Follow-Up 112 HH in 3 villages	Change (± %)	Remarks
Education	42	65	+55	% of respondents mentioning this factor
Health care	41	38	–7	
Technology	36	56	+56	
Agricultural development	35	32	–9	
Business opportunities	20	35	+75	
Village market	9	5	–44	
Microcredit	3	1	–67	

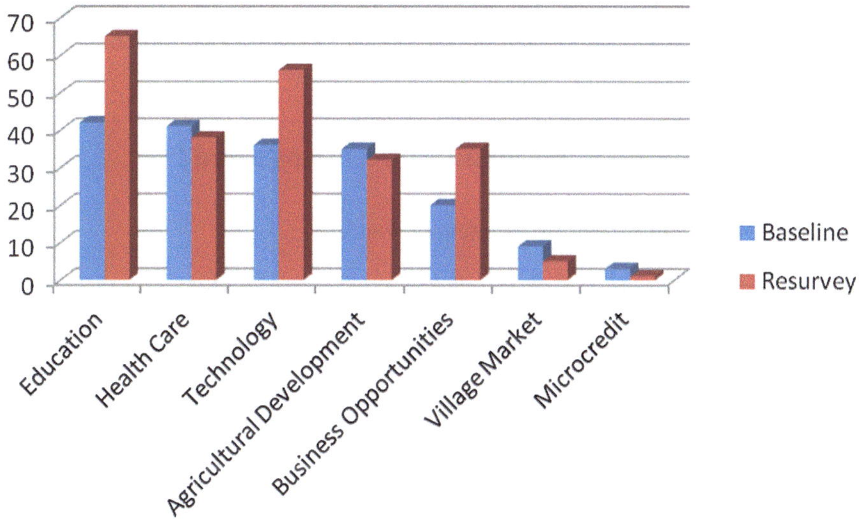

Introducing a contextualised technology that is principally in the women's realm helps to begin the process of empowerment through improving their access to resources and a community developed support network and non-formal education. Over time, with improved health outcomes and education, the community can develop a sustainable future according to their physical, mental and social needs. These expected outcomes will be measured in the years ahead.

7.2.3 What Other Changes are Needed

As part of both, the Base-Line and Follow-Up surveys, a question was included, asking the villagers what other changes they thought were needed in their village and community beside the "Family of 4". This question aimed to determine how much awareness the local people had initially of their overall circumstances and needs, and how the implementation of the "Family of 4" has changed their awareness.

Table 7.15 shows that people have become more aware of several, previously less recognised needs. Thus, the implementation of the "Family of 4" has not only improved people's living and health conditions in general, but significantly more people now identify education (more awareness, skills and knowledge through NFE classes), (contextualised) technologies, such as the SMS, and business opportunities (to grow vegetables in greenhouses or orchards to process, sell or trade) as important factors in their quest for sustainable development. The increased emphasis in responses between the surveys shows that the people are more aware that their active participation in the development process is needed. Such activities include education, implementation and maintaining of technologies in their local context, and that for all endeavours they need more financial resources which can be generated through better business opportunities.

7.3 Have the Research Questions been Answered?

In Sect. 1.3.1 the two research questions asked were:

- *What is the appropriate role of renewable energy technology in holistic community development?*
- *Does the newly developed holistic community development concept of the "Family of 4" and the "Family of 4 PLUS" provide a sustainable way of improving the quality of life for remote and impoverished mountain communities in Nepal?*

As stated in Sect. 1.1.9, today it is widely agreed that access to improved energy services is a key factor for better living conditions, which can be achieved through development activities. Further, in Sect. 1.1.7 it was highlighted that basically all the studies undertaken over the last decade agree that there is a direct relationship between poverty alleviation and access to clean, efficient energy sources and improved energy services. Nepal's rural communities' heavy reliance on traditional biomass in particular is an indicator of their impoverished status, due to limited, or no access to improved, modern energy services such as electricity. It is also widely accepted that "poverty alleviation and development depend on universal access to energy services that are affordable, reliable and of good quality" (Reddy 2002 and Saghir 2005), as mentioned in Sect. 1.1.6.

With a HDI of 0.458 (UN 2011) for Nepal as a country, and an HDI of 0.244 (Sect. 3.13.1) for the remote district of Humla, it is evident that there is a lot of development work required to alleviate the burden of poverty on the people of Nepal. Further, with Nepal's abundant, locally available renewable energy resources such as hydro power (Appendix 11.4) and solar energy (Appendix 11.5), the necessary resources for change are available.

The question initially asked in Sect. 1.3.4 was whether the conversion of the locally available renewable energy resources could meet the need for energy services of impoverished communities, and thus enable the improvement of their overall living conditions. This has been strongly affirmed.

Moreover, the subsequent question: "do RETs play an important role in HCD?" has also been strongly confirmed. Each chapter of this thesis provides ample examples and conclusive justifications that RETs can play a pivotal role in community development. It has to be highlighted, that RETs have to be developed locally, based on the locally available renewable energy resources. Due considerations must be given to the environmental and geographical context, in order to be relevant to the communities' living circumstances. Further, it is axiomatic that contextualized RETs are considered and implemented alongside the important development activities integrated into the "Family of 4" and the "Family of 4 PLUS".

This leads to the discussion of the second research question. Most of Chap. 2 aimed to provide the necessary understanding that the communities' identified needs to improve their livelihood and health, are multi-sectoral and not just single issues. In order to do justice to these needs, it is essential to think of multi-pronged community development programs and projects rather than single issue projects. A reductionist approach cannot address the multi-faceted needs of a disadvantaged community as effectively as HCD can. Further, to pay due respect to the meaning of the word "sustainable" and "development", as discussed in Appendix 10.3, a long-term perspective of community development and a long-term engagement with communities within their cultural context is essential. Thus it is the writer's view that it is only justifiable to talk about the "sustainability" of a community development program if it starts out with a multiple-year project cycle approach. Even better, it should have an active community interaction time frame, or project cycle, of up to two generations, according to the local community's circumstances.

With these ideas in mind, the "Family of 4" and the "Family of 4 PLUS" HCD concepts, as explained in Sect. 2.10 and Sect. 2.11 respectively, have been developed and applied since 2002 in 13 villages in Humla. This approach pays heed to the above criteria for holistic community development. It takes the identified needs of the end-users seriously and sets out with the long-term perspective and engagement with the local communities at the grass-roots level.

So far, no definite conclusion can be made whether a village based "Family of 4" and/or "Family of 4 PLUS" program, or a component of it, is truly sustainable, as they have been running only since 2002. The word "sustainable" has no single, clear cut definition or criteria to measure it against, resulting in a verdict saying "failed" or "passed". But with now a decade of experience since the start of the HCD programs, it can be said with some confidence, that the "Family of 4" and the "Family of 4 PLUS" HCD concepts bring substantial long-term, context related and culturally accepted, life improvements.

Chaps. 3, 4, 5 and 6, presented and discussed various projects of the "Family of 4" and the "Family of 4 PLUS" in detail. Their aim was to make apparent that the projects and programs, integrated in the "Family of 4" and the "Family of 4 PLUS", bring clear and definable improvements to the quality of life for the end-users. Further, Sects. 7.1 and 7.2 discussed some quantitative and qualitative improvements measured through the various numerical household surveys and anecdotal data collection.

Thus it can be concluded, that the newly developed and applied holistic community development concepts of the "Family of 4" and the "Family of 4 PLUS" provide a good starting point in regard to improving the quality of life for remote and impoverished mountain communities in Nepal. It is reasonable to say that a high degree of sustainability[15] can be reached if all stakeholders participating in the HCD programs are aware of their responsibilities and contribute to the projects accordingly. This is because community development is not the work of an individual or a single organisation, but of a community of stakeholders, who agree under defined rules and conditions to join hands to improve the lives of the disadvantaged in a joint, committed long-term response.

7.4 Have the Research Objectives been Met?

In Sect. 1.3.2 the primary objective was defined as:

- *To design, implement and evaluate an integrated approach to holistic community development, called the "Family of 4" and the "Family of 4 PLUS", in high-altitude areas of Nepal.*

As discussed in Sect. 2.8, the author's years of experience of living and working with remote village communities, led him to define the four most identified needs: a latrine, a stove, electric indoor lighting and clean drinking water . These are all deeply interlinked and should be implemented together, as a set of projects for whole villages. Thus the name for this new HCD concept, the "Family of 4", came into being. Once the "Family of 4", was implemented and in use in villages, the users started to identify over time more needs they wanted to address, so the "Family of 4 PLUS", came into being. In 2011 in 13 villages in Humla and one village in Jumla either one of these HCD concepts are being implemented in high-altitude villages.

Alongside the implementation over two generations, or ~20 years in the context of the average Humla life-expectancy, periodic Follow-Up surveys, in the 2nd, 5th, 10th and 20th year, are carried out/planned. The data and information gathered, which are based on the end-users' feedback of their changing life

[15] Sustainability, understood in today's context as discussed in Appendix 10.3 and in the definitions of "sustainable development".

Table 7.16 SWOT analysis of the HCD concepts the "Family of 4" and the "Family of 4 Plus"

Strengths:	**Weaknesses (Hard/Difficult):**
Takes the multi-faceted, locally identified needs of people into consideration	Difficult to find long-term committed donor
Takes the learning of decades of community development into consideration	Time and effort demanding to manage and keep the overall view of all HCD projects
Enables holistic development of people	Hard to trace back and identify the individual source(s) of synergistic benefits
Enables significant synergistic benefits	Needs highly motivated and dedicated staff
Equal concern for all people/communities, independent of gender, caste or background	Needs external donors
Each individual of a community takes part	Long-term financial dependence
Developed for a defined context	Long-term commitment demanded from all stakeholders
Can be contextualised for other areas/nations	Demands long-term skill and leadership training of local staff
High potential for sustainability	Often financially intensive programs
Allows stringent evaluation procedure	Demands intensive communication among the leadership and management team
Supports a project life-cycle approach	Great time demand to learn the local context
Empowers local people/communities	Is vulnerable to political instabilities and changes due to the long-term, intimate local engagement with the village communities
Local people are on staff	Danger of losing well trained staff to better paid jobs in other organisations
Local people are trained in new skills	
Enables development of new technologies	
Provides new skills/jobs to the local industry	
Economic improvement for local industries	
Lowers the migration rate	
Opportunities:	**Threats:**
Long-term life-changing potential	Donor(s) may pull out
Improves overall living conditions	Government puts restrictions on projects
Reproducible in another context	Other organisations head hunt trained staff
Local people's skills improve	Increased local competition among NGOs
More locally earned income	Exposed to local corruption
Local industry gets empowered	Exposed to the local political structures
New business opportunities through new, locally developed technologies	Old traditions hamper changes and growth
Income generation possibilities for locals	Village based "power" struggle
Community based improvements possible	Caste differences in each village
The HCD concept is adaptable to the context	The HCD project scope is too broad
The HCD concept is expandable	The HCD projects are hard to manage
	Political instability, war or natural disasters

conditions, allow a constant evaluation of the HCD programs. Thus their approach, impacts and synergistic benefits are monitored on an ongoing basis, allowing timely adjustments if needed. That pays due respect and consideration to people's own pace of coming to terms with changes in their living conditions, their culture and environment. All these are important factors for an integrated community development approach, besides including all the community development learning over the last few decades.

Using the SWOT analysis approach to the HCD concept, the following interim observations can be made (Table 7.16):

Further, Sect. 1.3.3 defined also three additional specific objectives:

- The **first specific objective** is *to understand the initial circumstances and context, in order to identify the needs and define possible projects together with all stakeholders.*
- The **second specific objective** is *the evaluation and identification of the quantitative and qualitative impacts of the implemented project activities.*
- The **third specific objective** is *to demonstrate that the holistic long-term approach to community development brings significant synergistic benefits.*

The **first specific objective** has been achieved through the detailed, 56 question Base-Line Survey, which is carried out in every village. This has demonstrated a sincere interest to work in partnership with RIDS-Nepal in the development and implementation of a long-term HCD program. This Base-Line Survey was developed by a professional task force, as explained in Sect. 1.3.3. It provides the initial data and information on each family in the village in question. The data from the later Follow-Up Surveys is compared to the Base-Line data, with some initial results after ~2 years presented in Sect. 7.2. The Base-Line Survey questionnaire is included in Appendix 14.1.

The Base-Line Survey provides an understanding of each family's actual situation in regard to their status within their village, their economic condition with regard to any debts, land ownership and "wealth" accumulated, as represented in their animal stock. It provides details such as how they traditionally meet their various energy demands for cooking and heating, as well as the level of education achieved. The health conditions of each family member over the past year(s) as well as any current health problems they face are recorded based on the interviewers' feedback. In this way a rather detailed "contemporary picture" emerges of the family, within their wider village community context. This is important in order to understand the reasons why the local people identify particular needs and how they would like them to be addressed as part of a long-term holistic community development program. Thus the initial Base-Line Survey is an important tool, against which all life improvements and changes, identified in the subsequent Follow-Up Surveys, will be measured.

The **second specific objective** is addressed through the periodic Follow-Up Surveys carried out/planned in the 2nd, 5th, 10th, and 20th year of a HCD program in a village community. This, 52 question Follow-Up Survey has also been developed by a professional group, as explained in Sect. 1.3.3. The full set of questions included in the Follow-Up Survey are presented in Appendix 14.2. The Follow-Up Surveys provide the latest information on how far the HCD program has come since its inception. They ensure that the overall direction towards the long-term development goals and achievements are within the acceptable range.

The **third specific objective** is by far the most difficult to address and to comment on. While each of the "Family of 4" and the "Family of 4 PLUS" projects involve life changing improvements which are increasingly identifiable and recognisable over time, it is harder to measure the extent of the potential and actual synergistic benefits that the fully integrated long-term HCD approach sets in motion. They are difficult to identify and highlight because all the "Family of 4" and "Family of 4 PLUS" project components are designed to bring about mutually enhanced results and direct benefits for individual end-users, which often take some time to appear.

While Sect. 7.2 shows some clear improvements through the current HCD projects, the complex nature of synergistic benefits cannot be easily identified in tables and graphs. In order to isolate and identify synergistic benefits in changed living conditions, requires not just a few months or years, but rather many years of intensive, ongoing follow-up research and data analysis. While RIDS-Nepal is committed to that, as part of the long-term approach to HCD, the following anecdotal summaries, by the author and the RIDS-Nepal staff illustrate some of the synergistic benefits. Thus, rather than equate synergistic benefits with numbers, the following provides a summary of enhanced benefits from each individual project embraced by the "Family of 4" and the "Family of 4 PLUS".

- People, in particular children, suffer from less intestinal disorder, such as worms and amoebic dysentery, due to the integrated implementation of a pit latrine for each family, access to clean drinking water from the village tap stand and a slow sand water filter for long-term clean water storage, in their home. While each of the individual projects will undoubtedly bring some improvements to people's health and living conditions, the overall impacts and benefits of all projects, jointly implemented, bring significantly more health improvement benefits than the sum of the individual projects.
- As highlighted and discussed in various chapters throughout the thesis, people suffer from heavy and life endangering indoor air pollution, in particular from particulate matter floating in the air, such as $PM_{2.5}$ and the poisonous, odourless gas, carbon monoxide. These indoor pollutants are generated by traditional open-fire place cooking, heating and indoor lighting. Thus it can often be seen that development organisations address these issues with either a stove to get rid of the indoor air pollution, or a basic electric indoor lighting system, often powered by a locally available renewable energy resource. Seldom, if ever, is it considered that a stove, and a basic electric indoor lighting system, should be implemented together. There is no doubt that each project individually brings improvements to the quality of life of the people under discussion. However, it can also be argued that if these two projects are implemented together, the final outcome or synergistic benefits is much greater than the sum of the individual improvements.

- As discussed in Sect. 1.1.8, Humla, the area where the HCD projects are implemented, is a permanent food shortage area. Thus the availability of more food is crucial for the local people. The HCD concept "Family of 4 PLUS" includes the building of greenhouses (Sect. 2.11.1), which were developed for this high-altitude alpine region, in order to grow more food locally. This has extended the traditional vegetable growing season from 3–4 months a year previously, to 10 months a year. In the years to come, this will undoubtedly improve the people's health conditions and reduce the severity of food/nutrition shortages. But there are still clearly identifiable periods of time over the course of the year during which certain food can not be grown due to the harsh climatic conditions, thus not allowing people to have the desired balanced nutrition throughout the year. To mitigate this, a solar drier was developed, enabling people with greenhouses to dry vegetables to the required moisture content in order to store them safely during times of plenty. Then the dried food (vegetables, fruits or meat), can be added to their daily diet in times when they cannot be grown, not even in the greenhouse, or are not available (goats or yaks for meat). This provides a much more balanced daily diet as well as added diversity to the otherwise rather monotonous daily food intake. The malnutrition alleviation program for children <5 years of age and their mothers, also part of the "Family of 4 PLUS" program, provides the right framework to teach the mothers (and fathers indirectly) about the problems of permanent food shortage and lack of food/nutritional variety. In addition, since the introduction of greenhouses and solar driers, several families have recognised the potential to grow more food and sell it at the local market in times of local scarcity, when vegetables fetch a much higher price. Thus, the greenhouse in particular, as well as the solar drier in a few cases, has become an income generation source, adding significantly to their meagre cash income. Again, it is clear that there are synergistic benefits from implementing the greenhouse, the solar drier and the malnutrition alleviation projects concurrently.
- As mentioned in Sect. 2.10, scabies is a very common skin disease that people suffer in Nepal's rural communities, due to unhygienic living conditions. This is no surprise, because of the rivers' cold water temperatures which, even in summer hardly ever reach 15 °C (Sect. 2.11.4). The remedy for this is the contextualised high-altitude solar water heater (HASWH) bathing centre for whole communities, utilising the locally available solar energy (Sects. 2.11.4, 4. 4.1.7 and 4.4.1.8). Now people have warm/hot water available, not heated by their dwindling biomass resources, but by the abundant locally available solar energy. In parallel, the NFE classes for women and children teach the participants how to take showers and how to achieve better hygiene within their own rural context, so that sicknesses such as scabies, over time, can be eradicated. Thus again, the joint implementation of an appropriate, locally manufactured HASWH, and the educational tool of NFE classes, produces synergistic benefits.
- Education is widely recognised as an important tool for sustainable development. Thus our, RIDS-Nepal NFE program and workbooks, all designed and written with the local context in mind, address all aspects and project

components of the "Family of 4" and the "Family of 4 PLUS". They are taught in visual ways, through posters (Appendix 22), brochures (Appendix 19), flip charts (Appendix 20) and topic specific videos (Appendix 16). They are also incorporated into songs (Appendix 24), written in the local language, or in active ways through games specifically developed and played in the NFE classes (Appendix 17.5, Appendix 6 and Appendix 7). Furthermore, up to four able young women and men from the project area, can apply every year to RIDS-Nepal for a scholarship to attend the Karnali Technical School (KTS) in the neighbouring district of Jumla for an additional 2½ years apprenticeship training program. Thus, while the NFE program brings people, who never had the chance to sit behind a school bench, from illiteracy to basic, functional literacy and numeracy, the KTS program leads them on to a professional education, enabling them to find a job and earn an income to support their families. Thus, these young men do not need to migrate to India, in the search of unskilled, mundane work, often under the hardest imaginable conditions. There is no shortage of stories of men who never returned from India or returned empty handed or of men who returned sick, even HIV/AIDS infected, bringing even more problems and suffering back home. Now, with basic skill training they can actively participate in the development of their own community, while at the same time earning a decent living. Thus, without leaving their family behind for months and years at a time they can remain together as a family unit. The long-term benefits to their families and their communities cannot be overestimated.

- The HCD programs do not just address the most urgent identified human needs, but also education and long-term income generation opportunities. But, as people and communities live in a particular environment, usually directly supported and fully dependent on it, it is also important to include HCD programs that impact positively on their environment. In the case of the "Family of 4" and the "Family of 4 PLUS", two of the more obvious and direct, positive environmental actions are mentioned here—the Smokeless Metal Stove (SMS) and the Solar Cooker (SK). Both were either, fully developed (by the author) and disseminated by RIDS-Nepal, such as the SMS, or partly, as the SK which was developed by a local Nepali company with feedback from the author. The SMS has been extensively tested in the field, as well as in the laboratory of the Kathmandu University. The data shows, that the local users consume 40 % less firewood, if they use the SMS properly, as taught by the RIDS-Nepal staff during the installation, and in the NFE classes (Appendix 17, Appendix 20, Appendix 21, Appendix 23, Appendix 24) and the follow-up visits. Further, it is obvious that the solar cooker does not need any other energy resource than the sun. The interventions taken, allow the local people to continue to cook their daily food the way they like it. Also, these new technologies enable the people/community to take more care of their precious local environmental resources, such as their forests, which thus far have been cut down mainly for firewood consumption in unsustainable ways, reducing their national forest resources by an average of 1.7–1.8 % per year (Appendix 11.3). Thus, an active fight against further deforestation and soil erosion (which causes the loss of land through landslides) are addressed in practical and active ways at the grass-roots level. So, another long-term benefit of these new, contextualised technologies is to allow the local forest and environment to recover from decades of over-cutting.

It is apparent that two or three projects implemented concurrently, as discussed above, bring synergistic benefits. Thus when all of the "Family of 4" and the "Family of 4 PLUS" projects, are simultaneously implemented as an integrated part of the long-term HCD concept, it is obvious that the potential for enhanced, or synergistic benefits of a holistic project approach, are enormous. While this is easy to grasp mentally, it is very difficult to scientifically track and identify the factors which have contributed over the years to the benefits. This is because of the number of factors and the length of time involved. Community benefits are easily measured, but their exact causes are often harder to determine.

7.5 Results and Benefits of Holistic Community Development

If the results and benefits of the HCD concepts of the "Family of 4" and the "Family of 4 PLUS" have to be summarised in one sentence, it could be: The whole is greater than the sum of its parts.

RIDS-Nepal's HCD concept has many actual and potential health benefits. These are achieved through directly and indirectly addressing the causes of disease in preventative and curative ways, while also fostering broader synergistic benefits, which are not always apparent. Respiratory disease and diarrhoea are listed by the WHO as two of the top causes of mortality in Nepal, with double the prevalence in children <5 years of age (WHO 2006b). The "Family of 4" directly affects the environmental conditions that cause the spread and proliferation of these diseases. It is a direct preventative health measurement, especially if considered over a long-term period.

The Smokeless Metal Stove (SMS) and indoor lighting significantly decrease the indoor air pollution, particularly the particulate matter and carbon monoxide, providing improved air quality inside the house. Fine particulate matter air pollution is a known cause of acute and chronic respiratory conditions, contributing to excess mortality in the USA (Dockery 1993). Our recent study (Sect. 7.2) has shown that the incidence of respiratory symptoms decrease following the installation of a SMS and indoor lighting in the remote region of Humla. Children, particularly benefit from this intervention due to the low oxygen environment at high-altitude and the added risk of hypoxia associated with a respiratory infection (Niermeyer 2008). Besides acute respiratory infections, there are also the potential long-term effects associated with air pollution exposure such as incidence of Chronic Obstructive Airway Disease (Stockley 2007) and Low Birth Weight children (Gouveia 2004). The RIDS-Nepal interventions through the HCD program have the potential to significantly alleviate these conditions, and in turn decrease the region's mortality and morbidity rate.

The improvement of water quality and quantity, the construction of a pit latrine and non-formal education directly address what the UNICEF (UNICEF 2012) identifies as water, sanitation and hygiene (WASH). Recent studies have estimated that almost one tenth of the global burden of disease can be attributed to "water, sanitation and hygiene" (WHO 2007a). These simple and achievable interventions help to decrease the transmission of diarrhoeal disease caused through worm infestation and bacterial/viral infection. This improvement in access to clean water for domestic and personal hygiene, combined with provision of sanitation facilities decreases diarrhoea morbidity and mortality (Esrey 1991). Pokhrel (2004) indicated that Nepal, with its unique geographical, socio-economic and cultural characteristics, would benefit from simple environmental interventions, such as improved water supply and sanitation to reduce diarrhoeal diseases. Besides the acute effects of diarrhoea, the repeated infections can cause malnutrition and other micronutrient deficiencies with their compounding health effects.

The "Family of 4 PLUS" scheme further supports the need for food security and micronutrient diversity through the greenhouse, solar drier and nutrition programs. Through the provision of an extended growing season and an increased range of food availability, the initiative can provide a stable nutritional base for people over a range of ages. Malnutrition, although not always a direct cause, is well known for its association with increased mortality through the endangerment of vital body functions (Van den Broeck 1995).

By providing an improved environment in which a community can develop, combined with the education and technical support provided by RIDS-Nepal, the HCD project not only provides direct, but also indirect health benefits. The synergistic nature of environmental health improvements is not easily quantifiable due to the very nature of scientific research methods, which by and large aim to control these confounding factors to give very specific results. When viewing each intervention individually there is research that shows benefit. It only stands to reason that by decreasing the disease burden of conditions like diarrhoea and respiratory ailments and increasing the nutritional status through improved food security and education, one would expect a decreased rate of morbidity and mortality and an increased quality of life. A study by Pandey (1991) conducted in Nepal showed an unexpected significant reduction in deaths from diarrhoea and measles when treating children <5 years of age for pneumonia.

The overall benefits to the lives of the people and their environment through the "Family of 4" and the "Family of 4 PLUS" community development projects demonstrate that the whole is greater than the sum of its parts. As the long-term project unfolds increasing evidence will become available about the synergistic benefits.

7.6 The Role of Renewable Energy Technology in Holistic Community Development

As previously mentioned, Nepal has no fossil fuel resources, but is rich in renewable energy resources such as hydro power, solar energy, as well as wind energy and biomass, in some areas. As shown in Chaps. 2, 3, 4, 5 and 6, these abundant, locally available and often equally distributed renewable energy resources can be tapped into with locally developed technologies that the community has helped to choose and adapt to the local conditions.

It has been shown, that RETs have to be considered as a fully integrated part of a development program. The direct link between access to improved energy services and poverty alleviation, discussed in Sects. 1.1.6 and 7, Appendix 8 and Appendix 9, further substantiates the assertion that applied RETs must take a central, crucial role in community development projects. Locally applied RETs enable a community to become energy independent of any external providers. That supports the communities in their quest for development, enabling them to become responsible, proud owners and managers of their own energy system(s). Further, a RET system contributes to the solution to global warming, making remote communities carbon emission free, or neutral in the case of a biomass based energy system. This is worth mentioning, considering the rather fragile and unique environment of the Himalayas.

Therefore, these are more than enough reasons to prove that contextualized RETs can meet the identified need for energy services of individual families as well as whole villages. This process empowers impoverished communities to go forward on the path of holistic development within their local context.

7.7 Conclusions

This chapter aimed to highlight some of the measurable changes that implemented HCD projects have thus far achieved. The various tools (Base-Line and Follow-Up surveys) developed to gather data and anecdotal information were used, to first understand in detail the initial circumstances and conditions of a village, as well as to monitor the ongoing changes and impacts the HCD programs will produce over the years to come.

While the HCD programs are a set of defined projects, enabling life improving changes, the outcome of each HCD project is strongly based on the active participation of the village partners. Thus there will not be a defined set of outcomes and benefits in each case. Their aim is to increase the understanding of the development process needed for the local village partner in its defined context, the benefits achievable and the concept of sustainability for each HCD program. The HCD programs of the "Family of 4" and the "Family of 4 PLUS" are defined to address the villagers' identified physical needs and a dynamic process of a close, active and long-term interaction and relationship between all the project's stakeholders is important. This is true for the end-users', the people of the village in question, and the partnering project organisation, in this case RIDS-Nepal. This relationship has to be established and nurtured, for the HCD program to succeed. In the rural communities it is still the case that unless relationships are intact, and trust among the parties is strong, that even the best of technologies will fail to bring the expected changes. This makes it essential that the necessary bonds between project partners are taken seriously and consequently intentionally nurtured. RIDS-Nepal's staff live in the area where the village based HCD programs run. They work hand in hand with the villagers during all the HCD project phases, from the initial Base-Line survey, during the long period of project implementation through to the periodic Follow-Up surveys. This facilitates the establishment and maintenance of a strong bond with the local people.

Without the technical components of a HCD program, the needed changes can not take place. As indicated above, the building and maintaining of relationships among project stakeholders takes effort, time and investment. Likewise, as seen throughout the thesis, energy and effort has to be invested into the design, development, implementation and monitoring of the projects. Sound engineering principles, in the context of an awareness of the anthropological background of the end-users, are mandatory in order to develop appropriate, contextualised designs of technologies.

Further, it can be concluded that the studies conducted over the last two decades (Sect. 1.1.7) about the direct relationship between poverty alleviation and access to improved energy services agree with the findings of the author during RIDS--Nepal's decade long project implementation and monitoring. There is a direct relationship between poverty and access to clean, improved energy services, and people and communities can actively start their process of liberation from poverty towards more dignified living conditions as a result of one to two decades of holistic community development in which improved access to energy services is highlighted and emphasised.

Consequently it can be said that for communities with ample renewable energy resources, such as the high-altitude villages of Nepal, contextualised RETs can play an important role, within a long-term HCD program, in providing the people with an opportunity to escape the vicious cycle of poverty. Improved health conditions, hygiene, increased food availability and nutrition as well as regrowth of their local biomass resources are all benefits the people directly experience in their family units and local community. Further, the local RET power plants empower the community to strive for energy independence and security, another important result, especially in the fragile and unique environment of the Himalayas, where the remote and high-altitude communities of Nepal are situated.

References

Amitai, Y., Zlotogorski, Z., Golan-Katzav, V., Wexler, A. & Gross, D. (1998). Neuropsychological impairment from acute low-level exposure to carbon, monoxide. *Archives of Neurology*, 55, 845–848.

Cook, D., & Strachan, D. (1997). Health effects of passive smoking. 3. Parental smoking and prevalence of respiratory symptoms and asthma in school age children. *Thorax,52*, 1081–1094.

Dockery, D. W., Pope, C. A., Xu, X., Spengler, J. D., Ware, J. H., Fay, M. E., Ferris, B. G. & Speizer, F. E. (1993). An association between air pollution and mortality in six US cities. *The New England Journal of Medicine*, *329*(24), 1753–175.

Edwards, R. (2004). ABC of smoking cessation. The problem of tobacco smoking. *British Medical Journal,328*, 217–219.

Esrey, S., Potash, J., Roberts, L., Shiff, C. (1991). Effects of improved water supply and sanitation on ascariasis, diarrhoea, dracunculiasis, hookworm infection, schistosomiasis, and trachoma. *Bulletin of the World Health Organization, 69*(5), 609–621. World Health Organization, Geneva.

Forbes W.H., Sargent, F., & F.J.W. (1945). The rate of carbon monoxide uptake by normal men. *American Journal of Physiology, 143*, 594–608.

Gouveia, N., Bremner, S.A., & Novaes, H. M. D. (2004). Association between ambient air pollution and birth weight in São Paulo, Brazil. *Journal Epidemiol Community Health, 58*, 11–17.

Hamer, M., Wolvers, D., & Albers, R. (2004). Using stress models to evaluate immunomodulating effects of nutritional intervention in healthy individuals. *Journal of the American College of Nutrition,23*(6), 637–646.

Hoffman-Goetz, L., & Klarlund Pedersen, B. (1994). Exercise and the immune system: a model of the stress response? *Trends in Immunology,15*(8), 382–387.

Lawther, P. J. (1975). Carbon monoxide. *British Medical Bulletin,31*(3), 256–260.

Longo, L. D. (1977). The biological effects of carbon monoxide on the pregnant woman, foetus and newborn infant. *The American Journal of Obstetrics and Gynaecology.,125*(1), 69–105.

Mann, J. (2009). *The impact on respiratory ailments through reduction of Indoor Air Pollution by smokeless metal stove and solar lighting: Results of a study in rural Nepal.* Masters of International Health Dissertation, Curtin University of Technology, Western Australia, and Alexander Zahnd RIDS-Nepal (Nepal based Supervisor).

Moore, David, S. (2007). *The Basic Practice of Statistics* (4th Edition). Purdue University, New York, USA: W.H. Freeman Company

Niermeyer, S., Mollinedo, P., & Huicho, L. (2008). Child health and living at high-altitude. *Archives of Disease in Childhood,2009*(94), 806–811.

Pandey, M. R., Daulaire, N., Starbuck, E., Houston, R., & McPherson, K. (1991). Reduction in total under-five mortality in western Nepal through community-based antimicrobial treatment of pneumonia. *The Lancet,338*(8773), 993–997.

Pokhrel, D. & Viraraghavan, T. (2004). Diarrhoeal diseases in Nepal *vis-a-vis* water supply and sanitation status. *Journal of Water and Health*, 71–81.

Pope, C., Bates, D., & Raizenne, M. (1995). Health effects of particulate air pollution: time for reassessment? *Environmental Health Perspectives,103*(5), 472–480.

Reddy, A. K. N. (2002). Energy technologies and policies for rural development. In: T. B. Johansson & J. Goldemberg (Eds.), *Energy for Sustainable Development: A Policy Agenda*. New York: UNDP, Ch. 4, pp. 115–36.

Saghir, J. (2005). *Energy and Poverty: Myths, Links, and Policy Issues.* Energy Working Notes, World Bank: Energy and Mining Sector Board, No.4 May, 2005

Schirnding, Y., Bruce, N., Smith, K., Ballard-Tremeer, G., Ezzati, M. & Lvovsky, K. (2002). *Addressing the impact of household energy and indoor air pollution on the health of the poor:*

implications for policy action and intervention measures. Retrieved July 19, 2012, from http://www.who.int/mediacentre/events/H&SD_Plaq_no9.pdf
Sopori, M. (2002). Effects of cigarette smoke on the immune system. *Nature Reviews Immunology,2*, 372–377.
Stockley, R. (2007). COPD and air pollution. *Chronic Obstructive Pulmonary Disease* (pp. 397–407). Blackwell publishing Ltd: Oxford.
UN. (2011). *Human Development Report 201—sustainability and equity: A better future for all*, p. 129. Retrived July 24, 2012, from http://hdr.undp.org/en/media/HDR_2011_EN_Complete.pdf
UNICEF. (2012). *Water, Sanitation and Hygiene*. Retrived July 24, 2012, from http://www.unicef.org/wash/
United States Environmental Protection Agency (USEPA). (2009). *National Ambient Air Quality Standards (NAAQS)*. Retrieved September 27, 2009 and July 24, 2012, from http://www.epa.gov/air/criteria.html
Van den Broeck, J. (1995). Malnutrition and mortality. *Journal of the Royal Society of Medicine,88*(9), 487–490.
Wald, N & Hackshaw, A. (1996). Cigarette smoking: An epidemiological overview. *British Medical Bulletin, 52*(1), 3–11.
WHO. (2000). *Air quality guidelines for Europe; Second edition*, World Health Organization, Regional Office for Europe, European series No 91, Copenhagen, Denmark, ISBN 13583. Retrieved July 24, 2012, from http://www.euro.who.int/document/e71922.pdf
WHO. (2000a). *Addressing the Links between Indoor Air Pollution, Household Energy and Human Health.* Washington D.C.:World Health Organisation 2002
WHO. (2006a). *Fuel for life: household energy and health.* Switzerland: WHO Geneva.
WHO. (2006b). Mortality Country Fact Sheet 2006. WHO website, http://www.who.int/whosis/mort/profiles/mort_searo_npl_nepal.pdf (accessed November 1, 2011, on July 24, 2012 not available any more).
WHO. (2007a). *Indoor air pollution: national burden of disease estimates*. Geneva:WHO press. Retrieved July 24, 2012, from http://www.who.int/indoorair/publications/indoor_air_national_burden_estimate_revised.pdf
WHO. (2007b). *Indoor air pollution and lower respiratory tract infections in children.* Geneva:WHO press. Retrieved July 24, 2012,from http://whqlibdoc.who.int/publications/2007/9789241595728_eng.pdf
Winrock (2004). *Household Energy, Indoor Air Pollution and Health Impact—Status Report for Nepal*, Winrock International Nepal, Kathmandu, Nepal. Retrieved July 24, 2012, from http://www.pciaonline.org/files/Nepal_Household_Energy_and_Health_Overview.pdf

Chapter 8
Conclusions

8.1 Summary of Findings

The **primary objective** of this work was explained in Sect. 1.3.2. It is to design, implement and evaluate an integrated approach to holistic community development, called the "Family of 4" and the "Family of 4 PLUS", in high-altitude areas of Nepal.

The **research questions** which I set out to address were discussed in Sect. 1.3.1 and were stated as:

- *What is the appropriate role of renewable energy technology in holistic community development?*
- *Does the newly developed holistic community development concept of the "Family of 4" and the "Family of 4 PLUS" provide a sustainable way of improving the quality of life for remote and impoverished mountain communities in Nepal?*

My **thesis** is that this holistic approach to community development in the high-altitude communities of Nepal provides significant, cost effective and sustainable benefits that go beyond those provided by conventional approaches to development.

In the preceding chapters I have attempted to describe the methods used in holistic community development and I have evaluated the outcomes of projects based on this approach. The main findings of this research are presented below.

(1) The importance of **learning and understanding of the local culture**, in the context of its history, has been discussed. A country such as Nepal, with its vast diversity of ethnicity, environment, climate, politics, poverty, energy resources and belief systems demands years of study from newcomers to come to grips with the way things work. In order to pay adequate respect to these issues, the thesis has highlighted in Sects. 1.1, 3.1, 6.1 and 6.2/ Appendices 5, 6, 7, 8 and 11 matters the author has learned by experience are important for understanding these people. In RIDS-Nepal, all foreign

A. Zahnd, *The Role of Renewable Energy Technology in Holistic Community Development*, Springer Theses, DOI: 10.1007/978-3-319-03989-3_8,

volunteers are put through an induction course in their first 2 weeks, which includes learning the basic language skills, reading Dor Bahadur Bista's book "Fatalism and Development" (Bista 1991) and watching a video[1] about the way of life and main belief system people adhere to in the area where they will participate in the project work.

(2) As a result of this research, it is clear that there is a **need to carefully define the meaning and practical implementation of "sustainability" and "development" in the context of Nepal.** As mentioned in Appendix 10.3, the concept of "sustainability" originated in the context of harvesting and managing renewable resources in such ways that they do not deplete, but will be available for future generations as well. Thus it is vital to maintain the ecological balance for survival, within the boundaries of the earth's ecosystem and the social and political framework of society. Also, it has become clear that equating "development" with ongoing economic and material growth is a physical impossibility within the realm of the earth's ecosystem, due to the earth's finite resources and limited boundaries. Thus we need to reconsider the definitions of these words, and to redefine them with an understanding that it is crucial to maintain the ecological balance, while striving for societal and political changes. A reorientation from abstract definitions of sustainability and development, worked out by international committees, to more practical project oriented forms, is needed. The HCD concepts of the "Family of 4"/"Family of 4 PLUS", with the contextualised renewable energy technologies embedded in them, has been shown to be a constructive step towards this goal.

(3) **Selective, or single component projects, though effective in achieving defined goals, cannot produce synergistic benefits** unless they are simultaneously and dependently implemented alongside each other. Thus, the comprehensive HCD approach, as presented in the thesis through the "Family of 4"/"Family of 4 PLUS" maximises each project's benefits and enables synergistic benefits. Chapter 2 argues and supports this from a historical and conceptual point of view, while Chaps. 3–5 provide details of

[1] In 1981 the 4½ h long film: "Shamans of the Blind Country", directed by the German filmmaker Michael Oppitz was filmed. The film is a remarkable anthropological documentary on Shamanism in the Nepal Himalayas and on their magical healing. The film explores in detail the séances and practices of Shamans in the secluded society of the Northern Magar in Central West Nepal, west of one of the >8,000 m high Dhaulagiri mountain group. It focuses on the sumptuous rituals performed by the Magar shamans during their night long séances. Their methods of diagnosis and treatment, their techniques of possession and their ritual journeys, undertaken to recover the fugitive souls of their patients, are all filmed and presented through their symbolic language of signs and gestures. It shows the transmission of the shaman's profession, following successive tests of aptitude and initiation rites, a shaman (male or female) is born on a conifer tree, the tree of life, during a lavish 3 day ceremony. These shamanistic rituals and séances are practiced in a very similar way by the shamans in Humla. The author has a copy of the film on DVD and thus it is part of each volunteer's introductory 2 week orientation program when they come to work with RIDS-Nepal for a defined length of time.

how this is done practically, through RIDS-Nepal, among the mountain communities in Nepal. Chapter 7 discusses some of the scientific and anecdotal data on HCD projects, their impact and changes thus far achieved. It concludes that **contextualised RETs play a key role, within a long-term HCD program**, in providing the people with opportunities to access improved health conditions, hygiene, increased food availability and nutrition. It also facilitates the regrowth of their local biomass resources, and thus provides opportunities for them to escape the vicious cycle of poverty.

(4) While many good community development and power generation projects are implemented, **there is almost a total lack of data in regard to a project's performance or impact achieved over time**, based on scientific monitoring or anecdotal records. This is a serious, widespread deficiency of most development projects. Sections 3.11 and, 3.12 are dedicated to addressing this question, presenting some examples of how, under the leadership of the author, RIDS-Nepal has developed **detailed scientific data monitoring** of some of its implemented village power generation projects (solar PV and pico-hydro). Recorded data, presented in graphical form in Chap. 3 (Figs. 3.110, 3.114, 3.119, 3.120, 3.121, 3.122, 3.127, 3.128, 3.130, 3.131 and 3.137) show the main environmental parameters, all the major technical parameters of the functional solar PV systems, as well as the losses incurred by a solar PV system due to non-ideal conditions. All of these parameters allow a detailed performance analysis. This provides an understanding of the systems' long-term functionality, down times and actual performance, and whether they can provide the end-users with their expected energy services. With real field data available, the initial design of a village power plant can be reassessed and necessary modifications of the design approach can take place. In this way, due respect is paid to the local environment and cultural context in which the power plants run. This in turn provides a learning experience which cannot be found in books or taught in academic institutions, as each context is unique. However, over time, theory and practice jointly provide the necessary experience and skills to design and teach about more sustainable power generation plants for a defined end-user's context. Since January 2010, after 2 years of development, the databank and adjustments of the open source graphical presentation software, detailed system information from twelve different monitored projects',[2] concepts,

[2] The RIDS-Nepal databank is available at: http://www.rids-nepal.org/databank/. A total of 12 functional projects are monitored in great detail to understand and determine their performance, short-comings and area of improvements needed. The 12 monitored projects are as follows:

(1) Simikot High-Altitude Research Station solar PV system, monitored since March 2004
(2) Simikot High-Altitude Solar Water Heater Bathing Centre, monitored since November 2009
(3) Simikot High-Altitude Greenhouse, monitored since April 2007
(4) Tulin Central Village 2-axis Solar PV Tracker system, monitored since November 2006
(5) Darapori village RIDS-Nepal field office solar PV system monitored since November 2008
(6) Darapori village Solar PV Home System, monitored since April 2007

pictures and recorded data are available for registered users of the RIDS--Nepal databank site (http://www.rids-nepal.org/databank/). The available data and information allows researchers and community development project designers of other local and international NGOs/INGOs to understand how these systems work in a defined context, user operation and management. The author also uses the databank as part of his renewable energy course (Appendix 27) at the Kathmandu University. It allows the teaching of students through actual field examples and the solving of assignment questions which are based on actual field data.

(5) **Without a detailed understanding of the local context, RETs usually fall short of their expected performance** (Sects. 3.6, 3.7 and 3.8). But with a thorough insight into a country's historical setting and cultural background, together with practical experience of applied RETs, the preconditions for the development of contextualised RETs is provided. The author, mainly through RIDS-Nepal and the Kathmandu University, has developed (since 1997) various, contextualised technologies, plus educational materials and infrastructure, with several of them now provided with a government subsidy to make them more affordable for the wider public. Among them are: the smokeless metal stove (50 % manufacturing cost subsidy); the solar drier (50 % manufacturing cost subsidy); the pico-hydro power plant, without the use of cement (context related government subsidy available); the high-altitude solar water heater thermo-siphon technology and bathing centre (context related government subsidy available); the slow sand water filter; the high-altitude greenhouse; various solar PV system concepts with WLED lighting (context related government subsidy available); and Non-Formal Education teaching and workbook materials. Table 7.1 provides a summary of the various installed projects up to June 2012. All of these newly developed technologies were first installed and field tested in RIDS-Nepal's High-Altitude Research Station (HARS), at 3,000 m altitude in Simikot. Only after the initial problems and shortcomings have been identified and overcome during the HARS field tests, are these technologies considered suitable for requested village projects. In this way, possible technical and non-technical problems are minimised. The necessary local maintenance and basic repairs training and the essential communication tools for the project implementer can also be defined at this stage. Through this joint effort and

(Footnote 2 continued)

(7) Darapori High-Altitude Solar Water Heater Bathing Centre, monitored since October 2009

(8) Darapori High-Altitude Greenhouse, monitored since April 2009

(9) Pamlathum village Solar PV cluster system, monitored since November 2006

(10) Kholsi village Pico-Hydro Power plant, monitored since March 2009

(11) Kathmandu RIDS-Nepal main office Meteorological (global solar radiation measurement on three different solar PV technologies), monitored since March 2009

(12) Kathmandu RIDS-Nepal main office Battery Testing Research project, monitored since July 2010, but suspended in September 2011 to improve the battery measurements.

time consuming consultation processes with the community, development projects can be successfully acculturated over time by the local communities. The GIS tool provides each village that RIDS-Nepal has been requested to partner with for a long-term HCD program, with information about the various projects implemented and in operation.

(6) **Local development and manufacturing of the required RETs** in the country where they are to be implemented, is important for the following reason. It **minimises imports and maximises local manufacturing**. It enables the local industry to learn new skills, expands their line of products, increases the local knowledge, creates local jobs and increases their income possibilities. In regard to the new technologies mentioned above in item 5, many new jobs have been created due to local manufacturing in Nepal. The author estimates that the following numbers of new jobs have been created:

- smokeless metal stove—10 government approved companies are qualified to manufacture the stove with a government subsidy nation wide: ~200 jobs;
- the solar drier: ~10 jobs;
- pico-hydro power plant: ~20 jobs;
- the high-altitude solar water heater system: ~20 jobs;
- the slow sand water filter: ~10 jobs;
- high-altitude greenhouse: ~10 jobs;
- various solar PV system concepts with WLED lighting: ~30 jobs.

Thus, all in all about ~300 new jobs have been created through the new, locally developed RETs, enabling local Nepali craftsman to earn a living to support their families and send their children to school. In addition, one RIDS-Nepal spin off company, Pico Power Nepal (PPN) has been established as a result of the various solar PV system developments. PPN manufactures whole village solar PV electrification systems and equipment, either under a contract for designed systems, as with RIDS-Nepal, or as a consultant for other local NGOs/INGOs. PPN employs on average 5 local staff.

(7) **RAPS systems (Sects. 3.2 and 3.3), powered by renewable energy sources (Appendix 11), are an appropriate and sustainable way to provide remotely located communities, with basic but improved energy services**. Minimal access to electricity is necessary for basic indoor lighting through WLED lamps (Figs. 2.4, 2.5, 2.6, 3.6, 3.15, 3.62, 3.96 and 3.99). This is available by tapping into and converting the local renewable energy resources, which provide a high degree of energy security and independence from outside authorities. This is demonstrated by the author's own load measurements of the Kholsi village pico-hydro power plant, inaugurated on the 6th December 2006, which generates ~600 W AC. The average load for all 230 WLED lamps (each consuming 1 W), in 73 households (3 WLED AC lamps per home), the local school's 7 WLED lamps and the 4 WLED "street" lamps (each consuming 5 W) during the night, was between 160 and

200 W in March 2007. However, the average consumption measured in May 2012 was between 460 and 480 W, or >2½ times the March 2007 power consumption. This shows that the local people have learned rather quickly what benefits even minimal access to electricity can provide. The consumption rose by 260 % within 5 years. Thus, today the WLED lights are no longer the only electricity load. Rechargeable batteries for torches and small radios are also charged, as well as an increasing number of mobile phones, which are plugged in for charging.[3] One man brought from his trip to nearby Tibet/China, to trade goods, a CD player and a small monitor, to run a home video theatre. He charges a minimal amount for people to come and watch a Hindi movie on CD, powered by the pico-hydro power plant. If asked, the people would not want to return to the pre-electrification times, when "*jharro*" (Figs. 2.3, 2.15, 3.3, 3.81 and 3.101) was the traditional and only means of lighting their homes. The energy independence, and energy security, obtained through a locally run and maintained pico-hydro power plant, means that their lives have changed for the better. Thus the energy security and independence gained through utilising and converting the available local renewable energy resource through a basic, simple to use and easy to maintain RET, is a vital catalyst for sustainable development and improved living conditions.

(8) Many single-pronged projects address the urgently needed health improvements of remote and impoverished communities through curative actions, such as vaccinations, yearly health camps for eyes (cataract operations), lung diseases, dentistry, de-worming or malnourished children. Important and good as they are, these curative actions address the consequences of an initial problem or cause, such as indoor air pollution, unhygienic conditions, consumption of contaminated water or the lack of nutritious food. They do not address or eradicate the source of the problem. **With access to improved energy services**, such as a stove or a solar cooker, basic electric indoor lighting, water heated through solar energy, more food grown through a greenhouse and dried with a solar drier, **the root causes of poor health, can be addressed in preventative ways**. Data collected and presented in Chap. 7 show that even 1–2 years after the installation of the "Family of 4", addressing root causes of the problems, people's general health conditions in regard to respiratory diseases (Tables 7.10, 7.11 and 7.13), eye diseases (Table 7.12), diarrhoea (Table 7.4) and intestinal worms (Table 7.5) have improved. People have become more aware of their problems (Tables 7.6, 7.8 and 7.15), and life has seen positive changes (Table 7.7). This confirms the importance of basic energy services for sustainable community development projects.

[3] Kholsi village is so remote that there are no mobile phone services available. However, cheap mobile phones have become increasingly a status symbol. Besides that, people do not have the financial capacity to pay for any mobile phone services. Thus the mobile phones are used mainly to display a status symbol and to play music.

(9) Most of the recently developed technologies (smokeless metal stove, solar drier, high-altitude solar water heater, slow sand water filter, high-altitude greenhouse, various solar PV system concepts) either emerged as an idea initially by the author, or were tested and further developed by the author through an academic institution such as the Kathmandu University (KU) in Nepal. This was because the author has been teaching a renewable energy course at KU since 2001, as well as developing and running students' renewable energy research projects. Since then, a stove and a solar energy (solar PV and thermal) testing laboratory have been built at KU, where students can convert their ideas into prototypes which can be tested for their functionality, efficiencies and appropriateness for the local context. All the prototypes are manufactured in partnership with the local industry. The students participate in the initial prototype manufacturing, bringing their ideas to the local producer, who with his practical experience turns a technical drawing into a prototype. The students, often unskilled in manual work, learn various new manufacturing procedures, while the manufacturer can add a new product to his line of work. As a result, an important and vital relationship between the student and the local industry is established, as from the day of the student's graduation onwards he is expected to work closely with the local industry. Now that more appropriate technologies are available on the local market, it is possible to place an order for the needed technology with a local manufacturer. In a joint effort, the local community and the NGO design and implement a HCD project, with the appropriate technologies and infrastructures in mind. Where the government has approved of the new technologies and has developed a policy to provide a subsidy,[4] projects become much more feasible with regard to finance, which would otherwise have to be raised partly by the local community and the NGO. Thus, **the link between the initial idea (University), the manufacturer (local industry), the NGO (RIDS-Nepal) as the project facilitator, the end-user community and the government (as subsidy provider), is established.**[5] In a country like Nepal, stricken by poverty,

[4] In 2011, one of the author's 7 research assistants completed a MSc dissertation in the field of: "Evaluation and Analysis of the Impact of Subsidies on Small Scale Renewable Energy Technologies Dissemination: A Case Study of Nepal", including and highlighting the importance of providing subsidies for holistic community development projects such as the "Family of 4" in the context of Nepal and its government.

[5] In this way, an initial idea, brought forth by an individual working in the field or at a University, may be explored, developed, tested and refined, with the final outcome being able to be manufactured by the local industry on a larger scale for quality and affordability. The NGO bridges the gap between the remote communities and the local industry, allowing the new technologies to have an entry point into the local communities. The local people, partnering with the NGO, design in mutual consultation the village based projects, which, if the government policies are in place, can be jointly implemented under a relevant subsidy scheme. In this way an important, thus far missing "link", the relationships between identified need assessments in the field, academic institutions (as the "brains" with the theoretical knowledge to develop new, contextualized technologies up to the prototype stage), local industries (to mass manufacture the

this conscious process of establishing relationships between the various stakeholders, in order to address their widespread needs, is of vital importance.

(10) Since 2002, the author has had 7 research assistants, all KU graduates. Each of them participated in the above process of developing and implementing new technologies for rural areas. Each also worked in the remote, high-altitude mountain communities with RIDS-Nepal and the local communities in the implementation of HCD projects as part of their 3 years research assistants' program. Following this, the author helped each of them to enrol in a Masters Degree program in RET and development in an international University. Experience shows that all of them, after having left Nepal for their further studies, have kept close contact, and have kept the vision alive, of coming back and participating in the development of their own country, despite the very difficult political and economic conditions in Nepal, since the mid 1990s. While 3 of them are still enrolled in their studies abroad (2 in an MSc and 1 in a PhD), three have expressed their desire to gain 2–3 years experience of working in a relevant industry abroad, and 1 is back in Nepal, working as a renewable energy and project consultant. Thus, the author's 3 year research assistant program, with 2 years working alongside him at the KU, teaching and running student research projects and 1 year working alongside the RIDS-Nepal staff and the local communities in the remote mountain areas, aims to overcome the common trend of Nepal's "brain drain". Many of today's undergraduates verbally express that they only intend to finish their basic studies in Nepal and then leave for further studies abroad, with the aim of never coming back. Considering all the difficult circumstances of Nepal's past and future, the author is convinced, based on the ongoing communication with his research assistants, that **a successful repatriation of Nepali professionals is possible** if their initial training at the undergraduate level is relevant, and a passion and a realistic vision for their own country has been implanted in their minds and hearts during their first few years of working as professionals.

(11) In order to highlight the long-term life and living condition improvements possible through improved access to basic indoor lighting, such as WLED lamps (Appendices 12, 12.5 and 12.6.1), the author examined (in Sect. 3.13) the possible **HDI improvement** through elementary indoor lighting over the course of one decade. The study concluded that a 60 % HDI improvement is

(Footnote 5 continued)
new approved technologies at affordable costs), indigenous NGOs (as direct project implementing partner of the local communities), INGOs (as main donor and funding partners), local communities as the end-users and the government (providing the available subsidies for the various technologies/project parameters), is established. Once this chain of relationships is established and active, it allows projects to take place within a set legal framework, which allows projects to take a life on their own from start to implementation and operation. In addition it assists the development of a country on a much broader basis. Projects are "vertically" (in regard to projects initiated from the bottom-up (community) and top-down (enabled through the government policies), as well as "horizontally" networked, uniting the different, and thus far very much socially and ethically dichotomized, communities.

feasible. While the base-line data used for the HDI calculation is based on the government data available at the time of calculation, the data for the possible changes after a decade are based on the author's and RIDS-Nepal's own data collection and experience. Considering the data presented in Chap. 7 (Tables 7.2, 7.4, 7.5, 7.7, 7.10 and 7.11), which were not available at the time of the HDI calculation, the average life-expectancy of the people in the communities with a running HCD project can be conservatively estimated to have increased by 10 years, from 54 to 64, or even more, considering the current data. This is because the **increased life-expectancy** is influenced not only by the electric indoor lighting, making the traditional smoky and dim "*jharro*" lighting obsolete, but as well by the cleaner indoor air through the smokeless metal stove; the cleaner environment in and around the home/village through the use of pit latrines; the cleaner water people consume from the tap stands in the village; and the slow sand water filters in their homes. In addition, the availability of warm/hot water for better personal hygiene and the increased availability of nutritious food from the greenhouses are also contributing factors. Further, we have seen a **doubling of both, the adult literacy rate**, especially among women who often enthusiastically participate in the NFE classes, **and the number of children enrolling in government schools** after their initial experience with the RIDS-Nepal NFE out-of-school children's classes. In some communities this figure is expected to be even higher. Also the assumed **per capita income increase of 40 %** over a decade, in the initial HDI calculation, is rather on the conservative side, as some families with current HCD projects have seen a tripling or more of their income through growing vegetables in their greenhouses and selling them on the local market during seasons when they were otherwise not available. In addition, the decreased amount of firewood families need to collect or purchase adds to their savings, as does the decreased expenditure for medicine for de-worming, recurring asthma attacks, diarrhoea and other indoor air pollution and malnutrition related diseases. Thus, with increased income and less expenditure, it is more likely than not that the per capita income increase is >40 %. Consequently it can be said with increased confidence, that the HDI improvement, among a community with a running and functional HCD project, after a decade, is at least 60 % or greater. Hence, it shows that making use of the locally available renewable energy resources, converted and applied through context related RETs, plays a central role in a long-term HCD project.

8.2 Contributions, Benefits and Beneficiaries of this Research

As previously mentioned, this thesis does not involve the formulation of a new theoretical community development concept in which applied RETs will have a central role. Rather, it involves the detailed evaluation of applied HCD concepts, developed by the author since 1996. With these new HCD projects in place, lives are being transformed and, over the course of the years, gradually more visible and measurable impacts will be seen. In the following, some of the direct benefits and beneficiaries are summarised and highlighted.

(1) The **local Humla communities** are surely the most direct beneficiaries of the HCD projects. As shown in Chap. 7, already 1–2 years after the "Family of 4" HCD projects have been implemented, the measurable scientific data, and the anecdotal information extracted from personal interviews with the end-users, show unambiguous improvements in their living conditions, due to the project's direct benefits. Undoubtedly, the long-term synergistic benefits of the HCD projects, which will become more apparent over the course of the years, will not only enhance and increase the benefits, but will be a significant part of the improvements in their overall life and living conditions.

(2) Since the author became intensively involved in community development in 1996, the following **new contextualised technologies have been developed, tested and are now available through the local market or local industry**, with some now supported by a government subsidy (GS):

- the smokeless metal stove (Sects. 2.10.2 and 4.1) with GS;
- three different solar PV system approaches to electrify rural homes (Sects. 2.10.3 and 3.3.2) with GS;
- the pico-hydro with no cement (Sect. 3.3.1.1) with GS according to the context;
- WLED lamps (Appendices 12.1, 12.5 and 12.6);
- solar PV system charge controllers (Sect. 3.9.4) with temperature adjustment, electronic fuse and MPPT (from 2011 onwards);
- solar PV system frames for the three different village electrification approaches (3.3.2);
- greenhouse for high-altitudes (Sect. 4.2);
- solar drier (Sect. 4.3) with GS;
- high-altitude solar water heater bathing centre (Sect. 4.4) with GS according to the context;
- slow sand water filter (Sect. 2.11.5);
- hand oil expeller (in development and testing stage in 2011/2012);
- hybrid solar PV/small scale wind turbine RAPS system for village electrification (in development and testing stage in 2011/2012).

(3) In contrast to the prevailing models of community development (Sects. 2.1, 2.3–2.5), which are most often individual, single project oriented, development projects, the new HCD concepts **of the "**Family of 4**" and the "**Family of 4 PLUS**" look at and address the wider issues of the local communities' own identified needs in a more holistic way**. The author's experience, having worked with 30 village communities in the remote high-altitude areas of north-west Nepal since 1996, shows that the "Family of 4", with its 4 pillar projects (pit latrine, smokeless metal stove, indoor lighting and clean drinking water), addresses 80–85 % of the four most often stated needs of the local

people. Thus the author has amalgamated these four projects and promotes them through RIDS-Nepal to be implemented through a joint effort with the local community. The time frame of a "Family of 4" HCD project is usually 2 years until each family has each project implemented in their home (such as the pit latrine, the stove and the indoor lighting). In addition, they participate in the designing and building of the village owned drinking water system, to which each family has access. While the "Family of 4" HCD concept is rather rigid, in the sense that each of the four individual projects has to be implemented alongside each other, the "Family of 4 PLUS" is more flexible and adaptable to the additional needs recognised by the local communities. Before July 2012 the "Family of 4 PLUS" included 8 project pillars (Sect. 2.11). However, only the ones the local village community identifies as their additional priority needs are implemented over the course of the following years. Thus, the "Family of 4 PLUS", in comparison to the "Family of 4", is a more adaptable concept, in that it can leave out projects if they are not seen as a priority need in the village in question. Also a new project, again identified and contextualised for its context, can be added if a new need is recognised. Thus **these two new HCD concepts provide novel ways of meeting the local community's needs**, with an understanding that people usually have multi-faceted needs. They enable us to hear the people's full stories and thus pay due respect to their circumstances and needs, as well as bringing forth synergistic benefits which will enhance the impact and long-term sustainability of a project. They permit the donors' financial support to be utilised to the maximum, while the end-users' voices are heard and their participation is active. Increased long-term sustainability of projects is likely, as the individual projects are contextualised, manufactured by the local industry, able to be operated and maintained by the local users and with increasing support and participation of the government. It provides opportunities at each level, enabling each project stakeholder to be a better steward and manager for the given tasks and responsibilities.

(4) Besides the new HCD concepts and the new contextualised RETs, an intensive effort has been devoted to developing and making available **new educational materials/tools and opportunities**,[6] such as: the interactive RIDS-Nepal web site at: www.rids-nepal.org, the NFE books (Sukhi Pariwar 1, 2 and Mahilo Ko Kalam 1, 2 as shown in Appendix 17; and interactive games (NFE Games) played during the NFE classes); various awareness raising and teaching videos (Appendix 16); posters (Appendix 22), flip carts (Appendix 20) and brochures (Appendix 19) for awareness raising and teaching; songs which address the HCD concepts the "Family of 4" and the "Family of 4 PLUS" as well as some social issues, such as the problems and effects of the misuse of alcohol and

[6] Most of the new developed educational materials, such as awareness and teaching videos, NFE teaching and reading books, conference papers and publications, dissertations co-supervised by the author are available and can be downloaded from the RIDS-Nepal web site at: http://www.rids-nepal.org/index.php/Downloads/ (accessed July 27, 2012)

cigarettes (Appendix 23). These are written in the local dialect so that the illiterate women can sing the songs during their field work. Other materials include: project related Bachelor and Masters Dissertations (Appendix 24); solar PV O&M teaching course and manual (Appendix 18); Base-Line and Follow-Up survey questionnaires (Appendix 14); a PowerPoint presentation explaining RIDS-Nepal's HCD and 3 Tier working principle with project examples (Appendix 29); data monitoring system documentation (Appendix 30). A RET course (developed for the context of Nepal and taught by the author) at the Kathmandu University since 2005 (Appendix 27). A scholarship scheme was established to enable up to 4 young men/women per year from Humla to learn a skill through apprenticeship training in the Karnali Technical School (KTS) (4–6 days walking distance from Humla in the district of Jumla). These educational tools are important alongside the more practical project implementations as they fulfil an essential HCD project supportive role in **providing increased awareness of the necessary skills** for the local people's village based projects. Further, through the **advanced theoretical and practical training** at the KTS, future local leaders are trained through relevant, context related and often hands on education.

(5) The manufacturing of the developed RETs and educational materials by the local industries has resulted in new jobs, new skills, wider and more context related product ranges, with increased income for local communities. These are all important factors for a long-term sustainable approach to development of a country with specific identified needs and shortcomings. RIDS-Nepal aims to enable **a development approach which is more rooted in the country's own culture and context in a sustainable way**, minimising the import of equipment and skills, by educating and empowering the local people and end-users. This approach provides an opportunity for the government to see, understand and support the development activities through relevant policies and subsidy programs. This allows a bottom-up development to take place, within the local context and with local people, working and striving for better living conditions in their own country and among their own people.

(6) The SMS, the solar PV home systems, the solar drier and the solar water heater have been recognised by the government as suitable and appropriate for the high-altitude remote mountain communities. Thus, **several recently developed and locally manufactured RETs have been able to receive subsidies from the Nepal government** to be implemented in HCD projects with mountain communities.

(7) The positive results of monitored and recorded changes in the local communities in Humla, through implemented HCD projects, convinced the author **to make this new approach of holistic community development more widely known among academics, community development project developers and practitioners, nationally as well as internationally.** Hence, in order to disseminate the experience gained thus far, twenty two peer reviewed conference papers and journal publication on these topics have been written since 2005. These publications have been written by the author in collaboration with

professional partners who have participated actively with the author and RIDS-Nepal in working with and monitoring of projects in Humla. In addition, the author gives on an average 8–12 public lectures and presentations per year around the world, often in academic institutions, philanthropic organisations or international development agencies. As a result of this experience, the author is convinced that **the principles of HCD and the concepts of the "Family of 4" and the "Family of 4 PLUS" are, if contextualised in professional ways according to the local context, highly applicable for many disadvantaged communities in countries such as Bolivia, Chile, Peru, Ecuador, Argentina, India, Bhutan and China.** Since the wider international dissemination of the HCD concepts through lectures, papers and journal articles, the author receives on an average 1–2 requests per month to advise development organisations, community development practitioners or students from around the globe, regarding issues related to HCD projects, or RETs for a particular country's context, or a student research project/dissertation.

8.3 Suggestions for Further Work

It has been highlighted throughout the thesis that one of the key parameters of the new HCD concept is the long-term involvement of the project implementing partner, often the local NGO, such as RIDS-Nepal, and the local village community. Long-term, in the context of the Humla communities, means 2 generations or ~20 years. The thesis' main focus includes the RIDS-Nepal HCD projects in Humla which commenced in 2002. Over the following years, more villages were added, according to the requests of local village communities. By 2012 the projects had expanded into 13 villages in the district of Humla. The partnership with the first villages is soon 10 years old, though only a part of the intended long-term period of two generations has passed, which means there is still a lot of work to be done. However, the following points highlight the recognised issues which the author deems to be relevant and important for RIDS-Nepal, and other organisations working in this field. This will be considered and addressed over the next few years, in order to remain faithful to the overall HCD philosophy as thus far developed and applied. Since RIDS-Nepal is a non-profit NGO, fully dependent on external funding for its work and projects, it will only be able to carry on this work to the extent that funding can be secured for it.

(1) Data from the Follow-Up survey, compared with the Base-Line survey, shows the positive impact of **pit latrines** on the overall environment in and around the village, resulting in fewer people suffering from diarrhoea and intestinal worms (Tables 7.4.and 7.5). However, some of the results **demand further investigation** due to the small sample sizes for some age groups, in order to draw definitive conclusions.

(2) In the remote, high-altitude mountain communities, the family roles in relation to gender are still very clear and strongly adhered to. The women are responsible for cooking and thus are exposed to the heavy IAP for far longer times per day, compared with the male population. They also take most of the responsibilities for the small children (aged <5 years), who consequently are also more exposed to the IAP. Thus it can be hypothesised that females and small children will gain more from a smokeless stove than the male occupants of the same household. While some of the measured data supports this hypothesis (Sect. 7.2.2.5), the initial follow-up surveys and anecdotal data, 1–2 years after the installation of the smokeless metal stoves, have not yielded unequivocal results. As anticipated, in view of the family roles and local culture, older females had a statistically significant improvement in regard to the decreased occurrence of respiratory symptoms with the stove and the indoor lighting. However, the small children had no significant change in reported respiratory symptoms, which was unexpected considering they are the most vulnerable to IAP. **A comparison of the indoor air pollution measurements for a traditional, open-fire cooking and heating place and a home with a smokeless metal stove shows clear numeric and visible improvements** (Fig. 7.2)**. Further investigations need to be carried out with increased numbers of households, the specifics of previous IAP exposure, and the incidence of health impairments as well as people's roles within the family, according to gender and age**.

(3) There have been unsatisfactory results in some households. It was recognised and measured, that people used their new stove during the winter more for room heating, and have decreased room ventilation, and some smoke more and spend longer periods indoors. Thus the season can have a confounding effect on the occurrence of respiratory symptoms and ailments. As there were significantly more data recorded before and after the SMS and indoor lighting interventions in the same households during the summer season, the study presented in Sect. 7.2.2 includes only the more consistent and reliable data for the summer season. It looks at the decrease of respiratory diseases and relative risk reduction among the various age groups after the installation of the SMS and WLED indoor lighting. However, the limited data recorded in the same households during winter shows higher occurrence of respiratory ailments before the stove and indoor lighting, but also a larger reduction in ailments during the follow-up surveys after the interventions (Table 7.13) when compared with summer. This was also noted in the prevalence of eye disease which had higher initial rates in winter (Table 7.12) and subsequently had a significantly marked drop in rate when compared to summer. Further, the decrease in respiratory ailments during winter occurred even though there was a recognisable increase of people smoking during winter (Table 7.14). Thus, **seasonality is a recognised determinant of respiratory ailments and further studies with this in view need to be conducted.**

(4) As shown in Table 7.2, the SMS consumes on an average, for the same energy services, $\sim$40 % less firewood. This reduces the time women need for

firewood gathering, helps to fight the rapidly increasing deforestation as well as slashing greenhouse gas emissions. Less time spent in the forests for firewood collection means that women's injuries due to carrying heavy loads, falls, miscarriages, exposure to violence (wild animals and men) and civil unrest are reduced. Thus, **the potential health improvement for the most vulnerable in society, the women and young girls, through improved access to resources and easier energy generation, community participation and non-formal education, all facilitate the process of decreased gender discrimination and women's empowerment.** Such changes bring forth cultural adaptations and transformations, which have to be monitored and recorded for years to come. **These issues have not yet been studied in relation to the SMS and the "Family of 4" and "Family of 4 PLUS" HCD projects.**

(5) As previously highlighted and discussed, the potential of enhanced, or synergistic benefits of a holistic project approach, such as the "Family of 4" and the "Family of 4 PLUS", are real and substantial. While this is easy to grasp mentally, it is very hard to scientifically identify synergistic benefits (Sects. 7.4 and 7.5), and even more difficult to identify and track the factors which have contributed over the years to the synergistic benefits a community enjoys. The overall benefits to the lives of the people and their environment through the ongoing "Family of 4" and the "Family of 4 PLUS" community development projects already demonstrate (Chap. 7) that the whole is greater than the sum of its parts. As the long-term project unfolds, increasing evidence will become available about the synergistic benefits. However, this is still an untouched field, which demands close, **long-term studies of HCD concepts versus single objective projects. Detailed scientific and anecdotal data monitoring and tracking is essential in order to uncover and identify in qualitative and quantitative ways the synergistic benefits obtained from an HCD approach.**

(6) Chapters 2–5 demonstrate how, with locally developed RETs, the abundant, locally available renewable energy resources can be tapped into. Thus **teaching the basic theories of RETs and RE resources at academic institutions** (such as the author's course described in Appendix 27), **in parallel with relevant student research projects, the local industry and INGOs/NGOs such as RIDS-Nepal, are compounding/reinforcing factors for long-term sustainable development of a country such as Nepal.** Thus there is ample scope and need to invest more effort in studies to develop additional educational tools to bridge the gaps in the national education curriculum in regard to the nation's development.

(7) The direct link between access to improved energy services and poverty alleviation, discussed in Sects. 1.1.6 and 1.1.7 and the first of the MDGs, also validates the statement that contextualised RETs have a vital role to play in community development projects. However, as part of understanding and establishing this central role more accurately, **further studies and investigations are needed into the "energy ladder" (Sect. 3.1). These should**

focus on communities who, for the first time in their history, gain access to improved basic energy services, such as electricity, better cooking facilities, hot/warm water and room heating. Also questions such as, what amount and quality of energy is accessible and economically realistic and sustainable for HCD projects, need to be tackled and answered satisfactorily.

(8) **Improved qualitative data monitoring systems are needed for technical projects**. They should be easy to learn and apply at affordable cost and thus easier to propose to project donors as an intrinsic part of HCD projects. Further, refined, anthropologically sound scientific and anecdotal data collection methodologies and procedures are indispensable in order to understand the impact, changes and adaptations taking place through long-term HCD projects at the deeper cultural and community grass-roots level.

(9) **Providing improved skills for NGO/project staff and local end-users** for more participatory and relevant project design, implementation, management (such as long-term project life-cycle management), operation and maintenance will allow greater sustainability of projects. Therefore, further work is **needed to develop relevant HCD and RET training and educational (formal and non-formal) courses/modules** in a manner and fashion similar to that used by the author and RIDS-Nepal (Appendices 16, 17, 18, 19, 20, 22, 23, 27, 29 and 30).

(10) Creating processes which **intensify the relationship and collaboration with the Government, in order to spearhead new HCD concepts and RETs nationwide through relevant policies** derived from grass-roots project experience and developed for defined contexts. Thus, for example, there is a need to develop new government subsidy policies for the "Family of 4" and the "Family of 4 PLUS" HCD concepts, in order to disseminate the HCD concepts nationwide. In this way, there is greater assurance that the necessary project processes are in place and adhered to.

(11) **Intensify the collaboration and partnership** through defined procedures, **between national academic institutions** (e.g. included as part of the compulsory curriculum), **the local industries** (e.g. enable them to provide government subsidised, work opportunities for student research projects) **and NGOs** (e.g. by presenting their field experience through practical lessons and workshops in academic institutions and offering "on the job" work/project opportunities for students and staff) **in practical and applied research and development** of context related RETs and HCD project implementation.

(12) RAPS systems, utilising the locally available renewable energy resources, are very relevant and applicable for remote communities with no access to the grid. However, due to the intermittent nature of renewable energy resources, a RAPS system needs an energy storage facility, in order to provide the required energy service, which is usually electricity, in times when the local renewable energy resources are not present or not sufficiently available. This is mostly done with batteries which pose some considerable danger to people and the environment (Sect. 6.1). This major technical downside will continue for years to come, as there is no real alternative, which is economically

feasible and technically reliable (Sects. 3.7.5 and 3.9.5). However, by utilising several locally available renewable energy resources, such as solar, wind and water, in combination with each other, in one so called **hybrid RAPS system, this problem can be addressed. These systems have significant potential to minimise the battery-bank capacity required.** However, there are few standard size hybrid RAPS systems available on the international market, especially not for the lower capacity range needed for rural development projects. In fact, RAPS systems are usually customised power generation systems for a specific context. Thus there is significantly more effort, **research and development needed to develop RAPS system design procedures which guarantee efficient and secure use of several renewable energy resources, converting them through contextualised RETs into energy services (such as electricity and heat) in a hybrid RAPS system for rural electrification.**

(13) There are several R&D projects which need to be considered to meet the overall aim of long-term sustainable HCD including the following:

(a) **Measurement of the intercepted global solar radiation at different altitudes** (e.g.120/1,350/3,000 m.a.s.l), to represent the three major altitude ranges of rural communities in the Nepal Himalayas. These measurements are done with different pyranometer technologies (e.g. mono-, poly- and amorphous silicon), in order to represent the three most common solar PV technologies used in Nepal (see RIDS-Nepal's first global solar radiation measurement station in Kathmandu at 1,350 m.a.s.l, at: http://www.rids-nepal.org/databank/date.php). This allows us **to determine which are the most appropriate solar PV technologies for the conversion of global solar radiation into DC electricity at these three different altitudes.**

(b) **UV radiation measurement at different altitudes** (120/1,350/3,000 m.a.s.l.) in order to measure, and **design effective protection from the harmful UVA and UVB radiations** the local people are exposed to throughout their daily field work at high-altitudes (Blumenthaler 1997).

(c) To **develop, design and propose mitigation measurements about how high-altitude communities can increase their resilience and how they can adapt to their fast changing environment**.

(d) With RAPS systems providing a great potential for rural village electrification, it is important that **increased effort in R&D is conducted into new energy storage technologies and facilities.** Possible technologies which could be considered, but need more R&D, are energy storage through flywheels, local generation and storage of hydrogen and its conversion into electricity on demand through fuel cells, capacitor banks for small scale applications, and water storage pumping during times of overproduction of energy (be it from a micro-hydro or a hybrid RAPS system), to cover peak demands and times of inefficient energy conversion capacity due to the variability of the renewable energy resources.

(e) WLED diodes are known for their extremely low power consumption (15 mA x 3.6VDC = 0.054 W per 5 mm diode) and their long life-expectancy (50,000–100,000 h). Since 1999 the author has electrified homes with WLED lamps for indoor village lighting. While at that time the best 5 mm WLED diodes provided a luminous intensity of 14 lm/w, the same brand and type (Nichia NSPW510) used in WLED lamps installed in 2009 in Humla villages provided 83 lm/w. However, with the potential of attaining 200 lm/w (Appendix 12.5.3) this technology has a great future and indoor lighting has come within reach, even for the remotest communities. Therefore, **there is need for increased development, testing and local manufacturing of high quality WLED lamps for indoor and street lighting applications for the rural village context**.

(f) From 1998–2002, the author installed and tested a self made, small scale 180 W horizontal axis wind turbine with wooden blades and a customised Bosch car alternator (Sect. 3.3.3). As a result of the positive experience and the increased world wide interest in harnessing the local wind energy resources, it is time to put **more effort into the R&D of Nepali made, 100–1000 W, horizontal axis small scale wind turbines for rural village electrification schemes.**

(g) The alarming deforestation in the high-altitude valleys of the Himalayas (Sects. 1.1.2, 1.1.8 and 4.1.1/Appendix 4) justifies the call for improved and efficient wood burning stoves (Sect. 2.10.2). More effort is needed to look into other, locally available energy resources, such as animal dung (cow, yak, mule, and horse), human faeces, kitchen waste and other minor biodegradable waste, which can be converted to generate heat. While these energy resources are already widely used in the low-altitude areas of Nepal to generate biogas with $\sim$60 % methane (CH_4) and $\sim$40 % carbon dioxide (CO_2), through an anaerobic process in biogas digesters (micro-organisms breaking down biodegradable materials in the absence of oxygen), there is no suitable technology and bio-digester technology available for high-altitude areas $\geq$3,000 m, where ambient temperature can drop below -15 °C during the long and cold winter nights. As the anaerobic digestion process takes place best at temperatures between 37 and 41 °C, which can never be reached in these cold high-altitude areas, new technologies are needed. Thus, **intensive R&D for high-altitude anaerobic digestion, including additional thermal heating systems, is needed.** Some family sized anaerobic biogas digester prototypes, providing the daily energy needed for cooking and room heating, should be installed and monitored in great detail.

(h) More **in-depth studies are needed for improved and more efficient biomass combustion processes for stoves.** A more detailed and accurate understanding is needed of the optimal biomass to air mass ratios and their consequences for the primary and secondary combustion processes and efficiencies (Sect. 4.1.5). In conjunction with practical combustion tests, under defined conditions in the laboratory, and with the support of

CFD (Computational Fluid Dynamics) these processes can be simulated, in order to understand how combustion stoves with a defined fuel, such as specified firewood or dung, need to be designed and manufactured.

(14) With the various RETs and infrastructures (such as the high-altitude greenhouse (Sects. 2.11.1 and 4.2) and the high-altitude solar water heater bathing centre (Sects. 2.11.4 and 4.4) thus far developed, and with more to be developed, there is an increasing demand to have these new contextualised technologies, developments and equipment tested under real, high-altitude field conditions in a more rigorous and scientific way before they are more widely disseminated and implemented in the local villages. This idea motivated the author in 2004 to build and run the first high-altitude research station (HARS) in Simikot, Humla, at 3,000 m.a.s.l. (Sect. 3.4). With the experience thus far gained we have realised that **an improved HARS is needed, which will include** greater opportunities for **expanded research and development of contextualised RETs** (such as hybrid RAPS systems, biogas digesters, ground source heat pumps, etc.). **Further R&D is needed in the field of new structural architecture for improved comfort in homes, schools, health posts, etc., as well as HCD programs.**. There is also a great need to include in the new HARS the ability to **provide hands on courses and skills training for the local users** of all the recently developed equipment and infrastructure. This will assist villagers to operate, maintain and competently repair some of the more common breakdowns. This will enhance the sustainability of all the new developments and HCD projects significantly, along with enabling the local end-users to become more responsible users and proud owners of these developments. This in turn is crucial for any new technology or infrastructure to be slowly taken up and included in their culture. Further, in order to make the knowledge and these experiences more widely available, nationally and internationally, **academic short (weeks) and medium length (3–6 months) courses, in the related fields, should be delivered by respected Professors in each field, and accredited degrees awarded** by one of the local Universities, such as the Kathmandu University. This could be done in conjunction with an international University. This would support the endeavour to disseminate the knowledge and experience to other countries, where the graduates can apply their new skills and knowledge in their own context.

(15) **RIDS-Nepal's web site,** www.rids-nepal.org**, is the gateway to all of the HCD and research projects**, as the working area is so remote. In fact, Humla is 17 days trekking away from the next road head (Sect. 1.1.8) and hardly any visitors ever come there due to the time required to get into and out of Humla, even though there is an airplane strip in Simikot, the main village of the Humla district. The small air planes fly only if weather conditions allow, as none of the aged air planes is equipped with radar or other modern navigation systems. Thus the author has invested significant effort, time and financial resources to develop the web site, to train a local Nepali RIDS-Nepal staff member, a

graduate of KU, to continuously develop, update and maintain the web site. It is also part of the vision of RIDS-Nepal to use the web site in the coming year(s) as a platform to provide the RE course (Appendix 27), presently only taught at KU, through the RIDS-Nepal Internet web site as an on-line course, though in a different visual format. Details of the implemented HCD projects of the "Family of 4" and the "Family of 4 PLUS" are available through RIDS-Nepal's GIS tool and databank. Also available The awareness raising and teaching videos (Appendix 16), songs (Appendix 23), the supervised dissertations (Appendix 24) and other educational, teaching and learning tools. This is the beginning of **the planned, long-term aim to have the web site as an Internet based learning platform for professionals, community development practitioners and students to learn from the grass-roots experiences of implemented projects.** This still needs to be planned, developed and designed and then made available through the web site over the coming years.

(16) With RIDS-Nepal being a non-profit NGO, the author needs to raise all the needed funds for the HCD village based projects and the R&D projects from donor agencies and organisations. This has been hard to sustain and has, since the recent worldwide economic downturn, become even more challenging. Thus, another important field **to improve RIDS-Nepal's long-term financial support basis, is the development of a professional portfolio for awareness raising and fundraising activities.** This will be among the major international governmental donor agencies, international development organisations and smaller, often privately managed, foundations for philanthropic works. A specially designed approach is intended to be developed to raise interest, awareness and financial participation among the thousands of Nepali professionals in the Diaspora in Australia, Europe and the USA. The project will help them to understand that they can play an active role in the development of their own country, despite the likelihood that they will not return to take an active hands on role. They can however, support RIDS-Nepal's endeavour to bring contextualised HCD and RETs to the remotest and most impoverished high-altitude communities of Nepal's Himalayas.

References

Back Matter Appendix Chapters referred to: 4, 5, 6, 7, 8, 10.3, 11, 12, 14, 16, 17, 18, 19, 20, 22, 23, 24, 27, 29, 30.

Bista, D. B. (1991). *Fatalism and development—nepal's struggle for modernization.* 5th Impression, Calcutta, India;Orient Longman, ISBN 0 86331 245 5, (2nd ed.) 1994.

Blumenthaler, M., Ambach, W., & Ellinger, R. (1997). Increase in solar UV radiation with altitude. Elsevier, Journal of *Photochemistry and Photobiology,* Biology *39,* 130–134, Retrieved July 28, 2012, Abstract available and paper to purchase at Doi:10.1016/S1011-1344(96)00018-8.

Appendix 1
Nepal Calendar—Western Calendar

1.1 Nepal Calendar: Bikram Sambat[1]

Bikram Sambat, (Devanagri: बिक्रम संवत, abbreviated "B.S.") is the calendar established by Indian emperor Vikramaditya. It is a popularly used calendar in India and the official calendar of Nepal. In addition to Bikram Samwat, the Gregorian calendar and the original Nepalese calendar, Nepal Sambat, are also used in Nepal. Nepal Sambat was officially used in Nepal till the era of Chandra Shamsher (~ 1929), but due to its limited use these days, it is not focused on in this study.

The Bikram Sambat was founded by the Tuar Rajput emperor Vikramaditya of Ujjain following his victory over the Sakas in 56 B.C. It is a solar calendar[2] based on ancient Hindu tradition. The Bikram Sambat calendar is 56.7 years ahead in count of the numbers of year of the solar Gregorian calendar. For example, the year 2068 BS began in middle of April 2011 and will end in middle of April in 2012. The calendar starts with the first day of the month Baisakh, which usually falls on the 13th or 14th of April in the Gregorian calendar.

1.2 Western Calendar: Gregorian Calendar[3]

"The Gregorian Calendar has become the internationally accepted civil calendar"..[4] It was first proposed by the Calabrian doctor Aloysius Lilius, and decreed by Pope Gregory XIII, after whom it was named, on 24 February 1582 by the papal bull *Inter gravissimas*. It is a reform of the Julian calendar and continues

[1] Bikram Sambat information is taken mostly from: http://en.wikipedia.org/wiki/Bikram_Samwat (accessed July 20, 2012).

[2] A solar calendar is based on the progression through the seasons as the Earth revolves around the sun.

[3] Gregorian calendar information is taken from: http://en.wikipedia.org/wiki/Gregorian_calendar (accessed July 20, 2012).

[4] From: http://www.usno.navy.mil/USNO/astronomical-applications/astronomical-information-center/calendars (accessed November, 2011, on July 20, 2012 not available anymore).

A. Zahnd, *The Role of Renewable Energy Technology in Holistic Community Development*, Springer Theses, DOI: 10.1007/978-3-319-03989-3,

Table A1.1 Comparison of the Nepali Bikram Sabat and the western Gregorian calendar

Gregorian calendar month			Nepali—Bikram Sambat calendar month			
No.	Name of month	Numbers of days	Nepali name in Devanagri	Colloquial Nepali name	Gregorian months equivalent	Numbers of days
1	January	31	माघ	Maagh	Mid January to mid February	29/30
2	February	28 or 29	फाल्गून	Falgun	Mid February to mid March	29/30
3	March	31	चैत्र	Chaitra	Mid March to mid April	30/31
4	April	30	वैशाख	Baisakh	Mid April to mid May	30/31
5	May	31	ज्येष्ट	Jesth	Mid May to mid June	31/32
6	June	30	आषाढ	Asadh	Mid June to mid July	31/32
7	July	31	श्रावण	Srawan / Saun	Mid July to mid August	31/32
8	August	31	भाद्र	Bhadau / Bhadra	Mid August to mid September	31/32
9	September	30	आश्विन	Aswin	Mid September to mid October	30/31
10	October	31	कार्तिक	Kartik	Mid October to mid November	29/30
11	November	30	मंसिर	Mangsir	Mid November to mid December	29/30
12	December	31	पौष	Poush	Mid December to mid January	29/30

the year numbering system of the Julian calendar, counting years from the traditional Incarnation of Jesus. Years after this date are given the designation "Anno Domini" (AD), or "Common Era" (CE); years before this date are labeled "Before Christ" (BC), or "Before the Common Era" (BCE). A Gregorian and the Nepali Bikram Sambat year are each divided into twelve months, each of irregular length. The following Table A1.1 shows how the two calendars relate to each other.

The basic rule of thumb for conversion, though this method provides only accuracy within 1–3 days and thus is suggested only for a general indication of the date, is as follows:

- Nepali Bikram Sabat to English Gregorian date: Subtract −56 Years − 8 Months − 15 Days.
- English Gregorian to Nepali Bikram Sabat date: Add + 56 years + 8 Months + 15 Days.

Or, for an even easier and quicker rough calculation for the year, the Nepali calendar is 57 years ahead of the Western calendar, with each Nepali year starting around the middle of April. All official and festival dates (including most of the marriage dates) in Nepal are determined by this calendar.

Appendix 2
Exchange Rates (Nepali Rupees per US Dollar)

In order to better understand the national, and to some extent as well, the local, economy of Nepal in relationship to the wider world better, it is important to understand the relationship between its own currency, the Nepali Rupee, or in short NRs, and the main foreign exchange currency, the USD. This provides an understanding of the purchasing power of a nation on the international market, for businesses dealing with foreign companies.

The USD has been and still is, despite its weakness demonstrated since around 2005, the leading international currency for trade, business and contract related international projects. Nepal, being a major receiving country of multiple international government and donor agencies is particularly dependent on the USD's value, trend and fluctuations. Thus the cost of imported goods and services is highly dependent on the USD's international level and the locally manufactured goods and markets are dependant on the USD's fluctuations.

One of the factors which has often not been given enough attention, because it is not so dependent on the USD's currency rate, is the salary scales and purchasing power, of the national people. Thus, in order to understand the direct and indirect reasons for the increasing struggles of the local people to make a decent living from their daily labour, it is also important to have an understanding of how the USD currency has related over the years to the local currency, the Nepali Rupee (NRs).

In view of the HCD concept and philosophy developed and presented throughout the thesis, it is clear, that development does not change over night, but rather over the course of years or even decades. Thus, in order to understand the full extent of the possible impact the USD currency, and its changes, on the lives and struggles of the local people, it is important that a more long-term time scale of the changes of the USD versus the NRs currency exchange rate is considered.

Thus a 26 year time span is chosen, representing approximately two generations of the present Nepali society. This allows us to demonstrate and "feel" the influence the USD currency exchange rate changes had on the local society.

In the following Table A2.1 and graphs the Nepali Rupee's (NRs) Exchange rate to the USD from 1986–2011 is provided.

A. Zahnd, *The Role of Renewable Energy Technology in Holistic Community Development*, Springer Theses, DOI: 10.1007/978-3-319-03989-3,

Table A2.1 USD to NRs exchange rates over the years 1986–2012 (July 7th)

Year	Average yearly NRs per USD	Year maximum NRs per USD	Year minimum NRs per USD	Average difference in % from year to year
1986	21.320	–	–	–
1987	21.819	–	–	2.341
1988	23.289	–	–	6.737
1989	27.189	–	–	16.746
1990	29.369	–	–	8.018
1991	37.255	–	–	26.851
1992	42.718	–	–	14.664
1993	48.608	–	–	13.788
1994	49.398	–	–	1.625
1995	51.890	–	–	5.045
1996	56.692	–	–	9.254
1997	58.010	–	–	2.325
1998	63.867	68.100	56.950	10.096
1999	68.328	69.050	67.400	6.985
2000	71.053	74.938	65.857	3.989
2001	75.474	80.811	71.822	6.222
2002	79.146	81.261	78.300	4.865
2003	77.162	80.163	74.040	−2.508
2004	74.864	75.702	73.415	−2.978
2005	73.985	76.435	70.823	−1.174
2006	75.238	78.213	72.354	1.694
2007	68.306	74.038	63.684	−9.213
2008	70.821	82.403	64.432	3.682
2009	79.513	80.232	78.506	12.273
2010	74.364	77.805	70.833	−5.390
2011	73.409	74.366	72.100	−1.285
2012 July	81.520	86.163	76.718	11.008
	Total change from 1986 to 2012 (July 7th)			382 %
	Average yearly change			14.7 %

Exchange rates from 1986 to 2003 are from: Asian Development Bank 2004, Key Indicators 2004. Poverty in Asia: Measurement, Estimates, and Prospect, Asian Development Bank, pp. 224–229. 1998–2009. Exchange rates for 1998–2012 are from the fxhistory web site: http://www.oanda.com/convert/fxhistory

The "Average Yearly NRs per USD" table is calculated from the following data sources:

1. Exchange rates for the averaged yearly values from 1986 to 1997 are from the Asian Development Bank data.
2. Exchange rates for the averaged yearly values from 1998 to 2003 are from the Asian Development Bank and the fxhistory web site based data. Maximum and minimum values are only from the fxhistory web site.

3. Exchange rates for the averaged yearly values from 2004 to 2011 are from the fxhistory web site based data.

Main findings and interpretations:

From the mid 1980s onwards to the early 1990s a high yearly USD to NRs exchange rate increasingly devalued the NRs internationally. Even more importantly, it made imported goods progressively more expensive, not proportional to the yearly remuneration and salary increase the local people have experienced. This is a counter productive experience towards the wide spread, international call for poverty alleviation Nepal made during these years. From around 1994–2000 a more or less steady, year by year rise of the USD currency over the NRs, of 2–6 % was encountered. The beginning of the economic downturn of the US, the massive yearly increase of their international debts and their ongoing war bills, caused the USD to lose ground from around 2003 to early 2008 against its value in NRs. While this decrease of value of the USD currency should be followed by a fall of costs of the imported goods and equipment, the local prices did not fall accordingly. As a consequence, the local people did not benefit from the USD's weakness. During 2008 a sharp increase in the USD's value was seen, followed by sharp increasing local prices, even for the more mundane, common daily needed goods. This trend was followed by the world wide economic crisis starting in the US, weakening the USD's value and making it again less stable and predictable. This state is still in effect in July 2012.

Thus, from this quick, and by no means in-depth or comprehensive, look at the past 26 years history of the USD—NRs currency exchange rate development, it can be seen, that the USD—NRs exchange rate has had dramatic changes, with an almost four fold value increase of the USD against the NRs.

Through personal interviews and discussions with local people, businesses and organizations one finds out quickly, that the payments and salaries of the local people and workmanship have not increased four-fold over the same time period. Neither have they on an average increased on a yearly basis by ~14 %, matching the average yearly value increase over the last 26 years of the USD against the NRs from 1986 (21.32NRs/USD) to 2012 (86.163 NRs/USD).

That means, that the local worker and the businesses and companies, involved in international trade and, contracts, had a very hard time to make a living over the past tow decades, due to external parameters which are not at all dependent on the Nepali economy, but rather the responsibility of the international economy and governments.

This point is important to mention as it shows, that any economic matter these days, and that includes the wide scope of development projects and activities, is not just restricted to national borders, but is linked and entangled in the most complicated ways at an international level. The issue of poverty alleviation is a major issue for Nepal and, most other developing nations and is one of the main points of the UN's MDGs. To address the local people's poverty realistically, a nation such as Nepal needs to look at a much wider perspective, including the international economy and markets.

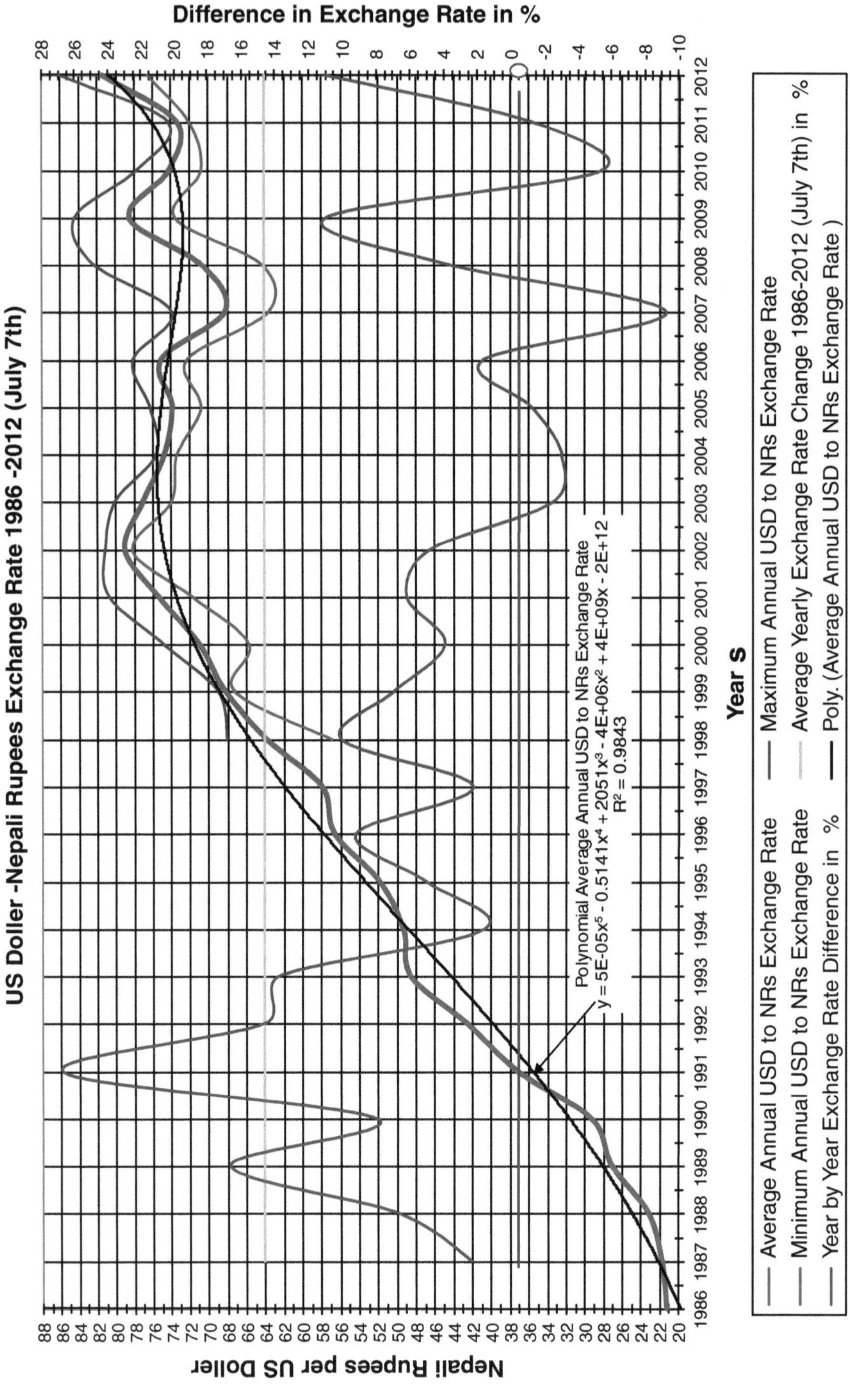
US Doller -Nepali Rupees Exchange Rate 1986 -2012 (July 7th)
Difference in Exchange Rate in %
Nepali Rupees per US Doller
Year S
Polynomial Average Annual USD to NRs Exchange Rate
y = 5E-05x⁵ - 0.5141x⁴ + 2051x³ - 4E+06x² + 4E+09x - 2E+12
R² = 0.9843
Average Annual USD to NRs Exchange Rate
Minimum Annual USD to NRs Exchange Rate
Year by Year Exchange Rate Difference in %
Maximum Annual USD to NRs Exchange Rate
Average Yearly Exchange Rate Change 1986-2012 (July 7th) in %
Poly. (Average Annual USD to NRs Exchange Rate)

Appendix 3
A Short History of Nepal

Nepal's earliest recorded history dates back to the beginning of the first millennium BC, when the Khas, an Indo-Aryan language speaking people group, arrived with their cattle, goats and sheep, settling in the western hill areas of Nepal. These nomadic pastoralists grew barley, millet and sesame (Bista 1991, 15). During the seventh and eighth century BC another people group, the Kirat, arrived in Nepal. Not much is known about them except that they are Mongols and spoke a Tibeto-Burmese language and came from the east with their domesticated pigs and buffaloes (Bista 1991, 15). It was during this period that Buddhism first came to the country. By 200 AD, Buddhism had waned, and was replaced by Hinduism, brought by the Licchavis, who invaded from northern India and overthrew the last Kirat king. They also introduced the caste system (which is still practiced today) and they were known for their art and architecture. By 879 AD, the Licchavi era was succeeded by the Thakuri dynasty. A grim period of instability and invasion often referred to as the "Dark Ages" followed.

Several centuries later, the Thakuri king, Arideva, founded the Malla dynasty, allowing another renaissance of Nepali culture. Despite earthquakes and battles between the three independent city states of Kathmandu, Patan and Bhaktapur in the Kathmandu valley, the dynasty flourished, reaching its high point in the fifteenth century under Yaksha Malla. Under the strong leadership of Prithvi Narayan Shah, the Ghorkha launched a campaign to conquer the Kathmandu valley. In 1768, after 27 years of war, they triumphed and the capital was moved to Kathmandu. Over the next 24 years the kingdom's power expanded, until progress was halted in 1792 through a short and punishing war with Tibet (Figs. A3.1, A3.2).

Some disagreement and hostilities followed in 1814, this time with the British over the territories of Sikkim (east of present Nepal) and the Terai region, which borders India to the south. Nepal had to hand over Sikkim and parts of the Terai plains as a result of the signing of the 1816 Sugauli Treaty. This established Nepal's present geographical border lines to the East and West. The Shah dynasty continued in power during the first half of the nineteenth century until the Kot Massacre of 1846, when Queen Rajendralakshmi failed in a bloody plot to kill Bir Narsingh Kunwar (known later as Jung Bahadur Rana), an influential warlord. Jung Bahadur seized control of Nepal by butchering several hundred of the most important men while they assembled in the Kot courtyard. He assigned to himself the more prestigious title Rana, and thus the Rana dynasty was formed. Jung

A. Zahnd, *The Role of Renewable Energy Technology in Holistic Community Development*, Springer Theses, DOI: 10.1007/978-3-319-03989-3,

Fig. A3.1 King Prithvi Narayan Shah (1723–1775), unified Nepal for the first time after conquering the Kathmandu Valley in 1769. He made Kathmandu the capital of Nepal

Fig. A3.2 Jung Bahadur Rana (1816–1877). Founder of the Rana dynasty. He was the first, self-appointed, Rana ruler and first Prime Minister of Nepal in 1846

Fig. A3.3 King Tribhuvan Bir Bikram Shah (1906–1955) was king of Nepal from 11 December 1911 until his death

Bahadur Rana proclaimed himself the first Prime Minister of Nepal, while the monarch was made a ceremonial figure. The post of prime Minister was the most powerful position and it was made hereditary, so that the Rana family could remain in power and live a luxurious life style in the palaces in Kathmandu, while the local people struggled to survive from what was left to them. The Rana leadership and regime, a harsh, strongly centralised autocracy, followed a policy of isolating Nepal from the surrounding world, so that no external or foreign influences were able to penetrate the society, purposely keeping people living in medieval conditions. The aim of the Ranas was to stay in power for as long as possible by suppressing any potential internal movement emerging from increased education and exposure of the local people to the outside world. On a more positive note, this policy kept Nepal free from any colonial powers and their influence, as was the case in the neighbouring India, through the British.

Thus Nepal remained a medieval society until 1950, when the Rana prime ministers and their antiquated regime, were overthrown and replaced by a democratic government, operating under the scrutiny of an actively interested and ultimately powerful monarch, King Tribhuvan (Fig. A3.3). He was appointed ruler in 1951 and established a new government comprising of Ranas and members of the newly formed Nepali Congress Party. After a failed democratic election, King Tribhuvan's son, who succeeded him, King Mahendra (Fig. A3.4), decided on a more appropriate "party less" panchaayat system for Nepal. He chose the Prime Minister and the cabinet and appointed most of the people sitting in the national assembly, making sure "his" people were sitting in the right positions, rubber

Fig. A3.4 King Mahendra Bir Bikram Shah (1920–1972) was king of Nepal from 1955 to 1972

stamping his ideas and policies. In this way he could remain in full control and power. King Mahendra was succeeded by his 27 year old son, King Birendra in 1972. Despite the new King, corruption, mismanagement and the diversion of foreign aid into the royal treasury, rather than into projects and programs they were meant for, continued until 1989.

The masses, by now much more informed and educated, were fed up with years of hardship and suffering. Finally they rose up in the popular protest called the "Jana Andolan" or "People's Movement" during 1989–1990. Months of violent and bloody encounters followed, leaving hundreds of people dead. King Birendra was forced to dissolve his cabinet and legalised political parties and invited the opposition to form an interim government, making the panchaayat system finally obsolete. Thus in May 1991, Nepal was able to hold its first free and fair parliamentary election. The Nepali Congress won the majority with the Communist Party as their main opposition. They jointly formed the first elected government in 32 years.

Since then, Nepal has discovered that the establishment of a workable democratic system in such a diverse and under-developed country is an enormously difficult task. Nepal has a weak and immature economy, with massive unemployment, in particular among the lower castes. It has widespread illiteracy, often enforced by the ethnic and religious diversity. It has also a high population growth rate. These problems continue to be the main challenges and stumbling blocks for Nepal as a country to overcome.

Through the early nineties constant small uprisings, demonstrations, clashes between students and the government (police and military forces) took place, mainly in the fast growing urban settlements.

Fig. A3.5 Dr. Baburam Bhattarai, born 26th May 1954 was the initial leader of the Maoist insurgency. He is a communist politician and was for the Maoist's 1st year in power (2008/9) the Minister of Finance of Nepal's government

Fig. A3.6 Pushpa Kamal Dahal, born 11th December 1954 is better known as Prachanda. He was the Maoist insurgency leader and then from 15th August 2008 to 5th May 2009, Nepal's Prime Minister from the ruling Maoist party

In February 1996 one of the Maoist parties started to raise its voice publicly with the call for a new people's democratic republic. Part of their plan was to replace the monarchy through a Maoist revolutionary strategy known as the people's war, which resulted in Nepal's civil war, led by Dr. Baburam Bhattarai and Pushpa Kamal Dahal (Figs. A3.5, A3.6), better known in Nepal as Prachanda.

Fig. A3.7 Nepal's last King (from 1st July 2001 to 20th June 2008), King Gyanendra, who was crowned after his brother, King Birendra and most of his family, were massacred on 1st June 2001

Fig. A3.8 Till the democracy movement, and the abolition of the monarchy, Nepal's Kings were always also the commanders in chief of Nepal's Army. Here is King Gyanendra in his army uniform

The insurgency initially began in the five, remote and impoverished, western districts of Rolpa, Rukum, Jajarkot, Gorkha and Sindhuli.

On June 1, 2001, Crown Prince Dipendra assassinated nine members of the royal family, including his father, the much beloved King Birendra, and his mother, Queen Aishwarya, before shooting himself. As per tradition, one of the few survivors of that massacre, the late King Birendra's brother, Prince Gyanendra (Figs. A3.7, A3.8) inherited the throne. Meanwhile, the Maoist rebellion escalated.

The government's inability to rule and bring peace and order to the country, and the siege of the Kathmandu Valley by the Maoists in August 2004, resulted in a steep and fast decline in support for the monarch and his government. Thus, King Gyanendra (Figs. A3.7, A3.8) dismissed the government and assumed executive powers on the 1st February 2005, declaring a "state of emergency". Politicians were placed under house arrest, all phone and internet lines were disconnected, and freedom of the press was severely reduced. But the King's new regime made no progress and the Maoist insurgency became stronger and increasingly widely supported by the local people. The boycotted municipal elections in February 2006 were "a backward step for democracy" (as described by the European Union). Ongoing and intensifying strikes and ever more violent street protests in Kathmandu, forced the King to restore the parliament in April 2006.

In November 2006, after 10 years of brutal and violent war and civil unrest, the peace agreement between the government and the Maoists brought the long hoped for framework for possible peace. The seven party alliance stripped the King of most of his powers after they assumed office. On 24 December 2007, seven parties, including the former Maoist rebels, agreed to abolish the monarchy and declared Nepal a Federal Republic. Elections followed on 10 April 2008, in which the Maoists emerged as the winners, securing a simple majority in Parliament and forming the new "Republic of Nepal" government.

On 28 May 2008, the newly elected Constituent Assembly declared, with a huge majority, that Nepal was a Federal Democratic Republic. It abolished the 240 year old monarchy, forcing the last King of Nepal, King Gyanendra to leave the Royal Palace and become an ordinary citizen. On 23rd July 2008, Nepali Congress leader, Ram Baran Yadav became the first President of the Federal Democratic Republic of Nepal. In a similar fashion, on 15 August 2008, Pushpa Kamal Dahal, popularly known as Prachanda, of the Maoist Communist Party of Nepal, defeated Sher Bahadur Deuba of the Nepali Congress, becoming the first Prime Minister of Nepal's "new era". However, this did not bring the hoped for changes and stability and Nepal continues to suffer in 2011 under its long-term, unstable and rapidly changing political power struggle.

Appendix 4
Geography, Climate and Geology of Nepal

Nepal, a landlocked country, bordered by China/Tibet to the north and India to the south, east and west, embraces one of the most diverse and unique climatic environments and physical ecosystems in the world. It lies within 26°22′–30°27′ northern latitudes, and 80°04′–88°12′ eastern longitudes. Nepal's overall land surface area is 147,181 km^2 (CIA 2000), and it has a roughly trapezoidal shape of $\sim$800 km length and an average $\sim$180 km width. Nepal lies between the Gangetic plain, bordering India to the south, and the Tibetan plateau to the north. Within a short north-south distance of merely 120–230 km, altitudes range from 70 to 8,848 m.a.s.l.

It is a country of unique and pristine natural beauty and has environments ranging from tropical and sub-tropical forests and fertile river plains to steep lush and barren valleys and high alpine peaks, including eight of the fourteen highest mountains in the world.

There are three distinct geographical zones, which cannot always be clearly separated as they "flow" into each other and, due to the extremely deep valleys cut by the rivers. They often exist in close proximity to each other. However, it is widely accepted that Nepal can be divided into three geographical zones. The mostly flat and fertile Terai to the south, the mid-hill area called the Siwalik and the high-altitude area to the north. The following provides a short summary of each of the three individual zones.

- The plains, ranging from 70 to $\sim$800 m.a.s.l., or Terai, bordering India to Nepal's south. This part is mostly flat and barely above sea level. It is a narrow strip of alluvial flatland from the many rivers flowing from the north to the south, with a hot and mostly humid climate. Since most of this area has good soil and can be irrigated, 38 % of the land is cultivated for food crops (Fang 2007). Some 46.7 % of Nepal's population live in the Terai (World Vision International 2003).
- Further north lies the Siwalik Range ($\sim$800 to $\sim$4,000 m.a.s.l.) and the Mahabharata Range, which generally run parallel, and sometimes merge, creating a chain of valleys which are called "Bhitri Madhes" (Bista 1991), a mid-montane area. This hilly region often has steeply sloping land with many small attractive and fertile valleys and basins between 800 and 1800 m.a.s.l., including such valleys as Kathmandu, Pokhara, and Surkhet (Fang 2007). Some

A. Zahnd, *The Role of Renewable Energy Technology in Holistic Community Development*, Springer Theses, DOI: 10.1007/978-3-319-03989-3,

45.5 % of the population live in this mid hill region (World Vision International 2003), with the majority today agglomerated in these heavy populated valleys. Only ~10 % of the land is suitable for cultivation (Banskota 1999).
- Marking a strong natural border with China/Tibet in the north are the well known high-altitude Himalayas (~4,000–8,848 m.a.s.l.), with permanently snow covered mountain peaks above 6,000 m.a.s.l. In between are the mountain valleys of the Inner Himalayas. Tony Hagen describes these Inner Valleys as "the real high mountain valleys of Nepal, surrounded on all sides as they are by ice clad giants" (Hagen 1980). The Himalayas contain eight of the world's fourteen highest mountains above 8,000 m, including the highest, Mt. Everest, or Sagarmatha in Nepali, or Chomolungma in Tibetan, with 8,848 m.a.s.l. Typically the populated regions between 1,800 and 3,500 m.a.s.l., where most of the "high-altitude villages" are situated, are called the mountainous regions of the Himalayas. Only ~2 % of the land is suitable for cultivation at these altitudes, since the higher elevated regions are mostly steep, rugged and cold. Thus, they are also the most sparsely populated. While the high-altitude area covers around a third of Nepal's area, it is inhabited by only 7.8 % of the nation's population (World Vision International 2003). The inhabited mountainous regions have an alpine climate with cool summers and freezing winters.

This enormous range and distinctive geographical environment is also the reason for Nepal's unique climates. It results in a variety of ecological zones that have a

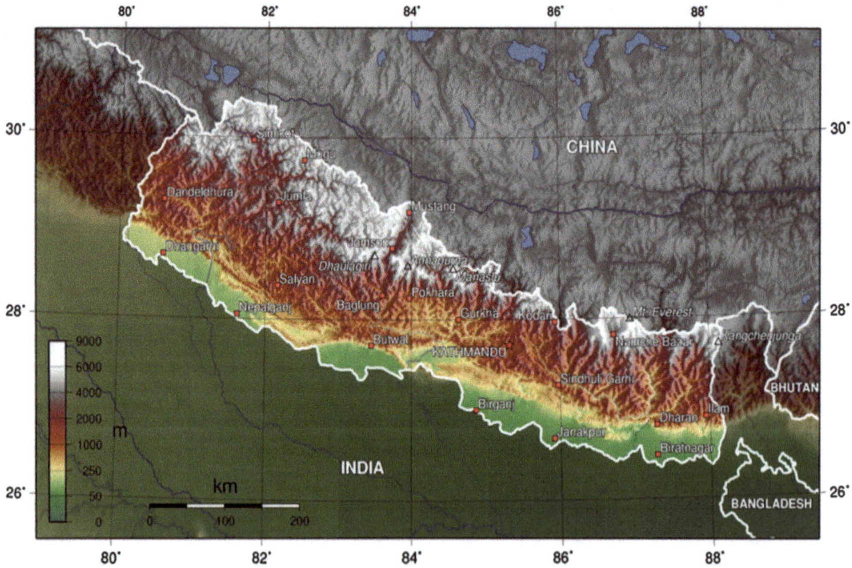

Fig. A4.1 Three distinct major geographical zones mark Nepal. The Terai to the south, merging with the Indian border, the varying mid region with its hills and mountains, and the high-altitude region with the permanent snow covered peaks to the north as well as the seasonal snow covered high mountains and valleys. (map from: http://en.wikipedia.org/wiki/File:Nepal_topo_en.jpg)

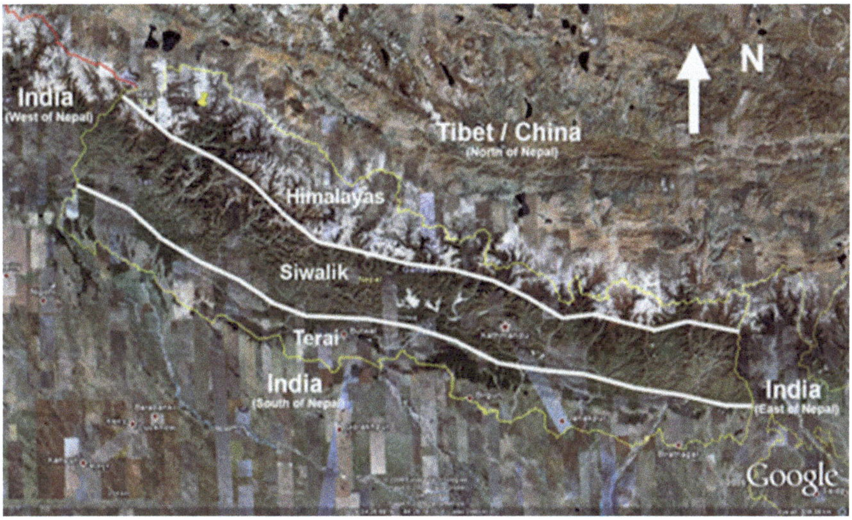

Fig. A4.2 Nepal map with national borders (*yellow line*), its neighbouring countries India to the south, west and east and China/Tibet to the north. Also, a rough, indicative area of the three main mountain ranges, the high-altitude Himalayas to the north, the Siwalik mountains/hills and the Terai, are included (by the author). (map from: Google Earth picture with author's additional explanations)

significant impact on the lives of the inhabitants. Five different climatic zones, corresponding to the north-south altitude changes, mark Nepal's climates. In the south, the tropical and subtropical zones dictate the year around conditions up to ~800 m.a.s.l. with its mostly warm to hot temperatures and very humid monsoon conditions. Further, the temperate zone, between ~800 and 2,400 m.a.s.l., provides overall the most comfortable year around conditions, with its pleasant spring and autumn conditions and rather chilly winters and quite hot and humid summers. Further up the altitude scale is the cold zone, with its typical weather pattern between ~2,400 and 3,600 m.a.s.l. (for highly elevated villages). The summers are short, though pleasant in temperatures and humidity, while the winters are cold and long, and in the higher parts, with snow covered fields and forest, sometimes for weeks at a time. The most inhospitable climate is the sub-arctic zone from ~3,600 to 4,400 m.a.s.l. and the arctic zone above 4,400 m.a.s.l. Very few villages have agglomerated in the sub-arctic zone, as the energy required to survive is so substantial that even subsistence farming is hardly possible.

No communities or villages, which are continuously occupied year around, exist in the arctic zone, as the conditions are too harsh and hostile. Temperatures drop down to −40 °C, with sub-zero temperatures during the day, for most of the year. The ground hardly thaws, making any farming or agriculture impossible. Further, the increasing lack of oxygen with increased altitude makes it impossible for human beings to live for extended time periods above 5,000 m.a.s.l. Thus the highest, permanent settlements in Nepal are found at a maximum altitude of around 4,000–4,400 m.a.s.l.

This wide variation of climates makes Nepal not just a world record holder with the world's highest mountain, but as well with regard to the most diverse climates within a country. In addition, with five different seasons a year, spring (March–April), summer (May–middle of July), monsoon (middle of July–middle of September), autumn (middle of September–October) and winter (November–February), Nepal again displays its distinctiveness and difference from other countries.

In summary it can be said that over 75 % of the country is rugged hills (42 %) or mountain terrain (35 %) (Karki 2002, 89–97), and only 17 % is suitable for cultivation (Fang 2007).

In the decades since 1960 there has been strong population growth throughout the country. That has taken its toll on the natural resources such as the biomass. The once picturesque, thick, green forested hill and mountain sides now show deep scars. Widespread deforestation is a major problem in all regions of Nepal, resulting in increased erosion, loss of topsoil, increased landslides during heavy rains, all resulting in the degradation of the ecosystem.

Nepal is prone to a variety of natural disasters and extreme weather, which are mostly related to the topography and climate. The hill ranges are mainly prone to landslides and hailstorms while the Terai is prone to floods, drought and fire, which damage crops and reduce yields. Freezing conditions and heavy snowfalls are characteristic phenomena in the mountains, which prevent crops being grown at all. Windstorms and lightening strikes also affect many areas of the country on a regular basis.

Another major potential for natural disaster is Nepal's "sandwiched" position between two major tectonic plates, the Eurasian Plate to the north and the Indian Plate to the south, as can be seen in Fig. A4.3. Several north-dipping thrust faults, lying to the south of the South Tibetan Detachment System (STDS) have been identified by Jouanne (2004):

- the Main Frontal Thrust (MFT, black line in Fig. A4.3)
- the Main Boundary Thrust (MBT, red line in Fig. A4.3)
- the Main Central Thrust (MCT, blue line in Fig. A4.3).

The Tectonic Plates, being in constant motion towards each other, cause an underthrusting of the Indian lithosphere beneath the higher elevated Himalayas and Tibet lithosphere, the Eurasian Plate. It is estimated that up to half of the continuing convergence between India and Eurasia is absorbed by underthrusting of the Indian lithosphere, beneath the Himalayas and Tibet, at a rate of an estimated 50 mm per year towards the North. Antoine (2004) states, that vertical displacement rates, expressed with reference to the Gangetic plain, indicate the current uplift of the high Himalayas at 6 mm per year.

The thrusts are generally younger from north to south and on the basis of these faults the structure of Nepal is generally subdivided into five tectonic zones (Fig. A4.3):

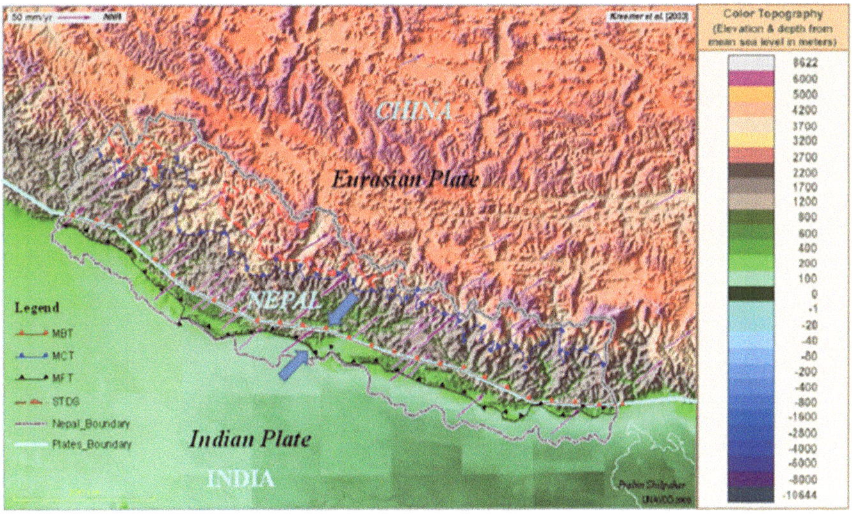

Fig. A4.3 Geological map of Nepal, showing the major Tectonic thrusts, from east to west between the Indian plate to the south and the Eurasian Plate to the north. *Map source*http://en.wikipedia.org/wiki/File:Nepal_Thrusts.jpg

1. Gangetic Plain in Nepal's Terai, the area south of MFT, bordering India
2. Sub-Himalayas, in the Siwalik hills and mountains, the area between MFT and MBT
3. Lesser Himalayas, the area between MBT and MCT
4. Higher Himalayas, the area between MCT and STDS
5. Tibetan-Tethys, the area north of STDS, bordering China/Tibet.

Three of the big Nepalese earthquakes (in 1905, 1934 and 1950, with magnitudes around 8 on the Richter Scale), were caused by a sudden under-thrusting of the Indian Plate beneath the Eurasian Plate and the resulting, abrupt release of, gigantic amounts of energy. Most of the earthquakes occur between the MCT and MBT, where some of the biggest urban and rural settlements are located. Thus Nepal, and most of its population, is under a permanent threat and fear of earthquakes.

The 6.9 magnitude earth quake in the East of Nepal, bordering Sikkim, on the 18th September 2011, causing the death of 112 people, was different, as it was along the north-south transverse of the main east-west Himalayan thrust faults. Thus, the ever-present fear of the long overdue "big" earthquake, which is expected to take place along the main East-West thrusts, continues. Or, as the Himalayan Times said: "The occurrence of this quake does not lower down the threat from the great earthquake which we expect in the Himalayas. The one (or more) which is expected will be a thrust type earthquake on the detachment which will be much more devastating than this one,.." (The Himalayan Times, 19th September 2011).

Appendix 5
Ethnicity and Religion in Nepal

Nepal is a unique country, not only because of its natural beauty and climates as described above, but also because of its people. Two main groups of people make up Nepal's population. The Aryans, with various sub-groups who live mostly along the Indian borders and the main urban settlements, and the Mongols, consisting of many different tribes, living mostly in the mid-hill and high-altitude mountain regions. Together they make up Nepal's population of ~30 million (estimated July 2010). Approximately 75 % of them live in remote, difficult to access rural and mountain areas, thus the majority of the population have no access to roads, clean drinking water, health facilities or electricity (Karki 2002). Fang (2007) also indicates that 42 out of the 75 districts are considered food deficit, acute food deficit and permanent food deficit areas. Further, Nepal is one of the only countries in the world with a lower female life-expectancy of 44–48 years in remote mountain areas, and 60.75 years (CIA 2000) in urban areas. Men live longer, 46–52 years on an average in remote areas,[5] and 61.12 years (CIA 2009) in urban settlements. High fertility rates (>4 children born per women in 2007[6]), feed the high population growth rate of >2 %, with an urbanisation rate of 4.9 % (CIA 2011) (Fig. A5.1).

Nepalese are in general tolerant and flexible in terms of religious beliefs and few religious tensions exist in Nepal among the accepted religious communities. Nepal also has a rich cultural and architectural history, with the Kathmandu Valley boasting the world's densest collection of world heritage sites.

The Nepal census of 2001 reveals that officially 92 different languages are spoken (although other sources quote over 100 different languages), by 103 distinct castes and ethnic groups (also enumerated by Nepal's 2001 census). It is not unusual that in a valley, divided by a stream all year around, the people on one side of the valley speak a language which the people on the other side of the valley can not understand. For centuries they could see each other across the valley, as the divide, the cold, furious mountain stream, is not wide. But there was no bridge

[5] The statistics for the remote areas are from the writer's own survey data over a 16 year time span (1996–2012) working in the remotest parts of Nepal.

[6] http://www.indexmundi.com/g/g.aspx?c=np&v=31, and http://www.indexmundi.com/nepal/total_fertility_rate.html (both accessed July 20, 2012).

A. Zahnd, *The Role of Renewable Energy Technology in Holistic Community Development*, Springer Theses, DOI: 10.1007/978-3-319-03989-3,

Fig. A5.1 Nepal, landlocked between China/Tibet to the north and India to the south, east and west lies along the Himalayas at 26°22′–30°27′ northern latitudes, and 80°04′–88°12′ eastern longitudes. With a total surface area of only 147,181 km^2, it has one of the most unique geographical, climatic and meteorological situations of any country in the world (map from: Magellan Geographix 1999)

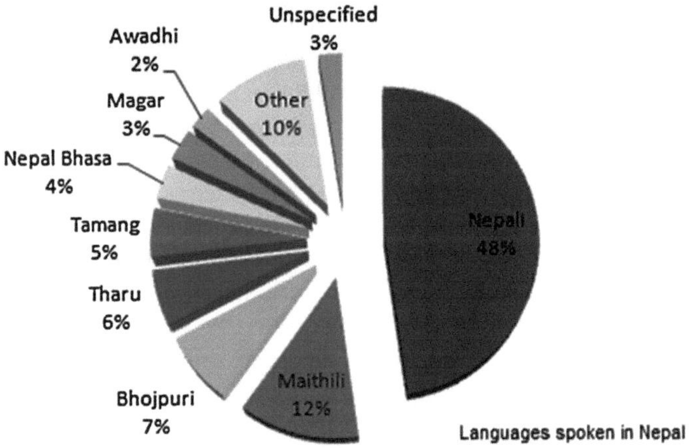

Fig. A5.2 Population distribution in per cent of the main languages spoken in Nepal. *Source*http://en.wikipedia.org/wiki/File:Languages_Spoken_In_Nepal.png

to cross or a rope to pull oneself across the stream. Thus there was until recent times no possible communication and thus different languages, cultural behaviour and habits developed.

The following pie chart provides an indication of the distribution of some of the major languages of Nepal's 92 officially identified languages (Fig. A5.2):

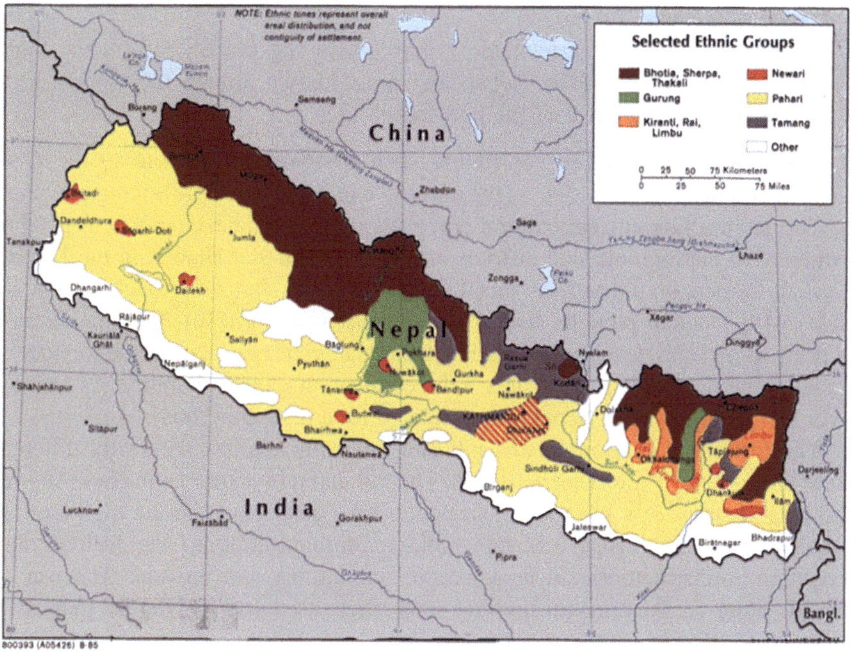

Fig. A5.3 Geographical distribution of the living areas of the main ethnic groups in Nepal, (map: http://en.wikipedia.org/wiki/File:Nepal_ethnic_groups.png)

The geographical uniqueness of Nepal with the Himalayas, the mountain and hill ranges, often covered with thick forests, and the hot and humid tropical region in the south, dictate that Nepal must have a great variety of ethnic groups inhabiting these different regions. As indicated above, there were 103 different ethnic groups identified in the 2001 countrywide census. For a rough indication of where the different major ethnic groups live, see Fig. A5.3.

The following list includes the major caste/ethnic groups among the 103 identified ethnic groups in the 2001 census:

- Chettri (middle caste, 15.8 %), with an European-Aryan ethnicity
- Bahun (high caste, 12.7 %), with an European Aryan ethnicity
- Magar (in the mid hill regions, 7.1 %), with a Mongol ethnicity
- Tharu (in the flat, subtropical Terai, 6.8 %)
- Tamang (in the mid-hill regions, 5.6 %), with a Mongol ethnicity
- Newar (mainly in the Kathmandu Valley, 5.5 %)
- Muslim (along the Terai belt bordering India, 4.3 %)
- Kami (low caste, 3.9 %),
- Yadav (Traditionally, classified under the Kshatriya, often pastoral occupations relating to cattle, 3.9 %)
- Rai (in the mid-hill regions, 2.7 %), with a Mongol ethnicity
- Gurung (in the mid-hill regions, 2.5 %), with a Mongol ethnicity

- Damai (low caste, 2.4 %).
- Of the remaining 91 caste/ethnic groups, including the Sherpa (in the high-altitude mountains in East Nepal, and most known around the world), each constitutes less than 2 % of the population.

Even though agriculture contributes such a significant amount to the GDP (38 %), and most of the people in the mid-hill and high-altitude hill/mountain areas are subsistence farmers, agricultural processes are not well developed in Nepal. For example, in the high-altitude districts of Jumla, Mugu and Humla, in the Karnali Anchal, farmers do not use any metal on their ploughs, but still plough their stony fields with wooden ploughs (Fig. 1.10). Since the majority of the population is dependent on subsistence agriculture, decreases in yields due to adverse weather or losses at the post harvest stage have a direct effect on food availability. As a result, over half the districts in Nepal are considered food insecure (2005).

The farming conditions in the hill and mountain regions are vastly different from those of the Terai (Khadka 1986). The hill and mountain regions have shorter growing seasons, less farm land and poorer soil, particularly in the high-altitude mountain regions. Crops are vulnerable to drought (due to the lack of any irrigation infrastructure), adverse weather conditions and erosion. The use of animal dung for heating means that it is often used for fuel rather than as a valuable fertilizer (Amatya 1998).

In the Mid-West Region, mountain areas such as Humla (Figs. 1.5, 1.6, 1.8, 7.2), food security and availability is also affected by additional food transport costs. Due to the lack of roads and rough terrain, markets are almost non-existent. Those that do exist, such as in Simikot, Humla's district centre, are extremely high priced as all the out of district food items, such as rice, sugar, salt, etc have to be flown in by airplane. Thus, staple food such as rice, costs up to 300 % more[7] compared to the Terai.

Agricultural work is different in the flat, low altitude, hot and humid, Terai. There, good soil and milder weather permit the year round cultivation of a variety of crops such as rice, wheat, maize, sugarcane and vegetables (Fang 2007). Reliable water resources, flat terrain and higher cropping intensity make it possible and cost effective to install irrigation systems. Food security issues in this surplus area are attributed to access to finance rather than scarcity (Fang 2007). These easier conditions (climate and availability of technical and agricultural tools) not only bring forth much larger crops, but also lower the cost of food.

Nepalese are commonly regarded as religious people and almost all tasks and events in their lives are believed to be connected with the spiritual world. There is a mix of religions followed by the local people. The main religious beliefs in the country are: Hindu 80.6 %, Buddhist 10.7 %, Muslim 4.2 %, Kirant[8] 3.6 %, Christians 0.5 %, other 0.4 %. Nepal was the only Hindu kingdom in the world

---

[7] These figures are from the writer's own survey data over a 16 year time span (1996–2012) working in the remotest parts of Nepal such as Humla and Jumla.

[8] The Kirant religion it is a blend of animism, such as ancestor worship and Shamanism, Shiva Hinduism (which is the oldest of the four main strands of Hinduism), and Tibetan Buddhism.

until the 18th of May 2006. On the 19th May 2006, Nepal became a secular state. Hindu temples and Buddhist shrines are scattered all over the mountains and valleys of Nepal.

The Nepalese are generally very family oriented, and modest. Due to their religiosity, fortune and misfortune are widely accepted as the will of fate or *karma*. Time is thought of more as a series of events or tied to seasons or ages than as a matter of minutes, hours or even years. The majority of the people believe to some extent that the *bhoot* (ghosts), *pret* (evil spirits), *bokhsi* (witches), and *graha dasha* (a bad position of the planets) can cause diseases in people and livestock, crop failures, or accidents. Incense, flowers, and food are offered to pacify the spirits and planets. It is very common that a rooster (for poor people) or a male goat (for richer people) may also be sacrificed to appease an evil spirit, ghost or deity.

Caste consciousness pervades this predominantly Hindu society and deeply affects the way people behave towards each other. Historically, Hinduism organised society into four main social castes:

- Brahman (priests)
- Kshatriya (nobles and warriors)
- Vaisya (traders and farmers)
- Sudra (servants).

These groups are further subdivided into many (literally thousands) of castes and sub castes, making it very difficult even for local people to fully grasp it.

In remote places such as Humla, people follow mainly two religious beliefs. Shamanism (a mixture of Hinduism, Animism and witch craft, with differences among them dependent on the people group and area), and Tibetan Buddhism (immigrants from Tibet). In each village *dhamis* and *jhankris* (witch doctors) are consulted as local healers and sorcerers. Nepal is believed to be the birthplace of Buddha, the founder of Buddhism (in the Terai city bordering India called Lumbini). There is a complex blending of Hinduism and Buddhism in the culture of Nepal.

Appendix 6
The Culture of Nepal

In the following a few indicative figures are provided to define Nepal a bit more with regard to its people and society. The figures are mostly taken from the CIA (2009) web site, as this web site is the one most frequently updated and matches closest with the author's knowledge. Further, where justified, the author's own experience is added. Following this section a more complex analysis is provided, to lead the reader "behind" the curtain which masks the culture, habits and some of the main reasons why things are the way they are in Nepal. It aims to bring more insight into patterns of the society which are still strongly at work today, dictating much of Nepal's culture, daily life, work and in particular its struggle for development (Fig. A6.1).

(1) By the middle of 2010 it is estimated that Nepal's population passed 30 million.
(2) Estimated population growth in 2008 was 2.095 % (CIA 2009)
(3) Estimated unemployment rate of 42 % (CIA 2009)
(4) Urbanisation rate of 4.9 % (CIA 2010) per annum (Fig. A6.2)
(5) Estimated sex ratio for 2008 for the total population: Male/Female=1.06/1 (CIA 2009)
(6) Estimated infant mortality rate for 2008 (CIA 2009). Male: 60.18 deaths/1,000 live births. Female: 63.91 deaths/1,000 live births
(7) Estimated total fertility rate for 2008: 3.91 children born/woman (CIA 2008)
(8) Major infectious and other diseases (CIA 2009): Food or waterborne diseases: bacterial diarrhoea, hepatitis A, and typhoid fever Vector borne diseases: Japanese encephalitis and malaria Other: Altitude sickness
(9) Estimated literacy rate (defined as age $\geq$15, and can read and write): total population: 48.6 % (CIA 2009) male: 62.7 % (in urban areas up to ~80 % and in remote areas as low as ~15 %) female: 34.9 % (in urban areas up to ~70 % and in remote areas as low as ~4 %)
(10) Political parties: 25 recognised political parties (CIA 2009)
(11) Manpower available for military service (CIA 2009): Males age 16–49: 7,322,965 Females age 16–49: 6,859,064.

While numbers are useful and helpful to give a clearer understanding of a country's people groups, size and main living conditions, they can not provide any

A. Zahnd, *The Role of Renewable Energy Technology in Holistic Community Development*, Springer Theses, DOI: 10.1007/978-3-319-03989-3,

Fig. A6.1 Women from the remote north-west Anchal of Karnali, with no road access within days of walking and a low population density of less than 100 people per km^2. They all live from the labour of their hands and what their fields are able to bring forth

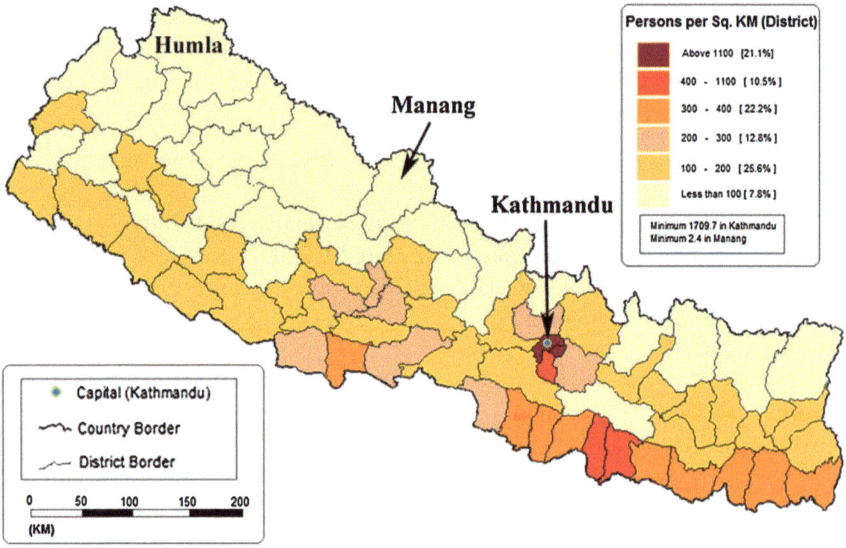

Fig. A6.2 Population density map of Nepal. The highest population density is found in Kathmandu with 1,710 people/km^2 and the lowest population density is found in Manang district with 2.4 people/km^2. The more fertile, flat, low-altitude land in the south of the country is much more densely populated compared to the high-altitude land to the north, which is meagre, rough and difficult to work. The still high population growth forces people to leave the higher lands for the south and urban settings, as the tradition of splitting the land among the sons has divided the arable land so much that it can not support the people anymore, thus they have to migrate to the lower terrain, or even to India, to find menial work. (map from: http://en.wikipedia.org/wiki/File:Population_density_map_of_nepal.png)

information on "why things work they way they do". These, more anthropological factors, are much more difficult to identify and even more difficult to understand, as to comprehend a culture in a defined context, is a time demanding and difficult task.

With Nepal having been isolated from almost any foreign influence from the middle of the seventeenth century to 1950, particular cultural habits and ways of life have developed. These are the unspoken behaviour patterns and to the eyes of a short-term visitor to Nepal, hidden reasons why things are the way they are. Thus in the following, an attempt is made to explain some of these less obvious, qualitative and hidden cultural factors of Nepali society. These are important to know if one intends to work and participate in the life of the Nepali people for some extended time period.

In this undertaking, besides the available literature, the author's experience of living and working in Nepal since 1983 provides some distinctive insights. However, in order not to lose track and to highlight the important, Nepali specific cultural issues, the book "Fatalism and Development—Nepal's Struggle for Modernisation", from Nepal's most famous anthropologist, Dor Bahadur Bista (Bista 1991), is indispensable. It provides a wealth of depth and insight into the mindset of the local culture and its people. In particular it provides an insight into the way people think and act, based on Nepal's deep rooted cultural background, which developed in isolation over the last two centuries. It addresses a neglected and publicly taboo field, revealing the way of thinking and acting of people born into, and shaped by, the Nepali culture. It provides valuable and important information for anyone involved in community development projects, by providing a better understanding of why things are done the way they are in Nepal.

Nepal is a heterogeneous society with a complex ethnic mix. The human element in Nepal is diverse and its contribution in defining the development potential of Nepal is complex. However, this diversity provides potentially valuable resources in regard to the social qualities which, with the right external and political environment could facilitate the development that the country needs. These qualities include a strong commitment to productive labour as today most of the communities are subsistence farmers who could not survive if their fields do not bring forth the life saving harvests. Other valuable qualities of the people include a high capacity for endurance (for which the Gorkha soldiers are famous), cooperative organisational styles (seen at the village community level, where the local villagers together build a house for one of the village members), and a high ability to adapt at individual and social levels (experienced for example during the frequent natural calamities such as droughts or landslides to which the people are regularly exposed). Further, over the centuries, the Nepalese have been known to be good merchants and craftsmen, using the natural resources of their land.

However, Bista (1991) points out, that the current prevailing nature of Nepalese society is such that some groups with positive elements in their value systems and in their social and communal practices are increasingly excluded from the mainstream of society and their values are endangered as another, essentially alien, culture becomes more pervasive. This other culture, the culture of fatalism, includes values and institutions that are inherently in conflict with, and unsupportive of development. Practical experience shows, that in Nepal, the culture of fatalism devalues the concept of productivity. This, for example, is reflected in the fact that most development and improvements in living standards has been attained through the long-term "heavy infusion of external aid". Bista (1991, 2) goes on to say that:

> ..the beneficiaries have not always been those who actually contributed to growth and production. But the irony is that those who enjoy the benefits continue to be increasingly unhappy and ask for more without contributing to the process of development.

He also points out that the inherent problem is not due to the great diversity of ethnic groups, but that conflict discloses a tension in the social structures whose origin lies in the class and caste systems. Bista (1991) defines them as having been "singled out as being at the base of many of Nepal's development difficulties". He continues that although the concept of caste was not unknown in ancient days, its rigid application and pervasiveness became apparent only in the twentieth century. He states (1991, 3) that:

> ...any rigidities associated with an ascriptive system will have an adverse effect on development and Nepal suffers from these problems.

Religion has always been a central part of Nepali life, but Nepal has never been plagued by any form of religious fanaticism. The Bahunistic principles, legitimised by the caste system, were introduced by the Brahmans arriving in Nepal during the medieval period. But it was the ruling elite, the Ranas in the twentieth century, who were responsible for rigidly applying it in order to secure and empower their own class status. The main consequence and adverse effect of this has been that since then there has developed an absolute belief in fatalism, with its emphatic sense that "one has no personal control over one's life circumstances, which are determined through a divine or powerful external agency". Bista (1991, 2–4) goes on to say that "the culture of fatalism devalues the concept of productivity", that it "includes values and institutions that are inherently in conflict with development", and that:

> ..this deep belief in fatalism has had a devastating effect on the work ethic and achievement motivation, and through these on the Nepalese response to development. It has consequences for the sense of time, and in particular such things as the concept of planning, orientation to the future, sense of causality, human dignity and punctuality.

Besides its relationship and connection with fatalism (as fatalism is strongly linked with various forms of dependency), dependency can be seen, though often in a more mature form, in the cooperative social behaviour of Nepalese indigenous groups. Nepalese are not strong individualists, but collectivism in itself is not necessarily an impediment to development. One example of the obvious, deep rooted and nationwide spread of collectivism in today's Nepalese society is the concept of *afno manche*, meaning one's own people, belonging to the "inner circle" of a group of people. While that does not make it inherently negative, it can be, and is, readily applied for unconstructive ends. It encourages the problem of inclusion—exclusion as group members gain favourable status and positions.

Being excluded from a chosen group (which could be a development project team) not based on professional reasons or lack of experience, can impede intended project goals and achievements. It is not uncommon to find that if one aims to belong to "the inner circle" that this can be acquired through the trading of privileges. Bista (1991, 4) summarises it in one sentence, saying:

> ..with *afno manche* one finds exclusionary tendencies, factionalism, failures in cooperation, and corruption in various forms leading to malfunctioning of development administration and dissatisfaction at every level.

Considering the two, dependency and fatalism, as being deeply rooted and combined in the society's belief and cultural systems, it explains why any undertaking or project has a hard time to bring forth sustainable accomplishments. It reveals that when combined, they exacerbate the tendency to impede development.

Bista (1991, 89–94) goes on to give another typical example found in practical daily life in Nepalese society, the practice of *chakari*, which had its origin in the religious ritual of obeisance, paying homage to a god to appease him or seek favour from him. This was first extended to the ruling classes and then in the later twentieth century, to all in certain positions of authority, influence or power. During the Rana period in the last century, people bragged about being a *chakariwal*, practicing *chakari*. In today's Nepal, where democracy is officially adopted, one does not brag about being a *chakariwal*, but those who have ambition and are interested in self-promotion know that this cannot be achieved without *chakari* and gifts (be it in the form of information or physical tokens of appreciation or gifts), in particular in governmental institutions.
Bista (1991, 5) explains *chakari* as:

> … a social activity which, in its most common form is simply being close to or in the presence of the person whose favour is desired. Instead of efficient fulfilment of duties and obligations, persistence in chakari is seen as merit, and with enough merit, favours may be granted. It is a passive form of instrumental behaviour whose object is to demonstrate dependency, with the aim of eventually eliciting the favour of the person depended upon.

As one can imagine, to practice *chakari* requires persistence and it can be seen often as a lifelong occupation. There are inherent vices in *chakari*. For example, once a *chakariwal's* "gift" is accepted, the patron feels obliged to respond positively to the *chakariwal's* terms. As the "services" of the *chakariwal* increase, the social obligation of the patron increases as well. This is a "one way path", and thus will lead the patron ultimately to a point where he is forced into decisions and actions in conflict with the decisions he would otherwise normally take, with the best interest and outcome for the project, the organisation or the business in mind. Thus the more highly ranked or strategically placed a patron is, the greater the ramifications for the society as a whole. In the long-term it can be seen that the whole *chakari* system can only play a negative role in any part of life, with particular impact on the political and public administrative life where it is still practised most often. Bista (1991, 93) puts it in the following way:

> Decisions are made, appointments determined because of pressing obligations formed through chakari and not as a result of an objective determination of what is best,

and:

> ..the entire social apparatus then suffers as critical positions are filled and governmental decisions made as a result of chakari. It is a built-in guarantor of incompetence, inefficiency, and misplaced effort.

It is said that families have been able over generations of practising *chakari* to even raise their family's class or caste status. Bista (1991, 5) comments on that saying:

> the strategy of chakari and the concept of productivity in those who practise it are alien to modern economic thought and systems, and can in no way support genuine development.

This practice is a vivid example of how deep rooted cultural behaviours, which are not readily visible or spoken about publicly, are still very much alive and practiced in the twenty-first century, in the midst of the striving of the nation to step out of the era of poverty, underdevelopment and insecurity. Bista (1991, 5) remarks that this practice is a sign that the country needs not only "to learn new things in order to progress, but must also unlearn old things". He is clear in his judgement when he says that hard as it will be to get rid of such a deep rooted cultural practice, unless it is purged from everyday life and from organisational life, Nepal's society in general and its economic development and hope for success will not become a reality.

Bista (1991) comes across rather strongly in his judgement of how these deeply embedded cultural behaviours and beliefs have an adverse effect on foreign development aid. He argues that the beliefs of fatalism and dependency "may allow perversion of development aid" (1991, 5). He goes on to explain that these beliefs can go so far that they blatantly manifest themselves in the "expectation of foreign aid as a divinely instigated redistribution" (1991, 5). Within that context foreign aid, meant for a defined development project, becomes merely something that is "justly due to Nepal and not a resource that is meant to be considered seriously and used productively" (1991, 5). Neither for that matter is education excluded from the influences of these beliefs. Bista states that "scholarship in Sanskritic tradition is associated with privilege and never with labour", and that "education is traditionally the prerogative of the upper classes" (1991, 5–6). What he means is that until recently, and even today if one looks at the names[9] of the ones with higher education or having had the privilege to study abroad with scholarships, almost all are from the highest caste and the upper classes. Thus to be educated is a powerful symbol of status and class identity, rather than a stepping stone to professionalism. In this way education is not perceived as a means of acquiring skills for future productive use in the society to secure more economic security, but is seen rather as

[9] Author's comment: In Nepal it is fairly easy to identify any person's caste according to his/her name.

an end in itself which, once achieved, signifies higher status. This in turn automatically brings more privileges.

Over the last two decades significant changes have started to occur. Educational opportunities increased for the lower classes through improved and adapted constitutional policies and educational scholarship programs. The lower class educated people see these new opportunities as their path for rapid social advancement, with which privileges and status are attached. This is believed because the acquisition of education is covered by the traditional beliefs. Thus once education has enabled one to reach the desired status, it is believed that one is "effectively removed from the workforce". Bista goes on to say (1991, 6):

> ….within such an atmosphere a genuinely productive professional workforce can never be developed and professional activities must continue to be performed by the untrained.

In order to have a more comprehensible sense of the impact of causality and fatalism on people's daily life in Nepal it is important to add a few explanations and illustrations. Even in the present generation of Nepalese from all castes it is widely believed that one's circumstances have been determined in advance and that one's life has been fated. If one is happy and or successful, it is because one must have earned this in a previous life. This is called *karma*. Therefore, one's life is guided at all times and it is not really possible to alter one's fate by any deeds, with the exception that some ritualistic deeds maybe could, or not, affect the powerful supernatural forces.

If this is believed, the course of future events is already determined and there is little sense to attempt to influence the outcome. This fatalistic belief greatly affects the way projects are planned, problem solving activities are addressed and how managerial responsibilities are exercised. Learned Hindus emphasise that the material world is *maya* and claim that this is not worth taking seriously. That results in people being made to bear poverty and deprivation only with a promise of a possible better rebirth.

Additionally, sometimes people are told by their priest, that they are chosen by a deity to be the ones to be poor and destitute so that they can remain devoted to the deity. Once people get used to these concepts and to their status of poverty, they are not really interested in working harder than is necessary for their subsistence. In addition, prosperity is often interpreted and presented as corruption. In any case, there is no dignity in labour as has been explained above. This concept is widely held amongst illiterate people, who express the view that any kind of work is equated with *dhuka* or pain and that they consider it an act of wisdom to avoid work. The opposite of dhuka is *sukha* or bliss, which means for them to live without having to work. People who can live without having to work are considered fortunate. This is supported by the testimonies of people on the search for a *jagir* (employment), which till recently even in the private industries, meant a job for life with a pay cheque at the end of each month but where one is not really expected to do any or much work. Further, many of the higher caste people do live without doing any visible work and seem to be living well and are certainly much better off than the poor, low caste people.

In contrast, low caste people have always been hard workers, and have developed high levels of endurance. They learned by experience, that when opportunities arose, they found that hard work led to better economic conditions. Therefore in many areas of Nepal, low caste people are fast improving economically when compared to people belonging to the higher castes who persist in carrying on with their traditional way of life.

The sense of a person's responsibility is also affected by fatalism, as responsibility is continually displaced to the external, typically the supernatural, over which individuals simply do not have any control. The sense of responsibility towards the society at large or a more specific given task is not supportive of the necessary development work at hand in Nepal. People know though that there is plenty of work which has to be done for Nepal's development, but they also know that means low status. Thus anyone who does not have to work, but can ask or demand others to work for them, is higher in status than those who work. For that reason, anyone who is educated, and thereby is in a position to identify with the traditional role of high caste, would never want to work. Supported by this belief and value system, people with a choice would not feel guilty for not working, and thus little work actually gets done.

In a hierarchical society, concepts of rights, privileges and obligations take on a different, limited meaning, as they exist only in connection with the caste. One's social circumstances are basically a matter of fate. Karma denies any possibility of holding oneself responsible for one's status in society and condition in life. If people are in less privileged positions this is a matter of their fate. To interfere with their lot is to interfere with the divine order and religious purpose. It is believed that people are in less fortunate positions in order to learn critical lessons in life, necessary for their spiritual growth and advancement, paralleling the caste hierarchy. And from a hierarchic perspective, to make things easier and to improve their lot is to interfere in their growth and jeopardise their life work.

Charity, not in the sense of alms-giving to a religious person, but as well known in non-profit NGOs and projects, is not valued under Hinduism. Fatalism also means, that one is never obliged to anyone for anything because everything occurs as it should, thus they do not try to interrupt it. Therefore no sense of obligation is instilled. This can be experienced by watching what happens when a priest receives a gift. He never thanks a giver. If at all, it is rather the giver who should be thanking the priest for accepting the gift, for in accepting it the priest bestows spiritual merit on the giver. The Nepali language also pays tribute to that. While there are ways to express gratitude, a word for "thank you" does not exist in the Nepali vocabulary, though in recent years a new word, *dhanyabad* has been coined to cover the translation of the English idiom "thank you".

If things are fated, it becomes difficult to attribute success, or to that extent as well, failure to individual action. Believing that one's life circumstances are fated, then personal success based on what is clearly individual effort and capability runs counter to these social expectations, thus threatening the widely accepted social order. This does not provide any incentives for people to give increased effort in order to get a job done or to achieve set goals or targets. It does not teach them to take initiative and responsibility to mitigate recognised failures or shortcomings, even less to accept defeat with dignity or grace. Thus there is a great lack of personal motivation, as it is perceived that personal effort would have little or no effect in altering the destined path of fate.

In today's world of globalisation it is difficult to imagine how one can get by without more complex social relationships, be it at work or on a private basis. For example, a marked outcome of strongly fatalistic beliefs is that people do not feel bound to keep any form of contractual relations, as there is anyway no free will or choice in decision making. Thus there is also no influence on how and whether the initial agreed contract will unfold over time. Plainly speaking, if a contract is completed it is because it was fated to be completed, not because the stakeholders of the contract felt bound to it and worked hard with determination to see it done within time and agreed scope.

An interesting aspect of the widespread sense of fatalism, that needs to be mentioned here, is the Nepali sense of time. Bista (1991, 84) states that:

> For Nepalese, time is like a flowing river,..., without a sense of beginning or end or particular divisions.

He goes on saying:

> Nepali culture does not have a sense of time as past, present or future; it is always an endless present

Thus as a rule, people living within this belief system do not plan the future. A busy person is usually running around for someone higher up, consequently he can not be of high caste or high status. That results in the traditional thinking that the more free time a person has on hand the higher and the more important he is considered to be.

Further, to live "constant" in the "present" reflects in the behaviour of the people as they are more oriented to consume now rather than considering how they are going to meet their needs in the days to come. Thus food, water and money are often spent right away without thought for the future. Consequently many people suffer from unnecessary deprivation due to the lack of planning for the future and due to the mismanagement of resources under their control.

This excursion aimed to lead the reader under the "first few layers" of Nepali culture, invisible to the short-term visitor or to the one unable to speak and understand the language. It is meant to provide some deeper understanding for people intending to live and work in Nepal. It is important to invest time and effort to start learning the language, gestures and cultural habits, in order to become more acquainted with the way life runs and, over time, to get to grips with more and more pieces of the whole puzzle. It will start if one enters the country and it will stop only when one leaves for good. Understandably, this is not just the case for foreigners who come to live and work in Nepal, but for any foreigner in any country, as the willingness to learn the new culture is also a sign of the basic respect one pays towards the host country's culture and ultimately its people, the wealth and most precious resource of any country. One may not agree with everything learned and experienced in a new culture, but this is the course of life and the privilege of those fortunate enough to be able to expose themselves to new worlds and cultures, and thus remain a learner throughout life.

With regard to what has been written above about Nepal's culture, the author wants to add that these are just a few cultural issues relevant to this thesis. It is believed they provide an important insight for people interested in development work in Nepal, which can only be conducted alongside the local communities and in a close relationship with the local people. Thus it is important for one entering the new sphere of a culture to give effort, interest and time, to learn and to come to know more about the "hidden reasons" of a culture, which are often unconsciously lived out by the local people. They are the key causes which help to explain why things run the way they do. Further it has to be added that the Nepali society is changing fast, with many signs of opening up to the wider world and with increasingly more international relationships and joint projects. That can only be positive in the long run for all stakeholders as one can learn the good things from each other and discard the bad things, or as Bista (1991, 9) succinctly says: "any culture needs to be scrutinised and its negative elements purged".

Appendix 7
Politics and Government in Nepal

Nepal's government (end of 2011) at a glance:

1. Country's name: Federal Democratic Republic of Nepal, or "Sanghiya Loktantrik Ganatantra Nepal" in Nepali language.
2. Capital: Kathmandu with the geographical coordinates 27°43′ Northern latitude and 85°19′ Eastern longitude, with an estimated population of 800,000–1,200,000.[10] The annual growth rate over the last decade was an average of 4.7 %.
3. Time zone: UTC + 5.75 (or 5 h 45 min ahead of Greenwich, UK time, or 10 h 45 min ahead of Washington, DC, USA, during Standard Times, or 3 h 45 min behind Perth, WA time). Nepal has no summer or winter daylight saving schedules.
4. Five Development Regions: Eastern, Central, Western, Mid Western and Far Western.
5. Fourteen Zones (in Nepal called Anchal): Bagmati, Bheri, Dhawalagiri, Gandaki, Janakpur, Karnali, Kosi, Lumbini, Mahakali, Mechi, Narayani, Rapti, Sagarmatha, Seti.
6. Seventy five districts.
7. Unified and called Nepal for the first time by Prithvi Narayan Shah in 1768.
8. Officially, (according to census 2001) 92 different languages are spoken in Nepal.
9. November 2006 peace agreement between the Government and the Maoists.
10. Democratic Elections: April 2008.
11. Cabinet formed: August 2008 by a majority coalition made up of the Communist Party Nepal (Marxist), Communist Party Nepal-United Marxist-Leninist, and Mahdesi Jana Adhikar Forum.

[10] Various estimations exist which vary significantly. Most are likely to be rather inaccurate as they only include the registered people in Kathmandu and do not include the mass of unregistered migrants. Thus it can easily be that the actual population in Kathmandu is 1.5 million or even more. This puts an enormous strain on the old insufficient infrastructure. See some figures at: http://population-of.com/en/Nepal/00/Kathmandu/, and http://apha.confex.com/apha/129am/techprogram/paper_28357.htm, and http://www.unep.org/Documents.Multilingual/Default.asp?DocumentID=498&ArticleID=5500&l=en&t=long (all accessed July 20, 2012).

A. Zahnd, *The Role of Renewable Energy Technology in Holistic Community Development*, Springer Theses, DOI: 10.1007/978-3-319-03989-3,

12. Prime Minister: Pushpa Kamal Dahal, better known with his Maoist name Prachanda, in office from the 18th August 2008 to 5th May 2009, then followed by Madhav Kumar Nepal from the Nepali-United Marxist-Leninist Party from the 25th May 2009 to 6th February 2011, and the Jhala Natha Khanal from the 6th February to 29th August 2011, and then Dr. Baburam Bhattarai since the 11th Sept 2011 from the United Communist Party, the Maoists.
13. Constituent Assembly: 601 seats, with a majority held by the two main Communist parties CPN-M (Maoists) with 220 seats and the CPN-UML (United Marxist-Leninist) with 103, and the main opposition the Nepal Congress (NC) with 110 seats.
14. Judiciary: Three tiers of Court: Supreme Court or Sarbochha Adalat (in Nepali), the Court of Appeal and the District Court. There is no distinction between Criminal and Civil Courts.

A more detailed account of the history of Nepal and its political leaders and Kings has been given in Appendix 3, thus only a short summary of the earlier political history is given and the focus will be more on the present constitution of the government.

After the unification of Nepal in the eighteenth century, the concern of the government was with consolidating its rule and battling against rebellious groups. With this emphasis on control, its form was largely military. The Sugauli Treaty, the peace agreement between the British East India Company and Nepal, ratified on the 4th March 1816, fixed the political boundaries of the nation. This ended the second British invasion of the Himalayan kingdom during the Anglo-Nepalese War (1814–1816).

This helped provide the sense of security that allowed the government to be less preoccupied with military affairs and to concentrate as well on other matters. Subsequently much of the effort of the ruling class turned towards personal ambitions, as the nobles built up a regime of intrigue, fostered by the in—and exclusive principles of *afno manche* and *chakari*.

In 1847, it was Jung Bahadur Rana, who came to power in the bloody Kot Massacre, when all his potential rivals were murdered. From then on the Ranas reigned with a strong hand without interruption till 1950. The throne was inheritable and so was the Prime Minister's rank, thus assuring that any potential danger and potential uprising could be averted at an early stage. The Ranas reigned in their own right, aiming only towards furthering the fortunes of their family and not the country, nipping any patriotism in the bud as it was regarded as suspicious and thus a threat to their personal interests. Thus at the time of their overthrow, Nepal was still a country living in its medieval stage.

King Tribhuvan, who came to power in December 1911, inspired by the Indian Independence, brought urgently needed changes to Nepal in 1951. For the first time, political parties, which were more like lobbying bodies, were formed placing some authority in the hands of the people. A parliament was formed, with members nominated by the King. This political system was inherently unstable, as the real power to rule still lay with the King and his Cabinet and Ministers. Further, fatalistic values also compromised the possibility of an effective political

Fig. A7.1 The 14 geographical regions called Anchal (in singular and plural form) of Nepal. (map from: http://www.un.org.np/reports/maps)

system. Then in 1955 King Tribhuvan's son King Mahendra came into power. Unable to elevate the quality of life in the country and incapable of putting the proper strategic activities in place to correspond to the increasing desire for material benefits of modernism, the system collapsed in 1958.

While the general election of a democratic parliament in 1959 brought forth the Nepali Congress as the ruling party, they could not deliver and thus King Mahendra, who remained at all times the ultimate power, declared in 1960 all parties to be illegal and arrested all cabinet members. Thus the government was once again in a monarch's hand.

In 1962 the constitution was amended and the first National *Panchayat* formed, with representatives from all 75 districts. King Mahendra's son, King Birendra came into power in 1972. In 1980 a referendum resulted in a party less system, and under the amended constitution, the ultimate power resided again with the King. By 1990 the country's political situation had once more deteriorated so much that the two main political parties, the Nepal Congress and the Communist Party rebelled against the party less Panchayat System, which finally collapsed under the pressure of the strong popular upraising in April 1990.

Subsequently the country was divided into five development zones (Fig. A7.2), and 14 political zones (Fig. A7.1) for purposes of planning and development projects. Additionally, the country was sub-divided into 75 administrative districts (Fig. A7.2), each headed by a Chief District Officer (CDO), a regular civil servant. He is responsible and endowed with all authority, for the district's related building

Fig. A7.2 Geographical regions of Nepal. Nepal is divided into 5 zones from east to west, the eastern, central, western, mid-western and far western zones, 14 Anchal (a cluster of districts) and 75 districts. (map from: http://www.un.org.np/reports/maps/npcgis/NatBio00002.jpg)

and maintenance of infrastructure such as roads, bridges, hospitals, schools, as well as the management of the natural resources such as forest and water. Additionally, he is responsible for the management and distribution of government grants and subsidies.

At the district level, each of the 75 districts has a number of VDCs or Village Development Communities, usually including 2–5 villages, dependent on their geographical location, remoteness and size. Each village has chosen community members representing their village at the VDC level. Each village, dependent on the village's size in regard to numbers of households, is sub divided into 3–6 Wards, with each Ward having 1–2 representatives. This complex political and administrative structure has been put in place to facilitate a more efficient development and modernisation of the country. But time showed that this arrangement of government is so massive in size and slow in functioning, that it developed into a certain kind of organism with a life of its own. It is self perpetuating and, due it its shear size, largely insensitive to the external environment and the functions that it should be fulfilling for the society at large.

Overseeing the general development of the country is the National Planning Commission (NPC), directly under the Prime Minister, who is its chairman. The NPC's purpose is to develop the five years plans that would methodically and efficiently develop and modernise Nepal. As it is written mainly by city based people, Bista comments (1991, 107):

> ...its five or six member, mostly economists, statisticians and geographers,..., In practice, the five years plan document is worth very little. The NPC also tends to ignore the people, local institutions, and even the Ministries once they have made their initial recommendations.

These instruments of government are still in place in 2011.

Since February 1996, when the Maoists publicly announced the beginning of the "people's war", till the peace agreement between the government and the Maoists was signed in November 2006, these government structures remained in place, though in remote areas in particular, they were most of the time even more crippled and inactive than before. The bloody and brutal civil war demanded a high price of the people. Many paid with their lives. Estimates of up to 20,000 and more killings were talked about and reported. The fact is that most of the remotely located families either have lost a son, the father or even a daughter in the ongoing conflict between the government troops and the Maoists' fighting units. The real figures of dead, crippled and wounded people will probably never be known.

In between, King Gyanendra, who came to the throne after his brother, King Birendra, was assassinated on the 1st June 2001, dissolved the parliament in February 2005, only to re-instate it again in April 2006 after massive popular street demonstrations. This led later to the declaration of a ceasefire between the Maoists and the government, and a Comprehensive Peace Accord (CPA) with the inclusion of the Maoist party in an interim government. The new parliament reduced the King to a figurehead, and later brought to an end the monarchy in Nepal, diminishing Gyanendra, Nepal's last King, to the role of an ordinary citizen, forcing him to leave the Royal Palace in June 2008. This ended the powers enjoyed by the royal Shah lineage over the Himalayan nation for 240 years (1768–2008).

The first democratic elections took place in April 2008, with the Maoist Party emerging as the leading party, with their leader later being selected as the new Nepali Prime Minister. Pushpa Kamal Dahal, better known by his Maoist name Prachanda (Fig. A3.5), took office in August 2008. The country's official name is now the Federal Democratic Republic of Nepal. Since then, the political struggles for survival and democratisation have not diminished and by September 2011 already the fourth Prime Minister is in place. Thus, rather than stability, the opposite has occurred. The country went through a very tough economic time, with people having no access to the essential energy resources such as petrol, diesel or kerosene for weeks at a time. In February 2009 the first violent uprising from within the Maoist's party, killing the Prime Minister's bodyguard, shows that the struggles for power are still going on. Also, by the end of 2011 the country still has no new constitution.

Only time will tell how many years it will take to heal the deep wounds, inflicted through the decade long civil war. This is not possible within a short 1–2 year period, as the opponents who previously fought one another to the last drop of blood, are now meant to lead the country towards increased development and prosperity. More realistically, to learn to forgive and go through the reconciliation process, to bury the grief and create a realistic, long-term vision for the country will take one or two generations.

Appendix 8
Poverty and Economy in Nepal

Poverty in Nepal is a deeply entrenched and complex phenomenon. (IFAD n.d.)

This section does not aim to be in any way comprehensive in defining what "poverty" means or has been defined as in the literature over the last few decades. Neither does it attempt to explain in detail what "poverty" is and/or can be in a defined context. However it is part of the thesis discussion because this thesis aims to be practical and rooted in the day-to-day life of people and communities who are extremely poor and will continue to be for several generations. They have struggled and will continue to struggle with poverty and its consequences for years to come. Within this scope it is helpful to have some definitions of previous thinkers, such as Mr. Ravallion (1992, 4) from the World Bank who defines poverty in the following way:

> "Poverty" can be said to exist in a given society when one or more persons do not attain a level of material well being deemed to constitute a reasonable minimum by the standards of that society.

This interpretation of poverty cannot be applied like a mathematical formula without knowing the particular context and people/community group. Rather, if the level and degree of poverty of a person, a family, a group or a village as a whole are to be defined as indicated above, one has to understand the identified people's, group's or village's life. One must experience at first hand their living conditions, or "dhuka and sukha" (sufferings and blessings) as one would say in Nepal, in order to understand their daily activities, hardships, sufferings, joys and community life. In this way it is possible to respect the people/community under analysis and keep their dignity intact while experiencing poverty in its full range of brutality, darkness and agony. With this mind, the following part of this chapter has been written.

One of the most comprehensive government surveys in regard to the poverty status of the different groups in the various main regions of Nepal was carried out in 1996, just at the time the Maoist insurgency began. Thus the following "Poverty Incidence by region: 1996" table (Bista 2008) and data are a summary, presented in Nepal's National Planning Commission Report (1998), of one of the most recent detailed and wide ranging household surveys in Nepal (Table A8.1).

A. Zahnd, *The Role of Renewable Energy Technology in Holistic Community Development*, Springer Theses, DOI: 10.1007/978-3-319-03989-3,

Table A8.1 The incidence of poverty by region in Nepal

Region	Population below poverty line (in %)		
	Poor	Ultra-poor	Total
Mountain	29.3	26.7	56.0
Hills	21.3	19.7	41.0
Terai	28.7	13.3	42.0
Urban	13.2	9.8	23.0
Rural	26.4	17.6	44.0
National average	24.9	17.1	42.0

Source National Planning Commission (1998, 203)

By the end of 2006, Nepal emerged from over a decade of violent, brutal and bloody civil war between the Maoists and the Government. Thus the above figures, old as they are, still represent fairly accurately the present conditions of Nepal's poverty and its distribution. Today, Nepal is a developing country with a national Human Development Index (HDI) of 0.428 which ranks it as 138th out of 169 countries according to the UN's Human Development Report (2010). Forty percent of Nepal's population is aged 15 years or less, and ~75 % of the population lives in rural, difficult to access areas, often with no road access within days of trekking and no electricity grid connection.

IFAD, a UN financial institution dedicated to eradicating rural poverty in developing countries, states that approximately 40% of Nepalese live below the poverty line of US$12 per person/per month. IFAD goes on to say:

> ..while the overall poverty rate for Nepal is 31 %, this figure increases to 45 % in the Mid-Western region and 41 % in the Far-Western region.[11]

If IFAD defines the poverty line as US$12 per person/per month (IFAD 2009), which is US$0.40 per day, the question arises, what has happened to the more commonly known poverty line definition of US$1 per day as initially defined by Mr. Martin Ravallion in the 1990 World Development Report[12] and now since end of 2008 redefined to US$1.25?

It has never been easy or clear what "poverty" is and how it is best defined. There have always been different opinions and literature on poverty measurement indicates as well that different people at different levels of living tend to hold different views about what "poverty" means. So as Ravallion (2008, 2) points out:

> ..the critical level of spending that a poor person would deem to be adequate in order to escape poverty is likely to be lower than the level a rich person would deem adequate to avoid becoming poor.

[11] http://operations.ifad.org/web/guest/country/home/tags/nepal (accessed July 20, 2012).

[12] http://www-wds.worldbank.org/external/default/WDSContentServer/IW3P/IB/2000/12/13/000178830_98101903345649/Rendered/PDF/multi_page.pdf (accessed November 6, 2011).

In 1990, Ravallion and the World Bank proposed that global poverty should be measured by the standards of the poorest countries. Thus based on the available national poverty lines of the poorest countries at the time, they proposed an international poverty line of US$32.74 per month per person, or US$1.08 a day, at 1993 PPP (Purchase Power Parity).
With the world-wide economy's dramatic changes over the last 15 years, the legitimate question has to be asked whether the $1.08 a day figure is still appropriate, and if "poverty" has to be redefined and indeed the "poverty line" has to be adjusted with a new benchmark in the light of today's much more globalised and interdependent, W century, economy? Further, it was recognised in 2005 that the US economy has experienced inflation since the US$1.08 poverty line was defined in 1993. The US$1.08 of 1993 is worth about US$1.45 in 2005 money. That raises the question of whether the poverty line should be re-calculated in today's changed world?

Acknowledging this, Ravallion and his research team collected and redefined, from 2006 onwards, 75 national poverty lines, ranging from the lowest, severe status of Senegal with US$0.63 per day to Uruguay's highest level of slightly above US$9 per day. Out of that range, the 15 lowest values, such as Nepal, Tajikistan and 13 sub-Saharan countries, were chosen and split the difference between them, emerging with a new international poverty line figure of US$1.25 a day. This new value, defined by the director of the World Bank's Development Research Group Mr. Martin Ravallion and his research team in their World Bank Policy Research Paper 4620, "Dollar a Day Revisited" (2008), is now the official newly endorsed yardstick for the poverty line.[13]

This shows that identifying poverty, and even more numerically categorising poverty, is not a straightforward task and can not be responsibly and correctly done with nationwide averaged data, in particular for a developing country of great diversity, such as Nepal. In order to pay due heed to the facts and reality on the ground, there was a need for re-calculation and revision for this social index the "poverty line".

Further, more justice is paid to Nepal's overall complex conditions and circumstances in regard to the definition of poverty if a range of additional values, such as for example caste adherence, family size or remoteness and accessibility of their village are considered to define the number of people living under the poverty line. Also, as Lekhal (2005, 203) and Hagen (1996) identified:

> …poverty in rural areas in Nepal has a direct impact on environmental degradation and biodiversity loss. Alleviation of poverty in rural areas is therefore vital not only for economic reasons, but also for preserving the environment and biodiversity.

---

[13] http://www-wds.worldbank.org/external/default/WDSContentServer/IW3P/IB/2008/09/02/000158349_20080902095754/Rendered/PDF/wps4620.pdf (accessed July 20, 2012), and http://www.timesofmalta.com/articles/view/20080828/business/more-people-living-below-poverty-line-world-bank (accessed July 20, 2012).

Thus, considering further detailed definitions of people's particular context and circumstances, it can be said more realistically that in Nepal, 30–45 % of the nation's population do not earn more than the US$1.25 poverty line. Following this, in order to be more specific and identify the poverty status of a specific village community or group, one has to draw closer, defined concentric circles around the particular cultural, economic and geographical context of the target group. This allows us to come up with a more accurate figure of the poverty rate within the above identified range, paying justice and respect to the target community's context.

While the definition of poverty and the numerical evaluation and assessment is important in order to identify the status of poverty in comparison with other countries, it also carries the danger that it "remains" a value with no attachment to real life. Poverty has many faces, and cannot only be defined in averaged economic values and figures, with no compassion, no relationship and no contact with the ones who are identified and talked about.

Thus, beside the academic way of defining poverty, one of the best ways to understand what poverty is, is to experience it first hand. That is, because poverty manifests itself vividly across cultural settings and boundaries in an inability to meet basic human needs such as food, shelter, work, clothing, clean water, fuel, health care and education, within people's cultural context. It denies human beings the chance to live dignified lives, with choices and opportunities for change and development. Further, it provides the understanding that women are much more vulnerable to the suffering and indignities of poverty than men in most societies and cultures. This factor is too often neglected or forgotten in reports, as it is hard to identify if not experienced first hand.

Additionally, it is hard to identify the effect of limited access to improved, modern energy services such as electricity and the burden that the heavy reliance on traditional biomass among poor communities brings upon them. These are hallmarks of poverty in developing countries (IEA 2002).

In terms of energy resources and uses, traditional fuel (biomass) consumption in Nepal represents $\sim$86 % of total usage nationwide and 100 % in the remote mountain areas. With an average annual per capita electricity consumption of around 70 kWh, accessible only to about 25 % of the nation's population, Nepal ranks very low globally in terms of the population's access to the grid. Infant mortality, compared to developed nations, is high, ranging from 53 to 86 per 1,000 live births for the richest 20 % and poorest respectively.

Thus one clear way to improve upon such conditions and situations is to work toward providing people living in poverty with access to, and control over, sustainable energy producing projects. It is widely accepted that:

> ..poverty alleviation and development depend on universal access to energy services that are affordable, reliable and of good quality (Reddy 2002; Saghir 2005).

Access to energy services is a key in satisfying the daily needs of human beings, and there are clear linkages between access to energy and reduced infant mortality and fertility rates and increased literacy and life-expectancy, as the WEA report explicitly illustrates (WEA 2000).

There is a clear relationship between poverty and remoteness (WEA 2000). Further, life becomes harder and the fight for survival tougher the more remote people are. In the remote hill and mountain valleys the terrain is rugged, the soil is infertile and difficult to farm. Rainfall is lower and increasingly less predictable. Thus agricultural harvests per household are the lowest in the country for remote, high-altitude communities. People's access to health services are minimal or non-existent and education has still not yet been seen and understood as one of the promising tools to break out of the vicious circle of fatalism and poverty.

Schumacher (1973) in his famous book, "small is beautiful" indicates that some of the primary causes of extreme poverty "lie in certain deficiencies in education, organisation and discipline". He goes on to say that in the context of poverty alleviation (1973, 138):

> ..development does not start with goods; it starts with people and their education, organisation and discipline. Without these three, all resources remain latent, untapped, potential", and "if the primary causes of poverty are deficiencies in these three respects, then the alleviation of poverty depends primarily on the removal of these deficiencies.

While it is difficult to be accurate due to the above identified reasons, it can be said that Nepal has made some progress towards poverty alleviation over the last decade from the mid 1990s onwards, despite the political instabilities and civil war. Poverty rates declined across all of Nepal's development regions and ecological belts, and the World Bank, in its "Poverty Reduction—Poverty Nepal[14]" web site provides a summary of the poverty reduction achievements in numerical values of the main strands of the society's communities.

- Headcount poverty rate declined from 42 to 31 % between FY95/96 and FY03/04
- Urban poverty declined from 22 to 10 %
- Rural poverty declined from 43 to 35 %
- Hill and Terai Dalits (lowest caste and poorest segment of the population) poverty declined from 58 to 46 %.

The standard of living improved (qualitatively) between FY95/96 and FY03/04:

- Agricultural wages and ownership of durables increased
- The actual consumption of 'luxury' foods rose (though mainly in urban areas)
- The proportion of households reporting inadequate food consumption declined
- Self-assessments of adequacy of housing, clothing, health care and children's schooling improved.

The World Bank indicated as well a very important factor, which is often neglected in the numerical value changes of poverty over time, the Gini

[14] http://web.worldbank.org/WBSITE/EXTERNAL/COUNTRIES/SOUTHASIAEXT/EXTSAREGTOPPOVRED/0,,contentMDK:20574069~menuPK:493447~pagePK:34004173~piPK:34003707~theSitePK:493441,00.html (accessed July 20, 2012).

coefficient.[15] For Nepal it has changed over the same time period from 0.342-0.411,[16] which is over a 20 % increase in inequality of increased income and wealth distribution among the Nepali people. It indicates that the dichotomy of poor and "better off" or "new rich" people in Nepal has dramatically increased and will probably continue in this fashion, a not unknown phenomenon in developing countries.

The World Bank's figures for Nepal's different poverty declines are indicative of this fact. While the urban poverty rate is now around 10 %, the poverty rate among the hill and Terai Dalits, the poorest ethnic caste (the untouchables), with now 46 %, indicates that almost every second person is considered to be living under the poverty line. This enormous difference, which will most likely continue to increase over the next decade, carries a high potential to raise political tensions among the various ethnic groups and political parties.

> ...poverty alleviation is what economic development is all about" (World Bank 1990).

While CIA in 2007 ranked Nepal 197th out of 225 nations with regard to an estimated US$1,500 GDP per capita, it down scaled that value to US$1,000 GDP[17] for the year 2008 (CIA 2000). The estimated GDP at purchasing power parity[18] of the country was US$31.09 billion, while the GDP at the official exchange rate,[19] is estimated to have been US$12.64 billion for 2008 (CIA 2009). By sector, the CIA allocates 38 % of the achieved GDP to the agricultural sector, 20 % to the industrial and 42 % to the service sector, including transportation, communications and tourism. In contrast, UNDP ranks Nepal as 68th out of 102

---

[15] The Gini coefficient is a measure of statistical dispersion most prominently used as a measure of inequality of income distribution or inequality of wealth distribution. It is defined as a ratio with values between 0 and 1: A low Gini coefficient indicates more equal income or wealth distribution, while a high Gini coefficient indicates more unequal distribution. From: http://en.wikipedia.org/wiki/Gini_coefficient (accessed July 20, 2012).

[16] http://web.worldbank.org/WBSITE/EXTERNAL/COUNTRIES/SOUTHASIAEXT/EXTSAREGTOPPOVRED/0,,contentMDK:20574069~menuPK:493447~pagePK:34004173~piPK:34003707~theSitePK:493441,00.html (accessed July 20, 2012).

[17] https://www.cia.gov/library/publications/the-world-factbook/geos/np.html (accessed November 6, 2011).

[18] A nation's GDP at purchasing power parity (PPP) exchange rates is the sum value of all goods and services produced in the country valued at prices prevailing in the United States. This is the measure most economists prefer when looking at per capita welfare and when comparing living conditions or use of resources across countries. For many developing countries, PPP based GDP measures are multiples of the official exchange rate (OER) measure. The differences between the OER and PPP denominated GDP values for most of the wealthy industrialized countries are generally much smaller. https://www.cia.gov/library/publications/the-world-factbook/docs/notesanddefs.html?countryName=Nepal&countryCode=NP&regionCode=sas#2001 (accessed July 20, 2012).

[19] A nation's GDP at official exchange rates (OER) is the home currency denominated annual GDP figure divided by the bilateral average US exchange rate with that country in that year. https://www.cia.gov/library/publications/the-world-factbook/docs/notesanddefs.html?countryName=Nepal&countryCode=NP&regionCode=sas#2001 (accessed July 20, 2012).

developing countries with an annual GDP per capita of US$252 (UNDP 2006). The World Bank puts Nepal with an average per capita Gross GDP of US$260 for 2004[20] among the poorest countries in South Asia, ranking it the twelfth poorest country in the world. Further, the author's practical experience shows that urban Nepalese earn multiple times more than people in rural areas, and that there is no, or just minimal, cohesion between them. Even in the context of this diversity of living standards, poor areas such as the people in the Humla district, have a per capita GDP of only US$72 (KIRDRC 2002). This shows that while national averaged figures are indicative for a country, they more than often do not represent the various communities existing and living alongside each other.

Considering UNDP's estimated GDP for Nepal (US$252 per capita), the UNDP indicates that this is around 40 % lower than Bangladesh, 55 % lower than India, almost 60 % lower than Pakistan, 75 % lower than Sri Lanka and Bhutan and about 80% lower than China (UN 2007). It also has the lowest economic growth rate among its neighbours, estimated at 2 % from 1979/1981 to 2003/2005 (Fang 2007).

An interesting factor in Nepal's economic changes over recent years is the significant increase in remittances from all the estimated 4–5 million Nepalese workers living and working abroad, sending their earnings back home. The World Bank indicates that the proportion of households in Nepal receiving remittances increased from 24 % in the FY 1995/96 to 32 % in the 2003/04 FY, and that the actual amount of remittance increased over the same time span by more than 80 %.

This major financial income source, earned in foreign countries, has become a substantial component in Nepal's domestic economy. This carries potential danger for the country's long-term economic conditions. Sending such substantial amounts of money earned in foreign countries into Nepal where it is spent, creates an economic boom fuelled by an unstable, volatile and artificially inflated source. As soon as a world wide recession, such as the one which started in 2008, takes its toll on mass unemployment, foreign, low educated labourers are usually the first ones to be laid off and often with no additional payments or securities. Thus the flow of remittance dries up very fast. The workers are forced to return to Nepal, but not with the hoped for and expected savings.

At the same time there is no work awaiting them in Nepal as the unemployment rate is estimated to be 42 % in urban areas, with the only opportunity to be engaged in agricultural work in the rural places from where most of the labourers come. Further, most of the Nepalese going to work in foreign countries are young men, from rural areas, between 18 and 25 years of age and with minimal education. Once they have been abroad for two to three terms (usually 8–12 years), they are too old to go back to school once they return to Nepal. Thus a whole generation of young men, still considered as a substantial part of the country's "work force", has no chance to participate actively in the development of Nepal.

---

[20] http://web.worldbank.org/WBSITE/EXTERNAL/COUNTRIES/SOUTHASIAEXT/EXTSAREGTOPPOVRED/0,,contentMDK:20574069~menuPK:493447~pagePK:34004173~piPK:34003707~theSitePK:493441,00.html (accessed July 20, 2012).

These are problems which have not been thought through nor raised in public, neither by the government nor any other organisation within Nepal. They will, in the years to come, be multiplied as the trend of leaving Nepal, in order to work in foreign countries, has increased, year by year, with the highest rates per day (estimated 1,500 people) leaving Nepal during 2008.

Besides that, the Nepalese economy is determined largely by the growth of its agricultural sector, providing a livelihood for three fourths of the population. It accounts for 38 % of GDP (Fang 2007). But the underperforming agricultural sector, with the majority of the population dependent on subsistence agriculture (WFP/FAO 2007), does not match up in any way to the fast increase in remittances flowing in from foreign countries.

Industrial activities are still small and include mainly the processing of agricultural products such as jute, sugar cane, tobacco, and grain. Other important sub-sectors include trade with China and India, transport and communication. Security concerns relating to the recent (1996–2006) civil war between the Maoists and the government have led to a sharp decrease in tourism, a key source of foreign exchange. This branch of income is again picking up since the peace agreement, but will take years to return to its previous level.

Prospects for a significant increase in foreign investment in Nepal will remain limited, due to the small size of its economy, its technological backwardness, its remoteness, its landlocked geographic location, its political instability, civil strife, and its susceptibility to natural disaster. The year 2008 though brought some glimpse of hope with an increased economic growth rate of 4 % compared to the previous 25 year average of around 2 %. Increased security since the peace agreement, resulting in increased tourism activities, some bumper crops and improved transportation, were the main reasons for this improved result.

Energy is another increasingly important commodity which has to be included in any economic overview of a country, as without access to modern energy services there are very limited opportunities for development. Thus it is important to look briefly at Nepal's present energy supply and its potential energy resources.

Nepal imports all of its fossil energy needs from India. This makes it not only politically dependent on its neighbour, but there is always the danger of having an unstable supply, unforseen price hikes and low levels of reserves in the country. In fact, these circumstances brought Nepal's economy to the verge of collapse during the final stages of the civil war in 2006, when the government owed the Indian government several hundred million US$ for petrol, diesel and kerosene bills. This resulted in the unavailability of fossil fuels for transport and cooking for months at the time, forcing car and motor bike owners to queue in front of petrol stations for days to get 2 L of petrol for a motor bike or 10 L of petrol for a car. These imports were also the main reason why Nepal's financial balances were negative during seven out of the nine years from 1996 to 2007. Only more recently, however, the financial account has registered small surpluses due to increased remittances from Nepalese working abroad and continued inflows of foreign aid (Fang 2007).

However, Nepal has an enormous amount of largely untapped renewable energy resources, in particular water for hydro power and sunshine which can be converted into electrical power and thermal energy.

The rivers of Nepal are the major contributors to the Ganga Basin in the north of India. The annual discharge of out flowing rivers from Nepal to India is about 236 billion m^3 (WECS 1999) from over 6,000 rivers, with many rivers losing a potential height of about 4,000 m within a north–south distance of 100 km. As previously mentioned, this creates a theoretical hydro power potential of 83,290 MW (USAID 2003; Warnock 1989; WECS1999b). The harsh terrain and difficult access to many areas limits the theoretically exploitable hydro power potential to a more technically realistic and economically profitable estimate of 42,130 MW (USAID 2003; Warnock 1989; WECS 1999b). With this figure, and an assumed capacity factor of 80 %, the annual energy potential of Nepal's rivers can be estimated to be around 300 TWh.

Until July 2005 only 546 MW (Regmi 2005), or 1.3 % of the feasible hydro power potential had been exploited. The average capacity factor was only 47 % (Gautam 2005) indicating that only half of the installed hydro power capacity was at any time available throughout the year. That increased till August 2009 to 635.845 MW (NEA 2009a) installed rated power capacity, from 69 major and small hydro power stations. While that amounts to 1.5 % of the technically and economically feasible hydro power potential of Nepal, the calculated average capacity factor of 44 % over the fiscal year 2008–2009, does not indicate any major improvements.

In addition, long dry periods, from October to March, force the authorities to introduce yearly load shedding in the main urban areas. Since early 2000, Nepal, and in particular its capital Kathmandu, experienced daily load shedding, increasing year by year. This culminated in February 2009 with up to 20 h per day of not having access to electricity. This condition, varying with the seasonal availability of rain water, will continue for the next decade. This is because of population growth, the minimal or lack of maintenance of operating hydro power plants, and the raging civil war from 1996–2006 which have not allowed any new major hydro power construction, thus leaving the country increasingly in the dark in the years to come.

Thus the consequences of wrong political decisions taken 15 years ago, a long-term unstable government with great political insecurity and no clear and realistic long-term vision for the country, are now borne by the local people and industries. Already the newspapers have reported towards the end of 2008 that the industries can not even pay the workers' salaries, rents and bank loan interest, as they cannot manufacture enough in the few hours remaining each day when they have electricity available, to cover even their basic costs.

Further, Nepal is also blessed with plenty of solar energy. It lies in the ideal 30° North "solar belt"[21] with about 300 sunny days a year (Renewable Energy South

[21] The convergence of heated air from the equator area in the Tropopause increases the air mass aloft around the 30° northern latitude (the Hadley cell). This in turn causes the air pressure at the surface to increase. Hence, at northern latitudes of around 30° a high pressure belt called "subtropical highs" (C. Donald Ahrens, Meteorology Today, 7th edition 2003, Brooks/Cole, Chap. 11, p. 286) is created, providing good favorable weather to utilize the incoming solar radiation.

Asia 2002; CRTN 2010). Solar irradiation measurements, conducted by the author in 2004–2008, in three different geographical locations in Nepal over the course of 4 years, showed that around 1,950–2,200 kWh/m^2/year (with an average of 5.342–6.027 kWh/m^2/day) of solar irradiation had been intercepted. The extreme values were 3.5 kWh/m^2 per day on a rainy overcast monsoon day in July in Kathmandu (at 1,350 m.a.s.l.), and 9.9 kWh/m^2 per day on a sunny late autumn day in November, in Simikot in Humla (at 3,000 m.a.s.l.).

Nepal is not a very windy country in general, as it is landlocked, and due to its unique geographical conditions, it has very few large flat areas of land. But there are various areas where rivers have cut deep north–south valleys into the massive Himalayan mountain range, stretching from the east to the west (Fig. A4.1, A4.2, A4.3, A11.9). Thus many valleys do have strong winds blowing, and villagers could benefit enormously from using their local wind energy resource. There are no national wind resource maps for Nepal available and therefore the country's potential wind resource can still not be assessed sufficiently accurately. Therefore locally available wind resources would need to be first evaluated/measured, which is time consuming and expensive, before potential wind projects can take place. For additional information on Nepal's renewable energy resources biomass, hydro, solar and wind, see 11.

Appendix 9
Development in Nepal

To provide a detailed picture of the development of Nepal within a reasonable amount of space, is impossible, considering what has been happening on all different fronts of development since the early 1980s. Thus, in order to set the scene and enable the reader at the end of the chapter to come to terms with some of the key features of the state and conditions of development in Nepal, it is important to include in this chapter, along with the more factual information, also some central anthropological and cultural mindsets of the Nepali society with regard to development. This will emphasise once more the importance of a deeper understanding of the culture's background, which determines many of the daily behavioural patterns.

Nepal was a roadless country in 1950. The extreme topographical and meteorological conditions in Nepal make development of infrastructure, such as drivable roads, exceedingly difficult. As can be seen in Fig. A9.1, indicated with the red lines, currently mostly only urban areas, along with the Terai and lower Hills, have serviceable, black topped roads. However, since 1950 an estimated 17,282 km (World Bank 2010a) of paved roads have been built by 2007 all over the country (Appendix Fig. A9.1). Nonetheless only about 40 % of Nepal's people have access to this road network, with the others within days of walking distance. The poor condition of the road network holds back the delivery of social services in the more remote regions, thus slowing down any kind of development. Additionally, the high transport costs into the remote areas add to the hindrance of Nepal's development.

Currently Nepal has 42 domestic and one international airport, in Kathmandu, connecting Nepal with the Middle East, Europe and the other major countries in South East Asia (India, Thailand, Hong Kong and Singapore), which themselves act as hubs to connect to all other continents. Most of the domestic airports, especially in the remoter hill and mountain regions, are still green fields with minimal navigation systems. All the airplanes serving these airports are now equipped with two engines, though none of them has modern navigational systems on board, thus all are flown by visual sight only. Consequently every year there are one or more detrimental outcomes, in particular during the monsoon season, when the lower and mid hills and valleys are covered with clouds.

A. Zahnd, *The Role of Renewable Energy Technology in Holistic Community Development*, Springer Theses, DOI: 10.1007/978-3-319-03989-3,

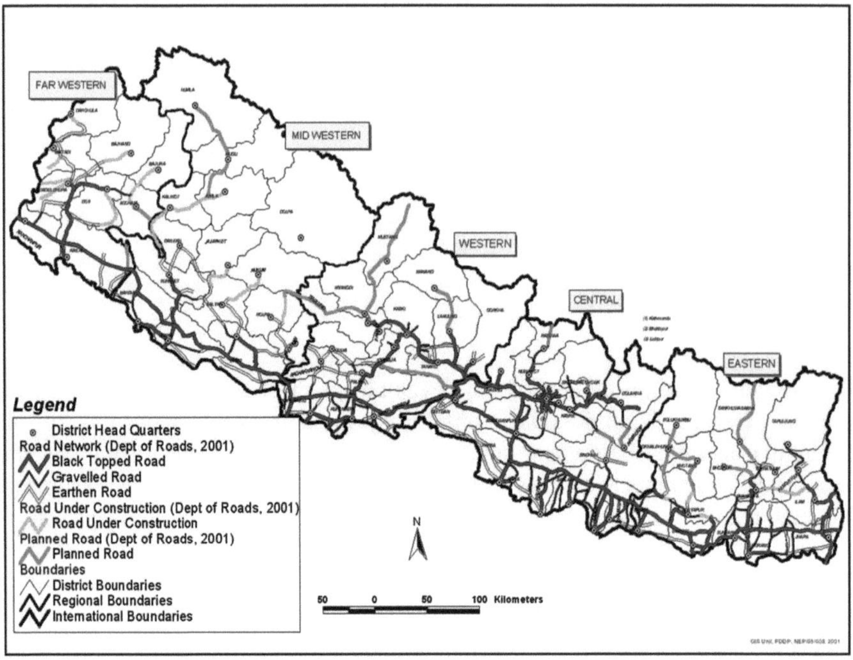

Fig. A9.1 Nepal's Road Network of paved roads (*red lines*) in 2001. Further, gravelled, earthen, under construction and planned roads all over the country, with the latter two probably difficult to provide any date of completion or start of construction. For example, the road from Surkhet to Jumla, in the mid Western region's Karnali zone, has been under construction (indicated with a *blue line*) for 30 years, and was initially planned to reach Jumla in 1986. (from: http://www.un.org.np/reports/maps/npcgis/NatInf00001.jpg, (available till May 2012, now the latest road map is available at: http://www.un.org.np/sites/default/files/2011-11-15-Nepal_Road_Distance_A3_22Aug2011.pdf, Retrieved August 10, 2012)

Nepal's GDP has grown by an average of 2.3 % in the early 1970s to an average of 7 % in the late 1980s, and was quite stable around 5 % in the early 1990s before the civil war. The insurgency, which started in 1996, had the effect of slowing economic growth to an average annual growth rate of 2 %. With the peace agreement in late 2006 and the new democratically elected Maoist government in place by August 2008, there is more chance of an increased GDP growth rate. But one has to consider that the decade long war and the global economic meltdown from 2008, will continue to have their toll on the country's ability to grow for years to come. The increased GDP rates from the mid 1980s till the mid 1990s were mainly due to the immense input of foreign aid and development projects, which, for understandable reasons, were withdrawn during the civil unrest (Fig. A9.2).

Historically, Nepal has been self-sufficient, as all its citizens were basically subsistence farmers for centuries, especially during the long period of the Rana regime which deliberately kept Nepal isolated from any external influence till 1950. The Nepalese may be poor by international standards but Nepalese peasants

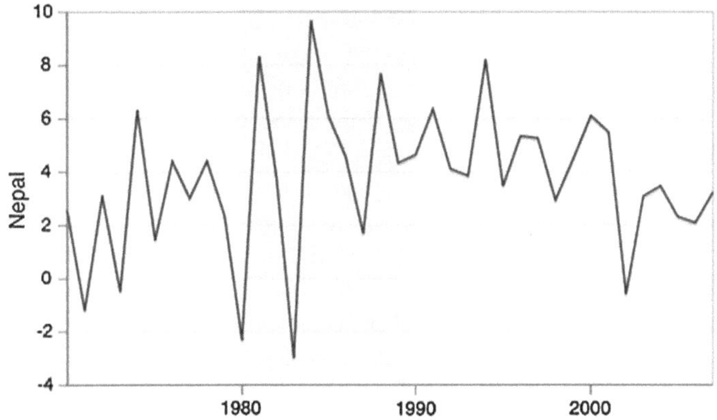

Fig. A9.2 GDP growth over the years 1970–2007 for Nepal. There is a clear trend, of higher GDP growth rates during the high foreign investment period from mid 80s to mid 90s, while the decade long civil war from 1996 to 2006 brought lower annual growth values. (source: http://www.swivel.com/graphs/show/11009802, in November 2011 no longer available)

have been, and many still are, self-sufficient and largely content. Thus the idea of foreign assistance is a novel one, and mainly due to their isolation from international affairs, the people had no idea that they were relatively impoverished compared with their neighbouring countries. That changed from the early 1980s onwards, when increasingly more foreign development projects started to take place. With an increased awareness of the relative affluence of the western world and the benefits of modern technology the desire for change and economic growth increased. Suddenly the awareness of being a poor country, compared to rest of the world, increased. Bista (1991) remarks that these changes began:

> …to destroy the gracefulness, charm, generosity and hospitality even among the rural people (1991, 133).

He goes on to say that the improved statistical and economic conditions in Nepal were not necessarily all positive. He concludes that:

> …people are paying their prices in terms of some positive human values which once lost will not be that easy to reinstate for generations to come (1991, 133).

As indicated above, the Ranas during their regime (1847–1950) were particular disinterested in foreign aid since it represented for them an alien intrusion into their private domain, and they intentionally kept the country and people isolated. They feared that any change in the social system represented a possible destabilising threat to their regime, and thus any process of change should be hindered as otherwise it might not be easily controlled. Thus it does not come as a surprise that Nepal, once the Ranas were overthrown in 1950, was ill prepared to meet the challenge of development. There was no national consciousness and no real national economy.

For the bulk of the people it can be said that their experience at the time was that the government was something at a distance, unconcerned with their welfare and preoccupied with its own self-centred interests. The concept of community was there, but it was limited to the local village society and context rather than beyond. Nepal did not share the legacies of past colonialism as did India, Bangladesh and Pakistan. Nepal had little or no sense of the outside world, had no established or functional infrastructure to support development and no trained manpower or professionals. Thus, while various foreign governments showed their willingness to provide aid, Nepal could not handle it as there was no effective indigenous administration to determine the country's needs, or to manage and distribute the aid.

From early days in Nepal's history, the idea of aid or assistance has been associated with religious activities, rather than any notion of giving charity or aid to the needy. Aid is not compatible with their fatalistic beliefs that all is determined, and thus a life of suffering is intended to either compensate for earlier misdeeds or prepare the self for some form of spiritual growth or progress. Bista (1991, 135) puts it in the following way:

> Charity to the poor and needy, outside of what is religiously prescribed, is not a genuinely altruistic act but might be seen more as an act of atheistic arrogance and as being ultimately negative and even destructive.

With this concept still around, though today maybe less significant than two decades ago, the attitude to the idea of aid is still affected by their opinions of begging and charity. Some of the more fundamental Hindu religious believers, are critical of foreign aid because they do not believe, as Bista puts it, "in the efficacy of the human hand" (1991, 135) in the process of development. In the theory of Hinduism it is thought that the present age, the age of Kali Yuga, is bad and any attempts to change it are looked upon as arrogance. Thus fundamentalists do not want foreign aid to interfere with what is determined for society. While this radical belief may not be shared any more by the majority of the people, it has still to be accounted for as a deeply rooted part of the culture and belief system and thus may not be verbally heard or uttered in this way. But certain actions and behaviours are still indicating that many subconscious decisions and working patterns are tainted with this train of thought.

Further, this is also having an impact on how the future is seen. Among the wider society it can be recognised that the sense of direction for the future is lacking. Even at the national planning level, the thinking has not reached the stage of pragmatism and empiricism (Bista 1991, 137). Within the Sanskritic tradition education is divorced from any practical use and seen and understood more for distinctive social positions. This in turn does not provide a very conducive environment for the needed pragmatic approach to community and economic development. In addition, the necessary planning and management skills for development, which means control and guidance, are placed in the hands of the planner and manager, run against the fatalistic principle.

Over the last three decades, numerous intelligent and able Nepali students, most of them from a high caste background, have studied abroad and returned with higher degrees and different new professional skills. They have been explicitly exposed, and were part of the culture, industry, education and way of life of the country they have chosen, or have been invited, to study in. They have learned new values and psychology necessary for success in the modern world. Upon their return many were placed, as they are now higher degree course graduates, in positions of authority and responsibility.

Many were initially very inspired when they came back, with lots of idealism, new ideas and ways they intended to implement and apply them. But too often, the new values and ideas they came back with, were abandoned once they were confronted with the persistent old ways and traditions, built upon the principle of fatalism. Thus, few of them could withstand that strong, invisible force of culture and tradition, and those who could were considered to be socially deviant. Most were compelled to accommodate their new ideas and intended innovations in such ways that they were rendered almost irrelevant, leaving behind great frustrations and disillusionment.[22] Those not willing to jeopardise their convictions will sooner or later depart to work abroad, and those who remain often become cynics and adapt to the chakari and afno manche culture.

Considering that a foreign aid or development worker, often specialised in his field of expertise, is new and inexperienced to Nepal, one can understand that he cannot be aware of the influences of culture at work within the society at large. He is utterly unaware and often insensitive to the peculiarities of the local cultural system and thus does not foresee any adjustments to the development model he adheres to and came with to Nepal.

A practical example of such a phenomenon, well known to all engaged in development work in Nepal, is the seminar. In its intended sense of meaning, a seminar is meant to show new ways, knowledge or examples of how things can be done, or are meant to be done, followed by practical application in a defined setting of a development project or business. In Nepal the seminar is much more commonly understood as an opportunity to demonstrate learning, scholarship and education level, and thus to raise one's stature and position. The activity of seminars is essentially a manipulation and a play on words. No real decisions are made and even less likely are effective actions and changes taking place afterwards. At such events it seems at times that it is more important that high quality food is served, in well known high class hotels. In this environment, seminars on poverty alleviation, rural community development, uplifting and empowerment of the poor are held. They are attended mostly by people who have never themselves lived or worked for an extended time period, in the situations the discussions are focusing on. Enormous financial resources are invested in such seminars, with very few useful outcomes.

Nepal is characterised by low levels of development and a high dependence on subsistence agriculture and imported goods. The combination of low agricultural

---

[22] The author, having worked with and in Nepali led companies and institutions for 16 years can testify that this experience is unfortunately still not unusual today. This issue is never verbally discussed or publicly dealt with, and often not even consciously realised by the one influenced, as the drift back into the mainstream cultural behaviours is subtle and slow, but in the long-term clearly recognisable by the surrounding people.

productivity, land constraints, natural hazards and high population growth causes short- and long-term food insecurity, particularly in rural areas. According to a recent FAO survey, 42 out of the 75 districts in Nepal were considered to be in food deficit (Fang 2007), where the amount of food available does not meet the needs of the population. Not only does the higher level of hunger affect the quality of life in these areas, it also fuels urban poverty by immigration of rural people to cities (Skoet 2006).

These conditions demand urgent and drastic development in the technology of the agricultural sector. With ~80% of the population directly dependent on agricultural production, the potential for improvement through agricultural development is enormous. But the people and institutions that have been recruited to lead this development are mainly academics with a high caste background, for whom agricultural labour is unthinkable. Most have never worked the soil as farmers, as they have used their academic education as an entry to a highly respected position. Thus what they learned never gets applied. Statistics show that Nepal has an impressive number of highly qualified agriculturists who could lead the country to bring to fruition its agricultural potential, but their talents and education are not being applied to practical problems.

Overall, agricultural production has not kept pace with population growth and average yields are low compared to neighbouring countries (Fang 2007). The use of fertiliser and pesticides is limited, due to limited availability and high cost, and where they are in use they are often applied in very inappropriate ways. Most of the people are not instructed in how to use these chemicals appropriately. There is an urgent need for investment in seed research and supply for the different climatic conditions, and in the extension of irrigation facilities. Food conservation techniques, such as drying and storing, are not well developed (Joshi 2002). The MOA reports (1997), that between 20 and 30 % of agricultural production is lost at the post-harvest stage (such as traditional sun drying of the crop).

Another important point to be mentioned is the issue of ownership. Often foreign aid is provided for a particular project. This can be a road, a bridge, a public building or a village based project. Often such projects are not defined and determined with local input and an initial base-line survey in partnership with the main end-users. Thus their participation is often limited to a paid labour job for a defined, often contracted, time period. After completion, and just before the main project implementation team moves on, the project is either handed over to the end-users, if it was a village based project, or to the public, if it was a public building project. But the connection, or better, ownership for the project handed over has not been established and thus is non-existent, or at its best is very weak and not understood in regard to the responsibilities of ownership for the end-users.

Neither initially, during the project's inception, design and planning, nor during the project implementation are enough practical steps undertaken towards participatory partnership and increased project ownership. Thus, while a small group of a village community may understand, within the village context, that they are the beneficiaries and thus need to be able to operate and maintain the new infrastructure put in place in their home or village, the intellectual leap between local community owned and public owned, is not made. Thus, the knowledge and experience that public projects such as roads, hospitals and schools, involving several communities, villages or even districts, need as well to be "owned" and cared for and maintained, is non-existent, even today. People, and the local government bodies, lack the sense of pride or ownership for such projects, compared to things they built through their own efforts.

This and the closely related and important issue of maintaining an implemented project go hand in hand. If no ownership feeling is present among the end-users, maintenance will not be done and thus the durability of the project is jeopardised. In that context, there is no justification to even talk about sustainability, a word which is being used these days too often as a buzz word. Further, forethought and funds for periodic maintenance of a completed project are rarely incorporated in the initial budget and thus not provided in Nepal. The general lack of support from the government for post-construction operation, maintenance and repair is still a major issue.

This spills over to Nepal's health sector, which is in particular affected by the above phenomenon, as all health facilities, except for a few new private hospitals in the major urban areas, belong to the public sector, and thus are publicly owned. Again this affects the people and communities in the rural and remote areas more than the urban dwellers.

Generally, the rural areas in Nepal are disadvantaged due to a lack of health services. As of 2005, health facilities in Nepal consisted of 84 government hospitals, 188 primary health centres, 698 health posts and 3,129 sub-health posts (WHO 2005). The hospitals and health centres are concentrated around urban and low-altitude areas. Remote mountain areas are mainly serviced by health posts and sub-health posts, where the availability of basic health equipment and drugs does not cover a whole year's needs (Emeriau 2006).

An FAO/WFP mission visited one sub-health post in the remote, high-altitude north-west district of Humla (1.1.8), that receives medical supplies once a year and generally runs out of medicine after 2 months (Fang 2007). Staff employed rarely include doctors or nurses, and services are limited to vaccinations, distribution of micro nutrients and basic first aid (Fournier 2007), if the drugs and medications are available and not out of stock.

Health outcomes, particularly for women and children, are negatively affected by living conditions in which people rely upon open indoor fire-places for heat, cooking and lighting. This is one reason why women in Nepal still have a lower life-expectancy than men, with average life-expectancy differences of up to 2 years in the remote areas. There is an urgent need for better health care facilities, with an emphasis on rural regions and in particular in the field of maternity health care and facilities.

Most of the women in remote rural areas give birth to their babies in the unhygienic environment of their mud home, in the cow shed or in a specially dedicated "bleeding hut" where the pregnant woman has to stay for up to 14 days, giving birth to her baby on her own, in total seclusion from anyone from the village, due to her "polluted" status. This status lasts for the first 2 weeks. Not

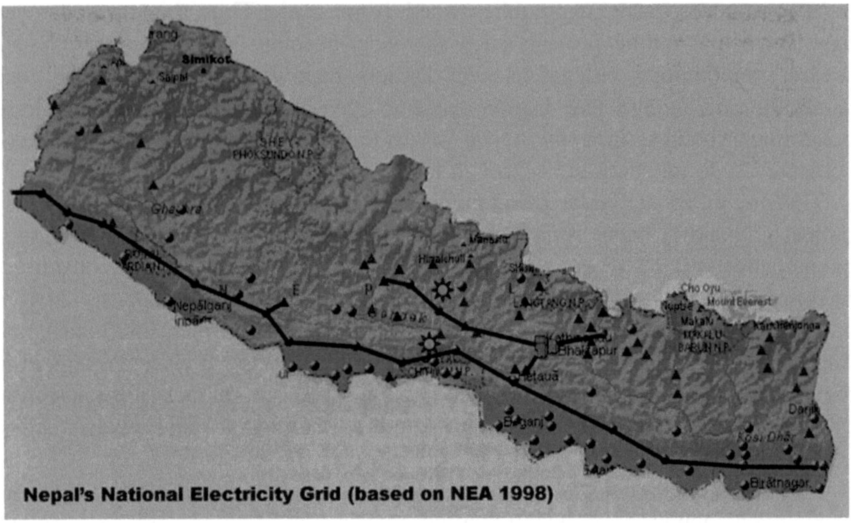

Fig. A9.3 Nepal's national Electricity grid (only the main transmission lines) in 1998. The map shows clearly that mainly the flat, low altitude Terai, bordering India, with the exception of Kathmandu the capital and Pokhara, Nepal's main tourist destination, have a major grid connection transmission line. (map: MSc RET Dissertation Alexander Zahnd, 2004)

even a locally trained birth attendant or midwife, who is often provided in other villages, is allowed to help.

As already indicated, improved access to more modern energy services and facilities, is a central issue in development endeavours. In Nepal this is a problematic issue. Karki (2002) says that only 18% of the population have access to the national electricity grid, most of which is concentrated in urban areas. In rural areas, rugged terrain makes grid expansion difficult and the combination of low population densities and low demand make it much less cost effective and economically unattractive for independent power producers (IPPs) and NEA.

As a result, only 4 % of the population in rural areas have access to the national grid (Karki 2002). The NEA forecasts the electricity needed up to the year 2020 (NEA 2009b). For the year 2009/10 the expected peak electricity demand was around 821 MW. The nationwide installed hydro power capacity in August 2009 was 635.845 MW_R (NEA 2009a). But with an estimated 500–520 MW only functional[23] and a capacity factor of 44 %,[24] the average power generation capacity is only about 230 MW, which covers just about 28 % of the maximum peak load (Fig. A9.3).

Many studies have been conducted over the last decade, outlining and highlighting the relationship between poverty alleviation and access to clean, efficient

---

[23] Personal experience of the author of the hydro power situation and conditions in Nepal.

[24] Calculated from the NEA electricity generation data for 2008–2009, in *A Year in Review—Fiscal Year 2008/2009*, Annual Report, Chap. 3.3.2 p. 4.

Fig. A9.4 Nepal's national Electricity grid (major and minor transmission lines) in 2009. In the legend the *green line* represents the 220 kV, the *red line* the 132 kV, the *brown line* the 66 kV and the *pink line* the 33 kV transmission lines. Even after a decade, the major remote, mid-hill and high-altitude regions, where approximately 75 % of the country' people live, is still not at all, or very sparingly served with electricity transmission lines. (map from: http://www.geni.org/globalenergy/library/national_energy_grid/nepal/index.shtml)

energy sources and improved energy services. They all basically agree that there is a direct link between improved access to energy services and increased development and better living conditions (Goldemberg 1995; Reddy 2000; Brew-Hammond 2004). Consequently it can be assumed that poverty in Nepal is partially linked to the lack of energy infrastructure, and thus lack of access to improved energy services.

The enormous, untapped potential of Nepal's renewable energy resources is discussed in 1.1.6/Appendix 11. It is clear that for Nepal, the building of the needed infrastructure, to harness the locally available renewable energy resources, must be at the top of the country's priority agenda, in order to provide a sound and stable foundation for long-term economic development (Fig. A9.4).

Another point of importance to highlight is the issue of ongoing relevant monitoring of projects beyond their implementation period. While there is no real shortage in Nepal of reports on all kinds of development activities and projects, with chapters on the results, impacts and lessons learned, there is no government system of (long-term) monitoring of the progress of development. Thus, how the people have been actually affected by such projects over time is simply not known. With no pressure by any government agencies for monitoring, the thousands of NGOs in Nepal have not given any attention to monitoring their projects beyond the time frame of donations and funds received (usually from abroad).

The more recent Maoist insurgency from 1996 to 2006, has further stalled development through destruction of existing infrastructure, forced migration and loss of life (Fang 2007). Although the main frustrations that fuelled the conflict are related to poor living conditions, experienced by the rural poor, ironically the insurgency has only worsened their plight. The functioning of normal economic activities deteriorated after 1996, reducing the ability of the economy to withstand shocks, such as loss of agricultural production due to natural disasters (Fang 2007). Much needed government funds have been diverted from urgently needed development to pay for the war bills. Foreign aid agencies have had to suspend their activities due to security concerns and inability to implement projects in the Maoist infested regions. The conflict has affected access to healthcare and education (UNICEF 2006), reduced agricultural production, and decreased the valuable income generated from tourism, previously one of Nepal's main foreign sources.

From the discussion in this chapter, it is clear that Nepal has overall low levels of development and high levels of poverty which are unevenly distributed throughout the country, the most extreme being in rural mountain areas. Indigenous natural resources are not well managed, with some being used unsustainably, while the potential of others is not fully exploited. Reliance on fossil fuels is not beneficial to the country economically or environmentally, as discussed in Sect. 1.1.5/Appendix 8 and 1.1.6. Food security is a major issue due to reliance on subsistence agriculture, poor farming conditions, lack of agricultural development, natural disasters and political conflict. Bista (1991, 148) summarised a discussion on the issue on foreign funded agriculture projects with foreign advisers in a succinct way by citing a UN expert, saying that "Nepal is over-advised and under-nourished".

More and more Nepalese are now highly educated professionals and possess technical skills with a better understanding of the problems and needs of Nepal. But one of the difficulties they face is being "enmeshed in the legacy of the hierarchic caste attitudes" (Bista 1991, 148). Nepal will need foreign aid and support for years to come, as this huge annual income has become part of the national economy over the last few decades. However, the doctrine of fatalism is hindering the international effort and the prevailing governmental structures and their management, incapacitating the country's social and economic advancement.

One promising step forward, escaping the vicious circle of fatalism, could be the focus on increased training and empowerment of local people. People who have thus far earned their living by putting their hands to work, be it on the land or in the expanding manufacturing or service industries, are a valuable human resource. Bista puts it in the following way: "The Nepali population that has remained untouched by the Hindu caste principles is Nepal's greatest treasure" (Bista 1991, 151). Most of these people though live in remote rural areas. This means, that increasingly foreign aid programs have to make the extra effort to go out into the most remote places to live and work together with these people, which make up a substantial part of Nepal.

One such area, where people live and have always struggled to survive, is the Humla district, where the non-profit NGO, RIDS-Nepal (co-founded by the author) established in 2002 a long-term holistic community development program. One of the aims of this NGO is to start out afresh, with all the above things in mind, on new paths, so that funds and support for village based projects get to those for whom they are intended.

Appendix 10
Discourse on Sustainability and Development

10.1 Introduction

This Appendix does not aim to propose a new definition of, or add significantly to the discussion on, "Sustainability and Development". Rather it intends to recall, highlight and summarize the train of thought and some of the main milestones in the four decade long debate. It seeks to remain clear and simple, emphasizing the need for ongoing applied research and well thought through grass-roots applications and evaluations.

Evaluation and the consequential "active learning", resulting in a redirected approach to sustainable development, is probably the least discussed topic and thus the weakest link, in this longstanding debate. This is because responsible assessment has to be based on real life experience, learned through collecting qualitative and quantitative data. Experience is ultimately linked with an extended time frame of exposure to and application in real life conditions. Consequently, this is also one of the main reasons why far more definitions and theoretical models of "Sustainability and Development" are available than practical examples rooted in real life which "walk the talk" and improve people's lives accordingly.

10.2 Alma Ata Conference 1978

In the 1960s and 1970s it was a common understanding among community development organisations, professionals and theorists of development that projects have to be multi-pronged and work in partnership with the local community. This was the climate in which the conference on primary health care took place in Alma Ata, in today's Kazakhstan, in September 1978.

People at the time built upon the idea that improvements in health are in various degrees subject to social, political and economic changes. Thus multi-sectoral projects and programs worked on both levels, preventative and curative, to improve primary health care conditions. They were part of the standard approach to community development for the improvement of the quality of life of the world's poorest and most marginalised people.

A. Zahnd, *The Role of Renewable Energy Technology in Holistic Community Development*, Springer Theses, DOI: 10.1007/978-3-319-03989-3,

In the Conference's Declaration (Alma Ata 1978) it is stated that:

> …health, is a state of complete physical, mental and social wellbeing, and not merely the absence of disease or infirmity", and that "a major social target of governments, international organisations, and the whole world community in the coming decades should be the attainment by all people of the world by the year 2000 of a level of health that will permit them to lead socially and economically productive lives.

The importance and relevance of multi-pronged projects came under increasing pressure and scrutiny in the 1980s and 1990s, when the various economic recessions, increasingly had taken their toll on international funding for development aid. Many development projects across sectors, and in particular those focused on supporting health programs of governments in developing countries, have failed. Countless people had lost confidence in the process of international development (LaFond 1995). Thus more and more the scope of projects narrowed down to single-pronged project approaches in order to suit the foreign governments' and donors' own agendas, with a constant demand to produce results within ever shorter time periods.

10.3 The Brundtland Report, "Our Common Future" 1987

Increasing awareness of trans-boundary air pollution problems triggered an UN-organised conference on the Human Environment in Stockholm, Sweden, in June 1972. This conference, mainly called by the industrialised Western countries, brought 113 nations together to discuss the necessary international actions to protect the global environment. It was one of the first times that it had been explicitly stated in a conference declaration that mankind is the most precious thing in the world and that we are also responsible for the widespread harm and pollution of the environment.

> Of all things in the world, people are the most precious. …through the rapid acceleration of science and technology, man has acquired the power to transform his environment in countless ways and on an unprecedented scale. ….We see around us growing evidence of man-made harm in many regions of the earth: dangerous levels of pollution in water, air, earth and living beings; major and undesirable disturbances to the ecological balance of the biosphere; destruction and depletion of irreplaceable resources; and gross deficiencies, harmful to the physical, mental and social health of man, in the man-made environment, particularly in the living and working environment. The Conference calls upon Governments and peoples to exert common efforts for the preservation and improvement of the human environment, for the benefit of all the people and for their posterity." (Excerpt from the Declaration of Stockholm Conference)

In 1983 the UN set up an independent commission to address the detailed issues discussed and agreed upon at the Stockholm Conference. Chaired by the Norwegian Prime Minister Gro Harlem Brundtland, the commission was directed to come up with new strategies for global environmental protection. Over the course of four years the Commission consulted developed countries as well as the fast growing developing countries. This was because the commission became aware that the increasing environmental problems will affect the poorer nations and communities much more than the more affluent ones.

The Commission, often called the "Brundtland Commission", presented their final report in 1987 with the title "Our Common Future" (Brundtland 1987). It is this report, which for the first time linked environmental problems to social and economic systems, introducing the term "Sustainable Development". Many definitions have been proposed and used since, but the one proposed in the Brundtland Commission report has prevailed as the most meaningful, commonly used and understood. In Chap. 2 it defines "Sustainable Development" as "meeting the needs of the present without compromising the ability of future generations to meet their own needs".

Further, the report countered the commonly held belief of the time that environmental protection can only come at the cost of economic growth. Rather, in Chap. 3 the report argues there needs to be a clear integration of environmental policies and development strategies through the concept of sustainable development.

With the fast growing popularity of the concept of sustainable development, environmental quality and economic development have come to be seen rather as interdependent and mutually reinforcing. Thus the discussions have shifted from debating whether the objectives of environment and development are compatible towards how to achieve environmentally sustainable development. The report emphasises that sustainable development requires the promotion of consumption patterns that are within the bounds of an ecologically determined framework. It highlights the priority of meeting the essential needs of the poor and that this responsibility lies foremost with the governments of the developed countries.

The achievement of sustainable development is intrinsically linked through major political, technical and social changes and adjustments. Among the essential changes are the elimination of poverty and exploitation, more equal distribution of global resources, drastic limitations of military expenditures, methods of just population control, switching to more appropriate technologies and renewable energy resource utilisation. Further, it includes lifestyle changes, institutional and governmental democratisation through active participation of each citizen in the decision making processes, and the removal of disparities in economic and political relationships between the South and North.

In addition, the report argues that sustainable development requires the conservation of the abundance and species diversity of plants and animals, because their loss can greatly limit the options of future generations and that this is a human moral obligation.

The report also draws attention to the fact that there is no blueprint for sustainable development and that despite the global agenda, each country needs to work out its own concrete policies, because the economic, ecological and social systems differ widely among different countries.

While this is all true and eloquently explained, it is also a weak point because these statements are too general for countries with limited experience and practice of sustainable development to implement. Thus to leave it in the hands of each country's policymakers, to interpret, contextualise and implement means that many countries have failed to act effectively on the Commission's recommendations.

During the 1990s the Brundtland Report shaped the mainstream thinking regarding sustainable development and some of the major organisations such as the European Union (EU), World Bank and UNEP have worked towards the implementation of the report's concepts. The report's ambiguity and generalisations have produced ongoing debates and the number of definitions of sustainable development is increasing. However this has compelled theorists to think about the original meaning of each of the words, "sustainable" and "development".

The word "development" has a distinct social and political connotation, often equated with growth and expansion. This can be physical, which is most often thought of such as in the context of economic growth or production of new materials, or non physical growth such as in measuring well being after a course of action. To reconsider the definition of development is important as it has become accepted over the last four decades that development is intrinsically equated with the western concept of materialistic economic growth.

However, since the beginning of the new millennium it has become more and more clear that equating economic and material growth simply with the concept of development is unrealistic, since ongoing exponential economic growth is a physical impossibility within the realm of the earth's ecosystem due to its finite resources and limited boundaries.

The word "sustainable", taken on its own, originated in a context of harvesting and managing renewable resources in such ways that they do not deplete but will be available for future generations as well. Thus it has its roots in the defined and limited ecosystem which supports human life and its endeavours. Sustainability is not like a project with an end in itself but an ongoing process of complex, interrelated and interwoven systems in nature and society (Hjorth 2006). It clearly indicates constraints within which mankind has to live in order to survive. Thus, the word "sustainable" takes on a much more serious meaning, as it becomes obvious that the need to maintain the ecological balance for survival is vital in order to allow mankind to continue to live within the boundaries of the earth's ecosystem. And all that takes place within a social and political framework of aspirations for advancement or growth (or development).

Hence, in joining the two concepts of "sustainability" and "development" to form "sustainable development", it becomes obvious that it emphasises the need to maintain the ecological balance while striving for societal and political changes. History has shown that these two aspects intrinsically influence each other and thus, the objectives of maintaining ecological balance or sustainability are paramount to the Brundtland approach to development.

10.4 Agenda 21, the Earth Summit, 1992

In Agenda 21's[25] preamble (point 1.1) it is stated:

> 1.1 Humanity stands at a defining moment in history. We are confronted with a perpetuation of disparities between and within nations, a worsening of poverty, hunger, ill health and illiteracy, and the continuing deterioration of the ecosystems on which we depend for our well being. However, integration of environment and development concerns and greater attention to them will lead to the fulfilment of basic needs, improved living standards for all, better protected and managed ecosystems and a safer, more prosperous future.

Half a decade after the Brundtland Report was presented, the UN held the largest environmental conference ever, the Rio Earth Summit in June 1992 in Rio de Janeiro, Brazil. The agenda included reviews on the progress thus far achieved and how to respond to global environmental problems and agreements on biodiversity, climate change and forest management.

While the Brundtland Report (1987) defined sustainable development as:

> …development which meets the needs of the present without compromising the ability of future generations to meet their own needs.

the Rio Earth Summit (UNEP 1992) affirmed that:

> …the right to development must be fulfilled so as to equitably meet developmental and environmental needs of present and future generations.

emphasizing that sustainable development is indeed about the economy and our society as well and not just about the environment. It represents the struggle, and to some extent the inability, to fully put into practice what the Brundtland Report demanded, a re-thinking and transformation of the mind set with regard to development.

The years since the Report's publication have shown that the concepts of development and economic growth are far more interlinked with each other than previously thought. The necessary paradigm shift will need more time and consultations among the nations in order to gain a better understanding of sustainability and development.

Agenda 21 is meant to be a blueprint for sustainability for the twenty-first century and was debated during the UN's Earth Summit Conference in 1992. Agenda 21 is a pledge, signed by all the governments of the nations committed to sustainable development. It is an outline for sustainable development for societies and economies as well as the conservation of the environment and natural resources in the twenty-first century. It aims to provide a framework to address today's urgent social and environmental dilemmas.

At the same time it seeks to deal with poverty, health, energy demands and consumption, waste, overpopulation and biodiversity loss. This blueprint is meant

[25] UN's Agenda 21, "Sustainable Development" website provides all the details and report's chapters: http://www.un.org/esa/dsd/agenda21/ (accessed July 20, 2012).

to be applied globally, nationally and locally not only by Governments and national and local groups, but also by organisations such as the United Nations and its agencies, in every area in which mankind has an impact on the environment.

Agenda 21 has recognized the centrality and importance of Government, NGOs, the general public and industry and the need for them to actively participate and engage in all activities as indicated in the Agenda 21 Report (Agenda 21 1992), Preamble 1.3 of the preamble:

> 1.3 Agenda 21 addresses the pressing problems of today and also aims at preparing the world for the challenges of the next century. It reflects a global consensus and political commitment at the highest level on development and environment cooperation. Its successful implementation is first and foremost the responsibility of Governments. National strategies, plans, policies and processes are crucial in achieving this. International cooperation should support and supplement such national efforts. In this context, the United Nations system has a key role to play. Other international, regional and subregional organisations are also called upon to contribute to this effort. The broadest public participation and the active involvement of the non-governmental organisations and other groups should also be encouraged.

Further, in Sects. 28.1 and 28.2(d) it states explicitly the centrality of the local authorities, women and the younger generation:

> 28.1 Because so many of the problems and solutions being addressed by Agenda 21 have their roots in local activities, the participation and cooperation of local authorities will be a determining factor in fulfilling its objectives. Local authorities construct, operate and maintain economic, social and environmental infrastructure, oversee planning processes, establish local environmental policies and regulations, and assist in implementing national and subnational environmental policies. As the level of governance closest to the people, they play a vital role in educating, mobilising and responding to the public to promote sustainable development.
> 28.2 The following objectives are proposed for this programme area:
> (d) All local authorities in each country should be encouraged to implement and monitor programmes which aim at ensuring that women and youth are represented in decision making, planning and implementation processes.

As the issues to be addressed are at the grass-roots level it has been recognised that national, regional and local Governments are key players in sustainable development. Local authorities, bestowed with the legal competence for the local area, are responsible for the stimulation of the local economy, environmental protection, waste disposal, fulfilling the growing energy service demands, local transport and infrastructure. Thus being the closest government authority to the grass-roots problems, they have a particularly important role to play in practical actions towards sustainable development. This role has been recognised and explicitly stated in the Agenda 21 strategies.

Another important theme, highlighted throughout Agenda 21, is the current high and unsustainable production and consumption of non-renewable energy resources. It draws attention to the fact that present energy production levels and consumption patterns, which do not seem likely to alter significantly over the next decade, will not allow any progress to be made in protecting mankind's health, the natural environment and the atmosphere.

Additionally, Agenda 21 recognised the need to distinguish between different countries' needs and agendas and thus called for all countries, to develop their own national strategies for sustainable development. Countries are called upon to contextualise the words, strategies and the commitments made at the Earth Summit in such a way that they can create their own, related policies and actions. It recognises that key decisions are needed at each country's national level, considering the needs of all the stakeholders. It was acknowledged that the gigantic task of sustainable development needed a specific strategy for each country. But what Agenda 21 failed to provide is any clear guidance, outline or definition of what constitutes such a strategy.

10.5 The Millennium Development Goals[26], New York, 2000

Following the various UN conferences called in the 1980s and 1990s, leaders of almost 150 countries met in September 2000 at the UN headquarters in New York to commit themselves to a new global partnership to tackle the major problems facing the world at the beginning of the twenty-first century.

The commonly agreed and signed Millennium Declaration set forth a series of targets, identified as the Millennium Development Goals (MDGs), which the world community commits to achieve by 2015 (UN 2007, 4).

The MDGs form a blueprint agreed upon by all the world's countries and leading development institutions and organisations. There are 8 MDGs covering the various dimensions of poverty, providing primary education for all, gender equality and the right to sustainable human development. Each of these goals has a set of targets associated with it.

Goal 1: Eradicate extreme poverty and hunger

Target 1.A: Halve, between 1990 and 2015, the proportion of people whose income is less than US$1 a day

Target 1.B: Achieve full and productive employment and decent work for all, including women and young people

Target 1.C: Halve, between 1990 and 2015, the proportion of people who suffer from hunger

Goal 2: Achieve universal primary education

Target 2.A: Ensure that, by 2015, children everywhere, boys and girls alike, will be able to complete a full course of primary schooling

---

[26] UN's main MDGs website: http://www.un.org/millenniumgoals/poverty.shtml, with the 8 MDGs individually listed and described in short at: http://www.unmillenniumproject.org/goals/index.htm (both accessed July 20, 2012).

Goal 3: Promote gender equality and empower women

Target 3.A: Eliminate gender disparity in primary and secondary education, preferably by 2005, and in all levels of education no later than 2015

Goal 4: Reduce child mortality

Target 4.A: Reduce by two thirds, between 1990 and 2015, the under five mortality rate

Goal 5: Improve maternal health

Target 5.A: Reduce by three quarters the maternal mortality ratio

Target 5.B: Achieve universal access to reproductive health

Goal 6: Combat HIV/AIDS, Malaria and other diseases

Target 6.A: Have halted by 2015 and begun to reverse the spread of HIV/AIDS

Target 6.B: Achieve, by 2010, universal access to treatment for HIV/AIDS for all those who need it

Target 6.C: Have halted by 2015 and begun to reverse the incidence of malaria and other major diseases

Goal 7: Ensure environmental sustainability

Target 7.A: Integrate the principles of sustainable development into country policies and programmes and reverse the loss of environmental resources

Target 7.B: Reduce biodiversity loss, achieving, by 2010, a significant reduction in the rate of loss

Target 7.C: Halve, by 2015, the proportion of the population without sustainable access to safe drinking water and basic sanitation

Target 7.D: By 2020, to have achieved a significant improvement in the lives of at least 100 million slum dwellers

Goal 8: Develop a global partnership for development

Target 8.A: Develop further an open, rule-based, predictable, non-discriminatory trading and financial system

Target 8.B:	Address the special needs of the least developed countries, landlocked countries and small island developing states
Target 8.C:	Address the special needs of landlocked developing countries and small island developing states
Target 8.D:	Deal comprehensively with developing countries' debt
Target 8.E:	In cooperation with pharmaceutical companies, provide access to affordable essential drugs in developing countries
Target 8.F:	In cooperation with the private sector, make available benefits of new technologies, especially information and communications

The MDGs indicate the clear trend towards a more holistic understanding of the dilemmas the world faces and their deeply rooted, complex relationship with each other. They demonstrate that unless the identified major issues are addressed and improved collectively, alongside each other, the necessary development cannot take place in sustainable ways nor within a set time frame.

Therefore a wide range of issues needs to be addressed, as the social, physical, political and environmental problems the world community faces at the start of the twenty-first century cannot be taken as individual challenges, but only as deeply interlinked and interdependent issues. These need to be tackled concurrently and in consultation.

The MDGs[27] present a much more realistic picture of the human condition within its environment and thus mark the most accurately defined outline of the progress towards sustainable development thus far. However, they highlight as well the complexity and interaction of all these issues at all levels. Thus, far from becoming easier after two decades of experience, the challenge the world community faces to accomplish the tasks at hand, has become much more difficult.

10.6 The Johannesburg Earth Summit 2002[28]

Two years after the MDGs were defined and signed the UN called another Summit, in 2002, in Johannesburg. An article, published shortly after the Summit (UN 2002a), said:

[27] A short UN video about the MDGs, within the context of identified needs, and the need for increased action, as till 2011 the world community of nations, committed to the MDGs, are NOT on track to fulfill their 8 identified MDGs by 2015. The 4 min 27 sec. long video, should be opened in the realPlayer video software, and can be watched through the following link: http://webcast.un.org/ramgen/ondemand/documentary/2008/mdg_makeithappen_081017.rm?start=00:00:11&end=00:04:38 (accessed July 20, 2012).

[28] The website with the main information and documents created at the Earth Summit 2002 in Johannesburg is at: http://www.earthsummit2002.org/ (accessed July 20, 2012).

> When the United Nations General Assembly authorised holding the World Summit on Sustainable Development, it was hardly a secret—or even a point in dispute—that progress in implementing sustainable development has been extremely disappointing since the 1992 Earth Summit, with poverty deepening and environmental degradation worsening. What the world wanted, the General Assembly said, was not a new philosophical or political debate but rather, a summit of actions and results. (http://www.johannesburgsummit.org/ accessed November 6, 2011)

Considering the tasks at hand identified in the MDGs it does not come as a surprise that only 2 years into the campaign no visible, satisfactory results could be shown. Thus one of the dominant topics addressed during this Summit was to promote and encourage more practical actions, especially on the most pressing issues which are poverty and environment.

Thus commitments were made by the participants to spur action to increase access to improved and cleaner energy services, clean drinking water and proper sanitation, to improve health conditions and to increase food availability and security.[29] Further, better protection for the world's biodiversity and ecosystems was agreed upon. For the first time the representative countries committed themselves, though without a time frame identified, to speed up the promotion and increased use of renewable energy resources and technologies.

An additional important step was taken during the Summit by establishing concrete partnerships between the countries' governments, different groups of people, businesses, NGOs and INGOs. This makes available new and important resources, expertise and capacities to achieve significant results in communities around the globe. UN Secretary General at the time, Mr. Kofi Annan, concluded the Summit (UN 2002b, 1–2) by saying:

> The Summit represents a major leap forward in the development of partnerships, ...with the UN, Governments, business and civil society coming together to increase the pool of resources to tackle global problems on a global scale. .....The true test of what the Johannesburg Summit achieves, are the actions that are taken afterward. ....We have to go out and take action. ....I am not saying that Johannesburg is the end of it. Johannesburg is the beginning.

---

[29] The commitments made by the participating countries at the Earth Summit 2002 in Johannesburg, to improve and to spur action to increase access to improved and cleaner energy services, clean drinking water and proper sanitation, to improve health conditions and to increase food availability and security, are all directly addressed, with awareness raising through NFE classes and practical project implementation through the author's newly developed Holistic Community Development (HCD) concepts of the "Family of 4" and the "Family of 4 PLUS". It shows that the world communities are aware of the urgency to increase the access to improved energy services for the poor, in order to aim for sustainable development. That shows that RETs do play a central and important role in long-term HCD programs.

10.7 Copenhagen Summit 2009[30]

From the 7–18th December 2009 the long awaited Climate Change Summit was held in Copenhagen Denmark. The representatives, Ministers and Heads of States of all the major countries and NGOs working in the field of climate change and the sphere of climate protection came together. Many arrived with great hope and expectations that together they could agree on new, legal binding emission cuts for the developed countries, either as part of the Kyoto Protocol (1998), or as a new agreed document.

There was hope that the developing countries, mainly China and India would agree to strengthen the Climate Convention by committing themselves to their respective countries' appropriate mitigation actions. Additionally, in order to support increased practical actions in the area of climate change/protection in developing countries, the negotiators sought to reach agreements on issues of adaptation, reducing deforestation and forest degradation (REDD), technology development and transfer, and financing to support climate actions in developing countries.

Behind closed doors, the negotiators and then the presidents, prime ministers and their senior advisors worked out a 12 point document, called the "Copenhagen Accord", between the USA, China, India, South Africa and Brazil, on the final day of he summit. While the European Community's endorsement followed, wider consensus with several other countries, however, was not achieved. Thus, for most delegates and countries the summit did not bring forth what they expected, despite the working out of the "Copenhagen Accord", as it missed a range of crucial elements, and is not a legally binding agreement.

The following is a summary of the 12 key points of the "Copenhagen Accord" (UNFCCC 2009):

The positive points in the "Copenhagen Accord":

- The USA, China, India and other major developing countries signed up to a joint climate agreement for the first time.
- These countries also made pledges to either cut or slow the growth of their emissions and/or implement specific measures to achieve this.
- Agreement on limiting average global temperature increase to ≤ 2 °C.
- Agreement for the monitoring, reporting and verification for developed countries' to deliver finance for developing countries and for mitigation actions in developing countries.

[30] The Copenhagen Accord 2009, a 12 key point, non-legally binding document, agreed by all the Heads of States, Governments and Ministers of the countries who participated at the United Nations Climate Change Conference 2009 in Copenhagen, can be consulted and studied at: http://unfccc.int/files/meetings/cop_15/application/pdf/cop15_cph_auv.pdf (accessed July 20, 2012).

- Developed countries to commit collectively 30 billion USD from 2010–2012 and USD 100 billion by 2020 for "meaningful mitigation actions and transparency of implementation".
- Establishment of a new "Technology Mechanism" to accelerate technology development and transfer for both adaptation and mitigation".

The "un-reached" issues include:

- Lack of defined, long-term global emission reduction goals.
- Lack of absolute emission targets for developed countries for 2020.
- Lack of relative emission reduction targets for developing countries for 2020.
- Lack of reference to a global emissions peaking date.
- Lack of a timetable for a legally binding agreement.
- Lack of reference/role for the carbon markets and CDM.
- Lack of clarity for the UNFCCC's and Kyoto Protocol's future and processes.

The "Copenhagen Accord" demonstrates a political step into the right direction, both at international and national levels. However, as the agreement is not legally binding it depends upon the commitment of all major economies, and thus in general, the summit did not bring the desired results.

10.8 Summary

This chapter's aim was to provide a short overview of the last four decades of discussions, summits and conferences which include the topic of "sustainable development" as it applies to developing countries. It aimed to highlight that development projects in the 1960s and 1970s had a multi-pronged approach. They tried to address mankind's multiple facets of needs but partly due to the economic recessions in the early 1980s and the need for greater accountability in the use of donors' funds, a more focused, single-project approach was applied by the main development agencies and organisations. In addition, development project agencies were put under increasing pressure to have shorter and shorter project time cycles, making it hard to focus on the actual village level needs of the end-users of the development projects. Not surprisingly, this raised the issue of sustainable development, which was for the first time highlighted in the Brundtland report. Ever since, all the major development summits and conferences have had debates about a more accurate definition of what sustainable development means.

The MDGs, agreed upon by all nations in September 2000, were one of the most important milestones reached in identifying mankind's needs and they are multi-pronged.[31] Further, they correctly recognise that the world's problems such as poverty, gender equality and equal opportunities for education are needs which can be addressed only over a longer time period. Thus the MDGs are targeted to be reached after a longer, 15 year time frame, in 2015. This significant development milestone initiated a new approach which is at the time of writing still developing.

[31] The shear fact that there are 8 MDGs, which break down into 21 quantifiable, defined targets that are measured by 60 indicators shows that the identified needs of the world community's poor and disadvantaged people are multi-pronged and therefore need a holistic community development project approach in order to be able to address their various needs. Further, the timeframe of 15 years, from 2000 to 2015, shows that the world community of nations understands that the MDGs cannot be reached within a short period of time but need a long-term project approach. Having set out with a 15 year time frame, the UN already recognized in 2010 that the set goals will most likely NOT be reached in 2015 but will still need a significantly longer time period to achieve the MDGs. For more details on the MDGs and the defined, quantifiable targets and indicators to be measured see: http://www.undp.org/mdg/basics.shtml (accessed July 20, 2012).

Appendix 11
Nepal's Renewable Energy Resources

11.1 Introduction

1.6–2 billion people in developing countries (Mills 2002a) live in dark homes, without access to electricity, and 2.4 billion rely on traditional biomass (IEA 2002) for their daily energy services, such as cooking, heating and lighting. Lack of electricity and heavy reliance on traditional biomass are hallmarks of poverty in developing countries (Reddy 2000; Saghir 2005), and women and children in particular suffer enormous health problems due to open-fire places.

The high internal migration and urbanisation rates in developing countries will continue, forcing governments to focus more on urban energy service provision and extension. That widens the gap between poor and rich, highlighting the relationship between poverty and access to electricity further.

Nepal, with the majority of its people living in difficult to access areas with no roads, is a typical example of that. Belonging to the poorest and most undeveloped countries, the per capita electricity consumption is among the lowest in the world. The geographical remoteness, the harsh climatic conditions, low population density with minimal energy demand and low growth potential, are some of the reasons why rural electrification costs in Nepal are prohibitive and the isolated rural mountain villages in Nepal will not be reached within the foreseeable future through grid extensions alone.

As previously mentioned, Nepal is not rich in fossil fuel resources but it has plenty of renewable energy resources, in particular water that is running down from the vast Himalayan mountain ranges in over 6,000 rivers. With 300 sunny days a year, the sun's freely available solar energy can also be converted into electricity.

In some of the most remote Himalayan valleys in Nepal (Fig. A11.1), among the poorest and most marginalised groups of people, some encouraging steps have been taken in regard to basic rural village electrification. The locally available, renewable energy resources have been tapped into, and through Remote Area Power Supply (RAPS) systems, miniscule amounts of power, in the "watt" range rather than "kilowatt", have been generated for basic rural village electrification. In this way several villages have been electrified, for lighting purposes only, through different types of village integrated solar PV systems and the smallest

A. Zahnd, *The Role of Renewable Energy Technology in Holistic Community Development*, Springer Theses, DOI: 10.1007/978-3-319-03989-3,

Fig. A11.1 Nepal map (with author's adjustment). From the "End of the Road" one either has to walk 15–17 days to Simikot (Humla district), or take a 1 h, adventurous airplane journey through the Himalayas from Nepalgunj (in the very south at the India boarder) to Simikot

kinds of hydro power plants, called pico-hydro. The lights considered most appropriate and sustainable are 1 watt energy consuming WLEDs, providing a minimum, but sufficient light output.

The author aims to highlight the urgency for appropriate and sustainable basic rural village electrification in Nepal and elsewhere, in order to create opportunities to bring light into the dark, smoke filled homes of the poorest of the poor. Thus in the following Nepal's main renewable energy resources: biomass, hydro power, solar and wind energy are discussed. A few practical examples of contextualised RETs such as solar PV systems, solar thermal applications and small scale wind turbines are highlighted.

11.2 Nepal's Renewable Energy Resources and Their Use

Nepal's lack of fossil fuel resources forces the government to import all its non-renewable energy resources, such as kerosene, diesel, petrol, liquefied petroleum gas (LPG) and coal, from its neighbouring country, India. The prices for these fossil fuels are strongly dependent on the global economic and political conditions.

Nepal's wealth is in its renewable energy resources, such as biomass, water, sunshine and also wind in some particular areas. The great benefits of these energy resources are, that they are free and renewable, and therefore do not incur ongoing fuel costs. Further, a great advantage is that these renewable energy resources are locally available, and thus can be part of the local community's economy and lifestyle.

11.3 Biomass

With still around 75 % of the population living in rural areas, the primary energy source used to provide most of the necessary daily energy services in Nepal for centuries, has been fuelwood, often supplemented by crop residues and animal manure, dependent on the prevailing local customs, caste, altitude and geographical zone. According to the Nepal Water and Energy Commission Secretariat (WECS), about 30 % (or 42,054 km^2) is covered with forest. About 11 % is covered with shrubs and bushes. Monitoring data shows that the forest areas are annually reduced by 1.7–1.8 %.[32] Further, forest and shrub areas are increasingly converted to cultivated land. In 1999 an estimated 15 million metric (MT) tons of air-dry fuelwood were consumed to provide the annual energy services. Ninety eight percent of that total fuel wood consumption is used in private households. But the annual sustainable fuelwood production from all of Nepal's accessible forests amounts only to an estimated 7 MT.[33]

Families in the remote areas of Nepal use precious trees as firewood for cooking, room heating and lighting. These activities, especially indoor cooking and lighting on open fire places, consume 20–40 kg of firewood a day,[34] with direct chronic impact on the health of women and children in particular, due to the enormous indoor smoke pollution. It is no surprise that they suffer from high rates of respiratory diseases, asthma, blindness and heart disease,[35] resulting in the low life-expectancy for women and the high death rate of children <5 years of age.[36]

The fuelwood consumption, through tree cutting, forest clearing and fuelwood collection, for the past 20 years, has been well above the rate of sustainable forest growth. Ever longer and more dangerous journeys for the women and young girls, of up to 7 h a day[37] are now required to collect the necessary fuelwood. This has forced the local people to increasingly utilise agricultural residues to meet their growing energy demands. This in turn results in decreased crop productivity, increased soil erosion and arable land loss. Further, as Prof. Smith[38] shows, the burning of dung and agricultural residues on open cooking fires, produces up to 3

---

32 Biomass energy uses: an experience and application of alternative energy technologies in Nepal, Bishwa S. Koirala Environment Advisor, Rural Energy Development Program, UNDP, Nepal, and WHO's booklet, *Little Green Data Book 2010*, p. 153, http://www.semide.net/media_server/files/semide/topics/water-data/little-green-data-book-2010-world-bank/GreenDataBook2010_WorldBank.pdf (accessed July 20, 2012).

33 Renewable Energy in South Asia, *Country Reports Nepal,* Chap. 2, 2002.

34 Alex Zahnd, *"Firewood consumption survey of 16 villages in the Jumla district"*, unpublished report 1999.

35 IEA, *World Energy Outlook 2002*, Chap. 13, pp. 367–368.

36 Warwick H. *Smoke—the Killer in the Kitchen,* ITDG Publishing 2002.

37 This agrees with data from India collected by UNDP in IEA, *World Energy Outlook 2002,* Chap. 13, p. 366.

38 *Health Impacts of Indoor Air Pollution;* Prof. Kirk Smith, University of California, Berkeley; workshop held in Kathmandu Nepal 7th June 2005.

times more indoor air pollution as burning fuelwood. Deforestation is widespread and the once picturesque, biodiverse forests and valleys are stripped of their resources in unsustainable ways.

11.4 Hydro Power

Nepal is the major contributor to the Ganga Basin in the north of India. The annual discharge of out-flowing rivers from Nepal to India is estimated to be 236 billion m^3 from over 6,000 rivers,[39] with many rivers losing a potential height of about 4,000 metres within a north–south distance of 100 km. This creates a theoretical hydro power potential of 83,290 MW.[40] The harsh terrain and difficult access to many areas limits the theoretically exploitable hydro power potential to a more realistic technically and economically profitable potential. Therefore, the realistic realisable, economically and technically feasible, hydro power potential has been estimated to be 42,130 MW.[41] With this figure, and an assumed average hydro power plant's generation capacity factor of 80 %, the annual national energy potential of Nepal's rivers can be estimated to be around 300 TWh (Fig. A11.2).

But the reality looks rather different, as until August 2009 only 635.845 MW[42] of hydro power generation capacity, from 69 major and small hydro power stations have thus far been installed. Thus, today a mere 1.5 % of the technically and economically feasible hydro-power potential of Nepal has been developed. During the fiscal year 2008/2009 the total annual energy generated by the government owned Nepal Electricity Authority's (NEA) 17 major hydro power plants (424.12 MW) and the two thermal, diesel power plants (53.4 MW) was 1841.636 GWh,[43] which implies a capacity factor of only 44 %. The national average electricity consumption per capita in the 2003/2004 fiscal year was just

---

[39] WECS (Water and Energy Commission Secretariat), 1999. WECS Bulletin No. 10., Kathmandu.

[40] WECS Annual Report, 1999.

[41] Regional Hydro power Resources: Summary and Analysis of Selected SARI Data, prepared for the USAID-SARI Energy program (ww.sari-energy.org), and available from: November 2003, Sect. 3.1, http://www.sari-energy.org/ProjectReports/RegionalHydroPowerReport.pdf (accessed July 20, 2012).

[42] Nepal Electricity Authority, *A Year in Review—Fiscal Year 2008/2009*, Annual Report, pp. 83–84, Kathmandu, Nepal August 2009. http://www.nea.org.np//reports/annualReports/YjqRDwQ9E3anrep09.pdf (accessed February 23, 2011, on November 6, 2011 not available anymore).

[43] Nepal Electricity Authority, *A Year in Review—Fiscal Year 2008/2009*, Annual Report, p. 15, Kathmandu, Nepal August 2009, http://www.nea.org.np//reports/annualReports/YjqRDwQ9E3anrep09.pdf (accessed February 23, 2011, on November 6, 2011 not available anymore).

Fig. A11.2 Some of the major rivers of Nepal, with their enormous north–south drop, from the 6,000 and 8,000 m altitude of the Himalayan range to the north, to the flat southern part, which is just above sea level. Map from: http://www.mapsofworld.com/nepal/nepal-river-map.html

under 70 kWh/year,[44] while it was around 66 kWh/year[45] during the 2008/2009 fiscal year, including the enormous energy loss of ~24 %,[46] accredited mainly to the large grid transmission line losses and power theft. That, compared to around 900 kWh/year[47] for most of the developing countries, and about 2,500 kWh[48] in

---

[44] Kathmandu Post. *1 billion Rupees "stolen" a year*, 23rd April 2005.

[45] The 2008/2009 per capita electricity consumption was lower than during the 2003/2004 fiscal year as Nepal has experienced, since 2005, increased load shedding due to too low power generation. Thus the annual average electricity growth demand of ~10% could not be provided. Rather people in urban settings had to learn to cope with increased load shedding. The 2008/2009 annual per capita electricity consumption was calculated by the author, assuming a population of ~28.5 million at the end of 2008, while using the electricity generation figures from: Nepal Electricity Authority, *A Year in Review—Fiscal Year 2008/2009*, Annual Report, p. 78, Kathmandu, Nepal August 2009, http://www.nea.org.np//reports/annualReports/YjqRDwQ9E3anrep09.pdf (accessed February 25, 2011, on November 6, 2011 not available anymore).

[46] Kathmandu Post. *1 billion Rupees "stolen" a year*, 23rd April 2005.

[47] Hunwick R.J. (2002), *The Rational Path To The Age Of Renewable Energy*, Hunwick Consultants Ptv Ltd, August 2002

[48] Hunwick R.J.(2002), *The Rational Path To The Age Of Renewable Energy*, Hunwick Consultants Ptv Ltd, August 2002.

OECD (Organisation for Economic Co-operation and Development countries), recorded in 2000, puts Nepal among the lowest electricity consumers in the world.

Thus one can say that the "luxury" of having electricity available at the flick of a switch is still a dream for about 75 %, or ~20 million people in Nepal. In NEA's long-term planning, 29 hydro power plants, with a total of 22,435 MW rated power output have been proposed for Nepal's future power development.

Under the Maoist led government, the Minister for Water Resources Bishnu Poudel announced on the 24th September 2008 that they plan to exploit and develop the first 10,000 MW within the next 10 years.[49] But considering the last 50 years of hydro power plant construction and operation (total built capacity of 635.845 MW), and the long-term political instability, it comes as no surprise that the public and the local industries do not have much faith and confidence in achieving that admirable goal.

11.5 Solar Energy

Nepal lies in the ideal 30° North "solar belt"[50] with about 300 sunny days a year.[51] Long-term solar irradiation measurements by the author with pyranometers on tracking frames for solar PV modules in three different geographical locations in Nepal since 2004, showed that 1,950–2,200 kWh/year (with an average of 5.342–6.027 $kWh/m^2/day$[52]) of solar irradiation had been intercepted. The extreme values were 3.5 $kWh/m^2/day$ on a rainy overcast monsoon day in July in Kathmandu, and 9.9 $kWh/m^2/day$ on a sunny late autumn day in November, in Simikot in Humla (at 3,000 m altitude).

According to recent studies, the PM_{10} (Particulate Matter <10 μm in diameter, and thus able to enter the respiratory system) values for Kathmandu during the day are as high as 495 $\mu g/m^3$ in core city areas, with TSP (Total Suspended

---

[49] http://www.nepalnews.com/archive/2008/sep/sep24/news06.php (accessed February 25, 2011, on November 6, 2011 not available anymore).

[50] The convergence of heated air from the equator area in the Tropopause, increases the air mass aloft around the 30° northern latitude (the Hadley cell). This in turn causes the air pressure at the surface to increase. Hence, at northern latitudes of around 30°, a high pressure belt called "subtropical highs" (C. Donald Ahrens, Meteorology Today, 7th edition 2003, Brooks/Cole, Chap. 11, p. 286) is created, providing good favourable weather to utilise the incoming solar radiation.

[51] Renewable Energy in South Asia, Chap. 1.2.2., report with the author and previously available at: http://www.worldenergy.org/wec-geis/publications/reports/renewable/overview/chap_1_2.asp ; Center for Rural Technology Nepal, http://www.crtnepal.org/new/technologies.php?mode=detail&technologies_id=12 (both accessed February 25, 2011, on November 6, 2011 both not available anymore).

[52] Alexander Zahnd measured and minutely recorded solar irradiation data for Kathmandu from 1.1.04–31.12.05, in Dhulikhel from January 2005 onwards and in Simikot Humla from April 2004 onwards.

Particulate) values up to 572 μg/m^3. Also 24 h average values were as high as 225 μg/m^3 for PM_{10}, and 379 μg/m^3 for TSP respectively.[53] In contrast, the WHO guidelines provide maximum average 24 h values of 50 μg/m^3 for PM_{10}[54], and 120 μg/m^3 for TSP.[55] These high values in Kathmandu are mainly produced by the inefficient combustion of fossil fuels from vehicles and the local brick kiln industry. Days with thick smog layers over the Kathmandu valley, reducing the intensity of the solar irradiation by ~20 % are frequent.[56] Considering this data, solar PV modules installed at an angle of 30° south (considered as the "Nepal standard"), can intercept 4.8–6.0 kWh/m^2 solar energy on a daily average in most sites in Nepal.[57]

Since 2002 the SWERA[58] (Solar and Wind Energy Resource Assessment) project under the leadership of NREL (National Renewable Energy Laboratory) in Golden Colorado USA, has developed (with the Tribhuvan University Nepal's national university- and the Ministry of Science and Technology) maps of the available solar resource in Nepal. Combined NASA satellite data with some measured ground data are presented in country maps with a 40 km grid for annual global horizontal, and at latitude tilt global solar irradiation. The maps and the actual ground measurement data recorded by the author at three different places agree within 10–15 %.

[53] *Kathmandu's Air Quality*, CEN (Clean Energy Nepal), http://www.cen.org.np/page.php?this=publications (accessed November 6, 2011, on July 20, 2012 not available anymore)

[54] http://whqlibdoc.who.int/hq/2006/WHO_SDE_PHE_OEH_06.02_eng.pdf , p. 9 (accessed July 20, 2012).

[55] *Towards establishing air quality guidelines for Pakistan, Table 1,*http://www.emro.who.int/Publications/EMHJ/1206/Article21.htm ; *Guidelines for air quality*. Geneva, World Health Organization, 2000 (WHO/SDE/OEH/00.02) (accessed November 6, 2011, on July 20, 2012 not available anymore).

[56] This estimation is based on recorded and averaged solar irradiation data on a 2-axis tracking frame for the following places: Kathmandu University in Dhulikhel, some 30 km east of the Kathmandu valley and thus free of its local pollution, from the 1.1.05 onwards; for Kathmandu (city area) from the 1.1.04 onwards and for Simikot in Humla from 24.4.2004 onwards, by Alex Zahnd.

[57] These values agree rather well with the available NASA data generated solar irradiation from http://eosweb.larc.nasa.gov/ (accessed July 20, 2012) for solar PV modules for each month for Nepal, for latitudes between 25°–31°North, and longitudes 79°–88° East, as presented in Zahnd A. 2004, *Case Study of a Solar Photovoltaic Elementary Lightning System for a Poor and Remote Mountain Village in Nepal,* M.Sc in Renewable Energy Dissertation, Chap. 8.3 p. 99, Figure 8-2, Murdoch University, Perth, Australia.

[58] SWERA Nepal project and its solar resource maps are available from http://swera.unep.net/index.php?id=metainfo&rowid=136&metaid=254 & http://swera.unep.net/index.php?id=metainfo&rowid=212&metaid=327 (both accessed July 20, 2012).

Fig. A11.3 Picture of the Simikot 50 kW Solar PV system, taken and prepared by Alex Zahnd. Simikot Humla 50 kW_P Solar PV System, after 18 years of operation

11.6 Overview of Solar PV Systems in Nepal

NEA has installed, as part of a French government development project in 1989, three solar PV array systems in three remote areas (3 0kW_P in Kodari, 50 kW_P in Gamghadi and 50 kW_P in Simikot), out of which the last two are still considered to be operational in NEA's 2009 annual report and thus part of the power plant generation development graphs and charts.[59] Personal investigations by the author confirmed, that since October 2003 the Gamghadi system has totally failed, leaving the people, who had simple, low power lighting in their homes for 14 years, again in the dark.

The Simikot system (Fig. A11.3, which was in the same village as the main RIDS-Nepal field office and HARS in Humla, (Fig. 3.59) has provided, from the late 1990s till it was fully shut down in 2009, only DC power (both the 5 and a 50 kW inverters failed due to lack of maintenance/repair funds, spare parts and skilled professionals). Thus for ~100 households, each with one to three 40–60 W incandescent bulbs not connected to an on-off switch, had only 1–2 h per day indoor lighting provided.

While one has to recognise that 20 years of operation in such a remote area, with such an extreme and harsh climate is a good track record, minimal maintenance, better initial planning and budgeting would have enabled the system to continue to provide its valuable lighting energy services for probably another 5 years. However, these central solar PV systems have prepared the way for the solar PV technology to have a good entry as an appropriate and sustainable technology, converting the locally available solar energy resource into electricity, mainly for lighting purposes. NEA also installed, over the years, smaller solar PV

[59] Nepal Electricity Authority, *A Year in Review—Fiscal Year 2008/2009*, Annual Report, p. 83, Kathmandu, Nepal August 2009, http://www.nea.org.np//reports/annualReports/YjqRDwQ9E3 anrep09.pdf (accessed November 6, 2011, on July 20, 2012 not available anymore).

systems for remotely located mountain air strips for civil aviation, for national telecommunication stations and some drinking water pumping systems in the lower flat areas, amounting to a total installed capacity of about 200 kW_P.

In 2001 the government of Nepal, mainly through the financial support of the Danish government development organisation DANIDA, established a solar PV home system (SHS) subsidy program.[60] That enables families living in the remote and poorer mountain regions to install a SHS mainly for lighting (10–40 W_P). Up to the end of November 2005, under the scheme's first phase (2001–2007), a total of 61,892 SHS have been installed in 73 out of the 75 districts of Nepal, with a total peak power rating of 2,024.574 kW. That has increased by the end of 2007 to a total of 87,245 installed SHS.[61] Phase 2, (2008–2012) aims to install another 150,000 SHS, of which up to March 2009 AEPC reports to have installed 76,619 SHS with an additional capacity of 1,818 kW.[62] Through personal communication with AEPC the author came to know that under all these subsidy programs up to the end of 2011 a total of 219,818 SHSs were installed. This growing demand for small scale SHS brought forth a mushrooming of solar energy companies in Nepal. While this is encouraging, not all companies have sound infrastructure and technical bases to provide quality products and after sales services. The latter in particular are very poor, as no subsidies are available for them. Therefore many systems, once they are installed, are listed in the books and statistics, but are not followed-up or properly maintained. Thus while the statistics look encouraging, with an appropriate renewable energy technology starting to make a difference in the lives of the poorest, the actual facts present a different picture.[63]

11.7 Solar Water Heaters for Domestic Use in Nepal

A solar energy conversion technology more widely used and installed in urban areas is the solar water heater (SWH). The thermo-siphon SWH technology was transferred from Switzerland to Nepal in the early 1970s and has now around 220 local Kathmandu based SWH manufacturers. Today, most of the locally manufactured SWHs are of the same old technology and size, as the focus was mainly on economic competitiveness among the fast growing number of

[60] SHS subsidy policy see: http://aepc.gov.np/downloads/SubsidyPolicyEnglish2006.pdf , pp. 3–4 (accessed October 25, 2010, on November 4, 2011 not available any more).

[61] http://aepc.gov.np/index.php?option=com_content&task=blogcategory&id=20&Itemid=49 (accessed February 25, 2011, on November 4, 2011 not available any more).

[62] http://aepc.gov.np/index.php?option=com_frontpage&Itemid=1 (accessed July 20, 2012).

[63] This is the experience of the author's 16 years of field experience with solar PV systems for SHS and whole villages in the remotest and poorest parts of Nepal, mostly in Jumla, Mugu and Humla, as reported in: Case *Study of the Tangin Solar PV Home System,* in Case Study of a Solar Photovoltaic Elementary Lightning System for a Poor and Remote Mountain Village in Nepal, M.Sc. in Renewable Energy Dissertation, Chap. 13, pp. 151–156.

Fig. A11.4 First unit of the Nepali made High-Altitude Thermo-Siphon SWH, with opened reflectors, designed to act as night time insulation, beside the ability to empty the absorber and connection pipes. A horizontal and 40° south (as the absorbers) inclined pyranometer allows the monitoring and recording of the actually received solar irradiation on the total absorber surface for efficiency calculations. The full size (4 units) of the High-Altitude SWH bathing centre system is designed to provide hot water for 1,100 people, for each to take a shower every two weeks

manufacturers. The average locally made SWH consists of $2 \times 2\ m^2$ window glass covered absorbers with galvanised steel pipes and steel absorber fins. Each unit has a 150–200 L warm/hot water storage tank, insulated with 50 mm glass wool. The overall efficiency [energy gained in the hot water (MJ)/incoming solar irradiation (MJ)] is low (~25 %[64]). Though there are no detailed figures of installed solar water heaters available, it is believed that around 80,000 units are operational in Nepal.

Since 2002, the Kathmandu University, under the author's supervision, has been engaged in an applied research project for an improved SWH model for wider use. The research project focuses in particular on a locally manufactured high-altitude SWH technology to run under freezing conditions, as a thermo-siphon system. A first unit for a high-altitude village SWH bathing centre was installed in November 2005 in Simikot Humla (Figs. A11.4, A11.5) and is at the moment of writing still undergoing intensive long-term field tests at an altitude of exactly 3,000 m.a.s.l.

In May 2009, as part of the long-term participatory HCD in Darapori village in Upper-Humla, the first village based high altitude hot water village bathing centre (it is believed to be the first of its kind in the world) was inaugurated (Figs. A11.6, A11.7, A11.8).

[64] Under the supervision of the author, and as part of the high-altitude solar water heater student research project at the Kathmandu University from 2002 to 2005, the students measured the performance and efficiency of solar water heaters installed on rooftops in Kathmandu.

Fig. A11.5 The High-Altitude Thermo-Siphon SWH in closed, or after sunset, position. The insulation all around the absorbers and the hot water storage tank consists of 100 mm polyurethane foam. Four thermocouples on one absorber measure the water temperatures at different heights, and 4 thermocouples, installed at different heights, measure the hot water storage tank temperature and stratification. Since November 2005, the various parameters have been recorded every minute with a DT605 dataTaker data logger. Overall efficiency values of 45–55% are attained

Fig. A11.6 Darapori village's High-Altitude Bathing Centre (BC) for the local community, built from March-May 2009. The BC provides 6 shower cabins, each with a hot and cold water tap to mix the water accordingly. It also has a clothes washing unit (see in the upper right corner in the Figure) with two cold water taps to wash clothes. Thus the BC meets the important need for improved personal hygiene. The wastewater is led through a reed bed wetland, in order to clean the water before it is drained into a sloping piece of land that is unsuitable for agriculture

Fig. A11.7 The BC consists of two units, each with 4 absorbers and each with a 300 L stainless steel warm/hot water storage tank. A 1,000 L cold water polyethylene reservoir tank (in the middle on top of the BC's roof) ensures that there is always enough water at the necessary pressure in the piping system. That enables 50–55 people a day to take a warm/hot shower on an average day (with an average solar irradiation of ~5.5 kWh/m^2 per day for Darapori village). Thus each of the 420 people of Darapori village can take a warm/hot shower every 10–14 days

Fig. A11.8 The absorbers are installed with an azimuthal angle of 30° towards the west, in order to compensate for the high horizon, caused by the surrounding high mountain chains and thus allow maximum capture of the solar radiation during the colder winter months from the middle of October to the middle of March. Each absorber has an adjustable cover, which acts as insulation protection during the cold winter nights and, as seen in the picture, during the day as an additional reflector, thus increasing the intercepted solar radiation

11.8 Wind

Nepal is not very windy, as it is landlocked, and due to its unique geographical conditions, it has very few large flat areas of land. But there are various areas where rivers have cut deep north–south valleys into the massive Himalayan mountain range, stretching from the east to the west. Thus many valleys do have strong winds blowing through them, and thus the local communities in these values could enormously benefit from using their local wind energy resource to generate their energy/electricity demands.

The SWERA project has not yet obtained sufficient wind resource data to create any national wind resource map for Nepal. Thus the country's potential wind resource can still not be assessed sufficiently accurately, and therefore locally available wind resources would need to be first evaluated/measured (time demanding and expensive) before a project can take place. Further, being directly related to the solar radiation and the local geographical conditions, local wind conditions in regard to strength, continuity and direction, can change rapidly. Thus, with the huge water resource, a maturing hydro power industry, and increasing awareness and implementation of solar PV projects under governmental subsidy programs, it comes as no surprise that hardly any wind generation projects have been implemented in Nepal.

The few wind generator projects in Nepal can be summarised in the following:

1. A first 30 kW wind generator was funded by the Danish government and installed by NEA in 2001 in Kagbeni, Mustang district, which is a wind rich village at the bottom of the Annapurna valley, with almost 6,000 m depth, called the world's "deepest valley". But already after a few months a strong gust damaged the wind turbine's blades so badly that it was beyond repair.
2. In 2001 an imported Synergy S5000DD wind turbine was installed as part of a small hybrid (750 W_P solar PV array and wind turbine) system as power generation for a high-altitude rescue clinic for alpinists, in the Mt. Everest region in the village of Pheriche. Apart from the installation, the local industry and community had no part and benefit from it.
3. In 2003 ITDG installed six small scale wind home systems with battery charging across Nepal. Emphasis was given to local manufacturing, and to train some local people. While this is the right approach to developing a future local manufacturing industry, no feedback or news has been made available on the outcome or results of the project.
4. The author has installed a self-made small wind turbine with $180W_P$ output power in 1998, and run it for 4½ years in the remote area of Jumla (Figs. 3.54, 3.58), north-west Nepal.[65] That wind turbine ran without any failure for 4½ years, generating small amounts of energy, just enough to charge the batteries for one home, even though the actual installation place was not ideal.

[65] http://www.rids-nepal.org/index.php/Wind_Energy.html (accessed July 20, 2012).

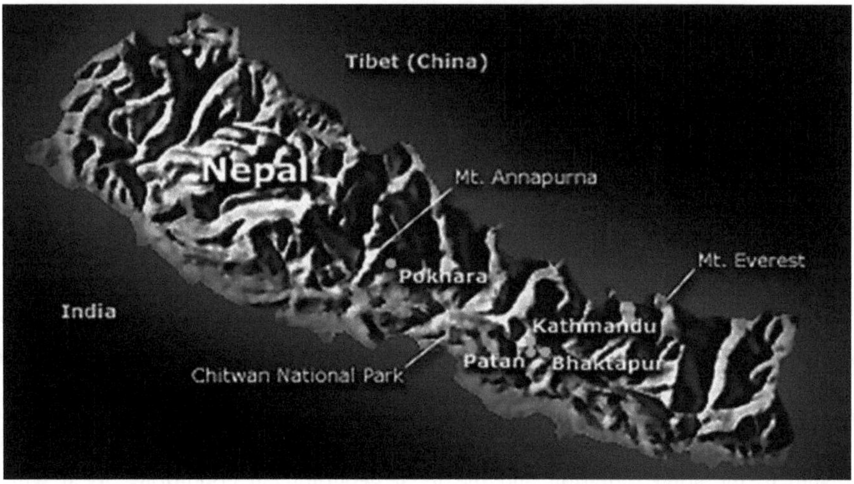

Fig. A11.9 Nepal relief map with some of the major deep north-south valleys, creating a natural venturi effect. (map from: http://www.bugbog.com/maps/asia/nepal_map.html)

5. Since 2004, KUPEG (Kathmandu University Power Electronics Group) started a small scale wind turbine design and manufacturing project under the guidance of Dr. Peter Freere. Part of their strategy is to apply locally available technologies, equipment and resources, alongside the training of local craftsmen.

These examples show that the application of small scale wind turbines for rural village electrification in Nepal is still very much in its infancy and that no project has thus far made headlines.

Appendix 12
Discourse on Light Emitting Diodes

12.1 Replacing Traditional Fuel Based Lighting with Light Emitting Diodes in Rural Nepali Homes

Since the beginning of the human civilisation, artificial lighting has been an important requirement for human social and economic activities. In 2005 the worldwide annual use of electric light was 133 Pl mh with an average of 20 Ml mh per person, and this consumes about 19 % of total global electricity use (IEA 2006). On the other hand, two billion people in the world are still living without access to electricity (Mills 2002a). Almost all of these people live in developing countries, with four out of five living in rural areas (IEA 2006). About 14 % of the urban population and 49 % of the rural population in the developing countries were without electricity in 2000, and in the least developed parts of Africa (e.g. Ethiopia, Uganda) only 1 % of households had access to electricity (Mills 2005).

People in the very remote parts of Nepal use precious trees as firewood on open fireplaces in their homes for cooking, room heating, and lighting. This has a direct impact on their chronic health problems due to the rooms filled with smoke, resulting in low life-expectancy, and high death rates of young children (IEA 2006). The extremely harsh and rugged environments of these communities in the midst of the Himalayan mountain region isolate them completely from the urban areas, and therefore most villages will not be reached by electric grid extensions within the foreseeable future. Thus the efficient use of available renewable energy resources and the adoption of appropriate contextualised technologies are essential for sustainable development of these rural communities.

12.2 Lighting in the Developed and in the Developing Countries

In the industrialised countries electric lighting is taken as a basic necessity. On the other hand, in the developing countries, particularly in rural areas, having basic lighting for the daily activities indicates a privileged status. The estimated per

A. Zahnd, *The Role of Renewable Energy Technology in Holistic Community Development*, Springer Theses, DOI: 10.1007/978-3-319-03989-3,

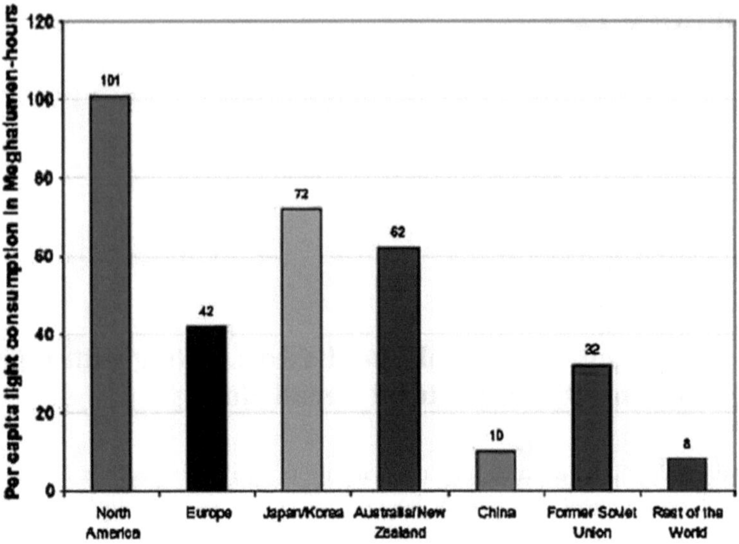

Fig. A12.1 Estimated per capita consumption of electric light in 2005 (IEA 2006)

capita consumption of electric light in 2005 in different regions of the world is given in Fig. A12.1. Many studies have shown that there is a relationship between the household annual lighting electricity consumption and economic wealth (ACMR 2004; Mills 2002b).

Most parts of the developing countries are still out of reach of electrical networks. Less than 5 % of the rural, and less than 40 % of the urban, households in Africa have access to an electric network (ABB 2003). Furthermore, even if the houses are electrified, many homes have only intermittent access to power as electricity blackouts are frequent, hence producing the need for alternative energy sources. For example, in the Indian state Madhya Pradesh, over 90 % of rural electrified households use kerosene as a backup fuel for lighting (IEA 2006). Lighting is the primary consumer of the total electricity production of the developing countries, where it accounts for a significantly higher share (up to 86 %) of residential electricity consumption compared to the industrialised countries (5–15 %) (Mills 2002b).

Due to very large differences in the household incomes and in the availability of energy sources, there are enormous differences in the quality of lighting encountered in the developed and developing countries. Furthermore, in the developing countries, there exist large differences in the quality and quantity of lighting used between the urban and rural areas.

In most of the developing countries, there are no guidelines on recommended lighting levels and lighting systems are usually installed, without proper calculations by technicians with insufficient professional training. The lighting levels applied are usually lower than required because of the lower affordability of electric lighting. In contrast, in the industrialised countries, there exist detailed

Fig. A12.2 "*Jharro*", a resin soaked pine tree stick for indoor lighting in Humla

codes and recommended levels for each specific application (CIBSE 1997; ISO 2002; CIE 2001; IEA 2006).

12.3 Fuel Based Lighting in Rural Nepali Villages

The primary energy source used to provide the necessary daily energy services in Nepal has for centuries been firewood, often supplemented by crop residues and animal manure.

The families and communities in upper Humla have to use "*jharro*", a resin soaked high-altitude pinewood stick, to get minimum, but smoky indoor lighting (Figs. A12.2, A12.3). "*Jharro*" is gathered by inflicting a deep "wound" on the pine tree, forcing it to produce a high amount of resin in order to "cure" the wound (Figs. 2.15, 6.3). This high resin content wood layer is cut away after a week and burned in small sticks to generate light. The author carried out illuminance measurements in houses burning "*jharro*" sticks in several villages.

Burning "*jharro*" sticks are typically placed on an elevated stone or mud pile or on a hanging metal plate (Figs. 3.3, 3.82, 3.101) at a height 40–50 cm from the floor. The average illuminance on the floor up to a horizontal distance of 1 m from the source was 2 lux. In the room corners (floor level), which were more than 1 m from the burning *jharro* sticks, the illuminance was less than 1 lux. These low lighting levels make it just possible to move around the room and to do some general work close to the light source, but the lighting is not adequate for any visually oriented tasks such as reading.

Fig. A12.3 Open-fire place for cooking and heating and light through "*jharro*" create hazardous indoor air pollution conditions

12.4 Defining Basic Lighting Levels for Remote Communities in Developing Countries such as Nepal

There are many factors that affect the definition of appropriate light levels for the homes of remote villages in developing countries. The availability of local energy resources, the cost of the lighting technology, and the local people's prevailing lighting practices should be considered in order to make the implemented lighting projects and programs sustainable. The defined lighting levels should be suitable and affordable for the rural people's activities and needs. The rural electrification projects are often first time electrification projects for a community, and thus have to aim to provide just minimal lighting for defined tasks, however in an affordable and sustainable way.

The primary function of any home lighting system is to provide a safe visual environment for movement around the space, to make it possible to perform visual tasks, and to provide a comfortable and pleasant visual environment. On the other hand, the lighting system has to be cost effective, efficient, non-polluting, and easy to clean and maintain. The standards and guidelines for the recommended lighting levels in the developed countries often categorise the household in different areas and give recommendations on lighting levels according to the specific need of each area. However, the homes in rural Nepali villages are made up of flat muddy roofs with stones and wooden beams, each home consisting of two rooms with low ceilings. There are no separate rooms for specified tasks. The whole family resides in the two rooms and these rooms serve them as kitchen, bedroom, study room, dining room and living room. In these homes, lighting is needed for cooking, orientation, security, and also for the children to do their homework in the evenings.

The illuminance measurements under fuel based lighting in the villages in Humla showed that the average illuminance levels available on the floor were around 2 lux. These lighting levels were just enough to move around the room and see the people and items in it. It was not possible to read at this lighting level, and any reading task had to be done very close to the light source. It was possible to read texts of a book when the illuminance level was around 25 lux, which was obtained by bringing the book close to the light source. This was tested by having the local school children (less than 18 years of age) perform reading tasks. Cooking was easier, because the light source is usually located near the cooking place in the houses. The illuminance level on the cooking stove varied between 5–15 lux depending on the height of the cooking place.

It is not feasible to recommend different lighting levels separately for orientation, cooking, cleaning, and reading for the rural homes that require a basic electric lighting system. Rather, two levels of illuminance can be recommended, one for reading purposes and another for all other activities. Based on the illuminance measurements under the traditional "*jharro*" lighting, it was concluded that a lighting level of 2 lux is not sufficient for general purposes. An improved illuminance of about 5–15 lux is recommended for general purposes. As it was possible to read at a lighting level of 25 lux, and this is recommended for a basic lighting service for home lighting in these communities.

12.5 Lighting with WLED Lights in Rural Nepali Villages

12.5.1 WLEDs and Renewable Energy Sources for Lighting

RIDS-Nepal uses solar PV systems and pico-hydro power plants with WLED lamps to implement lighting in villages as part of long-term community development projects. Until December 2011, RIDS-Nepal has electrified 10 villages, one school and one health post in remote Upper-Humla through basic village electrification projects. Nine villages, the school and the health post generate their energy through solar PV systems and one village uses a 1.1 kW pico-hydro power plant. In these villages, a total of 564 homes with ~3,400 people now have minimal indoor electric lighting for about 7–8 hours a day. In all three solar PV approaches (3.3.2), the central, the cluster, and the solar home system, as well as in the pico-hydro power plant approach, each home is provided with three luminaries with WLEDs (Fig. 3.83, A12.4).

12.5.2 Luminare Measurements at the Helsinki University of Technology in Finland

Illuminance measurements were carried out at the homes of Humla villages under WLED lighting. The luminaries used in the villages were also measured in the

Fig. A12.4 WLED luminaries manufactured in Nepal using Nichia NSPW510BS/DS WLEDs (*left*) and Lumiled's Luxeon Star LED (*right*)

Lighting Laboratory of Helsinki University of Technology to test their performance (A13.1). There are two different types of WLED luminaries manufactured for RIDS-Nepal village illumination programs, one consists of 9–12 Nichia NSPW510BS/DS WLEDs (5 mm low-power WLEDs); and the other consists of a single WLED, Luxeon Star from Lumiled (high power WLED), as shown in Fig. A12.4.

All of the luminaries are manufactured in Nepal by Pico Power Nepal (PPN), a local, RIDS-Nepal spin-off, manufacturing company. The luminare with low-power WLEDs is driven by using series resistors and a reverse protection diode. Three sets of WLED strings, each string containing three WLEDs, are connected in parallel and with one resistor. Since the current through each WLED is small (15–20 mA), the losses in the resistors are minimal. The rated current of the high power WLED is 350 mA and it is thus uneconomical to drive it with a resistor due to high-power losses. Hence it is driven by a driving circuit based on a switching regulator.

12.5.3 Characterisation of Nepali Solid State Lamps in Vilnius University in Lithuania

Solid state lighting technology based on injection electroluminescence devices, light emitting diodes (LEDs), has already found numerous applications in traffic lights, automotive signage, full colour video displays, liquid crystal display backlighting, optical measurements, phototherapy, and other niche applications (Žukauskas 2002; Harbers 2007). The development of WLEDs based on partial or complete conversion of short wavelength radiation in phosphors and phosphor blends and recent progress in compound semiconductor and phosphor materials quality and photonic design of the chips resulted in WLEDs that are also considered as promising candidates in general lighting. General lighting based on solid state lamps has a high potential in energy saving, offers vast environmental benefits, and features unsurpassed longevity, reduced maintenance, and functionality (Harbers 2007; Schubert 2006).

In industrialised countries where electrical lighting is very common, solid state lighting meets high competition from conventional sources of light, such as incandescent, fluorescent, high-pressure sodium, and metal-halide lamps. Meanwhile in developing countries, especially in rural areas, WLEDs encroach on the lighting market occupied by fuel based sources, such as firewood and kerosene lamps.

Because of their high efficiency and compatibility with off network and rechargeable battery based DC sources of electrical power, such as photovoltaic, hydro, wind, and pedal generators, WLEDs are the most appropriate sources for such substitution with huge benefits in energy consumption and cost, as well as in resolving education and health issues (Shailesh 2006; Bhusal et al. 2007). In particular, fuel based sources have luminous efficiencies below 0.1 lm/watt comparing to 15–80 lm/watt of present commercial WLEDs (Shailesh 2006; Bhusal et al. 2007) with a potential of attaining 200 lm/watt efficiency in the near future (Tsao 2004; Narukawa 2007). To this end, penetration of solid state lighting technology into general lighting markets might be even faster in developing countries than in industrialised ones, especially with international support taken into account.[66]

Solid state lamps are a novel source of general lighting with no well established standards developed so far. However, because of cost pressure, fabrication of solid state lamps for developing countries is drawn to local manufacturing facilities, which lack instrumental characterisation. Meanwhile the conditions of exploitation of lighting devices in developing countries are much harsher than in industrialised ones and involve wide ranges of temperature and humidity, variation in driving voltage, and smoke from organic fuel based stoves. This includes a need for international cooperation on characterisation of solid state lamps manufactured in developing countries. At the Vilnius University in Lithuania the temperature, directional, and chromatic characterisation of solid state lamps manufactured for the RIDS-Nepal village illumination programs were tested and in more details are presented in A13.2.

12.6 Energy Efficient Innovative Lighting and Energy Supply Solutions in Developing Countries

Fuel based lighting is not only inefficient and expensive compared to electric lighting, but is also a cause of severe respiratory and cardiac health problems (IEA 2002; Mills 2002b; Bhusal et al. 2007). Thus the use of different renewable energy sources in combination with efficient lighting technology to provide clean lighting

[66] Light Up The World Foundation. http://www.lutw.org/ (accessed July 20, 2012).

services to developing countries is an urgent need. There is a range of innovative and sustainable technology solutions which can meet the energy needs in developing countries (Gustavsson 2004; Doig 1999; Richards 2006). The technologies involving wind power, solar power, and small scale hydro power exploit the local resources, operate on a small scale and have an advantage of meeting the needs of widely dispersed rural communities. The efficient use of electrical energy is a very important issue in these situations because of the low-level of power production capacity from these technologies and also due to the associated costs.

LEDs are rapidly evolving light sources. As discussed in the previous chapters, the application of WLEDs, powered through the conversion of a renewable energy sources is considered to be a sustainable solution to the basic lighting needs of rural people. Further, increasing luminous efficiency, longer lifetime, and low power requirements make them suitable to be used for lighting in the rural villages. Cost analysis of WLED based lighting systems driven with renewable energy sources in different parts of developing countries have shown them to be cost effective in comparison with the conventional options (Bhusal et al. 2007; Shailesh 2006; Jones 2005).

12.6.1 Case Study: Light Up the World Activities in Developing Countries

Light Up the World (LUTW) is a non-profit organisation devoted to the development of renewable energy based innovative lighting technologies for the developing countries. Founded in Nepal in 1997 by Canadian Prof. Dr. Dave Irvine Halliday, LUTW is the first humanitarian organisation to utilise WLEDs in order to replace fuel based lighting in developing countries (LUTW 2009). The goal of the organisation is to bring safe, healthy, reliable, environmentally friendly and affordable home lighting to more than one quarter of the world's population that lives without access to electricity (Peon 2005).

LUTW has become the platform for the adoption of solid state lighting technology in the most remote and needy parts of the world. LUTW has lit up more than 14,000 homes in 42 countries, improving the life of more than 100,000 people (LUTW 2009).

12.6.2 LEDs and WLEDs for Illumination

LEDs are solid state lighting devices that convert electricity directly into light. LEDs were developed more than 40 years ago but were limited to emitting only coloured light. The possibility of using LEDs for illumination became a reality after Shuji Nakamura from the Nichia Corporation developed the high brightness blue LED in the mid 1990s (Nakamura 1994). After this invention it was possible to produce LEDs emitting white light. The two most common methods of generating white light with LEDs are: (1) the use of blue LEDs with a yellow phosphor, and (2) the mixing of two or more coloured LEDs in appropriate proportions.

Technical advances have greatly enhanced the performance of LEDs in recent years. According to Agilent Technologies, the lumens per package value of red LEDs has been increasing 30 fold per decade whereas the price is decreasing 10 fold per decade (Haitz 2001). Some of the currently available WLEDs have luminous efficacy of 139–150 lm/watt,[67] which is ~9 times greater than an incandescent lamp. The luminous efficacy of prototype WLEDs reached 150 lm/watt in 2006 and 161 lm/watt by 2008.[68] Although the theoretical maximum for luminous efficacy of white light is 400 lm/watt, half of that value is estimated to be feasible with LEDs (Haitz 2001). The optoelectronics industry development association (OIDA) roadmap has a target to achieve the value of 200 lm/watt by 2020 (OIDA 2002). The other important advantages of LED light sources making them suitable for rural lighting are their lifetimes measured in tens of thousands of hours, ruggedness, compact size, and low operating voltage.

12.6.3 Providing Basic Lighting Needs

Currently many homes in rural areas of developing countries without access to electricity are illuminated by the use of biomass or petroleum fuel, which cause health problems due to the smoke produced. On the other hand, the homes in the developing countries have very low income. The designed electric lighting systems have to be affordable, simple, and reliable. The aim of the basic lighting services is to provide a safe visual environment for movement, to enable people to see each other and to do daily indoor tasks such as cooking, cleaning, socialising and school homework. This is the first step on the path charted from a traditional lighting (use of biomass and petroleum for lighting) to cleaner, convenient, and modern lighting. The illumination level is chosen as previously defined, to provide minimal but sufficient lighting in an affordable and sustainable way.

---

[67] Cree announces Lighting Class Performance for Cool White XLamp® LEDs of 139 lm/watt, http://www.cree.com/products/xlamp (accessed October 12, 2009, on November 4, 2011 not available any more), while Nichia announced in October 2009 their commercial available WLED, NSPWR70CS-K1, has 130 lm/watt luminous efficiency, http://www.nichia.co.jp/specification/en/product/led/NSPWR70CS-K1-E.pdf (accessed November 4, 2011).

[68] Nichia achieved 150 lm/W with a WLED in 2006, http://www.nichia.co.jp/about_nichia/2006/2006_122001.html (on November 4, 2011 not available any more), CREE achieved 161 lm/W with a WLED in November 2008, http://www.cree.com/press/press_detail.asp?i=1227101620851 (accessed July 20, 2012).

12.6.4 Lighting Systems for Villages

For the Thalpi village in Jumla only lighting systems had to be provided, as already an experimental 200 W (producing 165 W) pico-hydro generation system was installed by the author in 1998 (Figs. 3.9, 3.10, 3.11, 3.12, 3.13, 3.14, 3.15). The power from the pico-hydro was used to provide lighting to two houses fitted with incandescent lamps. But 30 homes in the Thalpi village were illuminated in 1999 by replacing these lamps with two WLED light sources for each home.

12.7 Life-Cycle Cost Analysis of Energy Supply Systems

A simple life-cycle cost analysis is used to compare the costs of different energy supply systems used for lighting in the rural Nepali villages. The costs of generating capacity are calculated for pico-hydro and PV solar systems over their entire lifetime by taking into consideration the characteristics of each individual case.

12.7.1 Pico-Hydro Power System

The cost calculation is made for the pico-power system with generating capacity of 1.1 kW installed in the most remote village in Kholsi, Humla (Figs. 3.3.1.13.16, 3. 17, 3.18, 3.19, 3.20) in western Nepal. The system is used to provide lighting for 73 homes and a primary school, for ~450 people.

The high cost of equipment compared to that in the other parts of Nepal is due to the transportation costs. All the equipment must first be carried by airplane and then by Yak or porter to reach the installation site. Some parts of the construction work were carried out voluntarily by the villagers. The local labour costs are estimated in calculating the cost of the construction work. The cost components of the pico-hydro system include the cost for the equipment, dam, canal, power house, and installation work (Table A12.1). The costs are given in NRs.

Table A12.1 Characteristics of the pico-hydro power system

Generating capacity	1.1 kW
Cost of generation equipment (generator, turbine, accessories)	NRs200,000
Cost of transformer	NRs70,000
Cost of construction work (power house, dam, canal, etc.)	NRs150,000
Cost of installation	NRs100,000
Estimated lifetime of generation equipment	10 years
Estimated lifetime of transformer	10 years

Table A12.2 Characteristics of the solar PV system

Generating capacity	75 W_P
Cost of solar photovoltaic module	NRs30,000
Cost of battery (75 Ah)	NRs6,000
Cost of charge controller	NRs2,000
Estimated lifetime of photovoltaic module	25 years
Estimated lifetime of charge-controller	8 years
Estimated lifetime of battery	6 years

12.7.2 Solar PV System

The PV systems used in the villages consist of a PV panel, a battery, and a battery charging unit. The system with a generating capacity of 75 W_P and battery of capacity 75 Ah is considered in the calculation. Depending on the geographical location and distances between the houses, this kind of system is used to illuminate up to 15 homes (Table A12.2).

12.7.3 Results and Discussion

A 20 year life-cycle cost (LCC) analysis period is used for each system. Using a discount rate of 4 %, discount factors are calculated for each cost occurring year and the costs are converted into present value. The life-cycle costs are then converted into cost per kilowatt of generating capacity to enable the comparison of the two systems. The results of the calculations are presented in Table A12.3.

The LCC calculation over a 20 year service life does not show any significant difference in costs per kW generating capacity between the solar PV and pico-hydro. However, the cost calculations depend greatly on the assumptions made and the cost varies depending on the systems and condition and context of the villages.

The maintenance and operating costs were not considered in the LCC analysis. There are no/minimal operating costs associated with the solar PV system. The maintenance costs of a PV system are low including the costs for periodic inspection and cleaning of the solar panels, battery, and circuits. On the other hand, the pico-hydro system needs trained manpower for the operation and maintenance. Special training has to be given to the local people for operation and minor maintenance work. In cases where major maintenance is needed, the situation becomes more complicated because of transportation problems. The operation and maintenance costs for pico-hydro systems can be partly covered by making use of its power in daytime for other purposes, e.g. grinding grain and pumping water.

An energy supply system for rural village electrification has to be cheap, easy to maintain, and sustainable. Energy technologies that require low maintenance are suited for the remote areas due to unavailability of skilled labour. Thus, overall the

Table A12.3 Calculation of life-cycle cost (LCC) of pico-hydro power and solar PV systems

Pico-hydro system			
Year	Base year cost	Discount factor	Present value
0	NRs520,000	1	NRs520,000
11	NRs370,000	0.65	NRs240,500
LCC			**NRs760,500**
LCC per kW generating capacity pico-hydro system			*NRs*691,364
Solar PV system			
0	NRs38,000	1	NRs38,000
7	NRs6,000	0.76	NRs4,560
9	NRs2,000	0.703	NRs1,406
13	NRs6,000	0.601	NRs3,606
17	NRs2,000	0.513	NRs1,026
19	NRs6,000	0.475	NRs2,850
LCC			**NRs51,448**
LCC per kW generating capacity solar PV system			*NRs*685,973

solar PV system is found to be more favourable in regard to its LCC as well as a more reliable and appropriate technology for small loads and remote rural areas.

12.8 Summary

Connecting the rural and remote areas of developing countries with electric networks is a challenging task and it is not expected to occur in the near future. The application of locally available renewable energy resources and efficient lighting technologies is the best way to provide safe and clean lighting to the villages. It will improve the quality of life in the rural villages while reducing the environmental emissions. The availability of electric lighting will contribute to improving the health, education, and income generation of the rural environments.

In choosing the appropriate renewable energy technology for rural lighting, the reliability and sustainability of the technology in the local context are important issues for the successful implementation. Solar photovoltaic systems are economically competitive and are suitable for the widely scattered rural areas.

Appendix 13
Testing and Evaluation of WLED Lamps Manufactured in Nepal

13.1 Measurement of the RIDS-Nepal WLED Luminaries at Helsinki University of Technology in Finland

The luminous fluxes of both, the Nichia NSPW510BS WLEDs (left in Fig. A13.1) and Lumiled's Luxeon Star LED (right in Fig. A13.1) luminaries were measured in an integrating sphere at Helsinki University of Technology.

The operating current and voltage were also measured for calculating luminous efficacy. In order to make a direct comparison of the WLED light source with the "*jharro*" light source, the luminous efficacy of "*jharro*" was also calculated. The energy content of the resin soaked pine stick was measured using a calorimeter at the University of Jyväskylä and the value was converted into equivalent electrical power. The luminous flux of the "*jharro*" was measured in a dark room at Helsinki University of Technology. Table A13.1 shows the characteristics of the two WLED luminaries and the pine stick "*jharro*". The measurements indicated that the luminous efficacy of the pine stick lamp (0.04 lm/watt) is half of the efficacy of kerosene fuel based lamp (0.08 lm/watt (Mills 2005)) and more than 300 times less than that of the WLED luminare used in the villages.

Each home in the villages is provided with two luminaries with 9 WLEDs, and one luminare with a single WLED. The two low-power WLED luminaries are installed in the bigger room used for cooking and the luminare with a single high power WLED is installed in the smaller room. The luminaries in the homes are installed on the ceiling of the room at a height of about 1.8 m from the floor. The average illuminance in the bigger room with the two luminaries at the floor level was 5 lux, while in the other room under the high-power WLED luminare it was 3 lux. Based on the people's reaction to the lighting (interview), an average illuminance of about 5 lux seemed to be adequate for general purposes, while the average illuminance of 3 lux was found to be inadequate.

The illuminances measured in the laboratory dark room under both WLED luminaries at different horizontal and vertical distances are shown in Figs. A13.2, A13.3.

A. Zahnd, *The Role of Renewable Energy Technology in Holistic Community Development*, Springer Theses, DOI: 10.1007/978-3-319-03989-3,

Fig. A13.1 WLED luminaries manufactured in Nepal using Nichia NSPW510BS WLEDs (*left*) and Lumiled's Luxeon Star LED (*right*)

Table A13.1 The measured values of power (W), luminous flux (lm), and luminous efficacy (lm/W) of the WLED luminaries and "*Jharro*", and rated luminous efficacy of the WLEDs as given by the manufacturers

Light source type	Power (W)	Luminous flux (lm)	Luminous efficacy (lm/W)	Rated luminous efficacy of WLED (lm/W)
Luminare with 9 Nichia WLEDs	0.730	11	15.0	29
Luminare with 1 Luxeon WLED	1.070	14	13.1	38
"*Jharro*" (pine stick)	2167	88	0.04	–

When the luminare with low-power WLEDs was installed at 0.5 m above the illuminated plane, the illuminance on the plane directly under the luminare was 112 lux. Thus the WLED luminaries can provide sufficient light to read and to perform other visual tasks. The appropriate installation height of the luminare depends on the type of illumination needed. While the illuminance directly under the luminare is high with the luminare installed at lower heights, the illuminance at the adjacent areas decreases sharply. In contrast, the reduction of illuminance on a wider horizontal plane is not so sharp under the high-power WLED luminare because of the wide viewing angle (110°) compared to the WLEDs (50–55°) of the low power WLED luminare. A wider angle provides more evenly distributed horizontal illuminance on a wider area though with lower illuminance level, while a narrower angle provides higher illuminance directly under the luminare but on a very narrow area.

13.1.1 Technical and Economic Aspects

The performance and lifetime of the lighting system is dependent on all the associated auxiliary components. Routine checking of the equipment is needed to

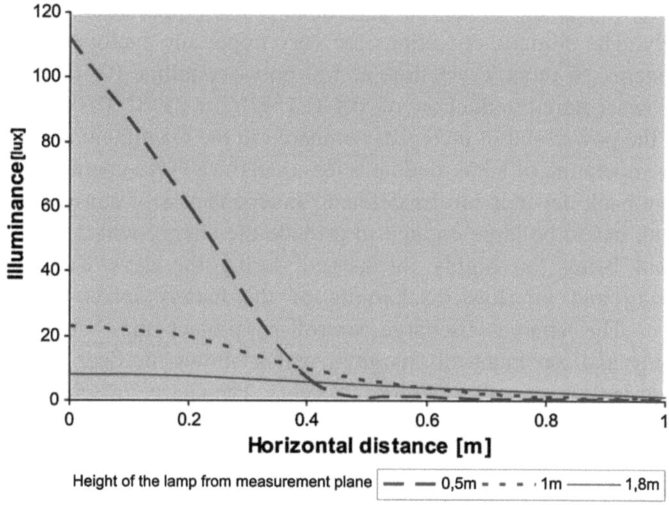

Fig. A13.2 Illuminance at floor level under the low-power WLED luminare as a function of horizontal distance and at three different luminare mounting heights

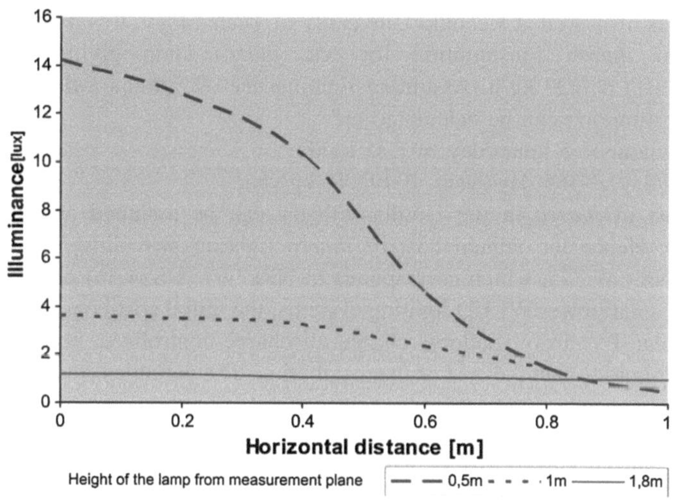

Fig. A13.3 Illuminance at floor level under the high-power WLED luminare as a function of horizontal distance and at three different luminare mounting heights

maintain the quality of the lighting. Cleaning of the PV panels, checking the battery voltage, topping up the batteries with rain water, and cleaning the glass of the luminare should be done regularly. The solar PV modules are the most expensive equipment in a PV system and they have the longest lifetime. The Si mono-crystalline PV arrays used in the Humla villages are guaranteed by the manufacturer to provide 90 and 80 % of rated power output after 12 and 25 years, respectively. The climatic conditions are very important factors when designing the PV systems. Si mono-crystalline and Si poly-crystalline PV modules have an average power output reduction of 0.4–0.5 %/°C > 25 °C (STC) temperature. Similarly, the power output increases compared to the manufacturer's rated power when the temperature of a PV module is less than the STC temperature. The design of a battery-bank depends on the defined "independence of sunshine" days. The battery-bank has to be large enough to provide the energy without being charged and without being too highly discharged during the days without sunshine. Overcharging and too low discharging of the battery leads to shorter life-expectancy. The charge-/discharge-controllers protect the battery-bank from overcharging and too much discharging, which allows the deep-cycle lead-acid battery used in the villages to last for 6–8 years. The charge-/discharge-controllers manufactured by PPN in Nepal have a lifetime of about 8–10 years. The whole system is protected by an electronic fuse for short circuits and overloading.

Cost analysis of two types of WLED lighting system and of *jharro* lighting was done to compare the costs in terms of NRs per lumen hours of light. Initial costs and variable costs of the lighting installations were converted into annual costs. In *jharro* lighting, there were no initial costs and the cost involved only the amount of *jharro* consumption. The amount of *jharro* consumption per hour in *jharro* lighting was measured at Helsinki University of Technology. It was found that the amount of *jharro* consumption for one *jharro* lamp giving 88 lumens (Table A13.1) is 0.27 kg/h. Assuming lighting use of 5 hours a day, the annual *jharro* consumption can be calculated as:

0.27 kg/hour × 5 hours/day = 1.35 kg/day

1.35 kg/day × 365 days/year = 492.75 kg/year.

The cost of *jharro* in the Humla villages can be assumed as NRs 100/kg (US$1.42). Hence the annual cost of *jharro* lighting providing 88 lm of light output is NRs 49,275, which corresponds to NRs 307 (US$4.36) per klmh.

For the solar power WLED lighting systems, the initial cost consists of the cost for the solar PV array, battery, charge-/discharge-controllers, wires, switches, WLED luminaries, and the cost of the installation. The variable costs consist of the cost for maintenance, and cost for the replacement of battery, controllers, and other auxiliaries. The cost analysis was done for a 25 year life-cycle, assuming the life of the solar panels to be 25 years. An example of a solar home system with a 12 W solar panel, two deep-cycle batteries, charge-/discharge-controller, luminaries, and switches was taken for the calculation. The cost of each component was assumed to be the cost at which they are available in the electrification project in Humla. We calculated from our analysis the cost of high power WLED (Luxeon) lighting to be NRs15.12 (US$0.21) per klmh, while the cost of low power WLED (Nichia) lighting was NRs15.59 (US$0.22) per klmh (kilolumen-hour).

Due to the development of WLED technology, the prices of WLEDs are decreasing and the luminous efficiencies of WLEDs are increasing. In the future, this will further increase the cost advantage of WLED lighting compared to the traditional *jharro* lighting.

13.2 Experimental Setup for the Temperature, Directional, and Chromatic Characterisation of Solid State Lamps at Vilnius University in Lithuania

Two kinds of lamps produced by PPN were investigated. The lamps were designed for the 12.6VDC input voltage that is the rated voltage of the deep-cycle lead-acid batteries typically used in combination with solar PV charging systems. The luminaries contained WLEDs, driving circuitry, and a flat aluminium reflector assembled within identical stainless steel cases with a sealed glass lid. The luminaries were designed for almost similar luminous flux (11–13 lumens) and differed in the model and number of WLEDs and driving electronics used. When mounted on the ceiling of a room, such luminaries provide an illuminance of about 5 lux, sufficient for performing simple visual tasks. Meanwhile at a shorter distance, the illuminance can amount 100 lux in magnitude that is sufficient for common tasks, such as reading (Bhusal et al. 2007).

The first lamp, with nine low-power white phosphor conversion LEDs (*Nichia* model NSPW510BS[69]) has a rated viewing angle of about 50–55°. The WLEDs were connected in three parallel strings each consisting of three WLEDs. Each string had a resistive ballast (100 Ω) connected in series for current regulation. Of nine WLEDs, eight were arranged in a ring (with a slight outside bending in order to provide a wider viewing angle of the lamp) and the ninth one was placed in the centre of the reflector. The rated input power of the lamp was 0.88 W. We designate this lamp as a low-power LED (LPLED) based lamp. The second lamp was based on a single white phosphor conversion high-power LED (*PhilipsLumiledsLighting* model LXHL-MW1D[70]) with the rated viewing angle of 110° ("batwing" radiation pattern). The WLED was placed in the centre of the reflector and driven by a switching converter based on a step down voltage regulator IC (*National Semiconductor* LM2575). The converter was controlled by a current regulator based on a measurement resistor and an operational amplifier (½LM358N). The lamp was also equipped with a comparator (½LM358N) for power cut-off at low input voltage in order to prevent battery overdrain. The rated

[69] Specifications for Nichia White LED. Model: NSPW510BS. Previous available at: http://www.nichia.com/specification/led_lamp/NSPW510BS-E.pdf (accessed March 15, 2010), http://pdf1.alldatasheet.com/datasheet-pdf/view/86901/ETC/NSPW510BS.html (accessed March 15, 2010, not accessible anymore by November 6, 2011, as this product has been superseded by the NSPW510DS in 2007/8, see at: http://www.nichia.co.jp/specification/en/product/led/NSPW510DS-E.pdf (accessed July 20, 2012).

[70] Power light source Luxeon® Star. Technical Datasheet DS23. http://www.luxeon.com/pdfs/DS23.pdf (accessed March 15, 2010, not accessible anymore by November 6, 2011, as this product has been superseded, see at: http://www.futurelightingsolutions.com/fr/Technologies/Semiconductors/Lighting-Solutions/High-Power-LED-Emitters/White/Pages/1391481-LXHL-MW1D.aspx?ManufacturerName=PHILIPS-LUMILEDS&isFLS=true (accessed July 20, 2012).

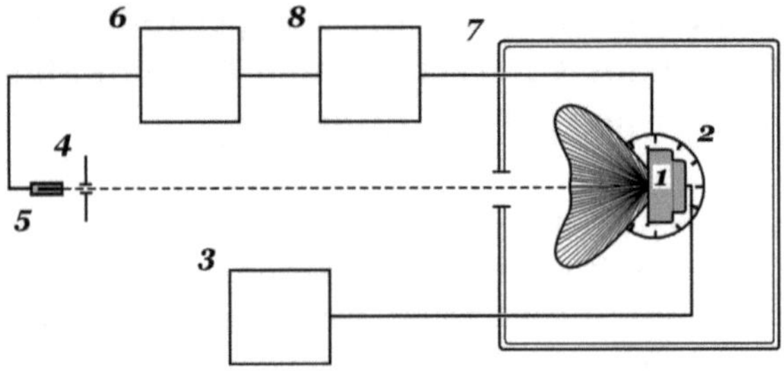

Fig. A13.4 Schematic view of the experimental setup. *1* solid state lamp; *2* rotating stage; *3* power supply; *4* pinhole; *5* optical fibre probe; *6* spectrometer; *7* climatic chamber; *8* computer

input power of the lamp was 1 W. Here we designate this lamp as a high-power LED (HPLED) based lamp.

The experimental setup used for the characterisation of the lamps is shown in Fig. A13.4. A lamp (1) was mounted on a computer controlled rotation stage (2) (*Standa* 8MR151 driven by a controller *Standa* 8SMC1-USBh). The lamp was driven using a power supply with regulated voltage (3) (*Keithley* 2430 *SourceMeter*), which is equipped with digital voltmeter and ammeter. The emission of the lamp was passed through a pinhole (4) with the aperture diameter of 2.5 mm and directed to the optical fibre probe (5) of a spectrometer (6) (*Hamamatsu* model C7473). The pinhole was located at a distance of 20*D* from the lamp, where *D* is the diameter of the lamp front window (77 mm). The lamp and rotation stage were placed within a climatic chamber (7) (*ILKA* model *Feutron* 3001-01). The temperature within the chamber was regulated with an accuracy of 2 °C. A desktop computer (8*)* was used for the control of the experimental setup. The CIE1931 chromaticity coordinates and the correlated colour temperature (CCT) were derived from the recorded spectra using spectrometer software.

13.2.1 Results and Discussion

The emission spectra of the lamps at two ambient temperatures are depicted in Fig. A13.5. The spectra were taken in the front view (along the axis of the lamps) at the input voltage of 12VDC. These spectra are typical of standard cool white phosphor conversion WLEDs. The narrow peak on the short wavelength side of the spectra (at about 460 nm) is due to electroluminescence of the InGaN chip of the LEDs. The broader long wavelength band is due to the photoluminescence of the Ce^{3+} doped yttrium aluminium garnet based phosphor converter. The mixture of blue light emitted by the semiconductor chip and yellow light emitted by the phosphor is perceived as white light.

When ambient temperature is increased from 23 to 50 °C, both spectra exhibit some broadening and long wavelength shift of the blue line typical of near band-

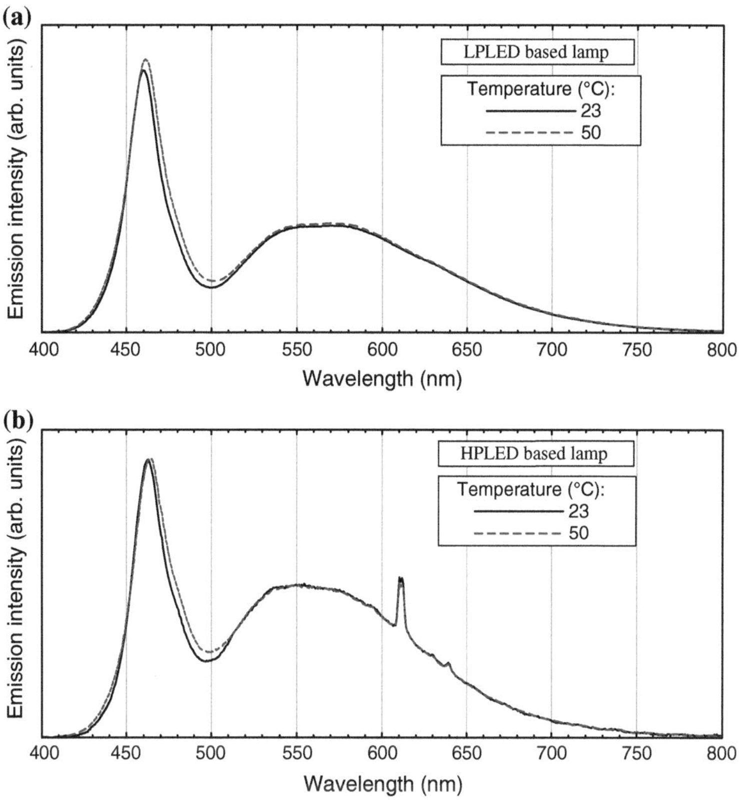

Fig. A13.5 Emission spectra of solid state lamps at front view at two ambient temperatures. **a** LPLED based lamp; **b** HPLED based lamp

gap emission from semiconductors, whereas the yellow line due to the phosphor, maintains almost constant shape and spectral position. Note that the overall emitted flux increases with temperature, although LEDs have negative temperature coefficients of output due to thermally activated non-radiative recombination. Therefore the observed increase of output with temperature must be attributed to the peculiarities of the driving circuits, which provide LEDs with a higher electrical power at elevated temperatures as discussed below. Figure A13.6 depicts the dependence of driving current on temperature for the lamps investigated at the input voltage of 12VDC. The LPLED based lamp exhibits an increase of current with temperature.

Such an increase is due to resistive ballast used and can be understood in terms of a negative temperature coefficient of the LEDs forward voltage, which has typical values of about −2 mV/°C. For three WLEDs connected in a string, this results in a decrease of the voltage drop over the string of about 160 mV, when the temperature is increased from 23 to 50 °C. A corresponding increase of the voltage drop over the ballast resistor results in an increase of the driving current and in a

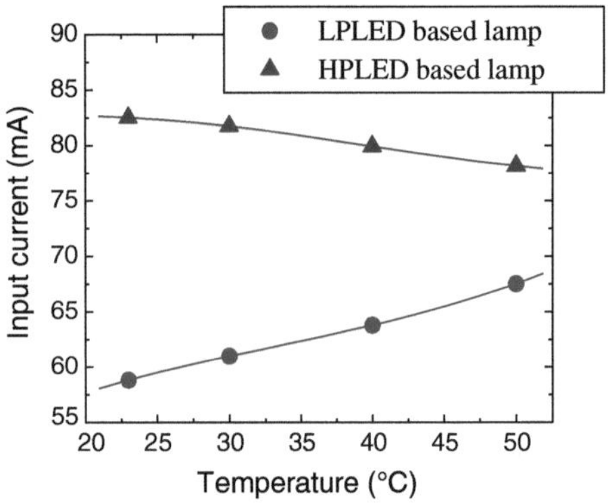

Fig. A13.6 Input current versus ambient temperature dependence for two types of solid state lamps. The input voltage is maintained at 12VDC

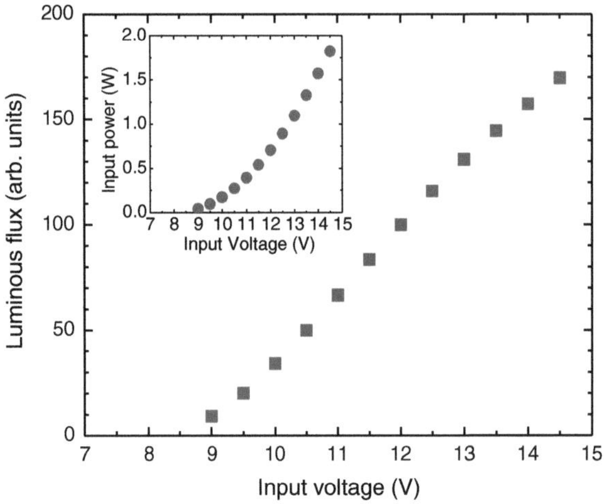

Fig. A13.7 Luminous flux versus input voltage dependence for the LPLED based lamp. The inset shows the electrical power versus input voltage dependence

small increase of the flux generated (Fig. A13.5a) despite the negative temperature coefficient of the LED output.

Although the resistive ballast used in the LPLED based lamp is sufficient for the LED forward voltage and output, such a driving circuitry is unable to ensure stability of the lamp against the variation of the input voltage. Fig. A13.7 shows luminous flux as a function of input voltage for the LPLED based lamp with resistive ballast. The flux is seen to almost linearly increase with input voltage with a coefficient of 30 %/V at the rated input voltage of 12VDC. The reason is a steep dependence of electrical power versus input voltage (inset in Fig. A13.7) that is due to the resistive ballast based driving circuit. Such a variation of the flux

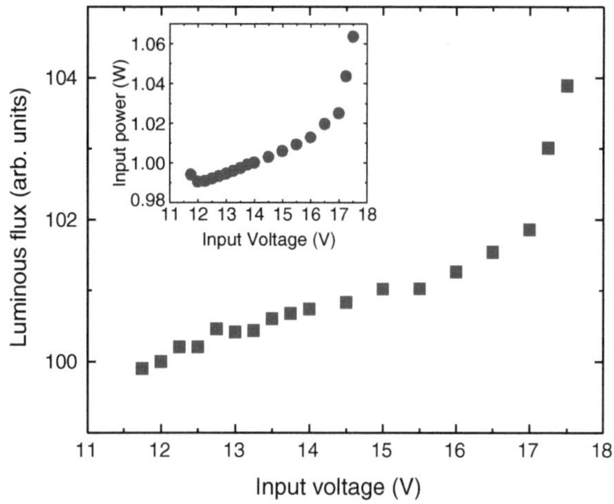

Fig. A13.8 Luminous flux versus input voltage dependence for the HPLED based lamp. The inset shows the electrical power versus input voltage dependence

independent output, due to the interplay of the temperature with input voltage, aggravates the use of standard lead-acid deep-cycle batteries that are very common in rural lighting installations. With a typical variation of the deep-cycle battery voltage from ~11VDC (uncharged) to 14.4VDC (fully charged), the output of such a lamp varies by about 250 %.

Triangles in Fig. A13.6 show the dependence of input current on temperature for the HPLED based lamp that is driven by the electronic ballast. The input current is seen to decrease with temperature. Since the electronic ballast has a circuitry for the regulation of the LED driving current (this results in almost temperature independent output of the lamp), the observed decrease of the input current with temperature can be attributed to an increased efficiency of the ballast at elevated temperatures.

However, the main advantage of the electronic ballast used in the HPLED based lamp is a high stability of the output flux against input voltage (Fig. A13.8).

For the input voltage ranging from 11.75VDC (start up voltage of the built-in battery overdrain protection circuit) to 17VDC, the output flux varies by less than 2 %. This is due to a small variation of the input power (inset in Fig. A13.8) provided by the electronic ballast. Since the corresponding variation in illuminance is almost indistinguishable by human vision, the electronically ballasted HPLED based lamp is much more appropriate for use in deep-cycle battery based lighting installations.

The measured angular distribution of the spectrally integrated luminous intensity for both of the two solid state lamps is depicted in Fig. A13.9. Solid and dashed lines present data obtained at 23–50 °C, respectively. The LPLED based lamp (Fig. A13.9a) has a narrow viewing angle of about 50°, whereas the HPLED based lamp (Fig. A13.9b) exhibits a "batwing" angular distribution with the viewing angle of 113° and two peaks at about 42° in both directions from the front view.

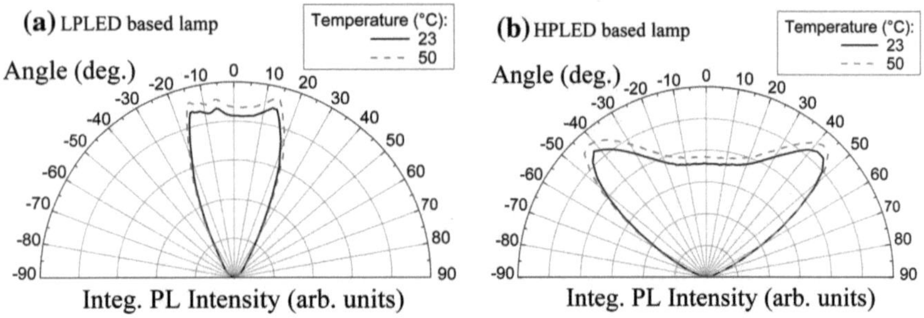

Fig. A13.9 Angular distribution of integrated Power Luminous (PL) intensity for solid state lamps based on LPLEDs (**a**) and HPLEDs (**b**). *Solid lines* and *dashed lines* correspond to 23 and 50 °C ambient temperatures, respectively

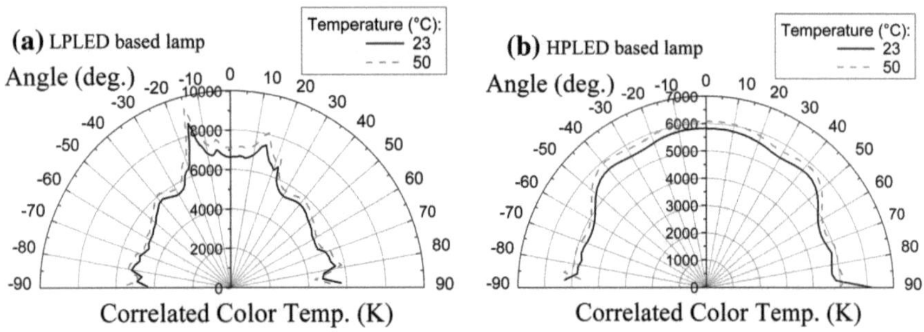

Fig. A13.10 Angular distribution of correlated colour temperature CCT for the solid state lamps based on LPLEDs (**a**) and HPLEDs (**b**). *Solid lines* and *dashed lines* correspond to 23 and 50 °Cambient temperature values, respectively

A small asymmetry can be traced in the angular distribution of the LPLED based lamp, probably, due to the deviations in the arrangement of the nine WLEDs. Basically, since the measured viewing angles are in line with the rated viewing angles of the WLEDs, one can conclude that the reflector and the front glass lid play a minor role in the formation of the angular distribution of intensity.

With increasing ambient temperature, both lamps are seen to exhibit a small increase in luminous intensity, while the angular distribution of luminous intensity remains almost unchanged.

The angular distribution of correlated colour temperature (CCT) is shown in Fig. A13.10. At 23 °C, the LPLED based lamp has CCT of about 6,610 K in the front view and a noticeable variation of CCT from about 5,700 to about 8,600 K within the viewing angle of ~50°. Such variation can be tolerated in performing simple visual tasks, such as orientation. However it can cause inconvenience in performing common visual tasks, such as reading. Differently, the HPLED based lamp has a CCT of about 5,800 K in the front view with a less important drop to about 5,000 K at the boundaries of the viewing angle.

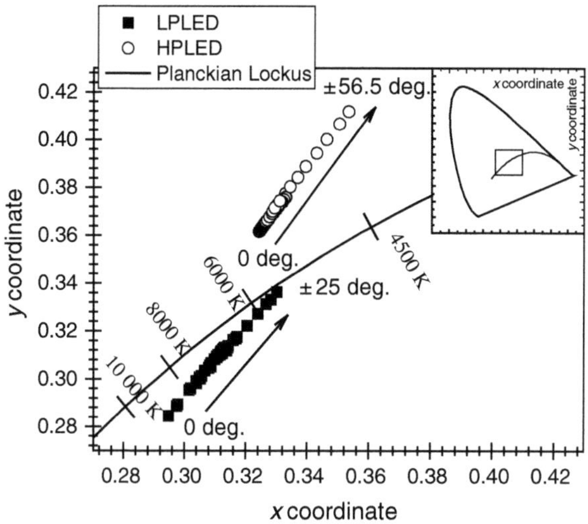

Fig. A13.11 Location of the chromaticity points on a segment of the CIE1931 diagram as a function of angular displacement within the viewing angle for LPLED and HPLED based lamps. The inset shows the position of the segment in the diagram

With ambient temperature increased to 50 °C, the shape of the angular distributions remains almost unchanged. However, an increase in CCT by up to 10 % is observed in all directions. Such variations in CCT are almost indistinguishable by human vision and can be tolerated in most visual tasks.

More details of the chromatic behaviour of the investigated lamps are shown in Figs. A13.11, A13.12 that depict a portion of the CIE1931 diagram with the Planckian locus, which contains chromaticity points of blackbody radiators at different temperatures.

Figure A13.11 shows the chromaticity coordinates of the two lamps as functions of the angular displacement from the in-axis position. One can see in Fig. A13.11 that with increasing angular displacement, the coordinates shift towards the upper right corner of the diagram, which corresponds to the aforementioned drop in CCT (Fig. A13.10). These angular distributions of CCT are due to peculiarities of the WLED design, in particular due to the position of the phosphor layer in respect of the semiconductor chip.

At high angular displacements the blue light passes a larger distance through the phosphor layer in comparison with that along the in axis direction. This results in a larger portion of the blue light being absorbed in the phosphor layer and the blue to yellow intensity ratio decreases. In terms of the chromaticity diagram, such a decrease means a shift outwards from the "blue" (lower left) corner of the diagram.

Note that in approaching the "blue" corner, the colour discrimination ability of human vision increases. Therefore, the chromaticity shift of the LPLED lamp and the corresponding drop in correlated temperature are more visually noticeable than those of the HPLED lamp.

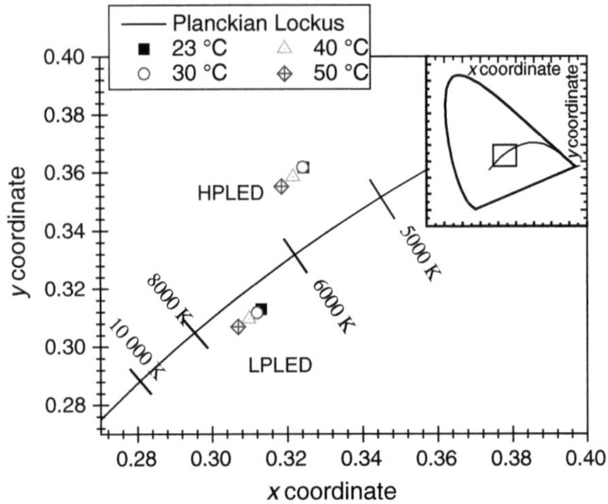

Fig. A13.12 Location of the chromaticity points on a segment of the CIE1931 diagram as a function of ambient temperature for front viewed LPLED and HPLED based lamps. The inset shows the position of the segment in the diagram

Figure A13.12 shows the chromaticity coordinates of the two lamps as functions of ambient temperature for front view. With increasing ambient temperature the chromaticity points are seen to shift towards the "blue" corner of the diagram, which corresponds to the aforementioned increase in CCT (Fig. A13.10). Such chromatic behaviour can be attributed to thermal quenching of the phosphor emission that results in an increase of the blue to yellow intensity ratio (such an increase can be seen in Fig. A13.5, where an increase of the spectrally integrated blue emission in relation to the yellow one can be observed). In terms of the chromaticity diagram, an increase in the blue to yellow ratio means a shift toward the "blue" corner of the diagram.

13.2.2 Summary

Solid state lamps manufactured in Nepal for village illumination programs were characterised by measuring the input current, optical output, radiation pattern, and chromaticity characteristics as functions of temperature. Also, the variation of the output flux with input voltage was examined. The lamps exhibited stability of output flux and chromaticity against temperature variation irrespective of the driving circuit and model of WLEDs used. However, the LPLED based lamp with resistive ballast was found to feature a high sensitivity of output flux to input voltage in comparison with the electronically ballasted HPLED based lamp. This implies that electronic ballasts are preferred in deep-cycle battery driven lighting installations based on WLEDs. At the time of writing, PPN has already upgraded

LPLED based lamps by the introduction of a switching regulator and battery overdrain protection circuit. Also, some models of WLEDs were found to have a large variation of CCT within the viewing angle which might cause visual inconvenience in performing some common tasks, such as reading.

Appendix 14
Base-Line and Follow-Up Survey Questionnaires

14.1 Base-Line Survey Data Form

For a copy of the Base-Line Survey Data Form please refer to the outline provided on the DVD, which is separately available from the author.

A. Zahnd, *The Role of Renewable Energy Technology in Holistic Community Development*, Springer Theses, DOI: 10.1007/978-3-319-03989-3,

Social and Health Changes from the Family of 4
Baseline Questionnaire
RIDS-Nepal

General

1. Name of household head: Year house was built:

2. Household head is (circle one): male female

3. Marital type (circle one): monogamous polygynous Amount of household debt:

4. Self-described socio-economic status (circle one): Low Middle High

5. Group-described socio-economic status (circle one): Low Middle High

6. Kind and numbers of livestock

Type:	Number:
Cattle	
Yaks/Naks	
Dzopa/Dzoma	
Goats	
Sheep	
Horses	
Tolba/tolbini	

7. Amount of land:

Type:	Days to plow	Describe fields (distance, quality of field, etc.)
Irrigated		
Non-irrigated		

8. Months of food security from their own crops per year:

9. What do they eat when the food they have grown is finished? Where do they get this food? What food do they buy?

10. Cooking method:
    a. Open fire with stone support
    b. Open fire with metal/steel frame (that strong expensive ~6,000 NRp metal frame)
    c. Open fire with stones and odhan
    d. Smokeless metal stove (define the kind)
    e. "Jumla" design smokeless metal stove
    f. Other:

11. Heating method:
    a. open fire
    b. smokeless metal stove (define the kind)
    c. "Jumla" design smokeless metal stove

12. Lighting method:
    a. jharro
    b. matitel (tupi)
    c. candle
    d. hydro power
    e. solar

13. Does this household have a latrine?

14.2 Follow-up Survey Data Form

For a copy of the Follow-Up Survey Data Form please refer to the outline provided on the DVD, separately available from the author.

Social and Health Changes from the Family of 4
Follow-up Questionnaire
RIDS-Nepal

General

1. Name of household head: Year that house was built:

2. Household head is (circle one): male female

3. Marital type (circle one): monogamous polygynous Amount of household debt:

4. Self-described socio-economic status (circle one): Low Middle High

5. Group-described socio-economic status (circle one): Low Middle High

6. Kind and numbers of livestock

Type:	Number:
Cattle	
Yaks/Naks	
Dzopa/Dzoma	
Goats	
Sheep	
Horses	
Tolba/tolbini	

7. Amount of land:

Type:	Days to plow	Describe fields (distance, quality of field, etc.)
Irrigated		
Non-irrigated		

8. Months of food security from their own crops per year:

9. What do they eat when the food they have grown is finished? Where do they get this food? What food do they buy?

10. Cooking method:
    a. Open fire with stone support
    b. Open fire with metal/steel frame (that strong expensive ~6,000 NRp metal frame)
    c. Open fire with stones and odhan
    d. smokeless metal stove (define the kind)
    e. "Jumla" design smokeless metal stove
    f. Other:

11. Heating method:
    a. open fire
    b. smokeless metal stove (define the kind)
    c. "Jumla" design smokeless metal stove

12. Lighting method:
    a. jharro
    b. matitel (tupi)
    c. candle
    d. hydro power
    e. solar

13. Does this household have a latrine?

Appendix 15
RIDS-Nepal Web Site

In order to visit the RIDS-Nepal website, developed by and updated under the leadership of the author, please connect to the Internet and go to: www.rids-nepal.org

RIDS-Nepal's Main Web Site.

A. Zahnd, *The Role of Renewable Energy Technology in Holistic Community Development*, Springer Theses, DOI: 10.1007/978-3-319-03989-3,

RIDS-Nepal's "Family of 4" Main Web site page.

Appendix 16
RIDS-Nepal Videos

To watch all RIDS-Nepal videos, please refer to the outline provided on the DVD, which is separately available from the author. Videos 17.1–17.4 are filmed and edited by the author and a trained RIDS-Nepal staff, while video 17.5 is a BBC video produced.

"Family of 4" Short Videos (PL, SMS, SHS, DWS)

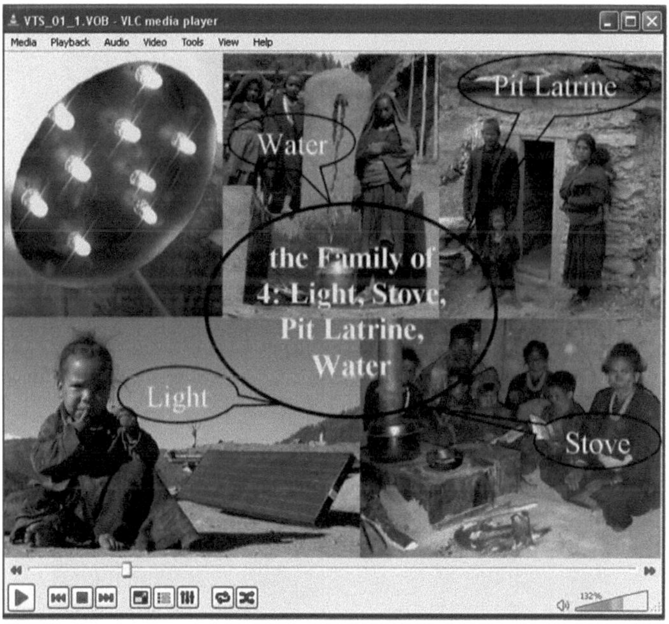

A. Zahnd, *The Role of Renewable Energy Technology in Holistic Community Development*, Springer Theses, DOI: 10.1007/978-3-319-03989-3,

RIDS-Nepal HCD in Humla

Local Battery Manufacturing

Sunshine In Shadows (Part 1, Part 2, Part 3, Part 4, Part 5)

Shamans in the Blind Country(Part 1, Part 2)

Appendix 17
RIDS-Nepal NFE Workbooks, Booklets, Cards and Games

To see, read (in Nepali language) and consult the various RIDS-Nepal NFE (Non-Formal Education) Workbooks, developed and written under the author's leadership, please refer to the individual files on the DVD, separately available from the author.

17.1 Sukhi Pariawar 1 (Healthy / Happy Family 1)

A. Zahnd, *The Role of Renewable Energy Technology in Holistic Community Development*, Springer Theses, DOI: 10.1007/978-3-319-03989-3,

17.2 Sukhi Pariawar 2 (Healthy / Happy Family 2)

17.3 Mahila Ko Kalam 1 (Women's Pen 1)

17.4 Mahila Ko Kalam 2 (Women's Pen 2)

17.5 Sukhi Pariawar (Health/Happy Family) 1 Booklets and Cards

17.6 Sukhi Pariawar (Health/Happy Family) 1 Games

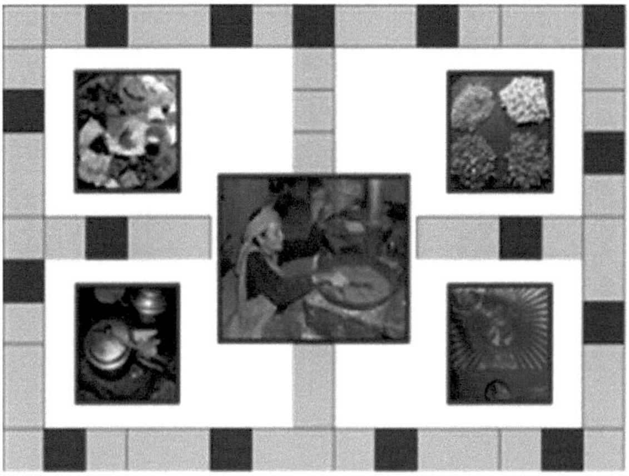

17.7 Sukhi Pariwar (Health/Happy Family) 2 Games

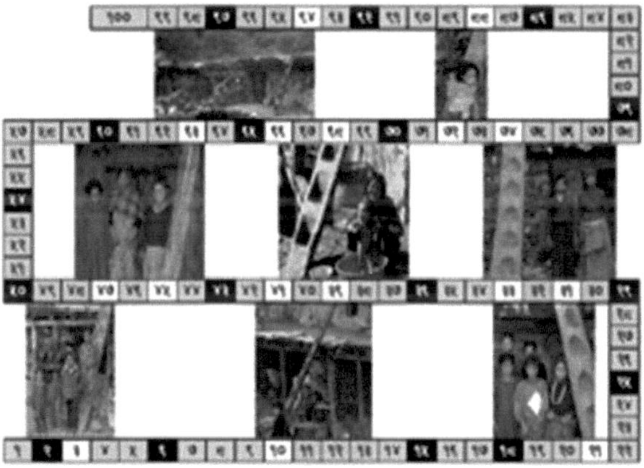

Appendix 18
RIDS-Nepal Solar PV System Training Manual

In order to read (in English), use and consult the RIDS-Nepal Solar PV System Training Manual, developed and written under the author's leadership, please refer to the Solar PV System Training Handbook file on the DVD, separately available from the author.

Solar Home System
Basic Maintenance & Repair
Training Workbook

RIDS-Nepal

Rural Integrated Development Services Nepal

www.rids-nepal.org

2009

A. Zahnd, *The Role of Renewable Energy Technology in Holistic Community Development*, Springer Theses, DOI: 10.1007/978-3-319-03989-3,

Appendix 19
RIDS-Nepal Brochures

To see, read (in Nepali) and print the various brochures in full pdf size, developed by and created under the author's leadership, please refer to the individual files on the DVD, separately available from the author.

A. Zahnd, *The Role of Renewable Energy Technology in Holistic Community Development*, Springer Theses, DOI: 10.1007/978-3-319-03989-3,

19.1 Pit Latrine

Pit Latrine Brochure page 1, pictures 1 – 5

Pit Latrine Brochure page 2, pictures 6 – 11

19.2 Smokeless Metal Stove

Smokeless Metal Stove Brochure page 1, pictures 1 – 3

Smokeless Metal Stove Brochure page 2, pictures 4 – 6

19.3 Basic Indoor Lighting

Basic Indoor Lighting with WLED lights page 1, pictures 1 – 4

Basic Indoor Lighting with WLED lights page 2, pictures 5 – 11

19.4 Clean Drinking Water

Clean Drinking Water System Building and Access page 1, pictures 1 – 5

Clean Drinking Water System Building and Access page 2, pictures 6 - 11

19.5 Greenhouse

Greenhouse Building and Increased Vegetable Growing page 1, pictures 1 – 5

Greenhouse Building and Increased Vegetable Growing page 2, pictures 6 - 11

19.6 Solar Water Heater Bathing Centre

Solar Water Heater Bathing Centre Building and Showering page 1, pictures 1 – 5

Solar Water Heater Bathing Centre Building and Showering page 2, pictures 6 - 11

Appendix 20
Flip Charts

In order to see the various Flip Charts, developed by and created under the leadership of the author with RIDS-Nepal, used for awareness raising, teaching and follow-up programs, please refer to the full-sized pdf files on the DVD, separately available from the author.

A. Zahnd, *The Role of Renewable Energy Technology in Holistic Community Development*, Springer Theses, DOI: 10.1007/978-3-319-03989-3,

20.1 Pit Latrine

 Pit Latrine Teaching Flip Chart 1	 Pit Latrine Teaching Flip Chart 2
 Pit Latrine Teaching Flip Chart 3	 Pit Latrine Teaching Flip Chart 4
 Pit Latrine Teaching Flip Chart 5	 Pit Latrine Teaching Flip Chart 6
 Pit Latrine Teaching Flip Chart 7	 Pit Latrine Teaching Flip Chart 8

20.2 Smokeless Metal Stove

Smokeless Metal Stove Flip Chart 1

Smokeless Metal Stove Flip Chart 2

Smokeless Metal Stove Flip Chart 3

Smokeless Metal Stove Flip Chart 4

Smokeless Metal Stove Flip Chart 5

Smokeless Metal Stove Flip Chart 6

Smokeless Metal Stove Flip Chart 7

Appendix 21
Humla Environment and Culture Pictures

In order to see pictures of Humla's environment, the geographical locations of the various village communities where RIDS-Nepal has started the long-term HCD projects of the "Family of 4 " and the "Family of 4 PLUS", as well as some pictures of the scenery and local culture, please refer to the individual files on the DVD, separately available from the author.

A. Zahnd, *The Role of Renewable Energy Technology in Holistic Community Development*, Springer Theses, DOI: 10.1007/978-3-319-03989-3,

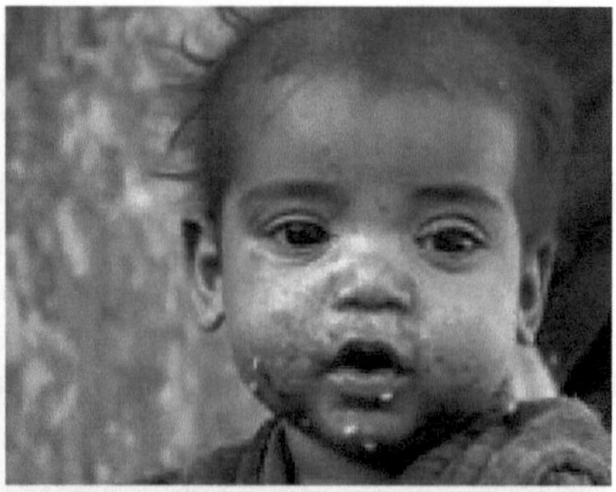

Appendix 22
RIDS-Nepal Posters

In order to see and print the various Posters, developed by and created under the leadership of the author with RIDS-Nepal, used for awareness raising, teaching and follow-up programs, please refer to the individual files on the DVD, separately available from the author.

A. Zahnd, *The Role of Renewable Energy Technology in Holistic Community Development*, Springer Theses, DOI: 10.1007/978-3-319-03989-3,

22.1 Smokeless Metal Stove & Pit Latrine

Sick and Suffering Life (with NO Pit Latrine and NO Smokeless Metal Stove)

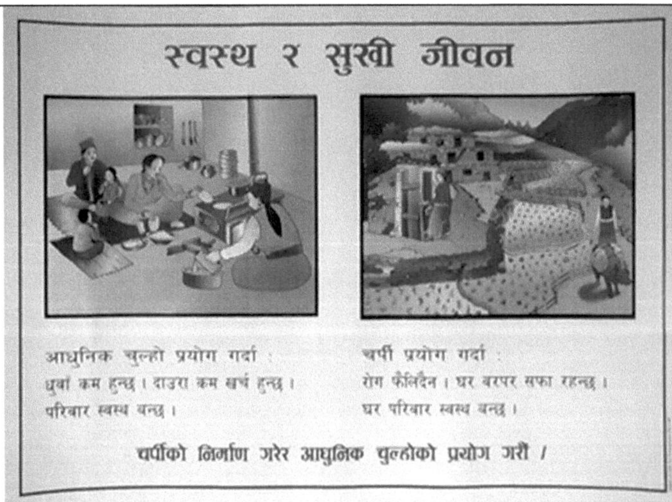

Healthy and Happy Life (WITH Pit Latrine and WITH a Smokeless Metal Stove)

22.2 Darapori Village Community Bathing Centre

Appendix 23
RIDS-Nepal Songs

In order to listen to the various songs (in local Jumla/Humla Nepali language), inspired by and written under the leadership of the author with RIDS-Nepal, and used for awareness raising, teaching and follow-up programs of the "Family of 4" and "Family of 4 PLUS" HCD programs, please refer to the individual song files on the DVD, separately available from the author.

23.1 Track 1: A Stove provides—Good Health—Good Heat—Good Cooking

23.2 Track 2: A Window into Jumla

23.3 Track 3: Working and Living Together

23.4 Track 4: Alcohol: The Road to Destruction

23.5 Track 5: Healthier Lives through Awareness and Action

23.6 Track 6: I am a Woman: Made to be Equal

23.7 Track 7: Life's Final Chapter—An Old Man's Sorrow

23.8 Track 8: Literacy for Life

23.9 Track 9: Mercy and Care

23.10 Track 10: Nutritious Food from our Fields and Gardens

23.11 Track 11: Steel Turbines for Faster Millstones

23.12 Track 12: Sunshine in Shadows

23.13 Track 13: Sweetness from Flowers

23.14 Track 14: The Dwindling Forest

23.15 Track 15: The Farmer's Burden

23.15 Track 16: The Yoke of Women

23.17 Track 17: Tragedy of Dependence and Poverty.

A. Zahnd, *The Role of Renewable Energy Technology in Holistic Community Development*, Springer Theses, DOI: 10.1007/978-3-319-03989-3,

Appendix 24
Supervised Dissertations

In order to read (in English (25.1/25.2/25.4/25.5 and in German 25.3)) and consult the various dissertations, co-supervised by the author, please refer to the individual files on the DVD, separately available from the author.

24.1 "Water Service System Concept Design Generation and Evaluation for a Remote Village in Nepal", Kris Robinson, Bachelor of Engineering, Australian National University, September 2007

24.2 "Technical Analysis of the Performance of a Small Scale, Centralised Village Photovoltaic System in Tulin, Humla Nepal", Alicia Ellul, Master of Science in Renewable Energy, Murdoch University 2008

24.3 "RIDS-Nepal—Projektbezogene Entwicklungshilfe", Samuel Glaser, Maturarbeit, Wirtschaftsgymnasium, Basel, Switzerland, December 2008

24.4 "Changing Well Being and Livelihoods: An Examination of the Impacts of Smokeless Metal Stoves on Individuals and Communities in Humla, Nepal", Gavin Naylor, B.Sc. Human and Physical Geography, University of Reading, February 2009

24.5 "The impact on respiratory ailments through reduction of Indoor Air Pollution by smokeless metal stove and solar lighting: Results of a study in rural Nepal", Jason Mann, Master of International Health, Curtin University of Technology, November 2009.

A. Zahnd, *The Role of Renewable Energy Technology in Holistic Community Development*, Springer Theses, DOI: 10.1007/978-3-319-03989-3,

Appendix 25
Pictures of the "Family of 4"

In order to see pictures of different implemented project examples included in the "Family of 4" HCD program, the pit latrine (PL), the smokeless metal stove (SMS), basic indoor lighting and access to clean drinking water, please refer to the individual picture folders on the DVD, separately available from the author.

25.1 Pit Latrine

Please open the Pit Latrine folder in the attached DVD for all pictures

25.2 Smokeless Metal Stove

Please open the Smokeless Metal Stove folder in the attached DVD for all pictures

A. Zahnd, *The Role of Renewable Energy Technology in Holistic Community Development*, Springer Theses, DOI: 10.1007/978-3-319-03989-3,

25.3 Basic Indoor Lighting

Please open the Basic Indoor Lighting folder in the attached DVD for all pictures

25.4 Village Drinking Water System

Please open the Village Drinking Water System folder in the attached DVD for all pictures

Appendix 26
Pictures of the "Family of 4 PLUS"

In order to see pictures of different implemented project examples included in the "Family of 4 PLUS" HCD program, such as the, high-altitude greenhouses, solar driers, Non-Formal Education programs for women and out-of-school children, malnutrition programs for children <5 years of age and their mothers, solar cookers, solar water heating bathing centres, slow sand water filters and Karnali Technical School Scholarships, please refer to the relevant picture folders on the DVD, separately available from the author.

26.1 Greenhouse

Please open the Greenhouse folder in the attached DVD for all pictures

26.2 Solar Drier

Please open the Solar Drier folder in the attached DVD for all pictures

A. Zahnd, *The Role of Renewable Energy Technology in Holistic Community Development*, Springer Theses, DOI: 10.1007/978-3-319-03989-3,

26.3 NFE (Non-Formal Education)

Please open the NFE folder in the attached DVD for all pictures

26.4 Mal-Nutrition for <5 years of age Children

Please open the Nutrition folder in the attached DVD for all pictures

26.5 Solar Cooker

Please open the Solar Cooker folder in the attached DVD for all pictures

26.6 Solar Water Heater

Please open the Solar Water Heater folder in the attached DVD for all pictures

26.7 Slow Sand Water Filter

Please open the Slow Sand Water Filter folder in the attached DVD for all pictures

26.8 Karnali Technical School Scholarship

Please open the Karnali Technical School Scholarship folder in the attached DVD for all pictures

Appendix 27
Renewable Energy Course

In order to read (in English) and use the Renewable Energy (RE) course, please refer to to the relevant course chapters' folders through the index provided on the DVD, separately available from the author. The RE course has been developed and written by the author, based in the Murdoch RE course 2003–2005 (used with permission). The public use of any material in these 9 chapters has to have the previous permission of the author. The course is contextualised for the context of Nepal's communities and lectured at the Mechanical Engineering Department of the Kathmandu University in Nepal to the 4th year Mechanical Engineering Bachelor and 2nd year Mechanical Engineering Masters Degree students from 2004–2011. The course is also accessible for registered RIDS-Nepal website users from the Download site at: http://www.rids-nepal.org/index.php?option=com_docman&task=cat_view&gid=91&Itemid=32 (accessed July 30, 2012).

A. Zahnd, *The Role of Renewable Energy Technology in Holistic Community Development*, Springer Theses, DOI: 10.1007/978-3-319-03989-3,

27.1 Chapter 1 Energy in Society and Environment

27.2 Chapter 2 Renewable Energy Resources

27.3 Chapter 3 Renewable Energy Conversion Devices

27.4 Chapter 4 Energy Systems

27.5 Chapter 5 Environmental Impact Assessment

27.6 Chapter 6 Renewable Energy System Design

27.7 Chapter 7 Energy Management

27.8 Chapter 8 Case Studies of Renewable Energy Technologies

27.9 Chapter 9 Renewable Energy and Sustainable Development

Appendix 28
Humla Virtual Tour

In order to provide the supporters of RIDS-Nepal's HCD programs in Humla with a more realistic "participation", a Virtual Tour has been developed under the author's leadership. In order to "come to Humla" and "walk through" Humla, its places and villages where RIDS-Nepal works and lives with the local communities, please refer to the relevant folder on the DVD, separately available from the author.

360° Virtual Tour around Simikot and its vicinity

A. Zahnd, *The Role of Renewable Energy Technology in Holistic Community Development*, Springer Theses, DOI: 10.1007/978-3-319-03989-3,

360° Virtual Tour around the RIDS-Nepal HARS office in Simikot and its vicinity

360° Virtual Tour around the RIDS-Nepal Darapori Field Office

Appendix 29
RIDS-Nepal Power Point Presentation, Vision, Mission, HCD Concepts, Project Examples, 3 Tier Working Principle

To see the RIDS-Nepal Power Point Presentation, explaining its Vision, Mission, Holistic Community Development Concepts, with project examples and the 3 Tier Working Principle, please refer to the Power Point file in the index provided on the DVD, separately available from the author.

A. Zahnd, *The Role of Renewable Energy Technology in Holistic Community Development*, Springer Theses, DOI: 10.1007/978-3-319-03989-3,

Experience Shows that the Communities Themselves Identify . .
Stove
Light
Latrine
Water
. . most of the time as the four MAIN NEEDS they want to address . .

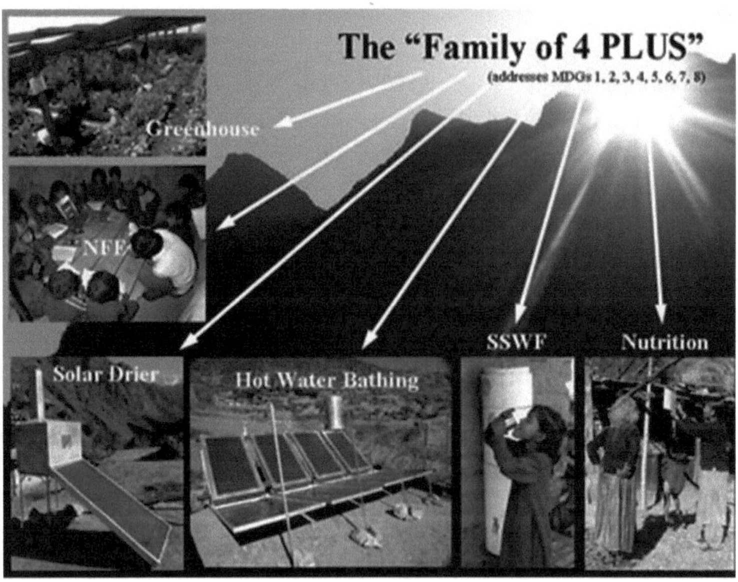
The "Family of 4 PLUS"
(addresses MDGs 1, 2, 3, 4, 5, 6, 7, 8)
Greenhouse
NFE
SSWF
Nutrition
Solar Drier
Hot Water Bathing

Appendix 30
Data Monitoring System Documentations

In order to read (in English) and consult the RIDS-Nepal Data Monitoring systems for several of its detailed energy generation projects for local villages and the RIDS-Nepal Humla field offices, please refer to the relevant folders and files according to the index on the DVD, separately available from the author.

A. Zahnd, *The Role of Renewable Energy Technology in Holistic Community Development*, Springer Theses, DOI: 10.1007/978-3-319-03989-3,

30.1 RIDS-Nepal Simikot Office Solar PV System High-Altitude Bathing Centre and High-Altitude Greenhouse

RIDS-Nepal Simikot Humla Data Monitoring Systems RIDS-Nepal

A description of existing power and monitoring systems

Date: 11th May 2009
Authors: Laurent, Solène, Xavier, Alex Zahnd

IMPORTANT NOTE

One of the major purposes of this document is to keep an up to date description of the existing systems in the High Altitude Simikot office. This includes the power systems as well as the monitoring systems. For this reason, **please update this report each time a change is made/observed**, and send a digital copy, with an increased index number and date of change, to the RIDS-Nepal's office, KTM.

Modifications Diary			
Date	Author	Sections(s)/page(s) modified	Brief description of the modification(s)
29th May 09	Adam F	Whole Document	English corrections
1st August 09	Alex Zahnd	Whole Document	Various remarks
11th September 09	Xavier, Laurent, Solène	Whole Document	Modifications according to Alex's remarks

RIDS-Nepal Simikot Office Data Monitoring Systems 1 RIDS-Nepal, 11th September 2009

30.2 RIDS-Nepal Darapori Office Solar PV System and High-Altitude Greenhouse

RIDS-Nepal Darapori Office Humla Data Monitoring RIDS-Nepal

Description of the existing power and monitoring systems

Date: 7th July 2009
Authors: Laurent, Solène, Xavier, Alex Zahnd

IMPORTANT NOTE

One of the major purposes of this document is to keep an up to date description of the existing systems in the RIDS-Nepal High Altitude Darapori Field Office. This includes the power systems as well as the monitoring systems.
For this reason, **please update this report each time a change is done/observed**, and send a digital copy, with an increased index number and date of change, to the RIDS-Nepal's office, KTM.

Modifications Diary			
Date	Author	Sections(s)/page(s) modified	Brief description of the modification(s)

30.3 Tulin Village Solar PV Central Tracking System

RIDS-Nepal TULIN Humla Data Monitoring Systems 24 V Tracking system – House Lighting	RIDS-Nepal
Description of the existing power and monitoring systems	
Date: 15th June 2009 **Authors:** Laurent, Solène, Xavier	

IMPORTANT NOTE

One of the major purposes of this document is to keep an up to date description of the existing system in Tulin village. This includes the power systems as well as the monitoring systems.
For this reason, **please update this report each time a change is done/observed**, and send a digital copy, with an increased index number and date of change, to the RIDS-Nepal's office, KTM.

Tulin system is about to be moved to another place in the village. This report describes it at the time we left (end of April 2009), but important modifications are to be made.

Modifications Diary			
Date	Author	Sections(s)/page(s) modified	Brief description of the modification(s)

30.4 Pamlatum Village Solar PV Cluster System

RIDS-Nepal Pamlathum Humla Data Monitoring Systems RIDS Nepal
12Volt PV Cluster Home Lighting System

A description of the existing power and monitoring systems

Date: 7th July 2009
Authors: Laurent, Solène, Xavier

IMPORTANT NOTE

One of the major purposes of this document is to keep an up to date description of the existing systems in the High Altitude Pamlathum village. This includes the power systems as well as the monitoring systems.
For this reason, **please update this report each time a change is made/observed**, and send a digital copy, with an increased index number and date of change, to the RIDS-Nepal's office, KTM.

30.5 Kholsi Village Pico-Hydro Power System

RIDS-Nepal KHOLSI Humla Data Monitoring Systems	RIDS-Nepal
Description of the existing pico hydro power and monitoring systems	
Date: 6th July 2009 **Authors:** Laurent, Solène, Xavier	

IMPORTANT NOTE
One of the major purposes of this document is to keep an up to date description of the existing system in Kholsi village. This includes the power systems as well as the monitoring systems. For this reason, **please update this report each time a change is done/observed,** and send a digital copy, with an increased index number and date of change, to the RIDS-Nepal's office, KTM.

Appendix 31
Humla Google Earth Tour Video

Google Earth provides high resolution pictures of the area in Humla where RIDS-Nepal works in close partnership with 13 village communities. In order to provide the supporters, and people interested in the RIDS-Nepal HCD programs in Humla with a more detailed understanding of how remote the village communities are where the project works take place, a Google Earth Video Tour has been developed under the author's leadership. In order to watch the Google Earth Video Tour, please refer to the relevant folder according to the index provided on the DVD, separately available from the author.

A. Zahnd, *The Role of Renewable Energy Technology in Holistic Community Development*, Springer Theses, DOI: 10.1007/978-3-319-03989-3,

RIDS-Nepal Google Earth Tour of Humla Main Web Site.

RIDS-Nepal Google Earth Tour showing Tangin Village.

RIDS-Nepal Google Earth Tour showing Chala Village.

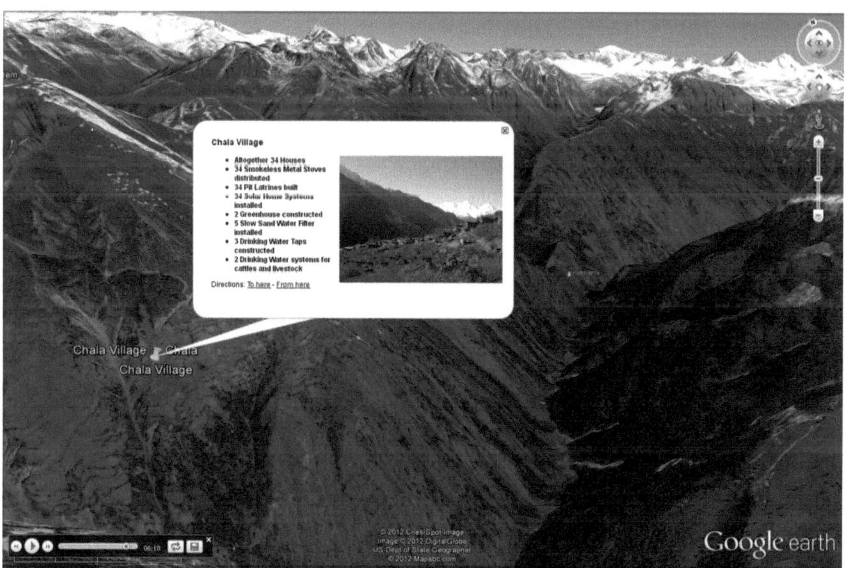

Short CV of the Author

I am Swiss, married to Sheila and we have two wonderful, adult children. I have a Mechanical Engineering degree (B.Sc.) from Switzerland, a Masters (M.Sc.) in Renewable Energy and a Ph.D. awarded through this thesis. My industrial experience ranges from development projects in extrusion technology for the food and plastic industry, to pharmaceutical production plants in Switzerland. I lived and worked in Nepal off and on since 1983. Since 1996 I lived and worked there full time developing and leading holistic community development (HCD) projects in the remotest and poorest mountain communities in the Himalayas. In 2002 I co-founded the not-for-profit NGO RIDS-Nepal (www.rids-nepal.org) and in 2013 the NGO RIDS-Switzerland (www.rids-switzerland.org) continuing to develop further HCD projects in the mountain communities. From 2001 to 2011 I have also been teaching a Renewable Energy course at the Kathmandu University as well as being active in applied research and application of renewable energy technologies for communities and industries in developing countries. Through my engagement in the field of applied HCD in remote Nepalese villages and my involvement in academic education of engineers in Nepal, I was elected as a member of the Swiss Academy of Engineering Sciences (SATW) in 2007.

Professional Experience from 1991–2013

2002—Current: International Project Director of the non-profit NGO RIDS-Nepal (www.rids-nepal.org) **and RIDS-Switzerland** (www.rids-switzerland.org)
Responsibilities/Activities:

Developing and implementing new concepts of contextualized long-term holistic community development (HCD) projects in partnership with remote, high altitude Himalayan communities in Nepal. Applied research and monitoring projects in the field of applied renewable energy technologies for improved access to energy services, such as rural village electrification. Working with local Nepali industries to manufacture and disseminate new developed technologies to make them more affordable and available for the rural communities.

2001–2011: Visiting Professor Mechanical Engineering Department Kathmandu University (KU), Nepal (I am still a member of the expatriate team at KU)

Responsibilities/Activities:

Lecturing of a renewable energy technology course for Mechanical Engineering Bachelor and Master Degree students that I had developed for the particular context of Nepal. Built and run a combustion (for stove development) and a solar energy (for solar photovoltaic and solar thermal applications developments) laboratory for student research projects and industry research and testing.

1996–2000: KCST Project Director (United Mission to Nepal project in Jumla, Nepal)

Responsibilities/Activities:

Developing and implementing holistic community development projects in partnership with remote, high altitude Himalayan communities in Nepal. Leading and training a team of 20 Nepali project, financial and administrative staff. Follow-up of village projects and developing teaching methodologies and educational tools, to teach local users how to use/operate and maintain their newly acquired technologies.

1991–1996: Chemical Process Engineer for Pharma Industry in Switzerland

Responsibilities/Activities:

Right hand of the CEO of a medium sized engineering consultancy company, responsible for a pharmaceutical production plant development, design and implementation team, working under competitive and clearly defined contract conditions for one of the major leading Swiss pharmaceutical companies.

Rewards and Honours

- Ph.D. scholarship (2009–2012) from the Australian Government and Murdoch University
- Swiss Academy of Engineering Sciences (www.satw.ch) member ship award "in honour and recognition of an active contribution towards a sustainable development of many Nepalese villages with a strong involvement in the education of engineers".
- M.Sc. scholarship from the Murdoch University for the 2nd and 3rd year (2002–2004), on-line "M.Sc. in Renewable Energy Technology", course.

Glossary of Nepali Terms

The following list contains short explanations of some of the more common and important Nepali terms frequently used in the field of community development and thus important to know and understand for anyone considering working in Nepal. Many of them are as well used in the thesis.

adhyaksha Chairman of a village or committee

afno manche A person, or people, belonging to one's own inner circle of associates, who can be approached whenever need arises

Anchal A geographical clustering of several districts to make a political region. Nepal has 14 Anchal. Example the Karnali Anchal has 5 districts, Kalikot, Jumla, Dolpa, Mugu and Humla

Anna Unit measurement for land measurement. One "anna" is one sixteenth of a ropani (508.74 m^2) which is equivalent to 31.79 m^2

Aryan The Indo-Aryan people migrated mainly from the South and West of Nepal's boarders with India

Bahun Nepali word for Brahmin. Brahmins represent the highest caste in the caste hierarchy. Also, used colloquially to mean priest

Bari Unirrigated sloping agricultural land

Bhitri Madhes Inner Terai. The Mahabharata hill range which sometimes merges with the higher elevated Siwalik mountain range

Bhoot Ghost

Bigha Unit of land in plain Nepal. 1 Bigha = 0.675 hectare = 6,760 m^2

Bikas The word used to mean "development" often for technical projects

Bokhsi Witch

A. Zahnd, *The Role of Renewable Energy Technology in Holistic Community Development*, Springer Theses, DOI: 10.1007/978-3-319-03989-3,

Chakari Originated from religious Hinduism, "*chakari*" is an essential concept which means to wait upon, to serve, to appease, or to seek favour from a god (Bista 1991). Since the nineteenth century though it has come to mean most often an institutionalised relationship between a patron and client based on information, gifts and services in return for favour. "*Chakari* is so ingrained in Nepali organizational life that to persist in attempting to exclude it is seen to be deviant—a demonstration of arrogance and conceit by those above and below" (Bista 1991)

Chettri The second highest Hindu caste

Chomolungma Mother goddess of the universe (Mt. Everest called in the local Tibetan culture). Mt Everest is called "Sagarmatha", meaning "on top of the world", in Nepalese language

dal bhat Nepal's national dish, usually eaten twice a day. Rice, lentils and a vegetable curry

Dhami Witch doctor, local healer and sorcerer, with a high status as religious and social leader in the village

dhanyabad Meaning for "thank you". This word was added to the Nepali vocabulary only more recent due to the increased interaction and work with English speaking foreigners. But previous to that there was no word in the Nepali language with the meaning of "thank you" as used in the English language

dhuka and sukha Suffering/pain and bliss/joys

graha dasha A bad position of the planets

hal Surface area of land ploughed with one ox in a day

jaat Belonging to a particular caste group

Jana Andolan Jana Andolan is the popular, violent and bloody protest movement called the "People's Movement" which brought thousands of people on the roads to protest during 1989–1990, leaving many dead

Jagir A salaried job where one does not have to (and often is not really expected to) work, but will receive a pay cheque. Till recently this was very common in all government lower and higher jobs as well as in many private industries. The job is guaranteed till one reaches the age to get retired and no one can get fired, even if the job or position is misused or one does not show up for weeks or months at the time

Jhankri Witch doctor, local healer and sorcerer (as the "dhami"), with a high status as religious and social leader in the village

IPP Independent Power Producer. Private organization or person who has been given the permission from NEA (Nepal Electricity Authority) to generate and sell to NEA electricity at a fixed price

jotnu To plough, as to plough the field with a plough and a pair of oxen

Kagyu Kagyu school of Tibetan Buddhism follows the Trantric teaching

Kali Yuga The Hindu teaching believes that there are four ages, called yugas. These are the Satya, Treat, Dwapar and Kali, the Golden, Silver, Copper and Iron ages respectively. After the end of the four ages, Satya Yuga comes back again automatically, and thus people do not need to and can not, do anything about this. Presently it is thought we are in the Kali Yuga, which is bad and accompanied with suffering. Only Satya Yuga afterwards, it is believed, will bring an abundance of happiness and bliss

Kami Metal smith worker, which is placed low in the caste hierarchy

karma Karma means literally action. But in Nepal in every day use people do not relate karma to their actions in the present life or in a previous one. The role of one's own personal actions in influencing karma is neglected and, instead, karma is thought to be predestined and as something which cannot be altered in any way

kattha A land measurement, equal to one twentieth, or 338 m^2, of a "bigha" (6,760 m^2)

kharsu Hard, small tree. The Humla farmers use this kind of wood to make every piece of their plough

khet (Usually) irrigated terraced agriculture land

khola Small river

kulo Canal (often used for drainage or irrigation canals)

Lamaism Lamaism, or Tibetan Buddhism, is the Buddhist belief system and religious doctrine adhered to in Tibet and in parts of several other Himalayan countries, notably Nepal, Bhutan and India

lakh Number for one hundred thousand (very common)

Licchavis A clan who migrated from today's Indian state of Bihar, in the north if India, establishing their dominance in parts of Nepal in the fourth century B.C.

Mahayana Mahayana Buddhism's main aim is to achieve spiritual enlightenment

maita Parental home of wife

mana One mana is 568 ml volume or approx. 400 g (~ 10 handful) of grain

maya An illusion and not real. Thus the world is looked upon as not realistic but as mere illusion and e.g. poverty has to be born with patience and humility without trying to improve or mitigate it

muri (1) A unit of land measuring 1,369 square feet or 127.18 m^2. 4 muri of land make 1 ropani (508.74 m^2)(2) A volumetric unit of equivalent to approximately 91 L 48.77 kg of paddy, 68.05 kg of wheat or maize or 65.78 kg of millet. One muri consist of 20 pathis

MDGs Millennium Development Goals. The eight Millennium Development Goals (MDGs)—which range from halving extreme poverty rates to halting the spread of HIV/AIDS and providing universal primary education, all by the target date of 2015—form a blueprint agreed to by all the world's countries and all the world's leading development institutions. (http://www.undp.org/content/undp/en/home/mdgoverview.html and http://www.un.org/millenniumgoals/) (accessed October 20, 2013)

NEA Nepal Electricity Authority. Nepal government's own and only electricity authority with the sole right to generate and sell electricity. Recently, under provided licenses IPPs have been given the permission to generate electricity and sell it to NEA

Newar One of Nepal's major ethnic groups, mostly Hindu but including some Buddhists, concentrated mostly in the Kathmandu valley. Newar's have their own occupational caste system. Very active in business, craftsmanship and are scattered throughout the country especially for trade purpose

NGO Non Governmental Organisation (such as e.g. RIDS-Nepal)

NPC National Planning Commission. Direct under the Prime Minister, with him as the NPC chairman, its main purpose is to develop each 5 years plan, to develop and modernise Nepal as a whole country

Nyingma Nyingma is the oldest tradition of the four (Nyingma, Kagyu, Sakya and Gelug) Tibetan Buddhist major schools, meaning "ancient" and thus often called the "old school" because it was founded on the first translation of Buddhist writings from Sanskrit into Tibetan in the eighth century AD

pani chalne Literally means "the ones who can accept water" and thus represent the caste groups that are touchable. Middle (Chettris) and high caste (Bahuns) people belong to this group

pani nachalne Literally means "the ones who can not accept water" and thus represent the caste groups that are untouchable. The low caste (Dalit) people belong to this group

panchaayat system "Panchayat" literally means assembly (*yat*) of five (*panch*) wise and respected elders chosen and accepted by the village community, settling disputes between individuals and villages. In Nepal it is a system of village governance in which chosen village leaders/elders are the basic units of political administration

papper Buck wheat, and part of the stable food of people in Humla

people's war Also called protracted people's war, is a military-political strategy invented by Mao Zedong. The basic concept behind People's War is to maintain the support of the population and draw the enemy deep into the interior where the population will bleed them dry through a mix of "Mobile Warfare" and Guerrilla warfare. The term is used by Maoists in Nepal for their strategy of the long-term armed revolutionary struggle http://en.wikipedia.org/wiki/People%27s_war (accessed July 30, 2012)

Prachanda Literally means: the fierce one, and is used for Nepal's Maoist leader

pret Evil spirit

Rana The Rana regime (1847–1950) held absolute power over the country, exploiting the people and resources mainly for their own benefit. Anyone with a concern for the society's development and the future of the country was considered their enemy and thus punished, jailed and even executed, on the pretext that it was high treason. The Ranas actively supported and encouraged the export of young and able men to British India, today known as the Gorkha soldiers

rakshi Local distilled alcohol usually made with millet

rin Loan, debt

ropani Unit of land measurement. 1 hectare = ~19.6 ropani (19.6 × 508.74 m^2)

sachib Village or committee secretary

Sagarmatha Goddess of the sky (Mt. Everest called in the local Nepali culture)

samiti Committee

samudai Community

Sanghiya Loktantrik Ganatantra Nepal Nepal, as a country name in Nepali language

Sarbochha Adalat Supreme Court

sarkaar Government

Sarki Leather worker. Placed low in the caste hierarchy

sau/sahu/sauji Landlord, often also used for the person who lends money

sarbotham pito "Super porridge" in English. Sarbotham pito is a mixture of the local available cereals of wheat, soya beans, corn and is fed to mal nourished children

Shifarish Recommendation

SLC School Leaving Certificate, received by Nepalese students after they successfully graduated from the government school class 10

Terai Southern plains of Nepal along the Indian boarder

theka/thekadar Contractor/contractor

VDC Village Development Committee: a geographical administrative unit below a district. Also a committee of elected members, the smallest local administrative unit

ward VDCs are made up of 2-5 villages. An average village is subdivided into 3-6 wards, with one ward consisting of up to 9 households, depending on the village's size. Representatives are elected from wards and sit on the VDC

References

ABB. (2003). *Access to electricity*. White paper on ABB's initiative for access to electricity. Retrieved July 20, 2012, from http://search.abb.com/library/Download.aspx?DocumentID=GFSA_ATE02&LanguageCode=en&DocumentPartID=&Action=Launch&IncludeExternalPublicLimited=True

ACMR (All China Marketing Research). (2004). *Survey report for annual follow-up evaluation of the promotion item of China green lighting project*. Beijing, China: China Greenlight Project Management Office.

Agenda 21. (1992). *Agenda 21*. http://habitat.igc.org/agenda21/ and Retrieved July 20, 2012, from http://www.un.org/esa/sustdev/documents/agenda21/english/Agenda21.pdf

Alm Ata. (1978). *Declaration of Alma-Ata—International Conference on Primary Health Care*, Alma-Ata, USSR, 6–12 September. Retrieved July 31, 2012, from http://www.searo.who.int/LinkFiles/Health_Systems_declaration_almaata.pdf

Amatya, V. B., & Shrestha, G. R. (1998). Chapter 4—Review on energy policies and their implications on renewable energy in Nepal. In K. Rijal (Ed.), *Renewable energy technologies: A brighter future* (pp. 85–115). Nepal: International Centre for Integrated Mountain Development Kathmandu.

Antoine, B., et al. (2004). Modelling the spatial distribution of present day deformation in Nepal: How cylindrical is the main Himalayan thrust in Nepal? *Geophysics International Journal,156*, 94–114.

Banskota, B. et al. (1999). *Pattern of energy use in Nepal: Energy use in mountain areas*. Kathmandu, Nepal: International Centre for Integrated Mountain Development (ICIMOD).

Bhusal, P., Zahnd, A., Eloholma, M., & Halonen, L. (2007). Replacing fuel based lighting with light emitting diodes in developing countries: Energy and lighting in rural nepali homes. *The Journal of the Illuminating Engineering Society of North America3*(4), 277–291. Retrieved July 22, 2012 http://www.iesna.org/leukos/Volume3/number4.cfm

Bista, D. B. (1991). *Fatalism and development—Nepal's struggle for modernization* (2nd ed. 1994, 5th Impression). Calcutta: Orient Longman, ISBN 0 86331 245 5.

Bista, Bir Raghu. (2008). *Poverty in Nepal: Causes, effects and alternative measures*. Nepal: Tribhuvan University.

Brew-Hammond, A. et al. (2004). *Reducing rural poverty through increased access to energy services: A review of the multi-functional platform in Mali*. United Nations Development Program, Environment and Energy, New York, USA. Retrieved July 21, 2012, from http://www.undp.org/content/undp/en/home/librarypage/environment-energy/sustainable_energy/reducing_rural_povertythroughincreasedaccesstoenergyservicesmppi.html

Brundtland, Gro Harlem. (1987). *Our common future*. New York: United Nations World Commission on Environment and Development (WCED), Oxford University Press, August 4, 1987, ISBN 0-19-282080-X. Retrieved July 30, 2012, from http://www.un-documents.net/wced-ocf.htm

A. Zahnd, *The Role of Renewable Energy Technology in Holistic Community Development*, Springer Theses, DOI: 10.1007/978-3-319-03989-3,

CIA, CIA. (2000–2012). *The World Factbook, Nepal*. Retrieved July 20, 2012, from, https://www.cia.gov/library/publications/the-world-factbook/geos/np.html

CIBSE (Chartered Institution of Building Services Engineers). (1997). *CIBSE Code for interior lighting 1994: Additions and corrections 1997*. London: CIBSE. ISBN 0900953 64 0.

CIE (Commission Internationale de L'Elairage). (2001). CIE S 008/E-2001.

CRTN, 2010. *Solar Parabolic Cooker*. Retrieved July 20, 2012 http://www.crtnepal.org/?option=technology&tid=3735333030

Doig, Alison. 1999. Off-grid electricity for developing countries, *IEE review* 45(1):25–28. http://dx.doi.org/10.1049/ir:19990104. Retrieved July 23, 2012 from http://ieeexplore.ieee.org/xpl/freeabs_all.jsp?arnumber=750256

Emeriau, S. (2006). *Nutritional exploratory mission—Humla and Mugu districts*. Kathmandu: Action Contre le Faim (ACF).

Fang, C., et al. (2007). *FAO/WFP food security assessment mission to Nepal*. Rome: Food and Agriculture Organisation / World Food Programme.

Fournier, A. (2007). *Action Contre la Faim: Anthropometric and mortality surveys, Mugu and Humla districts*, April 2007, Kathmandu, Nepal

Gautam, U. et al. (2005). Nepal: Thermal energy for export. *South Asia Journal,* July–September 2005. Retrieved April 13, 2010, from http://www.southasianmedia.net/Magazine/Journal/9-nepal_thermal_energy.htm, on November 1, 2011 not available any more.

Goldemberg, J., Thomas B. J. (1995). *Energy as an instrument for socio-economic development*. New York: UNDP/BDP Energy and Environment Group, UNDP. http://www.energyandenvironment.undp.org/undp/indexAction.cfm?module=Library&action=GetFile&DocumentAttachmentID=1025 (Retrieved November 1, 2010, on July 25, 2012 not available any more).

Gustavsson, M., & Ellegard, A. (2004). The impact of solar home systems on rural livelihoods. Experiences from the Nyimba energy service company in Zambia. *Renewable Energy,29*(7), 1059–1072.

Hagen, T. (1980). *Nepal—the kingdom in the Himalayas*. Berne: Kummerly + Frey, Geographical Publishers.

Hagen, T. (1996). *Environment and biodiversity seen with the eyes of the farmers* (pp. 1–11). Ecological Society, Nepal: Environment and Biodiversity in the Context of South Asia.

Haitz, R. (2001). Another semiconductor revolution: This time it's lighting. In *Proceedings of 9th International Symposium on the Science and Technology of Light Sources*, Ithaca, NY, USA.

Harbers, G., Bierhuizen, S. J., & Krames, M. R. (2007). Performance of high power light emitting diodes in display illumination applications. *Display Technology,3*(2), 98–109.

Hjorth, P., & Bagheri, (2006). Navigating towards sustainable development: A system dynamics approach. *Science Direct, Futures,38*(2006), 74–92.

The Himalayan Times. (2011). *Geologist say Sikkim quake unusual*. The Himalayan Times, September 20, 2011. Retrieved July 23, 2012, from http://www.thehimalayantimes.com/fullNews.php?headline=Geologists+say+Sikkim+quake+unusual&NewsID=303455

IEA. (2002). *Energy and poverty*. Chapter 13, World Energy Outlook. Paris: IEA.

IEA. (2006). *Light's labour's lost*. France: IEA Publications

IFAD (International Fund for Agricultural Development). (2012). n.d. *Rural poverty in Nepal*, UN agency. Retrieved July 19, 2012, from http://operations.ifad.org/web/guest/country/home/tags/nepal

ISO Standards. (2002). *ISO 8995:2002 lighting of indoor work places*. Retrieved July 19, 2012, from http://www.iso.org/iso/catalogue_detail.htm?csnumber=28857

Jones, R., Du, J., Gentry, Z., Gur, I., Mills, E. (2005). Alternatives to fuel-based lighting in rural China. In *Proceedings of Right Light 6*, Shanghai, May 9th–11th.

Joshi, C. B., et al. (2002). Development and dissemination of solar drying systems in Nepal: Problems and prospects. *International Energy Journal,3*(2), 53–73.

Jouanne, F., et al. (2004). Current shortening across the Himalayas of Nepal. *Geophysics, International Journal,154*, 1–14.

Karki, A., et al. (2002). Micro-hydro power in Nepal: Access to electricity for isolated rural population in the hills and mountains. *International Energy Journal,3*(2), 89–97.
Khadka, B. (1986). *Coarse grains and pulses in Nepal: Roles and prospects*. Bogor: UN/ESCAP CGPRT Centre.
KIRDRC (Karnali Integrated Rural Development and Research Centre), 2002. *Governance in the Karnali, an exploratory study*. August 2002, Jumla, Nepal.
LaFond, A. (1995). *Sustaining primary health care*. Save the Children Fund: New York.
Lekhak, Hari Datt, & Lekhak, Binod. (2005). *Natural resource conservation and sustainable development* (2nd ed.). Nayabazar: Kshitiz Publication.
LUTW, 2009. Retrieved July 20, 2012 from http://www.lutw.org/
Mills, E., S. Johnson. (2002a). *A dramatic opportunity for technology leapfrogging in the developing world,* Lawrence Berkeley National Laboratory.
Mills, E. (2002b). Why we're here: The $230-billion global lighting energy bill. In *Proceedings of the 5th International Conference on Energy-Efficient Lighting* (pp. 369–385). Nice, France.
Mills, E. (2005). The specter of fuel based lighting. *Science,308*(5726), 123–126.
MOA (Ministry of Agriculture), 1997. *Progress Report 1980/81–1995/96,* Lalitpur, Post Harvest Loss Reduction Division, Ministry of Agriculture (MOA).
Nakamura, S., Mukai, T., Senoh, M. (1994). Candela-class high brightness InGaN/AlGaN double-hetrostructure blue light-emitting diodes. *Applied Physics, 64*(13), 1687–1689.
Narukawa, Y., Sano, M., Ichikawa, M., Minato, S., Sakamoto, T., Yamada, T., et al. (2007). Improvement of luminous efficiency in white light emitting diodes by reducing a forward-bias voltage. *Japanese Journal of Applied Physics,46*(4), L963–L965.
NEA (2009b). *Electricity demand forecast,(Load Forecast) 2003–2020,* Chapter 4, Table 1. http://www.nea.org.np/corporateplan.php. Accessed March 23, 2010 (Since November 1, 2011 not available any more).
NEA (Nepal Electricity Authority) (2009a). A year in review—Fiscal year 2008/2009. Annual Report, Kathmandu, Nepal, pp. 9, 83–84. Retrieved March 23, 2010, from http://www.nea.org.np//reports/annualReports/YjqRDwQ9E3anrep09.pdf. (Since November 1, 2011 not available any more).
OIDA (Optoelectronics Industry Development Association) (2002). *The promise of solid state lighting for general illumination: Light emitting diodes and organic light emitting diodes*. Washington DC, USA. Retrieved July 23, 2012, from http://lighting.sandia.gov/lightingdocs/OIDALEDOLEDSummary200103.pdf
Peon, R., Doluweera, G., Platonova, I., Irvine-Halliday, D., & Irvine-Halliday, G. (2005). Solid state lighting for the developing world: The only solution. In *5th International Conference on Solid State Lighting*, San Diego, California, pp. 109–123.
Ravallion, M. et al. (2008). *Dollar a Day Revisited.* Policy Research Working Paper 4620. Development Research Group, World Bank, 2008. Retrieved July 25, 2012, from http://www-wds.worldbank.org/external/default/WDSContentServer/IW3P/IB/2008/09/02/000158349_20080902095754/Rendered/PDF/wps4620.pdf
Reddy, A. K. N. (2000). *Energy and social issues—world energy assessment: Energy and the challenge of sustainability, chapter 2* (pp. 39–60). New York: United Nations Development Program (UNDP).
Reddy, A. K. N. (2002). Energy technologies and policies for rural development. In T. B. Johansson, J. Goldemberg (Eds.), *Energy for sustainable development: A policy agenda* (Chap. 4, pp. 115–136). New York: UNDP.
Regmi, B. R. (2005). Sustainable development of hydro power potential in Nepal. *Kathmandu Hydro Power Conference June 2005,* Nepal.
Renewable Energy in South Asia. (2002). *Country reports Nepal* (Chap. 2). Previously Retrieved from http://www.worldenergy.org/wec-eis/publications/reports/renewable/country_reports/chap_2_5.asp
Richards, P. (2006). Renewable development: New strategies in rural electrification. *Renewable Energy World Magazine*, July/August edition

Saghir, J. (2005). *Energy and poverty: myths, links, and policy issues.* Energy Working Notes, World Bank: Energy and Mining Sector Board, No. 4 May, 2005.

Schubert, E. F., Kim, J. K., Luo, H., & Xi, J.-Q. (2006). Solid-state lighting—a benevolent technology. *Reports on Progress in Physics,69*(12), 3069–3099.

Schumacher, E. F. (1973). *Small is beautiful—a study of economics as if people mattered.* London: Blond & Briggs Ltd. ISBN 0 009 922561 1.

Shailesh, K. R. (2006). White LED based illumination systems for Indian villages. *LEUKOS, The Journal of the Illuminating Engineering Society of North America,3*(2), 167–173.

Sharma, S. (2005). Food security in Nepal. In S. C. Babu, & A. Gulati (Eds.), *Economic reforms and food security: The impact of trade and technology in South Asia* (pp. 395–411). New York: Food Products Press.

Skoet, J., Kostas S. (2006). *The state offood insecurity in the world 2006.* Rome, Italy: Food and Agriculture Organisation (FAO). Retrieved July 23, 2012, from http://www.ftp.fao.org/docrep/fao/009/a0750e/a0750e00.pdf

Tsao, J. Y. (2004). Solid-state lighting: Lamps, chips, and materials for tomorrow. *IEEE Circuit and Devices Magazine,20*(5), 28–37.

UN. (1998). *Kyoto Protocol to the United Nations Framework Convention on Climate Change*, United Nations 1998. Retrieved July 23, 2012, from http://unfccc.int/resource/docs/convkp/kpeng.pdf, http://unfccc.int/kyoto_protocol/items/2830.php

UN. (2002a). *The Johannesburg Summit Test: What Will Change?* WSSD Summit Coverage, New York. 25th September. Retrieved July 24, 2012, from http://www.un.org/jsummit/

UN. (2002b). *Sustainable Development Summit Concludes in Johannesburg: UN Secretary-General Kofi Annan Says It's Just the Beginning.* Johannesburg Summit 2002. Retrieved November 1, 2011, from http://www.un.org/events/wssd/pressreleases/finalrelease.pdf

UN. (2007). *The Millennium Development Goals Report 2007*, United Nations (UN), New York. Retrieved July 24, 2012, from http://www.un.org/millenniumgoals/pdf/mdg2007.pdf

UN. (2010). *Human Development Report 2010—The Real Wealth of Nations: Pathways to Human Development.* Retrieved July 24, 2012, from http://hdr.undp.org/en/media/HDR_2010_EN_Complete_reprint.pdf

UNDP. (2006). *Human Development Report 2006—Beyond scarcity; power, poverty and the global water crisis* (p. 294). United Nations Development Program, New York, USA. Retrieved July 24, 2012 from http://hdr.undp.org/en/media/HDR06-complete.pdf

UNEP. (1992). *Rio declaration on environment and development*, Principle 3, Rio de Janeiro 3 to 14 June, Brazil. Retrieved July 24, 2012, from http://www.unep.org/Documents.Multilingual/Default.asp?documentID=78&articleID=1163

UNFCC. (2009). *Fifteenth Session of the Conference of the Parties to the United Nations Framework Convention on Climate Change and Fifth Session of the Meeting of the Parties to the Kyoto Protocol*, December 7–18, 2009 Copenhagen, Denmark. Retrieved July 24, 2012, from http://www.pewclimate.org/docUploads/copenhagen-cop15-summary.pdf

UNICEF. (2006). *Situation of Women and Children in Nepal 2006*, Kathmandu, United Nations Children's Fund (UNICEF).

United Nations Environment Programme (UNEP). (1972). *Declaration of the United Nations Conference on the Human Environment*, 16th June 1972 Stockholm, Sweden. Retrieved July 24, 2012, from http://www.unep.org/Documents.Multilingual/Default.asp?documentid=97&articleid=1503

USAID. (2003). *Regional hydro power resources: Summary and analysis of selected SARI data*, USAID-SARI/Energy Program, Kathmandu, Nepal, November 2003. Retrieved July 24, 2012, from http://www.sari-energy.org/ProjectReports/RegionalHydroPowerReport.pdf

Warnock, J. B. (1989). *The hydro resources of Nepal.* UK: Water Power and Dam Construction.

WECS. (1999b). *Annual Report 1999.* HMG of Nepal, Kathmandu, Nepal.

WFP/FAO. (2007). *Food and Agricultural Markets in Nepal. Kathmandu*, Food and Agricultural Organisation (FAO)/World Food Programme (WFP), 42.

WHO. (2005). *Distribution of health facilities in Nepal*. Nepal: World Health Organisation Kathmandu.

World Energy Assessment (WEA). (2000). *Energy and the challenge of sustainability*, UNDP, New York, USA. Retrieved July 24, 2012, from http://www.undp.org/content/undp/en/home/librarypage/environment-energy/sustainable_energy/world_energy_assessmentenergyandthechallengeofsustainability.html

World Bank. (2010a). *Transport in South Asia, Nepal Transport Sector*. Retrieved July 24, 2012, from http://web.worldbank.org/WBSITE/EXTERNAL/COUNTRIES/SOUTHASIAEXT/EXTSARREGTOPTRANSPORT/0,,contentMDK:20560914 ~ menuPK:869038 ~ pagePK:34004173 ~ piPK:34003707 ~ theSitePK:579598,00.html

World Bank. (1990). *The World Bank Annual Report 1990: Poverty* (p. 57). Retrieved July 24, 2012, from http://www-wds.worldbank.org/external/default/WDSContentServer/IW3P/IB/2000/12/13/000178830_98101903345649/Rendered/PDF/multi_page.pdf

National Planning Commission (NPC), (1998). *The Incidence of poverty by region in Nepal*. HMG (His Majesty Government) of Nepal, Kathmandu, Nepal, p. 203

Ravallion, M. (1992). *Poverty comparisons—a guide to concepts and methods*, LSMS Working Paper No. 88, The International Bank for Reconstruction and Development/THE WORLD BANK, World Bank February 1992, ISSN: 02534517.

Water, Energy Commission Secretariat (WECS). (1999). *WECS Bulletin No. 10*, HMG of Nepal, Kathmandu, Nepal.

World Vision International. (2003). *Rapid Nutrition and Food Security Assessment Report*. Kathmandu, Nepal: World Vision International.

Žukauskas, A., Shur, M. S., & Gaska. R. (2002). *Intruduction to solid-state lighting*. New York: Willey, IBSN 0471215740.

MIX
Papier aus verantwortungsvollen Quellen
Paper from responsible sources
FSC® C105338
FSC
www.fsc.org

If you have any concerns about our products,
you can contact us on
ProductSafety@springernature.com

In case Publisher is established outside the EU,
the EU authorized representative is:
Springer Nature Customer Service Center GmbH
Europaplatz 3, 69115 Heidelberg, Germany

Printed by Libri Plureos GmbH
in Hamburg, Germany